Multiple Abiotic Stress Tolerances in Higher Plants

In the last 50 years, classical breeding has played a significant role in achieving higher crop productivity, but major crops have reached a plateau in their yield potential. Therefore, the current focus for sustainable intensification of agriculture is the use of biotechnological approaches to enhance the yield potential by combating the yield losses that occur due to abiotic stresses. The abiotic stresses are governed by multigenes, and therefore, a holistic approach is needed to get success in imparting stress tolerance to enhance the yield potential of our crops.

Plants face multiple stress conditions during their life stages and adopt several physiological, biochemical, and molecular strategies to combat that, which are sometimes not sufficient to survive, particularly crop plants. The climate change era has created a need to understand the abiotic stresses in a holistic way. Therefore, a deep understanding of multiple abiotic stress mechanisms is necessary to develop crops tolerant to climate fluctuation.

With this background, the outline of this book covers the following features:

- Agriculture sustainability and molecular understanding of multiple stress tolerance
- Systems biology for life-history strategies, conventional and genomic approaches above and underground
- Genetic resources and molecular understanding of seed priming
- Molecular signaling compounds, cell signal transduction, and crosstalk between plant growth hormones and regulators
- Roles of transcription factors, LEA proteins, reactive oxygen species, and alternative oxidase
- Genome editing, metabolomics, and "omics" technologies

Multiple Abiotic Stress Tolerances in Higher Plants
Addressing the Growing Challenges

Edited by
N. K. Gupta, Yuri Shavrukov, R. K. Singhal
and Nikolai Borisjuk

CRC Press
Taylor & Francis Group
Boca Raton London New York

CRC Press is an imprint of the
Taylor & Francis Group, an **informa** business

First edition published 2024
by CRC Press
6000 Broken Sound Parkway NW, Suite 300, Boca Raton, FL 33487-2742

and by CRC Press
4 Park Square, Milton Park, Abingdon, Oxon, OX14 4RN

CRC Press is an imprint of Taylor & Francis Group, LLC

Library of Congress Cataloging-in-Publication Data
Names: Gupta, Narendra Kumar, editor. | Shavrukov, Yuri, editor.
Title: Multiple abiotic stress tolerance in higher plants : addressing the growing challenges /
edited by Narendra Kumar Gupta, Yuri Shavrukov, Rajesh Kumar Singhal, and Nikolai Borisjuk.
Description: First edition. | Boca Raton : CRC Press, 2024. |
Includes bibliographical references and index. |
Identifiers: LCCN 2023013797 (print) | LCCN 2023013798 (ebook) |
ISBN 9781032292410 (hardback) | ISBN 9781032292427 (paperback) | ISBN 9781003300564 (ebook)
Subjects: LCSH: Crops–Effect of stress on. | Plants–Effect of stress on–Genetic aspects. | Crops–Physiology.
Classification: LCC SB112.5 .M85 2024 (print) | LCC SB112.5 (ebook) | DDC 632/.1–dc23/eng/20230628
LC record available at https://lccn.loc.gov/2023013797
LC ebook record available at https://lccn.loc.gov/2023013798

ISBN: 9781032292410 (hbk)
ISBN: 9781032292427 (pbk)
ISBN: 9781003300564 (ebk)

DOI: 10.1201/9781003300564

Typeset in Times
by codeMantra

Contents

Foreword

Plants face many adverse circumstances at different stages during their growth and development. The field environment presents multiple abiotic stresses to crops, the most widespread being drought and heat. These two conditions frequently occur together and are the cause of large yield penalties in the production of important crops. Global warming and climate change are increasing the incidence of drought and of extreme temperatures. Other stresses faced by crop plants in many countries of the world are soil salinity and sodicity, soil pollution by heavy metals, and flooding. Again, these conditions are not isolated but occur in combination. Breeding plants with increased tolerance to multiple stresses is essential to reduce the yield potential gap and secure global food security for the next generation.

This book presents a collection of 18 chapters written by leading scientists around the world who have carried out research into different aspects of stress tolerance: soil science, physiology, molecular genetics, and molecular biology. They come from as many as ten countries, including India, Israel, China, Ukraine, Russia, and Australia.

Researchers and plant breeders are continuously working with advanced methodologies to understand the limitations to crop production in adverse environments and to produce more stress-resistant varieties. This book shows how plants can tolerate and adapt to a combination of environmental stresses. Drought would usually be accompanied by high temperatures, and vice versa. Irrigation of crops may relieve drought, but it often increases salinity. Tolerance mechanisms to drought and soil salinity have much in common; however, the combined stress would have an additive effect on plant growth and yield. In these circumstances, modern physiological, biochemical, and biotechnological approaches are needed to accelerate the traditional crop improvement programs.

An understanding of the plant responses to soil salinity, drought, heat, nutrient deficiency, and toxicity will lead to the understanding of the mechanisms in common to improve plant tolerance to multiple stresses. This book shows that, for example, oxidative stress is the most common condition faced by plants under single, combined, and multiple stresses, and the molecular understanding of antioxidant compounds, secondary metabolites, and antioxidant enzymes will help to overcome the stress. Molecular chaperones are essential to repair damaged proteins, and the induction of heat shock proteins and LEAs are common to many stresses. Many chapters in this book emphasize innovative topics such as hormone signaling, speed breeding, integrative transcriptomics, and seed priming which focus on reducing the yield gap and securing future global security. Likewise, the modern biotechnological, molecular, and genetic tools such as GWAS (genome-wide association studies) and integrative "omics" enable a molecular understanding of plant reactions under unfavorable conditions.

This book entitled *Multiple Abiotic Stress Tolerance in Higher Plants: Addressing the Growing Challenges* edited by N. K. Gupta, Yuri Shavrukov, R. K. Singhal, and Nikolai Borisjuk and published by CRC Press, Taylor & Francis, 2023 provides new insights into the production of crops for multiple stress environments. I recommend it for its breadth of coverage and for a focus on new technologies.

Rana Munns
Professor
ARC Centre of Excellence in Plant Energy Biology,
School of Molecular Sciences
University of Western Australia
Perth, Australia

Preface

The field of plant stress biology has seen dramatic progress in the past two decades due to the development of new experimental techniques and the wide application of these innovative approaches to cultivated crops, not just a few model plant species. The recent advances in whole-genome sequencing, plant bioinformatics, genomics, proteomics, and metabolomics are used with increasing speed to understand plants responses and improve crop tolerance to various abiotic stress factors such as heat, drought, soil salinity, and soil acidity. These scientific actions aimed at addressing the effects of abiotic stresses on crop productivity have been motivated by the need to feed the growing human population, and the need to mitigate the effects of environment pollution resulting from human activities and climate change.

Instead of focusing on one particular problem or plant, our editorial team of four international experts in plant biology put together this book, collecting 18 chapters written by scientists specialized in various areas of plant stress biology, to give the readers a broad view of the modern problems, technologies, and plant systems used in the field. As a result, the book covers salinity, drought, heat, and cold stresses, emphasizing on the tolerance to combinations of different stresses and highlighting the physiological, biochemical, and molecular mechanisms underlying crops' reactions to abiotic stresses and paving the way for improving stress tolerance. The chapters overview many novel approaches from system biology to phenomics, genomics, metabolomics, transgenesis, and genome editing, examining their contributions to the field and providing perspectives for improving stress tolerance in a range of important crops including cereals, cotton, and potato.

The subjects and themes of the book are carefully indexed, helping the readers to easily find the information of interest. We believe this book will be interesting and useful not only for the scientists practicing in the related fields but also for crop breeders, biology students, and the broader audience interested in advanced science and technologies.

Our sincere thanks go to all the authors for their effort and time in preparing the chapters for their valuable contributions.

N. K. Gupta
Yuri Shavrukov
R. K. Singhal
Nikolai Borisjuk
March 2023

Editors' Biography

Dr N. K. Gupta obtained his PhD (1997) from prestigious Indian Agricultural Research Institute, New Delhi, India. He was awarded Biotech Overseas Associateship (2006–2007) by Government of India and worked as Post Doc at the University of Adelaide, Adelaide, Australia. He also served as Visiting Professor at the University of Nebraska, Lincoln, USA (2013) and University of Tottori, Japan (2023). At present he is working as Professor (Plant Physiology) & Director, Education at SKN Agriculture University, Jobner. Dr Gupta has teaching experience of more than 25 years and guided 15 PG students for dissertation. He is Fellow of many academic societies. Recently, he has been awarded prestigious J. J. Chinoy Gold Medal Award by ISPP, New Delhi (2019). He is recipient of first RS Paroda Best Scientist Award (2020). He has published more than 60 research papers in national and international journals including *Plant Cell Environment*, *Photosynthetica*, *Plant Growth Regulation*, *Journal of Agronomy & Crop Science*, *Functional & Integrative Genomics*, etc. He has four authored books and ten book chapters published by Nova Publication, CRC Press, etc. His H-Factor is 20.0.

Dr Yuri Shavrukov received his PhD degree (1991) from the Institute of Cytology and Genetics, Russian Academy of Sciences, Novosibirsk, Russia, with following employment in the same Institute and combined his research with teaching at Novosibirsk State University. He completed his STA Post-Doctoral Fellowship (1995–1996) at the Hokkaido National Agricultural Experiment Station (Japan); and was awarded JSPS Research Fellowship at the Hokkaido University (1998) and JSPS Senior Research Fellowship at Okayama University (2012). In 2001, Dr Shavrukov was selected and employed by CSIRO (Commonwealth Scientific and Industrial Research Organisation), Adelaide, Australia. After CSIRO, he received a position at the University of Adelaide, Plant Genomics Centre (2005–2016), and since that he moved to Flinders University, Adelaide. Currently, Dr Shavrukov is Adjunct Professor in the College of Science and Engineering (Biological Sciences), Flinders University. His H-factor is 23. He has more than 65 publications, and he is actively publishing especially in last years.

Dr R.K. Singhal is currently working as a scientist in the Division of Crop Improvement, ICAR-Indian Grassland and Fodder Research Institute, Jhansi (U.P.), India. He has published more than 40 research reviews and chapter papers in reputed journals such as *Frontier in Plant Science (IF-5.753)*, *Frontier in Agronomy*, *Physiologia Plantarum (If-4.5)*, *Plant Breeding*, *Russian Journal of Plant Physiology*, and many others focused mainly on abiotic stresses. Currently, he is working on multiple abiotic stress tolerance in different crops.

Dr Nikolai Borisjuk is Professor in the School of Life Sciences of Huaiyin Normal University. He has graduated from the T. Shevchenko Kiev State University (Kiev, Ukraine), majoring in microbiology and received his PhD in Molecular Biology in 1988. In 1990, he moved to Tuebingen University in Germany as a fellow of A. von Humboldt Foundation, and since then he continued working at various international research Institutions in Germany, USA, and Australia. He has published more than 60 research papers devoted to plant genome structure and evolution, plant genetic engineering, and stress resistance in prestigious scientific journals such as *Nature Biotechnology*, *PNAS*, *Plant Physiology*, *New Phytologist*, etc. with a current H-factor of 27. He taught molecular biology and plant genetic engineering courses and supervised undergraduate and postgraduate students at the Kiev State University, Institute of Plant Genetics (Gatersleben, Germany), Rutgers University (New Brunswick, USA), Thomas Jefferson University (Philadelphia, USA), and the University of Adelaide (Adelaide, Australia).

Contributors

Irfan Afzal
Department of Agronomy
University of Agriculture
Faisalabad, Pakistan

Mughair Abdul Aziz
Integrative Agriculture Department, College of Agriculture and Veterinary Medicine
United Arab Emirates University
Al Ain, United Arab Emirates

Satyaprakash Barik
Department of Agriculture and Allied Sciences
CV Raman Global University
Bhubaneswar, India

Tanzeel Bashir
Genome Engineering and Societal Biotechnology Lab, Division of Plant Biotechnology
Sher-e-Kashmir University of Agricultural Sciences & Technology of Kashmir (SKUAST-K)
Shalimar, India

Laxmipreeya Behera
Odisha University of Agriculture and Technology
Bhubaneswar, India

Neha Bhargavi
VIT School of Agricultural Innovations and Advanced Learning
VIT University
Vellore, India

Sougata Bhatacharjee
ICAR–Indian Agricultural Research Institute
Hazaribagh, India

Nikolai Borisjuk
School of Life Sciences
Huaiyin Normal University
Huaiyin, China

Stephen Chivasa
Department of Biosciences
Durham University
Durham, United Kingdom

Himaj Deshmukh
ICAR–National Institute of Abiotic Stress Management
Pune, India

Nardana Esmaeili
Department of Basic and Translational Sciences, School of Dental Medicine
University of Pennsylvania
Philadelphia, Pennsylvania

Saiyyeda Firdous
VIT School of Agricultural Innovations and Advanced Learning
VIT University
Vellore, India

Suresh Gali
Department of Botany and Plant Physiology
CCS Haryana Agricultural University
Hisar, India

N. K. Gupta
SKN Agriculture University, Jobner
Jaipur, India

Pernika Gupta
Department of Biochemistry, College of Basic Sciences and Humanities
CCS Haryana Agricultural University
Hisar, India

Syed Anam Ul Haq
Genome Engineering and Societal Biotechnology Lab, Division of Plant Biotechnology
Sher-e-Kashmir University of Agricultural Sciences & Technology of Kashmir (SKUAST-K)
Shalimar, India

S.G. Harsha
Division of Basic Sciences
Indian Institute of Horticultural Research
Bengaluru, India
and
Department of Plant Physiology
Indira Gandhi Krishi Vishwavidyalaya
Raipur, India

Umer Hayat
Department of Botany
University of Agriculture
Faisalabad, Pakistan

Vinay Hegde
ICAR–National Institute of Abiotic Stress Management
Pune, India

Akbar Hossain
Department of Agronomy
Bangladesh Wheat and Maize Research Institute
Dinajpur, Bangladesh

Amjad M. Husaini
Genome Engineering and Societal Biotechnology Lab, Division of Plant Biotechnology
Sher-e-Kashmir University of Agricultural Sciences & Technology of Kashmir (SKUAST-K)
Shalimar, India

Stanislav Isayenkov
Department of Plant Food Products and Biofortification, Institute of Food Biotechnology and Genomics
National Academy of Sciences of Ukraine
Kyiv, Ukraine

Rahul Jain
Department of Applied Sciences
Chandigarh Engineering College
Mohali, India

C.S. Kailash
Odisha University of Agriculture and Technology
Bhubaneswar, India

Sonika Kalia
School of Applied and Life Sciences
Uttaranchal University
Dehradun, India

Muhammad Kamran
Department of Agronomy
University of Agriculture
Faisalabad, Pakistan

Asha Kumari
ICAR–Indian Agricultural Research Institute
Hazaribagh, India

R.H. Laxman
Division of Basic Sciences
Indian Institute of Horticultural Research
Bengaluru, India

Khaled Masmoudi
Integrative Agriculture Department, College of Agriculture and Veterinary Medicine
United Arab Emirates University
Al Ain, United Arab Emirates

Pratikshya Mishra
Department of Plant Breeding and Genetics
Odisha University of Agriculture and Technology
Bhubaneswar, India

Ananya Anyatama Mohanty
Department of Plant Breeding and Genetics
Odisha University of Agriculture and Technology
Bhubaneswar, India

Rana Munns
ARC Centre of Excellence in Plant Energy Biology, School of Molecular Sciences
University of Western Australia
Crawley, Australia

Iteeshree Nayak
Crop Improvement Divisions
ICAR–National Rice Research Institute
Cuttack, India

Rudo Ngara
Department of Plant Sciences
University of the Free State
Phuthaditjhaba, South Africa

C. Parmeswaran
Crop Improvement Divisions
ICAR–National Rice Research Institute
Cuttack, India

Sanmay Kumar Patra
Department of Agricultural Chemistry and Soil Science
Bidhan Chandra Krishi Viswavidyalaya
Nadia, India

Krishnayan Paul
Division of Molecular Biology and Biotechnology
ICAR–Indian Agricultural Research Institute
New Delhi, India

K.A. Pawan
Odisha University of Agriculture and Technology
Bhubaneshwar, India

Zvi Peleg
The Hebrew University of Jerusalem
Rehovot, Israel

Ratneswar Poddar
Department of Agronomy
Bidhan Chandra Krishi Viswavidyalaya
Nadia, India

Aliza Pradhan
ICAR–National Institute of Abiotic Stress Management
Baramati, Pune, India

Pranaya Pradhan
Department of Nematology
Odisha University of Agriculture and Technology
Bhubaneswar, India

Smruti Ranjan Padhan
Division of Agronomy
ICAR–Indian Agricultural Research Institute
New Delhi, India

M.D. Prathibha
Division of Basic Sciences
Indian Institute of Horticultural Research
Bengaluru, India

Himani Punia
Department of Biochemistry, College of Basic Sciences and Humanities
CCS Haryana Agricultural University
Hisar, India

Karthika Rajendran
VIT School of Agricultural Innovations and Advanced Learning
VIT University
Vellore, India

Jagadish Rane
ICAR–National Institute of Abiotic Stress Management
Pune, India

Anju Rani
Department of Biochemistry, College of Basic Sciences and Humanities
CCS Haryana Agricultural University
Hisar, India

Maryam Razzaq
Department of Plant Breeding and Genetics, Seed Science and Technology
University of Agriculture
Faisalabad, Pakistan

Elena V. Rogozina
N.I. Vavilov All-Russian Institute of Plant Genetic Resources (VIR)
St. Petersburg, Russia

Oishika Sah
VIT School of Agricultural Innovations and Advanced Learning
VIT University
Vellore, India

Bijayalaxmi Sahoo
Crop Improvement Divisions
ICAR–National Rice Research Institute
Cuttack, India

Jyoti Prakash Sahoo
Department of Agricultural Biotechnology
Odisha University of Agriculture and Technology
Bhubaneswar, India

Kailash Chandra Samal
Department of Agricultural Biotechnology
Odisha University of Agriculture and Technology
Bhubaneswar, India

S. Sanghamitra
Crop Improvement Divisions
ICAR–National Rice Research Institute
Cuttack, India

Arindam Sarkar
Department of Agricultural Chemistry and Soil Science
Bidhan Chandra Krishi Viswavidyalaya
Nadia, India

Shraddha Bhaskar Sawant
Odisha University of Agriculture and Technology
Bhubaneshwar, India

Arup Sen
Department of Agricultural Chemistry and Soil Science
Bidhan Chandra Krishi Viswavidyalaya
Nadia, India

Lidor Shaar-Moshe
The Hebrew University of Jerusalem
Rehovot, Israel
University of California
Davis, California

Yuri Shavrukov
College of Science and Engineering
Flinders University
Adelaide, Australia

H.S. Sheela
Division of Basic Sciences
Indian Institute of Horticultural Research
Bengaluru, India

K.S. Shivashankara
Division of Basic Sciences
Indian Institute of Horticultural Research
Bengaluru, India

R. K. Singhal
ICAR–National Grassland and Fodder Research Institute
Jhansi, India

Madhavi Sonone
ICAR–National Institute of Abiotic Stress Management
Pune, India

Kathleen L. Soole
College of Science and Engineering
Flinders University
Adelaide, Australia

Crystal Sweetman
College of Science and Engineering
Flinders University
Adelaide, Australia

Tamilselvan A.
ICAR–National Institute of Abiotic Stress Management
Pune, India

Jayanti Tokas
Department of Biochemistry, College of Basic Sciences and Humanities
CCS Haryana Agricultural University
Hisar, India

Inosha Wijewardene
Department of Biological Sciences
Texas Tech University
Lubbock, Texas

Hong Zhang
Department of Biological Sciences
Texas Tech University
Lubbock, Texas

1 Molecular Understanding of Multiple Stress Tolerance in Higher Plants

An Overview

Anju Rani, Jayanti Tokas, Himani Punia, and Pernika Gupta
CCS Haryana Agricultural University

Rahul Jain
Chandigarh Engineering College, Chandigarh Group of Colleges

Sonika Kalia
Uttaranchal University

CONTENTS

1.1 INTRODUCTION

Plants are regularly exposed to changing environments and different types of biotic and abiotic stresses. Climate change and disease incidence really deter crop health and productivity; plants manage these stressful situations by adapting to harsh conditions or by inducing a defense system, which is an active part of every living being (Punia et al., 2019). These pathways are really complex and intricate, cross-talk among different components of defense pathways takes place to effectively prevent the crop plants (Punia et al., 2021a). Initially, stress is perceived by the specific receptors present on the plant, and then downstream signaling in the form of changed Ca^{2+} ion level,

DOI: 10.1201/9781003300564-1

TABLE 1.1
Abiotic Stress Regulation by Molecular Mechanisms

Plant	Genes Regulation	Reference
	Drought Stress	
Wheat	Under water-deficit condition 30% of genes were upregulated and the 18% were downregulated among 300 unique single expressed sequences tag (ESTs)	Way et al. (2005)
A. thaliana	Under drought condition 130 genes were upregulated	Bray (2002)
Drought-tolerant native Andeon potato	Under water-deficit condition, 186 upregulated genes were involved in drought tolerance among 1713 differentially expressed genes	Schafleitner et al. (2007)
	Metal Ion Stress	
A. thaliana	Under etal ion stress of lead (Pb) and cadmium (Cd) 19 and 76 genes up- and downregulated by Pb and 65 and 338 up- and downregulated by Cd	Kovalchuk et al. (2005)
A. thaliana	ABC transporters were differentially controlled upon Cd treatments, showing some ABC transporter play an important role in glutathione–Cd or phytochelatins–Cd complex transport both inside and outside the cell	Bovet et al. (2005)

membrane liquidity, reactive oxygen species (ROS), and disturbed ion homeostasis take place (Tokas et al., 2021), this further induces intracellular elements of defense system such as activation of specific proteins, enzymes for instance phosphatase/kinases, and regulation of phytohormones level to maintain plants' growth and development using various signaling pathways (Punia et al., 2020a,b). These enzymes finally regulate gene expression responsible for developing plant resistance and thus, protect the plant from stresses. Identification of these genes and targeted proteins having a key role in plant protection can help in the development of resistant varieties with improved resistance and high productivity traits. Table 1.1 shows the regulation of defense mechanisms by upregulation and downregulation of genes that respond to abiotic and biotic stress under drought and metal ion stress.

Biotic stress is caused by the infestation of insects, pests, bacteria, fungi, and virus in plants. Plants respond to biological stress using two major defense pathways: one is innate pathway and the other is systemic plant response as shown in Figure 1.1. Innate response and systemic plant response are further classified on the basis of specificity shown against pest attack. Different mechanisms are used for plant defense in these pathways. Innate immunity is classified as PTI and ETI as shown is Table 1.2. These are further classified based on the pathogen parts involved in the reaction (Table 1.3).

1.2 STRESS SIGNAL PERCEPTION

After encountering the stress, first step is the perception of signals related to respective stress, then signaling molecules transmit the signal further on in the form of phyto-hormones, H_2O_2, MG, and cytosolic Ca^{2+}, which activates certain transcription factors (TFs) that modify gene expressions of stress-related genes and aid in stress resistance manifestation (Hoque et al., 2012). Sensors are the first molecule that sense the stress. Temperature change results in the altered membrane fluidity (Örvar et al., 2000). Four sensors have been reported in plants, which initiated heat-shock reaction (Punia et al., 2021b). One such sensors is calcium ion channel that activates Ca^{2+} inflow inside cells, other are two unfolded protein preceptors in cytosol and endoplasmic reticulum and histone preceptor inside nucleus (Mittler et al., 2012). Therefore, further exploration in this field will be helpful in furthering insights in temperature sensing.

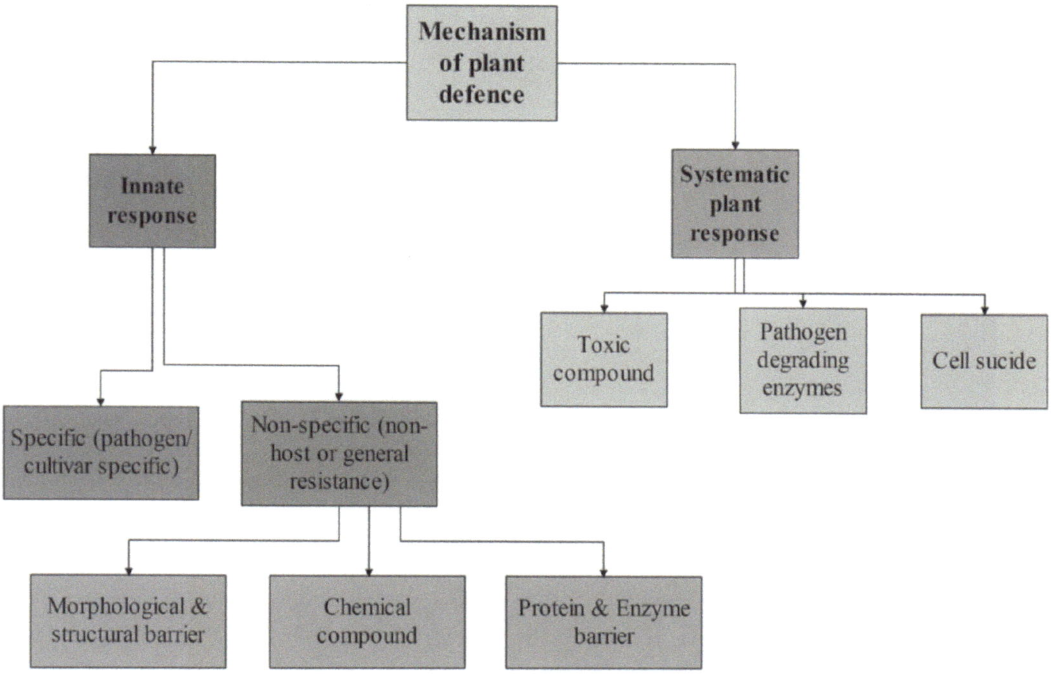

FIGURE 1.1 Plant defense mechanisms in response to pathogens.

TABLE 1.2
Classification of Different Innate Immunity

Innate Immunity	
Pathogen- or microbial-associated molecular-pattern (PAMP/MAMP)-triggered immunity (PTI)	Effector-triggered immunity (ETI)
Different classes of PTI immunity	Different classes of ETI
1. Flagellin-induced resistance	1. Direct NLR–effector interaction
2. Elongation factor (EF-Tu)-induced resistance	2. Indirect NLR surveillance of effector activities
3. Plant perception of PAMPs from fungi and oomycetes	
4. Plant perception of virus PAMPs	
5. Plant perception of insect PAMPs	
6. Infection self-perception DAMPs	

Sensors perceive the signal and start signaling by either primary signaling cascade or secondary cascade pathway. Primary signaling is performed by the receptor-like kinases (RLKs), it has three domains: one is extracellular domain, which binds with the ligand and is the example of protein–protein interaction, second domain is transmembrane domain, which crosses the cell membrane, and third domain is intracellular kinase domain, which phosphorylates the target protein and starts intracellular signaling in the cytosol or transmits the signal to the nucleus. First example of these types of signaling was histidine kinase signaling found in prokaryotes, which was later on reported in eukaryotes including plants. Herein, initially histidine is phosphorylated by histidine kinase and then transferred to aspartate residue of same protein or of other protein and further mitogen-activated protein kinase (MAPK) cascade was activated to transmit or amplify the signals. Protein phosphorylase and phosphatase are the key proteins involved in this sort of intracellular signaling. Finally, these cascades will regulate cell processes such as destining protein or degradation of proteins, complex biomolecule formation, and enhancing enzyme activity.

TABLE 1.3
sRNA Involved in Plants Functional Regulation

Plants	Type of Stress	sRNA and Function Regulated by sRNA	Reference
Brachypodium distachyon	Cold stress	The activation and repression of miRNAs	Zhang et al. (2009)
Rice	Oxidative stress	miRNAs regulated auxin amount, programmed cell death, and nutrient transport	Li et al. (2011)
	Abiotic stress	ABA signal transduction and resistance were positively mediated by siR441 and siR446	Yan et al. (2011)
Medicago truncatula	Drought	The synthesis and processing of miRNAs was controlled by MtDCL1 and MtAGO1, enzymes formed under drought stress and thus control small RNA species signaling	Capitão et al. (2011)
Transgenic *Arabidopsis*	Oxidative stress	Increased capacity to neutralize superoxide was observed as chloroplastic superoxide dismutase (CSD2) that is not suppressed by mi398	Sunkar et al. (2006)
Soybean	Water deficit and infection with soybean rust fungus	Under drought stress and rust fungus infection expression of 11 soybean miRNAs was observed	Kulcheski et al. (2011)
	Oxidative stress and biotic stress	Regulates mRNA decay and H_2O_2 metabolism and its signaling by activating peroxidase protein by MIR-Seq11 The pathogen infection downregulates the MIR-Seq11 during biotic stress and thus defends from the ROS	Kulcheski et al. (2011)
	Nat-siRNAs	Both biotic and abiotic stress are controlled	Ruiz-Ferrer and Voinnet (2009)
Arabidopsis	Salt	Antisense gene pair PC5DH–SRO5 results in nat-siRNA formation that control proline levels and provide resistance to salt	Borsani et al. (2005)
	Salt stress,	Nat-siRNA is produced, which targets the mRNA of the proline degradation gene PC5DH for destruction	Katiyar-Agarwal et al. (2006)
Arabidopsis	Nutrient deprived	miR398 (copper), miR399 (phosphate), and miR395 (sulfate) control these nutrients' balance	Buhtz et al. (2008); Buhtz et al. (2010)
	Inorganic phosphate limitation	PHO2, an E2 ubiquitin-conjugating proteins, are targeted by miR399 gene family, which represses uptake of P_i	Bari et al. (2006); Allen et al. (2005); Aung et al. (2006)

1.3 SIGNAL TRANSDUCTION

1.3.1 Phytohormones

A plant protects itself using different pathways involving different plant hormones such as salicylic acid, ethylene, and jasmonic acid (JA), these phytohormones instead have important role in plant development. Cross-talk among these hormones bring plant defense system in effect. Brassinosteroids (BR) are plant growth regulators and have a role in photomorphogenesis, cell elongation, stomata growth, flower development, and stress tolerance (Li et al., 2016). BR were first identified in *Brassica* pollens and so named Brassinosteroids. Many isoforms of BR are reported in plants and major classes are 28-homobrassinolide, 24-epibrassinolid,e and brassinolide, which are biologically active and have key role in plant growth and defense (Wu et al., 2017).

1.3.1.1 Role of JA Stress Signaling

The roots are the first part of plant encountering stress and then a signal is transmitted from the roots to upper parts of plant. Plants hormones such as JAs, salicylic acid, and ethylene are involved in defense mechanisms. JA is the hormone sensed by receptors COI1, which is a F-box protein and integral component of Skp1-CullinF-box (SCF) complex. It shares high similarity with TIR1 (transport inhibitor response 1) receptor, which is auxin's receptor, and it is the first receptor recognized as the component of SCF complex with F-box protein. JA is the hormone deciding the interaction between plant and microbes, which pass through the membrane and transmit long-term signals (Hause and Schaarschmidt, 2009). Many studies showed improved plant growth performance under stressed water conditions by regulating stomatal conductance (Punia et al., 2021c). JA induces Nod factor synthesis by upregulating the nod gene in bacteria via enhanced production of isoflavonoid in the plant cell (Zhang et al., 2007). JA produced under stressed conditions start signaling pathway by suppressing the binding of jasmonate-insensitive/jasmonate-zim (JAI3/JAZ) proteins to the different transcription regulators including MYC2 degrading the JAZ protein and thus, promote plant growth by developing defense against various biotic and abiotic stresses. Whereas in the absence of JA, JAZ protein binds the MYC2 transcription factor and regulates the transcription of various genes involved in plant defense (Chini et al., 2007), as depicted in Figure 1.2.

1.3.1.2 ABA Receptor and Downstream Signaling

ABA is a plant hormone synthesized naturally under stress conditions in plants. It influences plant growth, reproduction, and stress coping. Water stress and drought result into osmotic stress, which induces ABA synthesis and plant defense mechanism (Yamaguchishinozaki and Shinozaki, 2006). ABA is usually synthesized in the roots and transported to the upper part of plants via xylem (Kuromori et al., 2010). Stress signal perception and transduction in plants take place by either

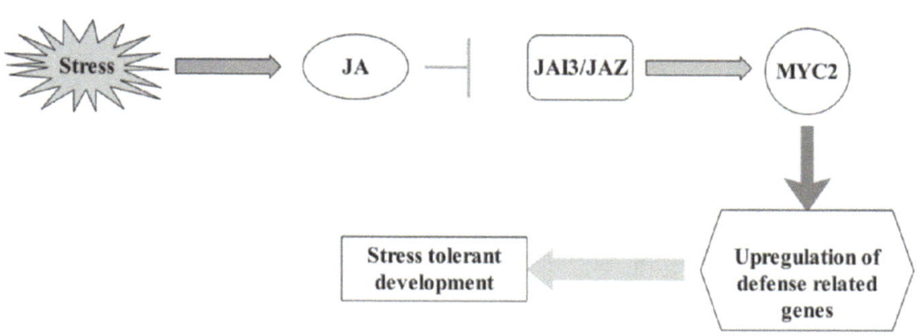

FIGURE 1.2 Role of jasmonic acid (JA) in stress tolerance.

ABA-dependent pathway or ABA-independent pathways, here ABA-dependent pathway plays a key role under most types of stresses. Receptors for ABA have been identified in different organelles, nucleus, cytosol, plasma membrane, and chloroplast envelope.

In *Arabidopsis*, receptor for ABA, which is responsible for regulating stress, senescence, plant growth, and seed dormancy, is synthesized primarily in the xylem and aerial parts of plants, and the receptors for perceiving this hormone are Gibberellin Insensitive Dwarf1 (GID1), Flowering Time Control Locus A (FCA), and Transport Inhibitor Response1 (TIR1). Signaling pathway of this hormone is independent of other proteins, short, and only has protein–protein binding using hormone receptors. Here, FCA receptor is similar to the C-terminus of ABA-binding protein (ABAP1) present in barley aleurone cell membrane. Thus, this receptor's affinity for RNA-binding protein regulates the transcription level of genes associated with different activities (Simpson et al., 2003).

1.3.2 Reactive Oxygen Species Formation under Abiotic Stress in Plants

ROS are oxygen entities formed by oxidation under normal as well as stressed environments, but accumulate in higher amount under adverse situations, decided by the scavenging and formation of the ROS. More commonly synthesized ROS are $O2^-$, H_2O_2, and $^{\bullet}OH$ (Hossain et al., 2012). Organelles such as chloroplast, peroxisome, and mitochondria are the major sites of ROS production, as these are involved in electron transport and oxidative metabolism (Gill and Tuteja, 2010). Chloroplast is the main site of ROS production. Here, electrons from photosystems (PS I and PS II) are transferred to oxygen that can be transformed to H_2O_2 by superoxide dismutase (SOD; Gechev et al., 2006). Under saline environment, stomatal conductance reduces and limits the CO_2 amount inside, thus reducing the CO_2 fixation by the Calvin cycle; and thus nicotinamide adenine dinucleotide phosphate ($NADP^+$) final acceptor is present in low amount and thus electrons siphoned to the O_2 produces $O_2^{\bullet -}$. Drought conditions lead to the stomatal closure and reduce water loss and CO_2 inside leaves, which again transfer electrons to oxygen with higher photorespiration rate (Sanda et al., 2011).

1.4 TRANSCRIPTIONAL ACTIVATION AND REGULATION

Plants' exposure to stress and changes in the gene expression could be known by profiling transcriptome. In response to stresses, transcription level changes differently and some sort of cross-talk takes place among different types of stresses. It has been experimented under laboratory conditions by applying single stress, whereas, in field, different types of stresses are encountered simultaneously. Thus, transcriptome of field crop is very different from the single stress exposure plants.

1.4.1 RNAi-Mediated Plant Defense

RNA interference is one of the emerging tools in crop improvement, which is based on sequence-specific inactivation of a particular gene using small RNA. Basically, two categories of RNA interference are employed: one using microRNA (miRNA) and other mediated by small interfering RNA (siRNA). Both these small RNA varies in their origin (Kamthan et al., 2015) and processing pathway, but mediates RNA or DNA interference ultimately. MiRNAs are small, noncoding RNAs produced endogenously against various stresses. These are involved in gene regulation posttranscriptionally, and thus regulate the quantum of various hormones and proteins involved in defense system. Phytohormones such as salicylic acid, abscisic acid, and ethylene mediate cross-talk to develop defense in plants (Huang et al., 2008).

miRNA functioning involves a series of steps, starting from admission of double-stranded RNA to the cell followed by splicing and cleavage, then 2-*O*-methylation and finally complex formation with target gene (Ruiz-Ferrer and Voinnet, 2009). RNA-induced silencing complex (RISC) forms and interacts with small RNA to bring sequence-specific interference. This technique is employed in improvement of plant resistance against herbivores and pests. Many such experiments have been done in the past. Double-stranded ribonucleic acid (dsRNA) development in engineered potato showed resistance against potato spindle tuber viroid (PSTVd) (Schwind et al., 2009). Posttranscriptional modification of DNA/RNA interference is the new technique in gene regulation known as virus-induced gene silencing (VIGS), which helps in plant defense against viruses (Simón-Mateo and Garcia, 2011). Using high-throughput sequencing, identification of small RNAs (Xin et al., 2010) was done, and several categories are now known to be involved in stress responses. Under low phosphate stress miR399 miRNA in *Arabidopsis* suppressed the expression of a ubiquitin-conjugating enzyme, thus facilitating the uptake and mobilization of phosphate and avoid phosphate scarcity (Fujii et al., 2005).

1.4.2 Transcriptional Factors

TFs are principally involved in gene regulation and regulates various activities by either downregulating or upregulating expression of responsible gene. Various TFs are associated with biotic and abiotic stresses and confer tolerance against respective stress. Many TFs were characterized and identified conferring salt tolerance. Such TFs include DREB, WRKY, bZIP, NAC, C_2H_2 zinc finger gene, and AP2 families. These TFs act on *cis*-acting element in the gene's promoter region (Figure 1.3). In rice (*Oryza sativa*) a salt-responsive TF has been reported, known as SALT-RESPONSIVE ERF1 (SERF1), which shows expression in roots on application of H_2O_2 and salt. Removal of *SERF1* leads to stoppage of MAPK pathway, and TFs responsible for salt tolerance. It shows that *SERF1* gets induced in response of H_2O_2 production during salt tolerance and is the site of phosphorylation of MAPK5. *SERF1* binds on the promoter region of various MAPK isoforms such as MAP3K6, MAPK5, DREB2A, etc., and it also stimulates its own production. Thus, *SERF1* overexpression in plants confer tolerance to salinity, whereas its inactivation promotes salinity sensitivity in plants and could be used as an effective target for improving salt tolerance in crop plants (Schmidt et al., 2013). Kinases regulate some of the TFs, which play a key role in salt tolerance. One such gene product is encoded by *OsRMC*, which negatively correlated with

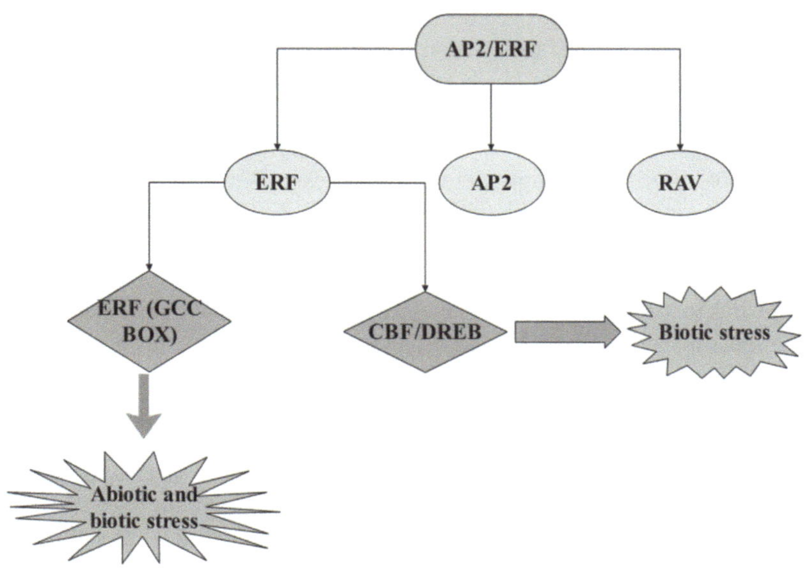

FIGURE 1.3 Highlights of transcription factors involved in abiotic stress tolerance.

salinity tolerance and code for a receptor-like kinase. The OsEREBP1 and OsEREBP2 are the two TFs belonging to AP2/ERF family, negatively regulate the *OsRMC* binding to the GCC-like DNA motif at promoter region (Serra et al., 2013).

1.5 DIRECT ACTION PROTEINS

1.5.1 Osmoprotection

Osmoprotection is provided by osmoregulants or osmolytes, which varies at chemical structures and biological activities having a key role in stress mitigation. These are uncharged, polar, and soluble, not disturbing the normal state of cell and confer resistance against abiotic stresses (Saito and Matsuda, 2010). Major compatible solutes are proline, sugar (fructans, sucrose, trehalose), glycine betaine, polyols (sorbitol, mannitol), amino acids (proline), and dimethylsulfonium compounds (Shulaev et al., 2008). Different plants accumulate different quantum of organic osmolytes for instance, proline accumulates in large number of classes (Saxena et al., 2013), whereas beta alanine betaine's storage is limited to certain members of Plumbaginaceae (Hanson et al., 1994). Amount of these osmolytes depends on the biosynthesis and degradation rate of respective osmolytes or irreversible synthesis of these organic solutes.

Glycine betaine is an organic compound with quaternary ammonium, amphoteric, electrically neutral, and water-soluble properties but having 3-methyl groups as nonpolar entity. These properties confer glycine betaine ability to bind with both hydrophobic and hydrophilic groups of enzymes and proteins. This increases the osmotic adjustment ability of cell under stress and thus ameliorate stress. It ensures plant growth by stabilizing protein (Mäkelä et al., 2000), defending photosynthetic machinery (Gadallah, 1999), maintaining osmotic balance, and eliminating ROS (Ashraf and Foolad, 2007; Saxena et al., 2013).

Polyols are organic compounds with multiple hydroxyl functional groups. Sugar alcohol constitutes its best example, act as an osmoprotectants, ROS cleanser, and chaperon, and thus, plays a crucial role in plant defense by countering stresses (Ashraf and Foolad, 2007). These polyols are studied under two categories – acyclic compounds such as mannitol and cyclic compounds for example pinitol. Proline accumulation has been reported under various abiotic stresses such as drought, salinity, etc. (El-Shintinawy and El-Shourbagy, 2001). Proline not only defends the plants from the stress but also acts as a nitrogen reserve under stressed situations. Accumulation of proline is an indicator of developing tolerance. Proline also acts as an antioxidant by quenching O_2 produced during different processes. In *Nicotiana tabacum*, antioxidants activity has been augmented by proline under salinity stress (Hoque et al., 2008).

Polyamines are small, polycationic aliphatic compound and ubiquitous in distribution. These have regulatory role in differentiation and morphogenesis, eliminating seed dormancy, fruit and flower development, and seed germination (Knott et al., 2007; Galston et al., 1997). In addition to this, it improves stress tolerance (Gupta et al., 2013; Kovács et al., 2010). Amine oxidases and diamine oxidases are the enzymes catalyzing polyamine oxidative catabolism and have important role in coping with stress (Takahashi and Kakehi, 2010; Cona et al., 2006). Synthesis and accumulation of polyamines such as SPM, SPD, and PUT in *Arabidopsis* and tobacco increased salt tolerance (Roy and Wu, 2002).

In analytical chemistry, methods such as MS-based methods and NMR, and sophisticated "data processing and mining" techniques, have allowed the plant biologists to venture into hitherto unexplored domains and generate extensive metabolic profiles due to various environmental stimuli including salinity. Results indicate that the metabolic processes are highly specific for given tissues, species, and plant–environment interactions. The clusters of identified compounds not only serve as base in the quest for novel defense compounds but also as markers for the characterization of the plants' defensive state.

1.5.2 Heat-Shock Proteins

Heat-shock proteins (HSPs) act as a chaperon that protects the protein from adopting wrong conformation and disassembly. As proteins are damaged during abiotic stress, so plants respond by synthesis of more HSPs under not only heat stress but most types of abiotic stresses. Proteins are prevented from unconventional folding and hence protect the protein and cell membrane with right folding, transportation, refolding, degradation, and oligomer formation under stress response (Hu et al., 2010). Five major classes of HSPs were reported in eukaryotes including plants. These are sHSPs, HSP60, HSP70, HSP90, and HSP100 localized in different organelles including endoplasmic reticulum, nucleus, chloroplasts, and mitochondria (de Ries et al., 2012). As chloroplast and mitochondria are major sites of ROS formation and hence of protein damage, so these HSPs are synthesized there to counter the loss. sHSPs are the HSPs expressed in mitochondria and defends the electron transport complex I by removing ROS from vicinity (Hamilton and Heckathorn, 2001). Various genomics and proteomics profiling revealed that HSP's higher synthesis is correlated with abiotic stresses. A positive correlation exists between sHSP synthesized and the amount of stress and confers defense to plant. Experiment using gain-of-function mutation in genetically engineered crop plants showed higher tolerance in plants.

1.5.3 Antioxidative Enzymes

Stresses, both abiotic and biotic, results in cellular and metabolic imbalance, which in turn results into ROS production. Plant proteomics study and analysis showed higher amount of ROS cleansing antioxidative enzymes such as catalase (CAT), SOD, ascorbate peroxidase (APX), glutathione reductase (GR), and glutathione peroxidase (GPX) (Barkla et al., 2013; Gill and Tuteja, 2010). ROS generation takes place due to acceptance of electrons by molecular oxygen, which results into production of ROS for their oxidizing potential and deters cell integrity. ROS thus produced are hydrogen peroxide (H_2O_2), singlet oxygen ($^1O_2$), the hydroxyl radical (OH^-), and the superoxide radical ($O^-{}_2$) (Groß et al., 2013). Antioxidants both enzymatic and nonenzymatic are produced in response to stress. Direct correlation between salinity stress and increased activity of antioxidative enzymes include CAT, APX, GPX, SOD, and GR with increased accumulation of antioxidative metabolites (Figure 1.4) (Gupta et al., 2005).

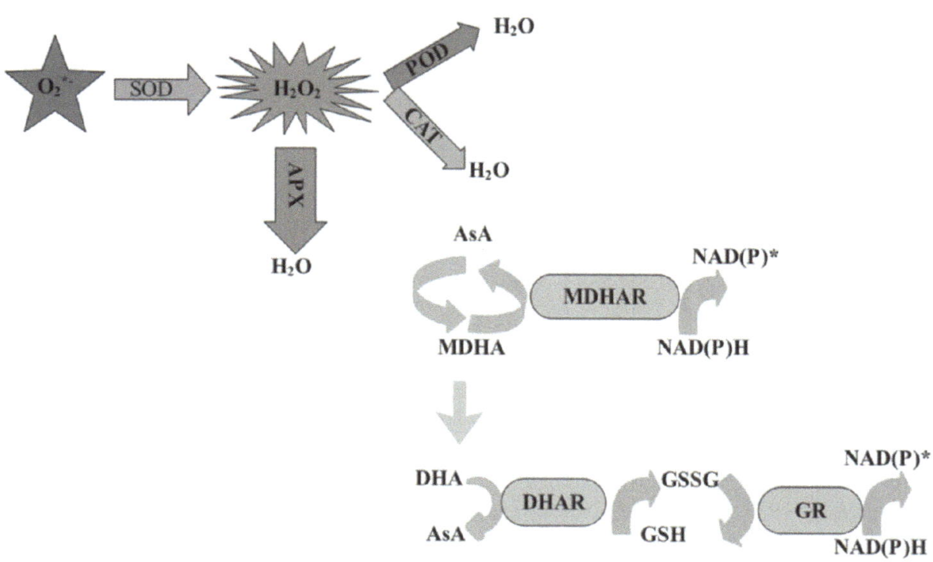

FIGURE 1.4 Different antioxidant enzymes and their mechanism of action in radicals scavenging.

1.6 CONCLUSION

In the era of climate change, the biochemical and NGS techniques ushered into the recognition of different TFs, small RNAs, and their regulatory role in stress tolerance. Understanding the molecular basis of stress tolerance in plants, which could find applications in development of stress-tolerant varieties of crop plant with enhanced yield using genetic engineering or through selective breeding.

ACKNOWLEDGMENTS

We would like to thank all the Indian funding agencies, which provided financial support (JRF, SRF and Merit Scholarship) to all the authors who have together contributed in this manuscript. The duly acknowledged funding agencies are University Grant Commission (UGC) and CCS Haryana Agriculture University (CCSHAU).

REFERENCES

Allen, E., Xie, Z., Gustafson, A. M., and Carrington, J. C. 2005. MicroRNA-directed phasing during trans-acting siRNA biogenesis in plants. *Cell* 121(2): 207–221.

Ashraf, M., and Foolad, M. R. 2007. Roles of glycine betaine and proline in improving plant abiotic stress resistance. *Environmental and Experimental Botany* 59(2): 206–216.

Aung, K., Lin, S. I., Wu, C. C., Huang, Y. T., Su, C. L., and Chiou, T. J. 2006. pho2, a Phosphate overaccumulator, is caused by a nonsense mutation in a microRNA399 target gene. *Plant Physiology* 141(3): 1000–1011.

Bari, R., Datt Pant, B., Stitt, M., and Scheible, W. R. 2006. PHO2, microRNA399, and PHR1 define a phosphate-signaling pathway in plants. *Plant Physiology* 141(3): 988–999.

Barkla, B. J., Castellanos-Cervantes, T., Diaz de Leon, et al. 2013. Elucidation of salt stress defense and tolerance mechanisms of crop plants using proteomics—current achievements and perspectives. *Proteomics* 13(12–13): 1885–1900.

Borsani, O., Zhu, J., Verslues, P. E., Sunkar, R., and Zhu, J. K. 2005. Endogenous siRNAs derived from a pair of natural cis-antisense transcripts regulate salt tolerance in Arabidopsis. *Cell* 123(7): 1279–1291.

Bovet, L., Feller, U., and Martinoia, E. 2005. Possible involvement of plant ABC transporters in cadmium detoxification: a cDNA sub-microarray approach. *Environment International* 31(2): 263–267.

Bray, E. A. 2002. Classification of genes differentially expressed during water-deficit stress in *Arabidopsis thaliana*: an analysis using microarray and differential expression data. *Annals of Botany* 89(7): 803–811.

Buhtz, A., Pieritz, J., Springer, F., and Kehr, J.2010. Phloem small RNAs, nutrient stress responses, and systemic mobility. *BMC Plant Biology* 10(1): 1–13.

Buhtz, A., Springer, F., Chappell L., Baulcombe, D.C., and Kehr, J. 2008. Identification and Characterization of small RNAs from the phloem of *Brassica napus*. *The Plant Journal* 53(5): 739–749.

Capitão, C., Paiva, J. A., Santos, D. M., and Fevereiro, P. 2011. In medicago truncatula, water deficit modulates the transcript accumulation of components of small RNA pathways. *BMC Plant Biology* 11(1): 1–14.

Chini, A., Fonseca, S. G. D. C., Fernandez, G., et al. 2007. The JAZ family of repressors is the missing link in jasmonate signalling. *Nature* 448(7154): 666–671.

Cona, A., Rea, G., Angelini, R., Federico, R., and Tavladoraki, P. 2006. Functions of amine oxidases in plant development and defence. *Trends in Plant Science* 11(2): 80–88.

de Ries, S.P., Lima, A.M., and de Souza, C.R.B. 2012. Recent molecular advances on downstream plant response to abiotic stress. *International Journal of Molecular Genetics* 13: 8628–8647.

El-Shintinawy, F., and El-Shourbagy, M. N. 2001. Alleviation of changes in protein metabolism in NaCl-stressed wheat seedlings by thiamine. *Biologia Plantarum* 44: 541–545.

Fujii, H., Chiou, T. J., Lin, S. I., Aung, K., and Zhu, J. K. 2005. A miRNA involved in phosphate-starvation response in Arabidopsis. *Current Biology* 15(22): 2038–2043.

Gadallah, M. A. A. 1999. Effects of proline and glycinebetaine on *Vicia faba* responses to salt stress. *Biologia Plantarum* 42(2): 249–257.

Galston, A. W., Kaur-Sawhney, R., Altabella, T., and Tiburcio, A. F. 1997. Plant polyamines in reproductive activity and response to abiotic stress. *Botanica Acta* 110(3): 197–207.

Gechev, T. S., Van Breusegem, F., Stone, J. M., Denev, I., and Laloi, C. 2006. Reactive oxygen species as signals that modulate plant stress responses and programmed cell death. *Bioessays* 28(11): 1091–1101.

Gill, S.S., and Tuteja, N. 2010. Reactive oxygen species and antioxidant machinery in abiotic stress tolerance in crop plants. *Plant Physiology and Biochemistry* 48(12); 909–930.

Groß, F., Durner, J., and Gaupels, F. 2013. Nitric oxide, antioxidants and prooxidants in plant defence responses. *Frontiers in Plant Science* 4: 419.

Gupta, K., Dey, A., and Gupta, B. 2013. Polyamines and their role in plant osmotic stress tolerance. *Climate Change and Plant Abiotic Stress Tolerance* 1: 1053–1072.

Gupta, K. J., Stoimenova, M., and Kaiser, W. M. 2005. In higher plants, only root mitochondria, but not leaf mitochondria reduce nitrite to NO, in vitro and in situ. *Journal of Experimental Botany* 56(420): 2601–2609.

Hamilton III, E. W., and Heckathorn, S. A. 2001. Mitochondrial adaptations to NaCl. Complex I is protected by anti-oxidants and small heat shock proteins, whereas complex II is protected by proline and betaine. *Plant Physiology* 126(3): 1266–1274.

Hanson, A. D., Rathinasabapathi, B., Rivoal, J., Burnet, M., Dillon, M. O., and Gage, D. A. 1994. Osmoprotective compounds in the Plumbaginaceae: a natural experiment in metabolic engineering of stress tolerance. *Proceedings of the National Academy of Sciences* 91(1): 306–310.

Hause, B., and Schaarschmidt, S. 2009. The role of jasmonates in mutualistic symbioses between plants and soil-born microorganisms. *Phytochemistry* 70(13–14): 1589–1599.

Hoque, M. A., Banu, M. N. A., Nakamura, Y., Shimoishi, Y., and Murata, Y. 2008. Proline and glycinebetaine enhance antioxidant defense and methylglyoxal detoxification systems and reduce NaCl-induced damage in cultured tobacco cells. *Journal of Plant Physiology* 165(8): 813–824.

Hoque, T. S., Uraji, M., Ye, W., Hossain, M. A., Nakamura, Y., and Murata, Y. 2012. Methylglyoxal-induced stomatal closure accompanied by peroxidase-mediated ROS production in Arabidopsis. *Journal of Plant Physiology* 169(10): 979–986.

Hossain, M. A., Piyatida, P., da Silva, J. A. T., and Fujita, M. 2012. Molecular mechanism of heavy metal toxicity and tolerance in plants: central role of glutathione in detoxification of reactive oxygen species and methylglyoxal and in heavy metal chelation. *Journal of Botany*. DOI: 10.1155/2012/872875.

Hu, X., Li, Y., Li, C., Yang, H., Wang, W., and Lu, M. 2010. Characterization of small heat shock proteins associated with maize tolerance to combined drought and heat stress. *Journal of Plant Growth Regulation* 29(4): 455–464.

Huang, Q. S., Wang, H. Y., Gao, P., Wang, G. Y., and Xia, G. X. 2008. Cloning and characterization of a calcium dependent protein kinase gene associated with cotton fiber development. *Plant Cell Reports* 27(12): 1869–1875.

Kamthan, A., Chaudhuri, A., Kamthan, M., and Datta, A. 2015. Small RNAs in plants: recent development and application for crop improvement. *Frontiers in Plant Science* 6: 208.

Katiyar-Agarwal, S., Morgan, R., Dahlbeck, D., and Borsani, O., et al. 2006. A pathogen-inducible endogenous siRNA in plant immunity. *Proceedings of the National Academy of Sciences* 103(47), 18002–18007.

Knott, J. M., Römer, P., & Sumper, M. 2007. Putative spermine synthases from *Thalassiosirapseudonana* and *Arabidopsis thaliana* synthesize thermospermine rather than spermine. *FEBS Letters* 581(16): 3081–3086.

Kovács, Z., Simon-Sarkadi, L., Szűcs, A., and Kocsy, G. 2010. Differential effects of cold, osmotic stress and abscisic acid on polyamine accumulation in wheat. *Amino Acids* 38(2): 623–631.

Kovalchuk, I., Titov, V., Hohn, B., and Kovalchuk, O. 2005. Transcriptome profiling reveals similarities and differences in plant responses to cadmium and lead. *Mutation Research/Fundamental and Molecular Mechanisms of Mutagenesis* 570(2): 149–161.

Kulcheski, F. R., de Oliveira, L. F., and Molina, L. G., et al. 2011. Identification of novel soybean microRNAs involved in abiotic and biotic stresses. *BMC Genomics* 12(1): 1–17.

Kuromori, T., Miyaji, T., Yabuuchi, H., and Shimizu, H., et al. 2010. ABC transporter AtABCG25 is involved in abscisic acid transport and responses. *Proceedings of the National Academy of Sciences* 107(5): 2361–2366. DOI: 10.1073/pnas.0912516107.

Li, Q.F., and He, J.X. 2016. BZR1 interacts with HY5 to mediate brassinosteroid-and light-regulated cotyledon opening in Arabidopsis in darkness. *Molecular Plant*, 9(1): 113–125.

Li, T., Li, H., Zhang, Y. X., and Liu, J. Y. 2011. Identification and analysis of seven H_2O_2-responsive miRNAs and 32 new miRNAs in the seedlings of rice (*Oryza sativa* L. ssp. indica). *Nucleic Acids Research* 39(7): 2821–2833.

Mäkelä, P., Kärkkäinen, J., and Somersalo, S. J. B. P. 2000. Effect of glycinebetaine on chloroplast ultrastructure, chlorophyll and protein content, and RuBPCO activities in tomato grown under drought or salinity. *Biologia Plantarum* 43(3): 471–475.

Malik, A., Mor, V.S., Tokas, J., Punia, H., Malik, S., Malik, K., Sangwan, S., Tomar, S., Singh, P., Singh, N. and Himangini, 2020. Biostimulant-treated seedlings under sustainable agriculture: A global perspective facing climate change. Agronomy, 11(1), p.14.

Mittler, R., Frinka, A., and Goloubinoff, P. 2012. How do plants feel the heat? *Trends in Biochemical Sciences* 37(3): 118–125.

Örvar, B. L., Sangwan, V., Omann, F., and Dhindsa, R. S. 2000. Early steps in cold sensing by plant cells: the role of actin cytoskeleton and membrane fluidity. *The Plant Journal* 23(6): 785–794.

Punia, H., Madan, S., Malik, A., and Sethi, S.K. 2019. Stability analysis for quality attributes in durum wheat (*Triticum durum L.*) genotypes. *Bangladesh Journal of Botany* 48(4): 967–972.

Punia, H., Tokas, J., Bhadu, S., Mohanty, A.K., Rawat, P., Malik, A., and Satpal, 2020a. Proteome dynamics and transcriptome profiling in sorghum [*Sorghum bicolor* (L.) *Moench*] under salt stress. *3 Biotech* 10: 1–10.

Punia, H., Tokas, J., Malik, A., Satpal, Rani, A., Gupta, P., Kumari, A., Mor, V.S., Bhuker, A., and Kumar, S. 2020b. Solar radiation and nitrogen use efficiency for sustainable agriculture. In: Kumar, S., Meena, R.S., Jhariya, M.K. (eds) *Resources Use Efficiency in Agriculture*. Springer, Singapore, 177–212.

Punia, H., Tokas, J., Malik, A., Bajguz, A., El-Sheikh, M.A., and Ahmad, P. 2021a. Ascorbate–glutathione oxidant scavengers, metabolome analysis and adaptation mechanisms of Ion exclusion in sorghum under salt stress. *International Journal of Molecular Sciences* 22(24): 13249.

Punia, H., Tokas, J., Mor, V.S., Bhuker, A., Malik, A., Singh, N., Satpal, Alsahli, A.A., and Hefft, D.I. 2021b. Deciphering reserve mobilization, antioxidant potential, and expression analysis of starch synthesis in sorghum seedlings under salt stress. *Plants*, 10(11): 2463.

Punia, H., Tokas, J., Malik, A., Sangwan, S., Rani, A., Yashveer, S., Alansi, S., Hashim, M.J., and El-Sheikh, M.A. 2021c. Genome-wide transcriptome profiling, characterization, and functional identification of NAC transcription factors in sorghum under salt stress. *Antioxidants* 10(10): 1605.

Roy, M., and Wu, R. 2002. Overexpression of S-adenosylmethionine decarboxylase gene in rice increases polyamine level and enhances sodium chloride-stress tolerance. *Plant Science* 163(5): 987–992.

Ruiz-Ferrer, V., and Voinnet, O. 2009. Roles of plant small RNAs in biotic stress responses. *Annual Review of Plant Biology* 60: 485–510.

Saito, K., and Matsuda, F. 2010. Metabolomics for functional genomics, systems biology, and biotechnology. *Annual Review of Plant Biology* 61: 463–489.

Sanda, S., Yoshida, K., Kuwano, M., Kawamura, T., Munekage, Y. N., Akashi, K., and Yokota, A. 2011. Responses of the photosynthetic electron transport system to excess light energy caused by water deficit in wild watermelon. *Physiologia Plantarum* 142(3): 247–264.

Saxena, S. C., Kaur, H., Verma, P., Petla, B. P., Andugula, V. R., and Majee, M. 2013. Osmoprotectants: potential for crop improvement under adverse conditions. In Tuteja, N., Singh Gill, S. (eds.) *Plant Acclimation to Environmental Stress* (pp. 197–232). Springer, New York, NY.

Schafleitner, R., Rosales, R. O. G., Gaudin, A., Aliaga, C. A. A., Martinez, G. N., Marca, L. R. T. and Bonierbale, M. 2007. Capturing candidate drought tolerance traits in two native Andean potato clones by transcription profiling of field grown plants under water stress. *Plant Physiology and Biochemistry* 45(9), 673–690.

Schmidt, R., Mieulet, D., Hubberten, H. M., Obata, T., Hoefgen, R., Fernie, A. R., ... and Mueller-Roeber, B. 2013. SALT-RESPONSIVE ERF1 regulates reactive oxygen species–dependent signaling during the initial response to salt stress in rice. *The Plant Cell* 25(6): 2115–2131.

Schwind, N., Zwiebel, M., Itaya, A., Ding, B., Wang, M. B., Krczal, G., and Wassenegger, M. 2009. RNAi-mediated resistance to potato spindle tuber viroid in transgenic tomato expressing a viroid hairpin RNA construct. *Molecular Plant Pathology* 10(4): 459–469. DOI: 10.1111/j.1364-3703.2009.00546.x.

Serra, T. S., Figueiredo, D. D., Cordeiro, A. M., Almeida, D. M., Lourenço, T., Abreu, I. A., ... and Saibo, N. J. 2013. OsRMC, a negative regulator of salt stress response in rice, is regulated by two AP2/ERF transcription factors. *Plant Molecular Biology* 82(4): 439–455.

Shulaev, V., Cortes, D., Miller, G., and Mittler, R. 2008. Metabolomics for plant stress response. *Physiologia Plantarum* 132(2): 199–208.

Simón-Mateo, C., and García, J. A. 2011. Antiviral strategies in plants based on RNA silencing. *Biochimica et BiophysicaActa (BBA)-Gene Regulatory Mechanisms* 1809: 722–731. DOI: 10.1016/j.bbagrm.2011.05.011.

Simpson, G. G., Dijkwel, P. P., Quesada, V., Henderson, I., and Dean, C. 2003. FY is an RNA 3′ end-processing factor that interacts with FCA to control the Arabidopsis floral transition. *Cell* 113(6), 777–787.

Sunkar, R., Kapoor, A., and Zhu, J. K. 2006. Posttranscriptional induction of two Cu/Zn superoxide dismutase genes in Arabidopsis is mediated by downregulation of miR398 and important for oxidative stress tolerance. *The Plant Cell* 18: 2051–2065.

Takahashi, T., and Kakehi, J. I. 2010. Polyamines: ubiquitous polycations with unique roles in growth and stress responses. *Annals of Botany* 105(1): 1–6.

Tokas, J., Punia, H., Malik, A., Sangwan, S., Devi, S. and Malik, S., 2021. Growth performance, nutritional status, forage yield and photosynthetic use efficiency of sorghum [Sorghum bicolor (L.) Moench] under salt stress. Range Management and Agroforestry, 42(1), pp.59-70.

Way, H., Chapman, S., McIntyre, L., Casu, R., Xue, G.P., Manners, J., and Shorter, R. 2005. Identification of differentially expressed genes in wheat undergoing gradual water deficit stress using a subtractive hybridisation approach. *Plant Science* 168: 661–670.

Wu, W., Zhang, Q., Ervin, E. H., Yang, Z., and Zhang, X. 2017. Physiological mechanism of enhancing salt stress tolerance of perennial ryegrass by 24-epibrassinolide. *Frontiers in Plant Science* 8, 1017.

Xin, M., Wang, Y., Yao, Y., Xie, C., Peng, H., Ni, Z., and Sun, Q. 2010. Diverse set of microRNAs are responsive to powdery mildew infection and heat stress in wheat (*Triticum aestivum* L.). *BMC Plant Biology* 10(1): 1–11.

Yamaguchishinozaki, K., Shinozaki, K. 2006. Transcriptional regulatory networks in cellular responses and tolerance to dehydration and cold stresses. *Annual Review of Plant Biology* 57: 781–803. DOI: 10.4161/psb.5.6.11398.

Yan, Y. S., Zhang, Y. M., Yang, K., Sun, Z. X., Fu, Y. P., Chen, X. Y., and Fang, R. X. 2011. Small RNAs from MITE-derived stem–loop precursors regulate abscisic acid signaling and abiotic stress responses in rice. *The Plant Journal* 65: 820–828.

Zhang, J., Subramanian, S., Zhang, Y., and Yu, O. 2007. Flavone synthases from Medicago truncatula are fl avanone-2-hydroxylases and are important for nodulation. *Plant Physiology* 144: 741–751.

Zhang, J., Xu, Y., Huan, Q., and Chong, K. 2009. Deep sequencing of Brachypodium small RNAs at the global genome level identifies microRNAs involved in cold stress response. *BMC Genomics* 10: 449.

2 Systems Biology Approach Unfolds Unique Life-History Strategies in Response to Abiotic Stress Combinations

Lidor Shaar-Moshe
The Hebrew University of Jerusalem
University of Haifa

Zvi Peleg
The Hebrew University of Jerusalem

CONTENTS

2.1 PLANT RESPONSES TO COMBINED STRESSES DEVIATE FROM RESPONSES TO THE CORRESPONDING SINGLE STRESSES

Abiotic stresses, such as drought, soil salinization, and heatwaves, are the major and most prevalent environmental constraints on crop productivity, causing devastating economic and social impacts worldwide (Vautard et al., 2007; Maggio et al., 2011). Combinations of environmental stresses are predicted to aggravate in the near future and surpass the adverse effects single stresses have on plant development and yield production (Mooney and Winner, 1991; Shaar-Moshe et al., 2017; Zandalinas et al., 2018, 2022; Rivero et al., 2022; Zandalinas and Mittler, 2022). Currently, most stress experiments focus on plant responses to single stresses. Several studies suggested that overlapping responses of single stresses, especially at the transcriptional level, act as a generalized stress mechanism (e.g., Seki et al., 2001, 2002; Kilian et al., 2007; Priest et al., 2014). Yet, stress combination studies demonstrated the limited ability to predict plant responses as different stresses are combined. These studies highlight the current knowledge gap of plant acclimation mechanisms to combined stresses (e.g., Rizhsky et al., 2002; Prasch and Sonnewald, 2013; Rivero et al., 2014; Sewelam et al., 2014; Shaar-Moshe et al., 2017; Zandalinas et al., 2018).

Natural environmental conditions are constantly changing (Figure 2.1). Thus, throughout their life cycle, plants are routinely exposed to different combinations of environmental constraints

DOI: 10.1201/9781003300564-2

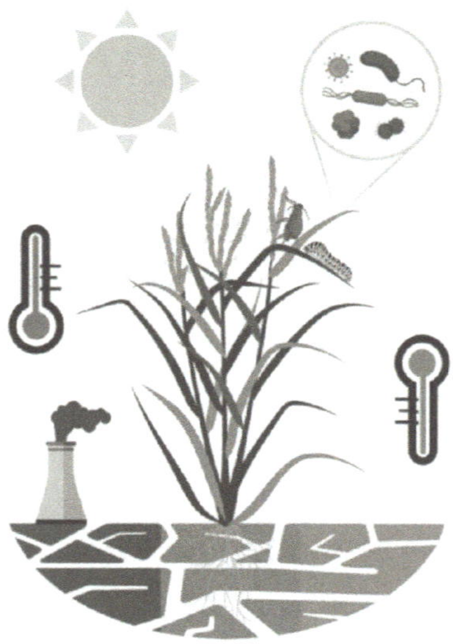

FIGURE 2.1 Major abiotic and biotic stresses that can occur individually or in different combinations and impact plant growth and production in natural and agricultural systems. The intensity, duration, and simultaneous co-occurrence of these and additional stresses are expected to aggravate and exceed the detrimental effect individual stresses impose on plant development, health, and productivity. Figure created with BioRender.com.

(Mittler, 2006). These stresses can develop progressively (i.e., salinity and drought) or rapidly (i.e., extreme temperature fluctuations) (Mickelbart et al., 2015). The occurrence of stresses at specific developmental stages, the nature of the stresses, and their concurrence can affect the acclimation mechanisms undertaken by the plants (Yamaguchi-Shinozaki and Shinozaki, 2006; Peleg and Blumwald, 2011; Claeys and Inzé, 2013; Mickelbart et al., 2015). These processes are classically divided into escape and resistance, and subdivided into avoidance and tolerance strategies (Levitt, 1980; Blum, 1988). ***The escape strategy***, which relies on successful completion of reproduction before the onset of severe stress (developmental plasticity), can be achieved by early flowering and/or short growth duration. ***The avoidance strategy*** can be defined as the plant's ability to retain a relatively higher level of 'hydration' under conditions of soil or atmospheric water stress. While ***dehydration tolerance*** is described as the plant's ability to maintain metabolism at low tissue water potential and continue growing despite dehydration of the plant or to recover after release from stress conditions. Under field conditions, plants combine different response types to cope with the dynamic environment (Ludlow, 1989). Response to a combination of environmental stresses involves multiple pathways and cross-talk between different sensors and signal transduction pathways (Mittler, 2006) that together elicit plastic phenotypic modifications. These modifications involve the mobilization of the whole metabolic machinery toward a new steady-state level, through a series of coordinated physiological, cellular, and molecular events (Chaves et al., 2009; Shaar-Moshe et al., 2017, 2019; Zandalinas et al., 2021). One of the major challenges of agricultural research is to decipher the molecular mechanisms that underlie plant acclimation to environmental stresses. In the past decades, plant biology studies have shifted from focusing on single-gene functions to exploring genetic pathways and regulatory networks, using genome-scale analyses (Moreno-Risueno et al., 2010). While high-throughput studies elaborated our understanding of molecular networks and their complexity, the degree to which transcriptional alterations shape the phenotype and the genetic variability that regulates plant stress tolerance are largely unknown (Sanchez, 2013). A possible means to overcome this challenge is to apply a systems biology approach that integrates large-scale measurements of different types of molecular components and regulatory levels (e.g. genomics, transcriptomics, epigenetics, proteomics, and metabolomics) (Fukushima et al., 2009) with measurements of plant physiology and noninvasive phenotyping (i.e., phenomics) at the tissue, organ, and whole plant level (Großkinsky et al., 2015). This complementary approach holds the potential to bridge the gap between genotype and phenotype and develop novel climate-resilient crops.

2.2 SHORT- *VERSUS* LONG-TERM PHENOTYPIC PLASTICITY UNDER STRESS COMBINATION

Exposure of plants to a combination of two stresses resulted in a specific physiological, molecular, and metabolic stress response that could not be inferred from exposure to the individual stresses (Rizhsky et al., 2004; Prasch and Sonnewald, 2013; Rasmussen et al., 2013; Sewelam et al., 2014; Shaar-Moshe et al., 2017; Cohen et al., 2021). These responses are affected by interactions between the different stresses, which can act in various modalities, including antagonistically, synergistically, additively, or neutrally. An antagonistic interaction resulting from a combination of heat and drought stress was demonstrated among several species. Although these studies involve various plant species, developmental stages, and stress procedures, the underlying physiological responses were largely similar and resulted in more severe damage under combinations of drought and heat than each of the single stresses (Zandalinas et al., 2018). *Nicotiana tabacum* and *Arabidopsis* plants that were subjected to heat stress opened their stomata to cool the tissue by transpiration. However, when heat stress was combined with drought, plant stomata remain closed, resulting in higher leaf temperature than under heat stress alone (Rizhsky et al., 2002; 2004). Reductions in leaf chlorophyll content and photosynthetic functions (i.e., photosynthetic rate and/or PSII activity), along with decreases in shoot dry weight and/or yield components, were detected under combinations of drought and heat in *Brachypodium distachyon* (Shaar-Moshe et al., 2017), *Triticum aestivum* (Prasad et al., 2011), *Leymus chinensis* (Xu and Zhou, 2006), *Cicer arietinum* (Awasthi et al., 2014), and citrus plants (Zandalinas et al., 2016). These results suggest the conservation of plants' responses to combinations of drought and heat across different species.

Rivero et al. (2014) showed that under a hydroponic system, *Solanum lycopersicum* plants exposed to a combination of heat and salinity exhibited a higher degree of tolerance compared with salt stress alone. This result, which suggests a synergistic effect, stands in contrast to the expected negative effect of heat and salt combination since enhanced transpiration to cool the leaf is projected to result in enhanced toxic ion uptake (Mittler, 2006). Another hydroponics study of combined salinity and heat stress in *S. lycopersicum* reported on aggravating effects of the combination (Chen et al., 2015). The discrepancy between the two tomato studies may arise from the different durations of the stresses and the developmental stages of the plants. While Rivero et al. (2014) focused on a fast response (24–72 hours after stress application), Chen et al. (2015) targeted long-term modifications or acclimation to salinity and heat combination (seven weeks after the stress was applied). Antagonistic or synergistic effects of the combination, compared with the single salinity or heat stress, were reported for other species (Zandalinas and Mittler, 2022). These studies exemplify the variability of plant responses, depending on the experimental design, and emphasize the need to impose realistic abiotic stresses that are similar to the conditions plants are likely to encounter in natural and agricultural systems.

For temperate grasses growing in the Mediterranean-like climate, the reproductive stage is one of the most sensitive stages in determining yield under stress (Beáta et al., 2008). The sensitivity of this stage is underscored considering that salinity and drought develop gradually and reach a high level of severity at the reproductive stages and that heatwaves occur during the spring, while grasses are flowering (Loss and Siddique, 1994). To survive the dynamic and unpredictable environment, plants must elicit a response that matches the environmental cues and their developmental stage.

An individual genotype can express a range of phenotypes in alternative environments, which are termed *norms of reaction*. These response patterns are attributed to plants' capacity to express phenotypic and/or developmental plasticity, which entails adaptive, maladaptive, or neutral outcomes on plant fitness and survival (Sultan, 2017). Cluster analysis of morphological and yield parameters, of *B. distachyon* plants subjected to single and combined stresses, highlighted the similar adverse effects of salinity treatments on plant development, including reductions in leaf area, shoot height, and yield. Under salinity treatments, leaf number per tiller and stomatal density increased, partly compensating for the reduction in plant growth and biomass accumulation. These long-term adjustments, which can be referred to as developmental plasticity, occurred during the early stages of the life cycle, parallel with the progressive exposure of the plants to salinity stress (Shaar-Moshe et al., 2017). Depending on the effect plasticity has on fitness, reaction norms can be characterized by similar phenotypes across environments (i.e., canalization) or by phenotypes that differ in specific functional traits (adaptive phenotypic plasticity) (Nijhout, 2003). Examination of physiological traits, which were clustered together under heat treatments, implies that plant acclimation to heat involves short-term phenotypic plasticity. These traits include stomatal conductance and leaf temperature (Shaar-Moshe et al., 2017), which exhibited similar values across different stresses (canalization). Other traits exhibited a variable response pattern that can be attributed to the intensity of the stress, for example, osmotic adjustment (Blum, 2017), or to antagonistic interactions between stresses that were found for transpiration under drought, heat, and their combination (Rizhsky et al., 2002; Shaar-Moshe et al., 2017).

2.3 FROM TRANSCRIPTIONAL CHANGES TO MORPHO-PHYSIOLOGICAL RESPONSES

The transcriptional response of plants to combinations of stresses is in part unique and cannot be deduced from plant responses to the corresponding single stresses (reviewed by Zandalinas et al., 2021). The limited ability to predict transcriptional changes is further augmented among partially overlapping stress combinations. *Brachypodium* plants subjected to a combination of salinity, drought, and heat showed a 37% increase in the number of differentially expressed genes (DEGs) that are assigned to expression modes that could not be predicted from transcriptional responses to double stress combinations. In addition, compared with single stresses, the number of DEGs increased under combinations of stresses, where 24%–30% of the transcriptome was modulated, and on average 13% of these genes were treatment-specific (Shaar-Moshe et al., 2017).

Transcriptional analysis of *B. distachyon* plants subjected to double and triple stress combinations assigned only a limited number of common stress genes (3%–9%) to antagonistic mode (Shaar-Moshe et al., 2017). A similar percentage of genes (5%–10%) were allocated to antagonistic responses in a transcriptional analysis of *Arabidopsis* plants subjected to six combinations of double stresses (Rasmussen et al., 2013). However, the number of genes assigned to the antagonistic mode peaked to 45% when genes uniquely expressed only under combinations of stresses were allocated to the different response modes (Shaar-Moshe et al., 2017). The increased number of genes exhibiting antagonistic responses, which mainly resulted from lack of expression under the triple combination of salinity, drought, and heat, implies the shutdown of unique combination pathways under this treatment. Furthermore, this finding suggests the importance of the core stress genes and processes to sustain plant growth and survival under such a harsh stress combination (Figure 2.2).

An increase in the number of DEGs under combinations of stresses was detected in other studies using different experimental procedures (Rizhsky et al., 2004; Prasch and Sonnewald, 2013). Yet, this massive transcriptional reprogramming was not manifested at the physiological level, where the phenotypes under combined stresses generally canalized in the same direction and magnitude as the single stresses (Vile et al., 2012; Ahmed et al., 2013; Shaar-Moshe et al., 2017). From a systems biology perspective, transcriptional regulation is one of several system regulatory mechanisms that

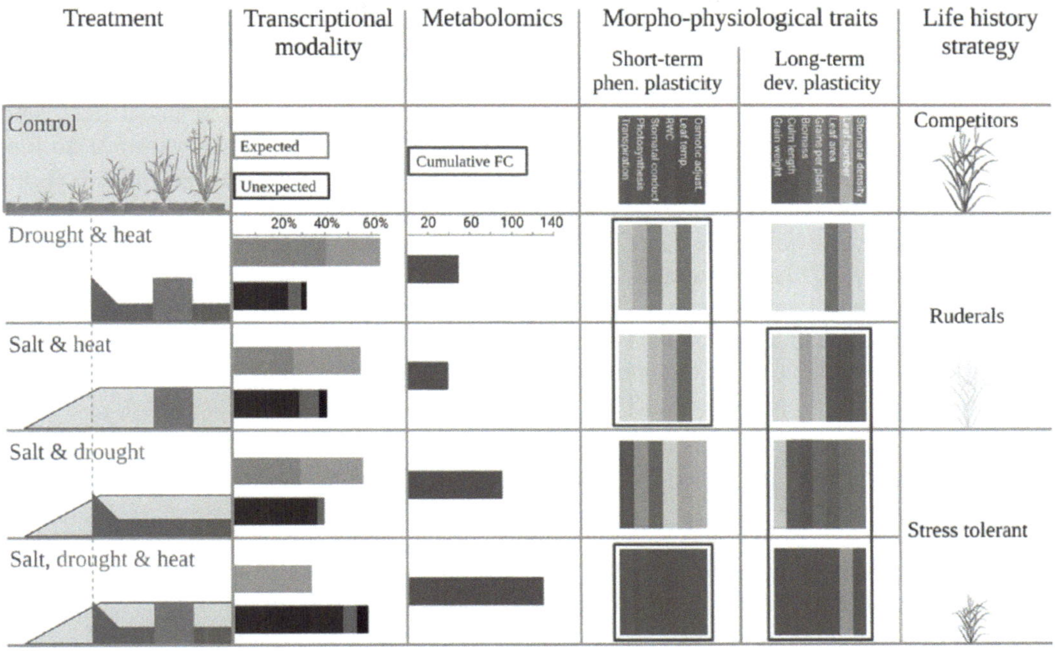

FIGURE 2.2 Plant plasticity and life-history strategies in response to combined stresses. *Brachypodium distachyon* plants were subjected to combinations of salinity, drought, and heat stress at distinct developmental stages to simulate the natural Mediterranean-like field conditions. Transcriptional patterns of DEGs shared among single and combined stresses were assigned into expected (i.e., neutral, additive, and equalization) and unexpected (i.e., synergistic, dominant, and antagonistic) modalities, indicated by different shades of light and dark gray, respectively. Compared with the double stress combinations, most DEGs in the triple stress combination were assigned to unexpected modalities (63%). Cumulative fold change (FC) of significant differentially accumulating compatible solutes demonstrated a gradual accumulation among the combined stresses, with the highest values under the triple stress combinations. Analysis of physiological traits revealed treatment clustering based on heat and non-heat stresses, implying that plant acclimation to heat involves short-term phenotypic (phen.) plasticity. Cluster analysis of morphological and yield parameters highlighted the dominancy of salt treatments on long-term developmental (dev.) plasticity that occurred at the early stages of plants' life cycle in accordance with the progressive exposure to salt stress. Red and blue colors indicate high and low values compared with the mean of each trait, respectively. The extent of transcriptional, metabolic, and phenotypic plasticity shaped the life-history strategies plants demonstrated in response to the different stress combinations (Shaar-Moshe et al., 2017, 2019). Figure created with BioRender.com.

act synergistically to suppress the propagation of variation to the phenotypic level, thus buffering the transcriptional alterations. For example, transcript abundance, and hence its biological significance, may not correlate with the activity of the gene product, especially under abiotic stresses, due to posttranscriptional regulation (Deyholos, 2010). In the absence of robust system architecture, minor expression alterations in thousands of genes would result in massive changes at the phenotype level (Fu et al., 2009). Such robustness functions as a capacitor against environmental changes and ensures normal development within suitable and relatively narrow boundaries of plant plasticity (Queitsch et al., 2002) and should be carefully considered and validated by additional regulatory levels and physiological responses (Blum, 2014).

Despite inherent dissimilarities, different stresses result in similar physiological effects, such as limited water availability, ion imbalance, impairment of the photosynthetic machinery, and growth inhibition. These similarities give rise to common acclimation processes, for example, osmotic adjustment, antioxidant production, membrane stability maintenance, growth modulation, and alteration in the source–sink relationship. Therefore, it is not surprising that in addition to the unique responses plants elicit under combinations of stresses, they also share common responses among single and combined stresses, at the transcriptional and metabolic levels (Rasmussen et al., 2013; Sewelam et al., 2014; Shaar-Moshe et al., 2017, 2019). Co-expression network analysis coupled with enrichment analysis of genes common to single and combined stresses (i.e., common stress genes) detected pathways and potential hub genes that are known to be involved in abiotic stress acclimation. These processes, which included degradation and transport of amino acids, metabolism of the raffinose family and trehalose, posttranslational modifications, and signaling, are suggested as core stress processes that are functionally conserved under both single and combined stresses (Shaar-Moshe et al., 2017). Importantly, osmotic adjustment plays a fundamental role in plant acclimation to single and combined stresses across species (Shaar-Moshe et al., 2015, 2017, 2019). Thus, genetic manipulation of these common stress genes may be an efficient strategy to achieve a higher tolerance across multiple types of environmental stresses, allowing the development of 'multipurpose' stress-tolerant crops (Sanchez, 2013).

2.4 ALTERATIONS IN LIFE-HISTORY STRATEGIES IN RESPONSE TO DIFFERENT ENVIRONMENTAL STRESSES

Plant life-history theory postulates that resource allocation is partitioned to maximize fitness in different environments, resulting in the occurrence of plants with diverse traits (Herms and Mattson, 1992). Grime (1977) defined three types of plant life-history strategies: competitors (C), stress-tolerant (S), and ruderals (R) in the C–S–R model. These categories represent trade-offs between growth, reproduction, and maintenance. Most *B. distachyon* accessions tend to favor ruderal strategy, yet competition and stress tolerance were also found in certain environments and seasons (Wilson et al., 2015). Focusing on *B. distachyon* plants from a single accession (Bd 21–3), Shaar-Moshe et al. (2019) found that this accession exhibited features of the different strategies under various treatments. Under control conditions, plants tend to competitors' characteristics, which were manifested by rapid growth with extensive resource acquisition. In contrast, under a combination of salinity and drought and the triple combination of salinity, drought, and heat, plants demonstrated classic stress-tolerant strategy, through slow growth, small stature, and low yield, as they allocated resources to defense processes. Ruderal strategy, which was characterized by small stature and relatively higher yield, was favored to some extent under the double combinations of salinity and heat or drought and heat (Shaar-Moshe et al., 2019). The plasticity and diversity in life-history strategies of *B. distachyon* may indicate an adaptive potential of this species to colonize variable and changing environments (Figure 2.2).

2.5 EVOLUTIONARY, ECOLOGICAL, AND AGRICULTURAL CONSEQUENCES OF STRESS COMBINATIONS FOR FUTURE FOOD SECURITY

It was assumed that under combinations of stresses plants elicit responses that are in part common with the individual stresses and in part unique to the combined stresses (Mittler, 2006). Derived from this hypothesis is the possibility that each combination has its unique responses. Shaar-Moshe et al. (2017) suggested that it is less likely that different combinations of stresses resulted in the development of new pathways and processes. Rather the expression of different gene subsets is altered according to the environmental cues. These findings coincide with previous reports on the limited ability to predict plant responses to combined stresses based on the corresponding single stresses (Suzuki et al., 2014), and further highlight combined stresses as discrete and autonomous stresses. The combined stresses resulted in differential impairment of plant fitness, predominantly combinations of salinity and drought, with or without heat, which diminished plant fitness (Shaar-Moshe et al., 2019). The reduction in fitness and genetic diversity can potentially reduce plants' ability to respond to environmental stresses (Reusch et al., 2005) and increase their extinction risk (Newman and Pilson, 1997).

In the face of the ongoing and projected climate change, breeding of novel crop plants with enhanced yield potential and improved resilience to stressful environments is crucial for global food security. The multifaced effects of stress combinations on crop production include both direct (e.g., climate-associated yield instability) and indirect factors (e.g., reduced soil fertility, microbial activity, and carbon sequestration). To meet the demands of the growing human population, a significant increase in crop-plant production will be needed (Sade and Peleg, 2020). Future food production will rely on our understanding of these various mechanisms and their integration, via rapid advances in crop phenomics and genomics technologies, into the development and cultivation of more climate-resilient crop plants. In areas prone to combinations of terminal drought and heatwaves, cereal crop varieties that have a rapid life cycle, efficient photoassimilate transport, and high sink strength would perform better and potentially produce higher yields. On the other hand, in regions characterized by high soil salinization and hot temperatures, varieties with improved osmoregulation that enables the preservation of normal water content through ion accumulation are expected to outperform. These mechanisms can potentially alter the trade-off between defense and reproductive output and prioritize yield production to ensure food security.

REFERENCES

Ahmed, I. M., Cao, F., Zhang, M., Chen, X., Zhang, G., and Wu, F. 2013. Difference in yield and physiological features in response to drought and salinity combined stress during anthesis in Tibetan wild and cultivated barleys. *PLoS One* 8:e77869. doi: 10.1371/journal.pone.0077869.

Awasthi, R., Kaushal, N., Vadez, V., Turner, N.C., Berger, J., Siddique, K.H.M., and Nayyar, H. 2014. Individual and combined effects of transient drought and heat stress on carbon assimilation and seed filling in chickpea. *Functional Plant Biology* 41:1148–1167. doi: 10.1071/FP13340. PMID: 32481065.

Beáta, B., Jäger, K., and Fehér, A. 2008. The effect of drought and heat stress on reproductive processes in cereals. *Plant, Cell and Environment* 31:11–38. doi: 10.1111/j.1365-3040.2007.01727.x.

Blum, A. 1988. *Plant Breeding for Stress Environments*. Boca Raton, FL: CRC Press. Inc.

Blum, A. 2014. Genomics for drought resistance - getting down to earth. *Functional Plant Biology* 41:1191–1198. doi: 10.1071/FP14018.

Blum, A. 2017. Osmotic adjustment is a prime drought stress adaptive engine in support of plant production. *Plant, Cell and Environment* 40:4–10. doi: 10.1111/pce.12800.

Chaves, M. M., Flexas, J., and Pinheiro, C. 2009. Photosynthesis under drought and salt stress: regulation mechanisms from whole plant to cell. *Annals of Botany* 103:551–560. doi: 10.1093/aob/mcn125.

Chen, T. W., Nguyen, T. M., Kahlen, K., and Stützel, H. 2015. High temperature and vapor pressure deficit aggravate architectural effects but ameliorate non-architectural effects of salinity on dry mass production of tomato. *Frontiers in plant science* 6: 887. doi.org/10.3389/fpls.2015.00887.

Claeys, H., and Inzé, D. 2013. The agony of choice: How plants balance growth and survival under water-limiting conditions. *Plant Physiology* 162:1768–1779. doi: 10.1104/pp.113.220921.

Cohen, I., Zandalinas, S. I., Huck, C., Fritschi, F. B., and Mittler, R. 2021. Meta-analysis of drought and heat stress combination impact on crop yield and yield components. *Physiologia Plantarum* 171:66–76. doi: 10.1111/ppl.13203.

Deyholos, M. K. 2010. Making the most of drought and salinity transcriptomics. *Plant, Cell and Environment* 33:648–654. doi: 10.1111/j.1365-3040.2009.02092.x.

Fu, J., Keurentjes, J. J. B., Bouwmeester, H., America, T., Verstappen, F. W. A., Ward, J. L., et al. 2009. System-wide molecular evidence for phenotypic buffering in *Arabidopsis Nature Genetics* 41:166–167. doi: 10.1038/ng.308.

Fukushima, A., Kusano, M., Redestig, H., Arita, M., and Saito, K. 2009. Integrated omics approaches in plant systems biology. *Current Opinion in Chemical Biology* 13:532–538. doi: 10.1016/j.cbpa.2009.09.022.

Grime, J. P. 1977. Evidence for the existence of three primary strategies in plants and its relevance to ecological and evolutionary theory. *American Naturalist* 111:1169–1194. doi: 10.1086/283244.

Großkinsky, D. K., Svensgaard, J., Christensen, S., and Roitsch, T. 2015. Plant phenomics and the need for physiological phenotyping across scales to narrow the genotype-to-phenotype knowledge gap. *Journal of Experimental Botany* 66:5429–5440. doi: 10.1093/jxb/erv345.

Herms, D. A., and Mattson, W. J. 1992. The dilemma of plants: To grow or defend. *Quarterly Review of Biology* 67:283–335. doi: 10.1086/417659.

Kilian, J., Whitehead, D., Horak, J., Wanke, D., Weinl, S., Batistic, O., et al. 2007. The AtGenExpress global stress expression data set: protocols, evaluation and model data analysis of UV-B light, drought and cold stress responses. *Plant Journal* 50:347–363. doi: 10.1111/j.1365-313X.2007.03052.x.

Levitt, J. 1980. *Responses of Plants to Environmental Stresses: Water, Radiation, Salt, and Other Stresses*. New York: Academic Press.

Loss, S. P., and Siddique, K. H. M. 1994. Morphological and physiological traits associated with wheat yield increases in mediterranean environments. In *Advances in Agronomy*, ed. D. L. Sparks, vol. 52, pp. 229–276. Academic Press. doi: 10.1016/S0065-2113(08)60625-2.

Ludlow, M. M. 1989. Strategies of response to water stress. In: Kreeeb, K.H., Richter, H. and Hinckley, T.M., Eds., Structural and Functional Responses to Environmental Stresses, SPB Academic Publishing, The Hague, Natherlands. pp. 269–281.

Maggio, A., Hasegawa, P. M., Bressan, R. A., Federica, C. M., Joly, R. J. 2001. Review: Unravelling the functional relationship between root anatomy and stress tolerance. *Functional Plant Biology* 28:999–1004. doi: 10.1071/PP01099

Mickelbart, M. V., Hasegawa, P. M. and Bailey-Serres, J. 2015. Genetic mechanisms of abiotic stress tolerance that translate to crop yield stability. *Nature Reviews Genetics* 16:237–251. doi: 10.1038/nrg3901.

Mittler, R. 2006. Abiotic stress, the field environment and stress combination. *Trends in Plant Science* 11:-15–19. doi: 10.1016/j.tplants.2005.11.002.

Mooney, H. A., and Winner, W. E. 1991. Partitioning response of plants to stress. In *Response of Plants to Multiple Stresses*, eds. H. A. Mooney et al., 129–141. San-Diego: Academic Press. doi: 10.1016/b978-0-08-092483-0.50011-6.

Moreno-Risueno, M. A., Busch, W., and Benfey, P. N. 2010. Omics meet networks - using systems approaches to infer regulatory networks in plants. *Current Opinion in Plant Biology* 13:126–131. doi: 10.1016/j.pbi.2009.11.005.

Newman, D., and Pilson, D. 1997. Increased probability of extinction due to decreased genetic effective population size: experimental populations of *Clarkia pulchella*. *Evolution* 51:354–362. doi: 10.1111/j.1558-5646.1997.tb02422.x.

Nijhout, H. F. 2003. Development and evolution of adaptive polyphenisms. *Evolution and Development* 5:9–18. doi: 10.1046/j.1525-142x.2003.03003.x.

Peleg, Z., and Blumwald, E. 2011. Hormone balance and abiotic stress tolerance in crop plants. *Current Opinion in Plant Biology* 14:290–295. doi: 10.1016/j.pbi.2011.02.001.

Prasad, P. V. V., Pisipati, S. R., Momčilović, I., and Ristic, Z. 2011. Independent and combined effects of high temperature and drought stress during grain filling on plant yield and chloroplast EF-Tu expression in spring wheat. *Journal of Agronomy and Crop Science* 197:430–441. doi: 10.1111/j.1439-037x.2011.00477.x.

Prasch, C. M., and Sonnewald, U. 2013. Simultaneous application of heat, drought, and virus to *Arabidopsis* plants reveals significant shifts in signaling networks. *Plant Physiology* 162:1849–1866. doi: 10.1104/pp.113.221044.

Priest, H. D., Fox, S. E., Rowley, E. R., Murray, J. R., Michael, T. P., and Mockler, T. C. 2014. Analysis of global gene expression in *Brachypodium distachyon* reveals extensive network plasticity in response to abiotic stress. *PLoS One* 9:e87499. doi: 10.1371/journal.pone.0087499.

Queitsch, C., Sangster, T. A., and Lindquist, S. 2002. Hsp90 as a capacitor of phenotypic variation. *Nature* 417:618–624. doi: 10.1038/nature749.

Rasmussen, S., Barah, P., Suarez-Rodriguez, M. C., Bressendorff, S., Friis, P., Costantino, P., et al. 2013. Transcriptome responses to combinations of stresses in *Arabidopsis*. *Plant Physiology* 161:1783–1794. doi: 10.1104/pp.112.210773.

Reusch, T. B. H., Ehlers, A., Hämmerli, A., and Worm, B. 2005. Ecosystem recovery after climatic extremes enhanced by genotypic diversity. *Proceedings of the National Academy of Sciences of the USA* 102:2826–2831. doi: 10.1073/pnas.0500008102.

Rivero, R. M., Mestre, T. C., Mittler, R., Rubio, F., Garcia-Sanchez, F., and Martinez, V. 2014. The combined effect of salinity and heat reveals a specific physiological, biochemical and molecular response in tomato plants. *Plant, Cell and Environment* 37:1059–1073. doi: 10.1111/pce.12199.

Rivero, R. M., Mittler, R., Blumwald, E., and Zandalinas, S. I. 2022. Developing climate-resilient crops: improving plant tolerance to stress combination. *Plant Journal* 109:373–389. doi: 10.1111/tpj.15483.

Rizhsky, L., Liang, H., and Mittler, R. 2002. The combined effect of drought stress and heat shock on gene expression in tobacco. *Plant Physiology* 130:1143–1151. doi: 10.1104/pp.006858.

Rizhsky, L., Liang, H., Shuman, J., Shulaev, V., Davletova, S., and Mittler, R. 2004. When defense pathways collide. the response of *Arabidopsis* to a combination of drought and heat stress. *Plant Physiology* 134:1683–1696. doi: 10.1104/pp.103.033431.

Sade, N. and Peleg Z. 2020. Future challenges for global food security under climate change. *Plant Science* 295:110467. doi: 10.1016/j.plantsci.2020.110467.

Sanchez, D. H. 2013. Physiological and biotechnological implications of transcript-level variation under abiotic stress. *Plant Biology* 15:925–930. doi: 10.1111/plb.12075.

Seki, M., Narusaka, M., Abe, H., Kasuga, M., Yamaguchi-Shinozaki, K., Carninci, P., et al. 2001. Monitoring the expression pattern of 1300 *Arabidopsis* genes under drought and cold stresses by using a full-length cDNA microarray. *Plant Cell* 13:61–72. doi: 10.1105/tpc.13.1.61.

Seki, M., Narusaka, M., Ishida, J., Nanjo, T., Fujita, M., Oono, Y., et al. 2002. Monitoring the expression profiles of 7000 *Arabidopsis* genes under drought, cold and high-salinity stresses using a full-length cDNA microarray. *Plant Journal* 31:279–292. doi: 10.1046/j.1365-313x.2002.01359.x.

Sewelam, N., Oshima, Y., Mitsuda, N., and Ohme-Takagi, M. 2014. A step towards understanding plant responses to multiple environmental stresses: A genome-wide study. *Plant, Cell and Environment* 37:2024–2035. doi: 10.1111/pce.12274.

Shaar-Moshe, L., Blumwald, E., and Peleg, Z. 2017. Unique physiological and transcriptional shifts under combinations of salinity, drought, and heat. *Plant Physiology* 174:421–434. doi: 10.1104/pp.17.00030.

Shaar-Moshe, L., Hayouka, R., Roessner, U., and Peleg, Z. 2019. Phenotypic and metabolic plasticity shapes life-history strategies under combinations of abiotic stresses. *Plant Direct* 3:e00113. doi: 10.1002/pld3.113.

Sultan, S. E. 2017. Eco-evo-devo. In *Evolutionary Developmental Biology*, eds. L. Nuno de la Rosa, and G. Müller, 1–13. Cham: Springer. doi: 10.1007/978-3-319-33038-9_42-1.

Suzuki, N., Rivero, R. M., Shulaev, V., Blumwald, E., and Mittler, R. 2014. Abiotic and biotic stress combinations. *New Phytologist* 203:32–43. doi: 10.1111/nph.12797.

Vautard, R., Yiou, P., D'Andrea, F., de Noblet, N., Viovy, N., Cassou, C., Polcher, J., Ciais, P., Kageyama, M., and Fan, Y. 2007. Summertime European heat and drought waves induced by wintertime Mediterranean rainfall deficit, *Geophysical Research Letters*, 34:L07711. doi:10.1029/2006GL028001.

Vile, D., Pervent, M., Belluau, M., Vasseur, F., Bresson, J., Muller, B., et al. 2012. *Arabidopsis* growth under prolonged high temperature and water deficit: independent or interactive effects? *Plant, Cell and Environment* 35:702–718. doi: 10.1111/j.1365-3040.2011.02445.x.

Wilson, P., Streich, J., and Borevitz, J. 2015. Genomic diversity and climate adaptation in *Brachypodium*. In *Genetics and Genomics of Brachypodium*, ed. J. Vogel, 107–127. Switzerland: Springer. doi: 10.1007/7397_2015_18.

Xu, Z. Z., and Zhou, G. S. 2006. Combined effects of water stress and high temperature on photosynthesis, nitrogen metabolism and lipid peroxidation of a perennial grass *Leymus chinensis*. *Planta* 224:1080–1090. doi: 10.1007/s00425-006-0281-5.

Yamaguchi-Shinozaki, K., and Shinozaki, K. 2006. Transcriptional regulatory networks in cellular responses and tolerance to dehydration and cold stresses. *Annual Review of Plant Biology* 57:781–803. doi: 10.1146/annurev.arplant.57.032905.105444.

Zandalinas, S. I., Balfagón, D., Gómez-Cadenas, A., and Mittler, R. 2022. Plant responses to climate change: metabolic changes under combined abiotic stresses. *Journal of Experimental Botany* 73:erac073. doi: 10.1093/jxb/erac073.

Zandalinas, S. I., Fritschi, F. B., and Mittler, R. 2021. Global warming, climate change, and environmental pollution: recipe for a multifactorial stress combination disaster. *Trends in Plant Science* 26:588–599. doi: 10.1016/j.tplants.2021.02.011.

Zandalinas, S. I., and Mittler, R. 2022. Plant responses to multifactorial stress combination. *New Phytologist* 234:1161–1167. doi: 10.1111/nph.18087.

Zandalinas, S. I., Mittler, R., Balfagón, D., Arbona, V., and Gómez-Cadenas, A. 2018. Plant adaptations to the combination of drought and high temperatures. *Physiologia Plantarum* 162:2–12. doi: 10.1111/ppl.12540.

Zandalinas, S. I., Rivero, R. M., Martínez, V., Gómez-Cadenas, A., and Arbona, V. 2016. Tolerance of citrus plants to the combination of high temperatures and drought is associated to the increase in transpiration modulated by a reduction in abscisic acid levels. *BMC Plant Biology* 16:105. doi: 10.1186/s12870-016-0791-7.

3 Advances in Phenomics and Its Implications for Crop Improvement under Multiple Stress Conditions through Conventional and Genomic Approaches

Himaj Deshmukh, Madhavi Sonone, Vinay Hegde, Tamilselvan A, Aliza Pradhan, and Jagadish Rane
ICAR-National Institute of Abiotic Stress Management

N. K. Gupta
Sri Karan Narendra Agriculture University

CONTENTS

DOI: 10.1201/9781003300564-3

3.1 INTRODUCTION

3.1.1 Definition, Cause, and Impacts of Stress on Crop Growth and Productivity

When environmental conditions aren't ideal for growth, plants are said to be stressed. Abiotic stresses are caused by deficiencies or excesses of environmental elements such as water, salt, light, temperature, and nutrients. They can significantly impair plant growth and productivity. It is not only a fundamental scientific subject about how harsh circumstances affect plant growth but it is also critical for agriculture and food security.

According to a 2007 FAO assessment, only 3.5% of the worldwide land area is unaffected by environmental constraints. Adverse environments were projected to result in a lower total yield of 70% in main agricultural crops, i.e., average yield accounted only for about a third of the genetic yield potential. While it is difficult to obtain precise estimates of the effects of abiotic stress on crop output, the percentage of land area affected and the number of scientific articles on diverse abiotic stresses indicate that abiotic stress continues to have a considerable impact on plants (Cramer et al., 2011).

Environmental stress can disrupt cellular structures and impair key physiological functions (Larcher, 2003). Drought, salinity, and low-temperature stress impose osmotic stress that can lead to turgor loss. Membranes may become disorganized, proteins may undergo loss of activity or be denatured, and often excess levels of reactive oxygen species (ROS) are produced leading to oxidative damage. As a consequence, inhibition of photosynthesis, metabolic dysfunction, and damage to cellular structures contribute to growth perturbances, reduced fertility, and premature senescence.

Abiotic stress responses in plants are dynamic and complicated, and they are both elastic and plastic. The majority of research focuses on individual stresses, despite the fact that plants in the field are subjected to a variety of stresses at the same time (Skirycz and Inzé, 2010; Cramer, 2010). Researchers observed a novel defense response pattern in *Arabidopsis* plants exposed to a combination of heat and drought conditions, which differed from that seen in studies looking at the effects of individual stress variables. A better understanding of the mechanisms in plants that allow them to acclimate to environmental fluctuations is needed in economically important crops in order to facilitate the selection or development of varieties that can tolerate environmental extremes and return to a new state of homeostasis when stressed.

The tissue or organ that is being injured determines the plant's response to stress. Transcriptional responses to stress in roots are tissue- or cell-specific and vary widely depending on the stress (Dinneny et al., 2008). In addition, the severity and duration of stress (acute vs. chronic) may have a significant impact on the complexity of the response (Tattersall et al., 2007; Pinheiro and Chaves, 2011).

Stress-induced systemic signaling and systemic acquired acclimation are critical for plant survival during periods of environmental stress. According to recent studies, plants mount a full stress-specific systemic response in response to a single abiotic stress, which comprises the accumulation of several different stress-specific transcripts and metabolites, as well as a coordinated stress-specific whole-plant stomatal response (Zandalinas et al., 2020).

Plants are frequently exposed to a combination of two or more different abiotic stresses in nature, each of which has the ability to activate its own stress-related systemic response. Plants can integrate two different systemic signals simultaneously generated during stress combination, and the way they sense the different stresses that trigger these signals (i.e., at the same or different parts of the plant) influences how quickly and efficiently they induce systemic ROS signals, transcriptomic, hormonal, and stomatal responses, and plant acclimation.

3.1.2 Addressing Abiotic Stress through Genomic Approaches

Due to the rapid growth of functional genomics, a growing number of crop genomes have been sequenced, and dozens of genes affecting essential agronomic traits have been discovered (Song et al., 2021). Next-generation sequencing technology has greatly accelerated progress in functional genomics (Li et al., 2018), allowing quantitative trait locus (QTL) mapping and genome-wide association studies (GWASs) (Xiao et al., 2017) to become powerful tools for elucidating the genetic architecture of complex traits (Wang et al., 2018; Zhang et al., 2019; Yao et al., 2018; Fernie and Yan, 2019; Shi et al., 2019). In rice, for example, only about 130 genes had been cloned prior to 2000. By the end of 2017, around 3,000 genes have been thoroughly characterized using a variety of standard phenotypic techniques (Yao et al., 2018). However, crop breeding and functional genomics investigations are still hampered by a lack of phenotypic data (Deery et al., 2016).

Traditional crop phenotyping techniques are labor-intensive, time-consuming, subjective, and often destructive. Current genome sequence information has not been effectively utilized for understanding the complex features of numerous genes due to a lack of crop phenotypic data. Technology and platforms that can capture phenotypic data and combine it with genetic information for crop development at all stages of development have become just as important as genotyping. As a result, high-throughput phenotyping (HTP) has risen to the top of the crop breeding (Houle et al., 2010).

3.1.3 Need for Phenotyping and High Throughput Phenotyping in Stress Management

Plant phenotyping is a set of techniques and protocols that are used to evaluate plant features with specified accuracy and precision under various environmental conditions at various sizes of organization, from organs to canopies (Fiorani and Schurr, 2013). Plant stress monitoring is critical for evaluating management tactics, plant breeding selection strategies, and the ability of novel cultivars to mitigate crop losses. As a result, improvements in the accuracy and automation of stress severity evaluations can boost genetic gain across crops and lead to new management options (Araus et al., 2018).

Phenomics is the collection of high-dimensional phenotypic data on an organism-wide scale (Fiorani and Schurr, 2013); it is a set of approaches and protocols for accurately evaluating plant development, architecture, and composition at various scales. Crop phenomics, on the other hand, is an interdisciplinary study of high-throughput accurate acquisition and processing of multidimensional phenotypes on an organism-wide scale throughout crop growth.

HTP and genomic selection (GS) select complicated features controlled by multiple alleles with little effects using genome-wide markers and statistical modeling. It helps forecast crop performance in a specific growth environment or agroecology. Vast volumes of genotype and phenotype data from individuals or lines are required to construct an accurate and robust predictive model. Markers may now be obtained quickly and correctly due to the advancement of next-generation sequencing technology. Phenotyping, on the other hand, is a significant bottleneck. Grain crops have been shown to benefit from HTP platforms. For example, for HTP of sorghum plant height and multiple genomic prediction models, an unmanned aerial vehicle (UAV) carrying a remote-sensing unit with either an RGB or a near-infrared, green, and blue (NIR-GB) camera were employed (Watanabe et al., 2017). Morphology and physiology may now be measured non-destructively and frequently across broad populations and throughout development because of advances in a variety of technologies ranging from sensors to software for data extraction (IT) as well as systems integration (Yang et al., 2020).

On the other hand, the technologies are still in the early stages of development. The subject of why and how to measure all aspects of genomics has been thoroughly addressed over the last decade, and technical advancements will now allow us to answer the question of why and how to measure whole-plant phenotypes in the future decades (Houle et al., 2010).

Improvements in automated phenotyping systems that incorporate artificial intelligence (AI) could be the answer to difficulties with present plant stress phenotyping techniques. The introduction of sophisticated sensors, HTP, and advanced data analytics through machine learning, as well as the widespread availability of computer infrastructure and resources, has paved the way for a hopeful future in plant stress phenotyping (Singh et al., 2016). The need to remove any lingering subjectivity, improve data gathering quality, and increase scalability and versatility are all challenges. High-throughput AI technologies for early stress detection before visual symptoms (Rumpf et al., 2010) could be the next step for precision agriculture, field-based use, and eventually, research in small test plots in this area (Singh et al., 2021).

Indeed, developing new varieties and focusing on cost-effective returns are key objectives of breeding companies and breeders. Many commercial plant breeding software tools, such as AGROBASE Generation II (https://www.agronomix.com/) and Plant Research Information Sharing Manager (PRISM, http://www.sytseed.com/), have been developed to improve the management of breeding data and optimize breeding programs. Most commercial breeding software, on the other hand, has limited capability. Thus, crop breeding requires the creation of multifunctional breeding software that incorporates genomes, phenotypes, environments, data management, multiomics data analysis, and genetic mapping.

3.2 ADVANCES IN PHENOMICS APPROACHES

3.2.1 Ground-Based Sensors for Phenotyping

A precise and in-depth understanding of phenotypes in controlled and field situations is required to describe the agricultural plants' adaptive mechanisms accurately. Understanding stress-associated phenotypic features and their interrelationships is always a scientific challenge. Image processing with visible to near-infrared spectrum light sources is used in advanced phenotyping approaches to generate nondestructive image datasets of the plant phenotypes (Rahaman et al., 2015; Rane et al., 2021a). The use of a robust software framework for the acquisition and analysis of image acquired in visible light, hyperspectral, infrared, and fluorescence imaging, as well as X-ray computed tomography, promise to generate unique, multilayer phenotyping data (Sozzani et al., 2014; Paproki et al., 2012). The use of sensors for HTP of crop plants is outlined in the following sections.

3.2.1.1 Visible Light (300–700 nm) Imaging

Shoot-related variables such as shoot biomass (Neilson et al., 2015), leaf morphology, shoot tip extension, panicle and seed morphology, root architecture, and other yield-related traits have all been measured using visible imaging approaches that rely on two-dimensional (2D) digital images (Fahlgren et al., 2015). Visible imaging sensors, such as silicon sensors (CCD or CMOS arrays), are sensitive to visible spectral ranges. Three-dimensional (3D) imaging was developed to provide more precise detail on complex phenotypes, and more recently, 2D and 3D imaging technologies have been combined to improve phenotyping accuracy (Rahaman et al., 2015). Plant height, leaf morphology, and shoot dry weight have all been measured using 3D imaging techniques that have been optimized (Paproki et al., 2012). Using a LemnaTec 3D Scanalyzer, Golzarian et al. (2011) correctly assessed shoot dry weight to examine wheat seedlings for salt stress.

Software has also been developed that allows for the superimposition of 2D to 3D images, allowing for the monitoring of root architecture, morphology, and growth (Clark et al., 2013). Phenotyping using RGB has been used for different abiotic stresses in several crops using various platforms, including PHENOPSIS (Granier et al., 2006) and WIWAM (https://www.wiwam.be/) for drought stress in *Arabidopsis*; LemnaTec for drought stress in barley (Honsdorf et al., 2014), mung bean (Rane et al., 2021a), and maize (Ge et al., 2016); and salt stress in rice (Hairmansis et al., 2014; Al-Tamimi et al, 2016), wheat, and barley (Meng et al., 2017), and PlantScreen and GROWSCREEN for chilling tolerance in *Arabidopsis* and peas, respectively (Humplík et al., 2015).

3.2.1.2 Thermal Imaging

Infrared radiation can be visualized through thermal imaging, which reveals the surface temperature of an object. Thermal cameras have a sensitive spectral range of 3–14 μ, and the most popular wavelengths for thermal imaging are 3–5 or 7–14 μ. Current thermal cameras have great thermal sensitivity to monitor plant canopy temperature, as well as more user-friendly interfaces and higher-resolution detectors. Furthermore, this method provides deeper insight into plant health under different stress conditions by combining NIR imaging with visible imaging, that is, visible and short-wave infrared (VSWIR; 0.4–2.5 m), because it provides well-defined spectral features for pigments, leaf water content, and biochemical components such as lignin and cellulose (Yang et al., 2013). Furthermore, by observing differences in canopy temperature, infrared thermography has been utilized to analyze stomatal responses under salinity and drought (Rahaman et al., 2015).

In dry conditions, the temperature difference between the canopy and the surrounding air can be utilized as a proxy for drought tolerance. Based on osmotic tolerance and Na+ exclusion, thermal infrared imaging can be used in the lab and in the field to characterize tolerance to diverse stresses such as drought and salinity (Kluitenberg and Biggar, 1992; Kumar et al., 2017; Price et al., 2002). To monitor salt tolerance in wheat genotypes, infrared thermal imaging devices have been utilized to measure stomatal behavior under various stress situations (Bayoumi et al., 2014).

3.2.1.3 Spectroscopy Imaging

Spectroscopy imaging is the result of multispectral and hyperspectral cameras capturing the radiation reflected by the plant surface. Hyperspectral imaging, unlike visible and infrared imaging, separates images into bands, resulting in images that contain a considerable portion of the electromagnetic spectrum (Yang et al., 2013). It is possible to engage this technology for (1) NDVI (normalized difference vegetation index), which compares red and near-infrared reflectance, (2) CRI (carotenoid reflectance index), which determines three wavebands in the yellow region, the 970 nm water index defining small water absorption area, and (3) PRI (photochemical reflectance index), which correlates functional status of non-photochemical energy conservation (Fiorani et al., 2012).

The vegetation indices are linked to a variety of traits, including water status, pigment content, and photosynthetically active biomass, and are used to calculate total green biomass, leaf area, chlorophyll content, and yield in a variety of crop species (Din et al., 2017). The use of spectroscopy imaging for field phenotyping is more effective when combined with aerial platforms; however, spectral cameras and their associated infrastructure are expensive.

3.2.1.4 Other Imaging Techniques

Light detection and ranging (LiDAR) (or laser scanner) sensors, stereo vision, photon mixer devices (PMD), time-of-flight cameras, and even the consumer-gaming interface Microsoft Kinect are among the imaging sensor technologies now employed for 3D mapping of plants. LiDAR is the most well-known and commonly utilized type of sensor for 3D canopy reconstruction due to its superior resilience, accuracy, and resolution. Structured light projection and laser range scanners are used to build accurate and detailed 3D models. However, it can be costly, complicated, and time-consuming. Chlorophyll fluorescence imaging can also be engaged in abiotic stress response of crop plants, as demonstrated in wheat (Rane et al., 2019), sorghum (Pradhan et al., 2022), and horticultural crops (Rane et al., 2021b).

3.2.2 Airborne Sensors

UAVs have been used in agriculture and general plant sciences for several years. To achieve HTP, more researchers are combining airborne remote-sensing techniques with plant science. A novel method of crop management and monitoring has emerged in the form of unmanned aerial systems (UAS). UAS-based phenotyping is a good replacement for earlier approaches, as they can deliver high-resolution measurements for field-based HTP (Haghighattalab et al., 2016). Furthermore, UAVs can collect data from the entire research site in a short amount of time, reducing the impact of meteorological conditions (e.g., solar angle, sun intensity, wind, and clouds). UAVs have benefits over satellites because of (1) low cost, (2) ease of flying and operating, (3) flexibility in fly time and altitude, (4) high-quality images, (5) high spatial and temporal resolution, (6) less influence by weather conditions, (7) timesaving, an (8) frequent and need-based data collection (Colomina and Molina, 2014; Vega et al., 2015).

Plant attributes such as yield, biomass, height, and leaf area index can be calculated non-destructively with this technology, which is crucial for HTP in agriculture. When compared to ground-based sensors, UAS can boost phenotyping throughput and frequency. UAS have recently paved the door for the rapid development of field HTP for crops. In field-scale trials, various sensors such as a regular RGB camera (red, green, and blue), a multispectral imaging camera (several wavebands), a hyperspectral imaging camera (hundreds or even thousands of wavebands), a thermal imaging sensor, and/or a LiDAR sensor can be mounted on a UAV.

3.3 IMPLICATIONS OF ADVANCES FOR MULTIPLE STRESS

Strategies to phenotype stress tolerance in plants should consider the associated mechanisms and plant traits, which have been well documented based on decades of investigations on plant responses to abiotic stress (Taiz and Zeiger, 2006). Any abiotic stress affects the growth of a plant at any stage of development though some of the growth stages are highly sensitive. It may reduce germination and lead to poor stand establishment, as in drought (Kaya et al., 2006) or cell growth due to water/osmotic stress caused by reduction in turgor pressure as in case of drought or salinity (Taiz and Zeiger, 2006). Leaf rolling is the prominent visible symptom and one of the survival tactics against drought, salinity, and heat stress, which reduces transpiration and canopy temperature (Joshi and Karan, 2013). Many of these changes can be quantified by phenomics as it is possible to derive shape features from image analysis and periodic observations can help to interpret abiotic responses of plants based on surrogate traits including digital volume of plants (Rane et al., 2021b; Neumann et al., 2015; Kumar et al., 2020). Surrogate traits generated based on the shape and color features of plant images can help in quantifying many of the plant responses mentioned earlier.

Leaf/plant senescence is one of the common responses to many of the abiotic stresses, including drought, high temperature, waterlogging, salinity, and nutrient deficiency. Abiotic stress-induced reduction in grain yield has often been related to delayed leaf senescence (Edmeades et al., 1999; Ziyomo and Bernardo, 2013; Wehner et al., 2015). Senescence of the leaves is a dynamic process that has a significant impact on plant output; this can be quantified by indices derived from RGB pixel information (Lyu et al., 2017).

Change in the phenology of plants to escape abiotic stresses is one of the common strategies of crop plants though it comes at the cost of grain yield penalty. Plant phenological variations are the most easily detectable consequences of severe environmental circumstances. Plants prevent yield losses by modifying the duration of their vegetative and reproductive growth when they are stressed (Golabadi et al., 2008; Alam et al., 2014) as reported in field crops (Liu et al., 2012), rice (Prasad et al., 2008), maize (Acevedo et al., 2002), lentils (Choukri et al., 2020), wheat (Yin et al., 2009; Grieve et al., 1994; Acevedo et al., 2002), and soybean (Füzy et al., 2019). Phenotyping protocols have been developed to capture the phonological changes (Dingkuhn et al., 2017).

The root system of the plants is highly sensitive to abiotic stress such as drought, salinity, waterlogging, and nutrient imbalance in the soil. Root system architecture (RSA) refers to the shape and spatial arrangement of root systems within the soil, including structural features such as root length, spread, number, and length of lateral roots, among others. It shows a lot of flexibility in response to environmental changes, and it could be important for developing crops with more efficient roots. Several strategies for phenotyping root systems have been published (Atkinson et al., 2019; McGrail et al., 2020), including both nondestructive and effective methods (Atkinson et al., 2019; Kitomi et al., 2018; Zhao et al., 2018; Kumar et al., 2020). Thus, assessment of common response to abiotic stress promises to help in phenotyping multiple abiotic stresses and to support genomic-aided genetic improvement in abiotic stress tolerance of crops.

3.3.1 Drought and Heat

Drought and heat stress typically coexist in tropical field environments. Drought stress and low atmospheric humidity raise the vapor pressure deficit (VPD), resulting in fast water loss from the soil and plants. As a precaution, plants close their stomata in reaction to increased VPD, but they frequently fail to fulfil the transpirational requirement to cool the surface, resulting in higher tissue and canopy temperatures. Similarly, high temperatures can cause drought stress by causing fast evapotranspiration, which results in a drop in soil and plant tissue water status. Under water-scarce situations, this direct heat effect is amplified (Prasad et al., 2008). As a result, drought and heat interact in the field to create a more formidable combined stress. Plants that are subjected to both drought and heat stress have physiological, metabolic, and molecular responses that are distinct from individual stress responses (Silva et al., 2010; Jin et al., 2016). As a result, unique combined stress tactics are being developed in order to improve crop tolerance (Pandey et al., 2015).

At the seedling and vegetative phases of rice, combined stress has negative impact on photosynthetic carbon metabolism and mitochondrial respiration. The current challenge is to develop high-yielding cultivars in both stressed and ideal settings. Understanding and enhancing rice tolerance to individual drought or heat stress relies heavily on plant breeding.

Our understanding of the genotype and environmental interaction from the individual stress may not be of much help while studying the combined stress. Besides, while selecting for only yield, the gene pool required for combined stress tolerance might have been lost in the presently grown cultivars. Understanding the mechanisms and pathways regulating plant responses to combined drought and heat stress through omics approaches is critical in an endeavor to develop combined stress-tolerant rice. Genetic engineering tools play vital roles in introducing these mechanisms and pathways. A large number of drought and heat stress-responsive genes have been discovered, and transgenic techniques have been utilized to modify these genes in rice to improve individual drought and heat stress tolerance (Yu et al., 2016). Furthermore, several of these genes are affected by drought and heat stress. When provided independently, similar genes have been found to improve drought and heat stress tolerance. Overexpression of the SNAC3 transcription factor, for example, confers resistance to individual drought and heat stress in rice by maintaining homeostasis of ROS (Fang et al., 2015).

In breeding, phenotyping for physiological features has two main applications. One is the characterization and selection of parents to guide strategic crossing, which is usually done on a hundred or more lines from crossing blocks. The other is the screening of a huge number of lines—typically tens of thousands—which could be the offspring of crosses or big panels of genetic resources (Rebetzke et al., 2019). An essential but often neglected truth is that the objectives and tactics used in each are extremely different. The former necessitates a more intensive strategy (Molero and Reynolds 2020), whereas the latter is better suited to high-throughput instruments such as proximal and remote sensing (Tattaris et al., 2016). Physiological breeding can also benefit from remote and proximal sensors to precisely identify crop growth.

The uses of phenotyping in breeding under three broad categories: handy traits that are by definition the most breeder friendly being inexpensive and easy to apply; high-throughput traits that as stated earlier are most suitable for pre-screening of potential parents within genetic resource panels, and selection of new genotypes among tens of thousands of progeny from breeding pipelines; and precision phenotyping to thoroughly characterize potential parents to inform strategic crosses.

3.3.2 Drought and Salinity

Various plant phenotyping platforms have recently been developed, aiming to detect the physiological status of plants exposed to stressful conditions. Most widely used methods for nondestructive studying of plant phenotypic traits under stressful conditions are chlorophyll fluorescence imaging (Brestic and Zivcak, 2013; Bresson et al., 2015). Employing those techniques gives valuable insights into plant performance under specific environmental conditions and estimates different traits such as light utilization by photosystem II (PSII) and underlying biochemical processes, leaf pigment content, leaf chemical composition, leaf and shoot morphological and architectural traits, etc. Due to often co-occurrence of drought and salinity, similar effects (symptoms) which they cause on plants, and the complexity of morpho-physiological adaptations to drought and salinity, it is difficult to differentiate between traits related to these two stresses.

When comparing drought-stressed wheat cultivars to salinity-stressed wheat cultivars, Paul et al. (2019) discovered a greater drop in biomass and leaf area. Furthermore, additional assessed features (chlorophyll fluorescence parameters) show that salinity, as well as drought, produced severe stress to basil plants after only 1 week. Drought and salinity both enhanced red, green, and blue reflectance while decreasing NIR reflectance, with severe drought treatment producing the quickest and most profound alterations on plants. Under stressed conditions, similar alterations in visible and NIR reflectance have been reported (Mulla, 2013). NIR reflectance, on the other hand, is determined by leaf optical parameters such as shape, thickness, water content, and light scattering, rather than leaf pigment content (Merzlyak et al., 2003). NIR reflectance and morphological features had a favorable connection, as did their simultaneous reduction in stress treatments.

3.3.3 Waterlogging and Salinity

A fundamental barrier for good crop production is salinity combined with waterlogging. Both of these stresses are key barriers to irrigated land long-term viability, effective crop production, and farmer subsistence. Waterlogging and salinity tolerance are both complex features governed by a variety of processes. Because direct selection of general tolerance is difficult, breeders rely on tolerance-related molecular markers. Li et al. (2008) and Xue et al. (2010) found that the majority of QTL for waterlogging/salinity tolerance are based on plant survival rate, plant healthiness, and leaf chlorosis under stress.

While these characteristics are useful for high-throughput screening, they are unrelated to the mechanisms that bestow tolerance. As a result of the huge number of QTLs involved, detailed mapping of these QTLs to offer reliable markers to breeders is extremely challenging, even if theoretically achievable. A method in which specific QTLs are connected directly to suitable mechanisms is far more promising. Because most mechanisms are believed to be controlled by only one or two QTLs, fine mapping them is significantly easier. Barley waterlogging tolerance, the key QTL for waterlogging on 4H, is a good illustration of this achievement (Li et al., 2008; Zhou, 2011). The plasma membrane is required for intracellular ionic homeostasis because it maintains ionic and electric gradients between the cytosol and external media. In plants under stress, it is also a crucial component of signal transmission (Kim et al., 2007). One of the common aspects of salinity and waterlogging stresses is plasma membrane depolarization, which causes a significant disruption in ionic homeostasis (Palmgren and Nissen, 2011; Shabala et al., 2016), which leads to metabolic disturbances and ultimately determines the cell's fate. The electrogenic HC-ATPase pumps are important for maintaining the PM's negative potential. Due to a considerable decrease in ATP availability to feed HC-ATPase, oxygen-depleted circumstances led in a large depolarization of the plasma membrane.

3.3.4 Waterlogging and Cold Stress

Depending on the depth of the water table, soil flooding causes a composite and complicated stress in plants known as submergence or waterlogging stress. These pressures are major determinants of the ecosystem's species composition in nature. They create economic damage on agricultural land, as well as long-term societal implications (Fukao and Bailey-Serres, 2008).

The adaptability to submergence and waterlogging stress is facilitated by root architecture and plasticity. Waterlogging-tolerant species have morphological characteristics such as the production of aerenchyma and adventitious roots. Internal oxygen transport from the aerial parts to the waterlogged roots is reported to be enhanced by aerenchyma, allowing the roots to maintain aerobic respiration (Armstrong, 1980). Plants are deprived of energy and carbohydrates in flooded circumstances due to reduced photosynthesis and aerobic respiration. When plants are subjected to prolonged darkness, they experience a similar energy state. As a result, it's not unexpected that plants that can withstand submergence can also withstand prolonged darkness. Rice resistance to submergence is increased by constitutive and stress-induced expression of SUB1A, which inhibits carbohydrate breakdown and elongation growth (Fukao et al., 2006).

3.3.5 Drought and Nutrient Stress

Drought has morphophysiological consequences for plants that are similar to nutrient stress in terms of metabolic disturbance and lower output. Drought stress inhibits photosynthesis by altering the quantities of chlorophyll and other photosynthetic pigments, causing plant growth to cease (Abedi and Pakniyat, 2010; Lee et al., 2016). When plants are exposed to both nutrient and drought toxicity, the situation becomes even worse, resulting in a critical loss of agricultural productivity. Ünyayar et al. (2005) found that nutritional and drought stress had a negative impact on the cellular antioxidative defense system, chlorophyll content, and growth of two tomato varieties. As a result, effective soil reclamation/remediation solutions are needed to deal with the negative impacts of nutrient and drought on crop plants for long-term agriculture and food security.

Drought stress and associated reduction in soil moisture can reduce plant nutrient uptake by reducing nutrient supply through mineralization (Fierer and Schimel, 2002; Schimel et al., 2007; Sanaullah et al., 2012), but also nutrient diffusion and mass flow in the soil (Chapin, 1991; Lambers et al., 2008). When drought stress is followed by rewetting, this often results in enhanced mineralization (Austin et al., 2004), which has been attributed to nutrient release from dead microbial biomass that has accumulated during the drying period (Borken and Matzner, 2009). The net effect of drought stress on nutrient supply through mineralization may depend on the duration and intensity (or severity) of these drying–rewetting cycles, which can be attributed to many factors such as frequency and occurrence of rainfall, soil type affecting water potential, and evapotranspiration demand (Farooq et al., 2009). Generally, negative effects on soil microbial activity and plant nutrient uptake become larger with increased soil moisture deficit.

Given the potential of nanomaterials as nano-fertilizers or nano-pesticides, their tailored application could deliver excellent results not only in terms of increasing crop productivity but also in terms of addressing metal and drought-related challenges at the soil–plant interface (Ragab and Saad-Allah, 2020; Prasad et al., 2014). Numerous studies have shown the use of artificial nanoparticles (NPs) as nano-fertilizers in resource-scarce polluted arable soils to boost crop yields (Venkatachalam et al., 2017; Rui et al., 2016). Biologically fabricated NPs are more sustainable than physiochemically produced NPs as an environmentally friendly and cost-effective nano-based technique. Nanotechnologists have been drawn to iron oxide nanoparticles (IONPs) because of their low toxicity, biocompatibility, and unique physiochemical and biological features (Xiao et al., 2020; Xu et al., 2020).

3.4 WAY FORWARD

Different combination of stress in diverse agroecologies needs specific strategies though the phenotyping tools may be the same. Hence, it is difficult to design a single solution for all the situations, and phenotyping protocols have to be customized for a given combination of stresses and agroecologies. The approach to initiate with a conventional approach of identifying distinct genotypes, including the desired genetic resource for a particular trait, can provide desired dividends if integrated with high-throughput phenomics and genomic approaches for molecular biology-aided crop improvement programs. It is always necessary to include the popular cultivar or those in the breeding pipeline to harness such dividends and translate the plant biology knowledge into products in the form of improved cultivars for multiple stress tolerance.

REFERENCES

Abedi, T., and Pakniyat, H. 2010. Antioxidant enzymes changes in response to drought stress in ten cultivars of oilseed rape (*Brassica napus* L.). *Czech Journal of Genetics and Plant Breeding* 46(1): 27–34.

Acevedo, E., Silva, P., and Silva, H. 2002. Wheat growth and physiology. Bread wheat: Improvement and production (No. CIS-3617. CIMMYT.).

Alam, M., Bodruzzaman, M., Hossain, M. M. and Sadekuzzaman, M. 2014. Growth performance of spring wheat under heat stress conditions. *International Journal of Agriculture Research* 4: 91–103.

Al-Tamimi, N., Brien, C., Oakey, H., Berger, B., Saade, S., Ho, Y. S., ... and Negrão, S. 2016. Salinity tolerance loci revealed in rice using high-throughput non-invasive phenotyping. *Nature Communications* 7(1): 1–11.

Araus, J. L., Kefauver, S. C., Zaman-Allah, M., Olsen, M. S., and Cairns, J. E. 2018. Translating high-throughput phenotyping into genetic gain. *Trends in Plant Science* 23(5): 451–466.

Armstrong, W. 1980. Aeration in higher plants. *Advances in Botanical Research* 7: 225–332.

Atkinson, J. A., Pound, M. P., Bennett, M. J., and Wells, D. M. 2019. Uncovering the hidden half of plants using new advances in root phenotyping. *Current Opinion in Biotechnology* 55: 1–8.

Austin, A. T., Yahdjian, L., Stark, J. M., Belnap, J., Porporato, A., Norton, U., ... and Schaeffer, S. M. 2004. Water pulses and biogeochemical cycles in arid and semiarid ecosystems. *Oecologia* 141(2): 221–235.

Bayoumi, T. Y., El-Hendawy, S., Yousef, M. S., Emam, M. A., and Okasha, S. A. 2014. Application of infrared thermal imagery for monitoring salt tolerant of wheat genotypes. *Journal of American Science* 10(12): 227–234.

Borken, W., and Matzner, E. 2009. Reappraisal of drying and wetting effects on C and N mineralization and fluxes in soils. *Global Change Biology* 15(4): 808–824.

Bresson, J., Vasseur, F., Dauzat, M., Koch, G., Granier, C., and Vile, D. 2015. Quantifying spatial heterogeneity of chlorophyll fluorescence during plant growth and in response to water stress. *Plant Methods* 11(1): 1–15.

Brestic, M., and Zivcak, M. 2013. PSII fluorescence techniques for measurement of drought and high temperature stress signal in crop plants: protocols and applications. In: Rout, G. R., Das, A. B., (eds.) *Molecular Stress Physiology of Plants*. Springer, Dordrecht, pp. 87–131.

Brestic, M, and Zivcak, M. 2013. PSII fluorescence techniques for measurement of drought and high temperature stress signal in plants: protocols and applications. In: Rout GR, Das AB (eds.) *Molecular Stress Physiology of Plants*. Springer, Dordrecht, pp. 87–131.

Chapin, FS III. 1991. Effects of multiple environmental stresses on nutrient availability and use. In Mooney HA, Winner WE, Pell EJ, eds. *Response of Plants to Multiple Stresses* (pp. 67–88). San Diego, CA, USA: Academic Press.

Choukri, H., Hejjaoui, K., El-Baouchi, A., El Haddad, N., Smouni, A., Maalouf, F., ... and Kumar, S. 2020. Heat and drought stress impact on phenology, grain yield, and nutritional quality of lentil (*Lens culinaris* Medikus). *Frontiers in Nutrition* 7: 596307.

Colomina, I., and Molina, P. 2014. Unmanned aerial systems for photogrammetry and remote sensing: a review. *ISPRS Journal of Photogram and Remote Sensing* 92: 79–97.

Clark, R. T., Famoso, A. N., Zhao, K., et al. 2013. High-throughput two-dimensional root system phenotyping platform facilitates genetic analysis of root growth and development. *Plant, Cell & Environment* 36(2): 454–466.

Cramer, G. R. 2010. Abiotic stress and plant responses from the whole vine to the genes. *Australian Journal of Grape and Wine Research* 16: 86–93.

Cramer, G. R., Urano, K., Delrot, S., Pezzotti, M., and Shinozaki, K. 2011. Effects of abiotic stress on plants: a systems biology perspective. *BMC Plant Biology* 11(1): 1–14.

Deery, D. M., Rebetzke, G. J., Jimenez-Berni, J. A., James, R. A., Condon, A. G., Bovill, W. D., ... and Furbank, R. T. 2016. Methodology for high-throughput field phenotyping of canopy temperature using airborne thermography. *Frontiers in Plant Science* 7: 1808.

Din, M., Zheng, W., Rashid, M., Wang, S., and Shi, Z. 2017. Evaluating hyperspectral vegetation indices for leaf area index estimation of *Oryza sativa* L. at diverse phenological stages. *Frontiers in Plant Science* 8: 820.

Dingkuhn, M., Pasco, R., Pasuquin, J. M., Damo, J., Soulié, J. C., Raboin, L. M., ... and Kretzschmar, T. 2017. Crop-model assisted phenomics and genome-wide association study for climate adaptation of indica rice. 1. Phenology. *Journal of Experimental Botany* 68(15): 4369–4388.

Dinneny, J. R., Long, T. A., Wang, J. Y., Jung, J. W., Mace, D., Pointer, S., ... and Benfey, P. N. 2008. Cell identity mediates the response of Arabidopsis roots to abiotic stress. *Science* 320(5878): 942–945.

Edmeades, G. O., Bolaños, J., Chapman, S. C., Lafitte, H. R., and Bänziger, M. 1999. Selection for drought tolerance increases maize yields across a range of nitrogen levels. *Crop Science* 39(4): 1306–1315.

Fahlgren, N., Gehan, M. A., and Baxter, I. 2015. Lights, camera, action: high-throughput plant phenotyping is ready for a close-up. *Current Opinion in Plant Biology* 24: 93–99.

Fang, Y., Liao, K., Du, H., Xu, Y., Song, H., Li, X., and Xiong, L. 2015. A stress-responsive NAC transcription factor SNAC3 confers heat and drought tolerance through modulation of reactive oxygen species in rice. *Journal of Experimental Botany* 66(21): 6803–6817.

Farooq, M., Wahid, A., Kobayashi, N., Fujita, D., and Basra, S. M. A. 2009. Plant drought stress: effects, mechanisms and management. *Agronomy for Sustainable Development* 29: 185–212.

Fernie, A. R., and Yan, J. 2019. De novo domestication: an alternative route toward new crops for the future. *Molecular Plant* 12(5): 615–631.

Fierer, N., and Schimel, J. P. 2002. Effects of drying–rewetting frequency on soil carbon and nitrogen transformations. *Soil Biology and Biochemistry* 34(6): 777–787.

Fiorani, F., Rascher, U., Jahnke, S., and Schurr, U. 2012. Imaging plants dynamics in heterogenic environments. *Current Opinion in Biotechnology* 23(2): 227–235.

Fiorani, F., and Schurr, U. 2013. Future scenarios for plant phenotyping. *Annual Review of Plant Biology* 64: 267–291.

Fukao, T., and Bailey-Serres, J. 2008. Submergence tolerance conferred by Sub1A is mediated by SLR1 and SLRL1 restriction of gibberellin responses in rice. *Proceedings of the National Academy of Sciences* 105(43): 16814–16819.

Fukao, T., Xu, K., Ronald, P. C., and Bailey-Serres, J. 2006. A variable cluster of ethylene response factor–like genes regulates metabolic and developmental acclimation responses to submergence in rice. *The Plant Cell* 18(8): 2021–2034.

Füzy, A., Kovács, R., Cseresnyés, I., Parádi, I., Szili-Kovács, T., Kelemen, B., and Takács, T. 2019. Selection of plant physiological parameters to detect stress effects in pot experiments using principal component analysis. *Acta Physiologiae Plantarum* 41(5): 1–10.

Ge, Y., Bai, G., Stoerger, V., and Schnable, J. C. 2016. Temporal dynamics of maize plant growth, water use, and leaf water content using automated high throughput RGB and hyperspectral imaging. *Computers and Electronics in Agriculture* 127: 625–632.

Golabadi, M., Arzani, A., Tabatabaie, B., and Mohamadi, A. 2008. Identification of SSR markers linked with genes controlling phenological characteristics of durum wheat under moisture stress and non-stress. In *Proceedings of the Tenth Congress of Genetics*, May 21–23, Tehran, Iran.

Golzarian, M. R., Frick, R. A., Rajendran, K., Berger, B., Roy, S., Tester, M., and Lun, D. S. 2011. Accurate inference of shoot biomass from high-throughput images of cereal plants. *Plant Methods* 7(1): 1–11.

Granier, C., Aguirrezabal, L., Chenu, K., Cookson, S. J., Dauzat, M., Hamard, P., ... and Tardieu, F. 2006. PHENOPSIS, an automated platform for reproducible phenotyping of plant responses to soil water deficit in Arabidopsis thaliana permitted the identification of an accession with low sensitivity to soil water deficit. *New Phytologist* 169(3): 623–635.

Grieve, C. M., Francois, L. E., and Maas, E. V. 1994. Salinity affects the timing of phasic development in spring wheat. *Crop Science* 34(6): 1544–1549.

Haghighattalab, A., Perez, L. G., Mondal, S., Singh, D., Schinstock, D., Rutkoski, J., et al. 2016. Application of unmanned aerial systems for high throughput phenotyping of large wheat breeding nurseries. *Plant Methods* 12: 15.

Hairmansis, A., Berger, B., Tester, M., and Roy, S. J. 2014. Image-based phenotyping for non-destructive screening of different salinity tolerance traits in rice. *Rice* 7(1): 1–10.

Honsdorf, N., March, T. J., Berger, B., Tester, M., and Pillen, K. 2014. High-throughput phenotyping to detect drought tolerance QTL in wild barley introgression lines. *PLoS One* 9(5): e97047.

Houle, D., Govindaraju, D. R., and Omholt, S. 2010. Phenomics: the next challenge. *Nature Reviews Genetics* 11(12): 855–866.

Humplík, J. F., Lazár, D., Fürst, T., Husičková, A., Hýbl, M., and Spíchal, L. 2015. Automated integrative high-throughput phenotyping of plant shoots: a case study of the cold-tolerance of pea (*Pisum sativum* L.). *Plant Methods* 11(1): 1–11.

Jin, R., Wang, Y., Liu, R., Gou, J., and Chan, Z. 2016. Physiological and metabolic changes of purslane (*Portulaca oleracea* L.) in response to drought, heat, and combined stresses. *Frontiers in Plant Science* 6: 1123.

Joshi, R., and Karan, R. 2013. Physiological, biochemical and molecular mechanisms of drought tolerance in plants. In Gaur, R. K., and Sharma, P. (eds). *Molecular Approaches in Plant Abiotic Stress*. CRC Press, Taylor & Francis Group, pp. 209–231.

Kaya, M. D., Okçu, G., Atak, M., Cıkılı, Y., and Kolsarıcı, Ö. 2006. Seed treatments to overcome salt and drought stress during germination in sunflower (*Helianthus annuus* L.). *European Journal of Agronomy* 24(4): 291–295.

Kim, B. G., Waadt, R., Cheong, Y. H., Pandey, G. K., Dominguez-Solis, J. R., Schültke, S., ... and Luan, S. 2007. The calcium sensor CBL10 mediates salt tolerance by regulating ion homeostasis in Arabidopsis. *The Plant Journal* 52(3): 473–484.

Kitomi, Y., Nakao, E., Kawai, S., Kanno, N., Ando, T., Fukuoka, S., ... and Uga, Y. 2018. Fine mapping of QUICK ROOTING 1 and 2, quantitative trait loci increasing root length in rice. *G3: Genes, Genomes, Genetics* 8(2): 727–735.

Kluitenberg, G.J., and Biggar, J.W. 1992. Canopy temperature as a measure of salinity stress on sorghum. *Irrigation Science* 13: 115–121. https://doi.org/10.1007/BF00191053.

Kumar, D., Kushwaha, S., Delvento, C., Liatukas, Ž., Vivekanand, V., Svensson, J. T., ... and Chawade, A. 2020. Affordable phenotyping of winter wheat under field and controlled conditions for drought tolerance. *Agronomy* 10(6): 882.

Kumar, M., Govindasamy, V., Rane, J., Singh, A. K., Choudhary, R. L., Raina, S. K., ... and Singh, N. P. 2017. Canopy temperature depression (CTD) and canopy greenness associated with variation in seed yield of soybean genotypes grown in semi-arid environment. *South African Journal of Botany* 113: 230–238.

Lambers, H., Chapin, F.S. and Pons, T.L. 2008. *Plant Physiological Ecology*. New York, NY, USA: Springer.

Larcher, W. 2003. *Physiological Plant Ecology: Ecophysiology and Stress Physiology of Functional Groups*. Springer Science and Business Media, New York.

Lee, B. R., Zaman, R., Avice, J. C., Ourry, A., and Kim, T. H. 2016. Sulfur use efficiency is a significant determinant of drought stress tolerance in relation to photosynthetic activity in *Brassica napus* cultivars. *Frontiers in Plant Science* 7: 459.

Li, H., Vaillancourt, R., Mendham, N., and Zhou, M. 2008. Comparative mapping of quantitative trait loci associated with waterlogging tolerance in barley (*Hordeum vulgare* L.). *BMC Genomics* 9(1): 1–12.

Li, Y., Xiao, J., Chen, L., Huang, X., Cheng, Z., Han, B., ... and Wu, C. 2018. Rice functional genomics research: past decade and future. *Molecular Plant* 11(3): 359–380.

Liu, J., Guo, T., Yang, P., Wang, H., Liu, L., Lu, Z., ... and Chen, S. 2012. Development of automatic nuclear magnetic resonance screening system for haploid kernels in maize. *Transactions of the Chinese Society of Agricultural Engineering* 28(1): 233–236.

Lyu, J. I., Baek, S. H., Jung, S., Chu, H., Nam, H. G., Kim, J., and Lim, P. O. 2017.High-throughput and computational study of leaf senescence through a phenomic approach. *Frontiers in Plant Science* 8: 250.

McGrail, R. K., Van Sanford, D. A., and McNear, D. H. 2020. Trait-based root phenotyping as a necessary tool for crop selection and improvement. *Agronomy* 10(9): 1328.

Meng, R., Saade, S., Kurtek, S., Berger, B., Brien, C., Pillen, K., ... and Sun, Y. 2017. Growth curve registration for evaluating salinity tolerance in barley. *Plant Methods* 13(1): 1–9.

Merzlyak, M. N., Gitelson, A. A., Chivkunova, O. B., Solovchenko, A. E., and Pogosyan, S. I. 2003. Application of reflectance spectroscopy for analysis of higher plant pigments. *Russian Journal of Plant Physiology* 50(5): 704–710.

Molero, G., and Reynolds, M. P. 2020. Spike photosynthesis measured at high throughput indicates genetic variation independent of flag leaf photosynthesis. *Field Crops Research* 255, 107866.

Mulla, D. J. 2013. Twenty five years of remote sensing in precision agriculture: key advances and remaining knowledge gaps. *Biosystems Engineering* 114: 358–371.

Neilson, E. H., Edwards, A. M., Blomstedt, C. K., Berger, B., Møller, B. L., and Gleadow, R. M. 2015. Utilization of a high-throughput shoot imaging system to examine the dynamic phenotypic responses of a C4 cereal crop plant to nitrogen and water deficiency over time. *Journal of Experimental Botany* 66(7): 1817–1832.

Neumann, K., Klukas, C., Friedel, S., Rischbeck, P., Chen, D., Entzian, A., ... and Kilian, B. 2015. Dissecting spatiotemporal biomass accumulation in barley under different water regimes using high-throughput image analysis. *Plant, Cell & Environment* 38(10): 1980–1996.

Palmgren, M. G., and Nissen, P. 2011. P-type ATPases. *Annual Review of Biophysics* 40: 243–266.

Pandey, P., Ramegowda, V., and Senthil-Kumar, M. 2015. Shared and unique responses of plants to multiple individual stresses and stress combinations: physiological and molecular mechanisms. *Frontiers in Plant Science* 6: 723.

Paproki, A., Sirault, X., Berry, S., Furbank, R., and Fripp, J. 2012. A novel mesh processing based technique for 3D plant analysis. *BMC Plant Biology* 12(1): 1–13.

Paul, K., Pauk, J., Kondic-Spika, A., Grausgruber, H., Allahverdiyev, T., Sass, L., and Vass, I. 2019. Co-occurrence of mild salinity and drought synergistically enhances biomass and grain retardation in wheat. *Frontiers in Plant Science* 10: 501.

Pinheiro, C., and Chaves, M. M. 2011. Photosynthesis and drought: can we make metabolic connections from available data? *Journal of Experimental Botany* 62(3): 869–882.

Pradhan, A., Aher, L., Hegde, V., Jangid, K. K., and Rane, J. 2022. Cooler canopy leverages sorghum adaptation to drought and heat stress. *Scientific Reports* 12(1): 1–11.

Prasad, P. V. V., Staggenborg, S. A., and Ristic, Z. 2008. Impacts of drought and/or heat stress on physiological, developmental, growth, and yield processes of crop plants. *Response of Crops to Limited Water: Understanding and Modeling Water Stress Effects on Plant Growth Processes* 1: 301–355.

Prasad, R., Kumar, V., and Prasad, K. S. 2014. Nanotechnology in sustainable agriculture: present concerns and future aspects. *African Journal of Biotechnology* 13(6): 705–713.

Price, A. H., Cairns, J. E., Horton, P., Jones, H. G., and Griffiths, H. 2002. Linking drought-resistance mechanisms to drought avoidance in upland rice using a QTL approach: progress and new opportunities to integrate stomatal and mesophyll responses. *Journal of Experimental Botany* 53(371): 989–1004. https://doi.org/10.1093/JEXBOT/53.371.989.

Ragab, G. A., and Saad-Allah, K. M. 2020. Green synthesis of sulfur nanoparticles using *Ocimumbasilicum* leaves and its prospective effect on manganese-stressed *Helianthus annuus* (L.) seedlings. *Ecotoxicology and Environmental Safety* 191: 110242.

Rahaman, M., Chen, D., Gillani, Z., Klukas, C., and Chen, M. 2015. Advanced phenotyping and phenotype data analysis for the study of plant growth and development. *Frontiers in Plant Science* 6: 619.

Rane, J., Raina, S. K., Govindasamy, V., Bindumadhava, H., Hanjagi, P., Giri, R., ... and Nair, R. M. 2021a. Use of phenomics for differentiation of mungbean (*Vigna radiata* L. Wilczek) genotypes varying in growth rates per unit of water. *Frontiers in Plant Science* 12: 1138.

Rane, J., Babar, R., Kumar, M., Kumar, P. S., Singh, Y., Nangare, D. D., ... and Minhas, P. S. 2021b. Desiccation tolerance of Photosystem II in dryland fruit crops. *Scientia Horticulturae* 288: 110295.

Rane, J., Sharma, D., Ekatpure, S., Aher, L., Kumar, M., Prasad, S. V. S., ... and Singh, N. P. 2019. Relative tolerance of photosystem II in spike, leaf, and stem of bread and durum wheat under desiccation. *Photosynthetica* 57(4): 1100–1108.

Rebetzke, G. J., Jimenez-Berni, J., Fischer, R. A., Deery, D. M., and Smith, D. J. 2019. High-throughput phenotyping to enhance the use of crop genetic resources. *Plant Science* 282: 40–48.

Rui, M., Ma, C., Hao, Y., Guo, J., Rui, Y., Tang, X., ... and Zhu, S. 2016. Iron oxide nanoparticles as a potential iron fertilizer for peanut (*Arachishypogaea*). *Frontiers in Plant Science* 7: 815.

Rumpf, T., Mahlein, A. K., Steiner, U., Oerke, E. C., Dehne, H. W., and Plümer, L. 2010. Early detection and classification of plant diseases with support vector machines based on hyperspectral reflectance. *Computers and Electronics in Agriculture* 74(1): 91–99.

Sanaullah, M., Rumpel, C., Charrier, X., and Chabbi, A. 2012. How does drought stress influence the decomposition of plant litter with contrasting quality in a grassland ecosystem? *Plant and Soil* 352(1): 277–288.

Schimel, J., Balser, T. C., and Wallenstein, M. 2007. Microbial stress response physiology and its implications for ecosystem function. *Ecology* 88(6): 1386–1394.

Shabala, S., Bose, J., Fuglsang, A. T., and Pottosin, I. 2016. On a quest for stress tolerance genes: membrane transporters in sensing and adapting to hostile soils. *Journal of Experimental Botany* 67(4): 1015–1031.

Shi, C., Zhao, L., Zhang, X., Lv, G., Pan, Y., and Chen, F. 2019. Gene regulatory network and abundant genetic variation play critical roles in heading stage of polyploidy wheat. *BMC Plant Biology* 19(1): 1–16.

Silva, E. N., Ferreira-Silva, S. L., de VasconcelosFontenele, A., Ribeiro, R. V., Viégas, R. A., and Silveira, J. A. G. 2010. Photosynthetic changes and protective mechanisms against oxidative damage subjected to isolated and combined drought and heat stresses in *Jatrophacurcas* plants. *Journal of Plant Physiology* 167(14): 1157–1164.

Singh, A., Ganapathysubramanian, B., Singh, A. K., and Sarkar, S. 2016. Machine learning for high-throughput stress phenotyping in plants. *Trends in Plant Science* 21(2): 110–124.

Singh, A., Jones, S., Ganapathysubramanian, B., Sarkar, S., Mueller, D., Sandhu, K., and Nagasubramanian, K. 2021. Challenges and opportunities in machine-augmented plant stress phenotyping. *Trends in Plant Science* 26(1): 53–69.

Skirycz, A., and Inzé, D. 2010. More from less: plant growth under limited water. *Current Opinion in Biotechnology* 21(2): 197–203.

Song, P., Wang, J., Guo, X., Yang, W., and Zhao, C. 2021. High-throughput phenotyping: Breaking through the bottleneck in future crop breeding. *The Crop Journal* 9(3): 633–645.

Sozzani, R., Busch, W., Spalding, E. P., and Benfey, P. N. 2014. Advanced imaging techniques for the study of plant growth and development. *Trends in Plant Science* 19(5): 304–310.

Taiz, L., and Zeiger, E. 2006. *Plant Physiology*. Sinauer Associates Incorporated, Sunderland, MA.

Tattaris, M., Reynolds, M. P., and Chapman, S. C. 2016. A direct comparison of remote sensing approaches for high-throughput phenotyping in plant breeding. *Frontiers in Plant Science* 7: 1131.

Tattersall, E. A., Grimplet, J., DeLuc, L., Wheatley, M. D., Vincent, D., Osborne, C., ... and Cramer, G. R. 2007. Transcript abundance profiles reveal larger and more complex responses of grapevine to chilling compared to osmotic and salinity stress. *Functional and Integrative Genomics* 7(4): 317–333.

Ünyayar, S., Keleş, Y., and Çekiç, F. Ö. 2005. The antioxidative response of two tomato species with different drought tolerances as a result of drought and cadmium stress combinations. *Plant, Soil and Environment* 51: 57–64.

Vega, F. A., Ramirez, F. C., Saiz, M. P., and Rosua, F. O. 2015. Multi-temporal imaging using an unmanned aerial vehicle for monitoring a sunflower crop. *Biosystems Engineering* 132: 19–27.

Venkatachalam, P., Priyanka, N., Manikandan, K., Ganeshbabu, I., Indiraarulselvi, P., Geetha, N., ... and Sahi, S. V. 2017. Enhanced plant growth promoting role of phycomolecules coated zinc oxide nanoparticles with P supplementation in cotton (*Gossypiumhirsutum* L.). *Plant Physiology and Biochemistry* 110: 118–127.

Wang, H., Xu, S., Fan, Y., Liu, N., Zhan, W., Liu, H., ... and Yan, J. 2018. Beyond pathways: genetic dissection of tocopherol content in maize kernels by combining linkage and association analyses. *Plant Biotechnology Journal* 16(8): 1464–1475.

Watanabe, K., Guo, W., Arai, K., Takanashi, H., Kajiya-Kanegae, H., Kobayashi, M., ... and Iwata, H. 2017. High-throughput phenotyping of sorghum plant height using an unmanned aerial vehicle and its application to genomic prediction modeling. *Frontiers in Plant Science* 8: 421.

Wehner, G. G., Balko, C. C., Enders, M. M., Humbeck, K. K., and Ordon, F. F. 2015. Identification of genomic regions involved in tolerance to drought stress and drought stress induced leaf senescence in juvenile barley. *BMC Plant Biology* 15(1): 1–15.

Xiao, C., Li, H., Zhao, Y., Zhang, X., and Wang, X. 2020. Green synthesis of iron nanoparticle by tea extract (polyphenols) and its selective removal of cationic dyes. *Journal of Environmental Management* 275: 111262.

Xiao, Y., Liu, H., Wu, L., Warburton, M., and Yan, J. 2017. Genome-wide association studies in maize: praise and stargaze. *Molecular Plant* 10(3): 359–374.

Xu, W., Hu, X., Lou, Y., Jiang, X., Shi, K., Tong, Y., and Lou, L. 2020. Effects of environmental factors on the removal of heavy metals by sulfide-modified nanoscalezerovalent iron. *Environmental Research* 187: 109662.

Xue, D. W., Zhou, M. X., Zhang, X. Q., Chen, S., Wei, K., Zeng, F. R., ... and Zhang, G. P. 2010. Identification of QTLs for yield and yield components of barley under different growth conditions. *Journal of Zhejiang University Science B* 11(3): 169–176.

Yang, W., Duan, L., Chen, G., Xiong, L., and Liu, Q. 2013. Plant phenomics and high-throughput phenotyping: accelerating rice functional genomics using multidisciplinary technologies. *Current Opinion in Plant Biology* 16(2): 180–187.

Yang, W., Feng, H., Zhang, X., Zhang, J., Doonan, J. H., Batchelor, W. D., ... and Yan, J. 2020. Crop phenomics and high-throughput phenotyping: past decades, current challenges, and future perspectives. *Molecular Plant* 13(2): 187–214.

Yao, W., Li, G., Yu, Y., and Ouyang, Y. 2018. funRiceGenes dataset for comprehensive understanding and application of rice functional genes. *Gigascience* 7(1): gix119.

Yin, X., Guo, W., and Spiertz, J. H. 2009. A quantitative approach to characterize sink–source relationships during grain filling in contrasting wheat genotypes. *Field Crops Research* 114(1): 119–126.

Yu, J., Lai, Y., Wu, X., Wu, G., and Guo, C. 2016. Overexpression of OsEm1 encoding a group I LEA protein confers enhanced drought tolerance in rice. *Biochemical and Biophysical Research Communications* 478(2): 703–709.

Zandalinas, S. I., Fichman, Y., Devireddy, A. R., Sengupta, S., Azad, R. K., and Mittler, R. 2020. Systemic signaling during abiotic stress combination in plants. *Proceedings of the National Academy of Sciences* 117(24): 13810–13820.

Zhang, H., Wang, X., Pan, Q., Li, P., Liu, Y., Lu, X., ... and Li, L. 2019. QTG-Seq accelerates QTL fine mapping through QTL partitioning and whole-genome sequencing of bulked segregant samples. *Molecular Plant* 12(3): 426–437.

Zhao, Y., Zhang, H., Xu, J., Jiang, C., Yin, Z., Xiong, H., ... and Li, Z. 2018. Loci and natural alleles underlying robust roots and adaptive domestication of upland ecotype rice in aerobic conditions. *PLoS Genetics* 14(8): e1007521.

Zhou, M. 2011. Accurate phenotyping reveals better QTL for waterlogging tolerance in barley. *Plant Breeding* 130(2): 203–208.

Ziyomo, C., and Bernardo, R. 2013. Drought tolerance in maize: Indirect selection through secondary traits versus genomewide selection. *Crop Science* 53(4): 1269–1275.

4 Improving Yield under Combined Salinity and Drought – Physiological and Molecular-Genetic Approaches

Rana Munns
University of Western Australia

Yuri Shavrukov
Flinders University

CONTENTS

DOI: 10.1201/9781003300564-4

4.1 INTRODUCTION

Soil salinity occurs under a wide range of climatic conditions, both under natural and human-induced conditions, but are especially widespread in arid and semi-arid climates where rainfall or irrigation is inadequate to leach accumulated salts below the plant's rooting zone (Hopmans et al., 2021). As both drought and soil salinity occur in arid zones, the two stresses commonly occur together. The responses of plants, and the mechanisms of adaptations to the two stresses, have much in common (Munns, 2011; Ahmed et al., 2015).

With dryland agriculture, where there is no water available for irrigation, concomitant decreases in matric and osmotic potentials determine plant water uptake (Rengasamy, 2006). The influence of soil texture and type of clay on plant-available water compounds the effect of matric and osmotic potentials, and increases the energy needed by plants to take up water as the soil dries and the salt concentration increases. For example, in a sandy loam soil with a field capacity of 35% water content, and with no salt in the soil, plants are able to take up water until the soil dries to 5% water content. Whereas, when the soil salinity is high, plants cease to take up water at 18% soil water content, leaving a lot of unused water in the soil (Rengasamy, 2006).

Generally, the occurrence of combined stresses is more detrimental to plant growth and crop production than a single stress (Langridge and Reynolds, 2021). In this chapter, we consider the responses and adaptations to the combined stresses of drought and salinity, and to the measurable traits that have been selected for identifying QTLs and candidate genes. We then speculate on two questions. Can we breed for drought tolerance and predict it will help plants when the soil becomes saline, and can we breed for salt tolerance and assume it will help in drought?

4.2 RESPONSES TO DROUGHT AND SALINITY STRESS

Soil salinity like drought reduces the soil water potential and the ability of plants to take up water, and this quickly reduces the rate of cell expansion in growing tissues. Leaves expand more slowly, and new leaves also appear more slowly. The numbers of lateral branches or tillers are reduced. Roots grow more slowly, although leaves are often more strongly impacted than roots (Munns, 2002).

The slower formation of photosynthetic leaf area in turn reduces the flow of assimilates to the meristematic and growing tissues of the plant. At the same time, both water and salt stress reduce stomatal conductance in the older leaves which limits their photosynthetic activity. It is difficult to gauge whether this causes an additional reduction in growth rate, or the reduction in growth rate is in balance with the reduced photosynthesis, as carbohydrate status of the growing tissue is usually not affected and sometimes is higher indicating unused assimilate.

Leaf senescence starts earlier, so leaves die more quickly. The rate of leaf death is crucial for the survival of the plant. If new leaves are continually produced at a rate greater than the old leaves die, there are enough photosynthesising leaves for the plant to produce flowers and seeds, although in reduced numbers. However, if old leaves die faster than new ones develop, the plant may not survive to produce seeds. For an annual plant, there is a race against time to initiate flowers and form seeds, while the leaf area is still adequate to supply the necessary photosynthate. For perennial species, there is an opportunity to enter a state akin to dormancy, and thus survive the stress.

The low soil water potential (due to drought or salinity) affects the growth rate of the younger leaves, whereas the toxic effect of NaCl causes premature death of older leaves and can be seen by the earlier starting of yellowing. These two effects can be quantitatively measured over time. In a laboratory experiment, when salt is introduced to a constant level, a two-phase effect on plant growth is evident. The first phase of the growth response results from the effect of NaCl outside the plant. The cellular and metabolic processes involved are common to drought-affected plants. Neither Na^+ nor Cl^- builds up in the growing tissues at concentrations that inhibit growth: This is supported by the evidence that Na^+ and Cl^- are below toxic concentrations in the growing cells themselves, in leaves and roots (Munns, 2011). Meristematic tissues are fed largely in the phloem from which salt is effectively excluded, and rapidly elongating cells can accommodate the salt that arrives in the xylem within their expanding vacuoles.

With time, salt may exert an additional effect on growth. If excessive amounts of Na^+ or Cl^- enter the plant, it may rise to toxic levels in the older transpiring leaves. This injury, added to an already reduced leaf area, will then further limit the flow of carbon compounds to meristems and growing zones in leaves. Therefore, the plant may die before seeds are produced.

The second phase of the growth response results from the toxic effect of salt inside the plant. Na^+ and Cl^- taken up by the plant concentrates in the old leaves: continued transport into transpiring leaves over a long period of time eventually results in very high concentrations of both Na^+ and Cl^-, and results in leaf death. The cause of the injury is probably the salt load exceeding the ability of the cells to compartmentalise, particularly Na^+, in the vacuole (Mujeeb-Kazi et al., 2019).

4.3 MECHANISMS OF ADAPTATION

4.3.1 Commonalities between Drought and Salinity: Growth of Roots and Shoots

Mechanisms controlling leaf and root growth are common to drought and salinity; they are due to factors associated with osmotic component related to the reduced water availability in cells.

When roots sense a saline soil, the responses are similar to those in a drying soil. Hormonal signals may coordinate assimilate production and use in the competing sinks of roots, leaves and lateral shoots. The hormonal regulation of source–sink relations during the osmotic phase of salinity, the phase when growth rate and development is reduced and before ions build up to toxic levels in leaves, affects whole plant energy balance, and is critical to delay the accumulation of ions to toxic levels (Munns, 2011). Productivity in saline as well as dry soil depends on maintaining a high growth rate of young leaves, while at the same time delaying the senescence of older leaves.

Hormonal control of cell division and differentiation is clear from the appearance of leaves, which are smaller in area but often thicker, indicating that cell size and shape has changed (James et al., 2002). Leaves from salt-treated plants have a higher ratio for 'weight/area', which means that their transpiration efficiency is higher (more carbon fixed per water lost), a feature that is common in plants adapted to both dry and saline soils.

Hormonal control of cell division and elongation is also evident in roots. Several studies have shown that salinity has differential effects on root elongation rates and lateral root initiation that is due to the osmotic effect of the saline soil (Rahnama et al., 2011).

4.3.2 Turgor Maintenance and Osmotic Adjustment

Turgor maintenance is essential for cell viability, as well as for providing the energy needed for elongating cells to expand, and stomata to open. Osmotic adjustment is of course essential for adapting plants to soils of low water potential, but may bring penalties in terms of carbon allocation for the cell expansion phase of rapidly growing plants such as wheat and barley (Munns, 1988). Compatible solute synthesis comes with an energy cost and hence involves a potential growth penalty. The synthesis of these compounds occurs at the expense of plant growth, but may allow the plant to survive and recover from the period with reduced water (Munns et al., 2020).

4.3.3 Salt-Specific Differences

The critical issue for tolerance of a saline soil is to keep Na^+ and Cl^- concentrations in the cytoplasm below toxic levels. Na^+ must be partitioned within cells so that concentrations in the cytoplasm are kept low, possibly as low as 10–30 mM (Munns and Tester, 2008). However, the concentration at which Na^+ becomes toxic is not well defined. In vitro studies showed Na^+ starts to inhibit most enzymes at concentrations approaching 100 mM, although some enzymes are sensitive to lower concentrations (Munns and Tester, 2008).

The concentration at which Cl^- becomes toxic in the cytoplasm is even less well defined, but is probably similar to that for Na^+ (Munns and Tester, 2008; Teakle and Tyerman, 2010).

Mechanisms for keeping cytoplasmic concentrations of Na^+ and Cl^- below toxic levels are of two main types: those minimising the entry of salt into the root and its transport through the plant, and those minimising the concentration of salt in the cytoplasm by sequestration (or compartmentalisation) in vacuoles.

Roots must exclude most of the Na^+ and Cl^- dissolved in the soil solution or the salt will gradually build up with time in the shoot to toxic levels. Plants transpire about 50 times more water than they retain in their leaves (Munns et al., 2020). If a plant lets in only 1/50 of the salt in the soil solution, i.e., it excludes 98%, the concentration of salt in the shoot as a whole would never increase over that in the soil and the plant could survive indefinitely in saline soil (Munns et al., 2020).

Na^+ exclusion from leaves is associated with salt tolerance in cereal crops including rice (Asch et al., 2000; Ul Haq et al., 2010), durum wheat (Munns and James, 2003; Shavrukov et al., 2009, 2011; Shamaya et al., 2017), bread wheat (Cuin et al., 2009, 2010), barley (Shavrukov et al., 2010), pearl millet (Krishnamurthy et al., 2007) and wild relatives such as *Hordeum* species (Garthwaite et al., 2005), tall wheatgrass (Colmer et al., 2006), *Triticum tauschii* (Schachtman et al., 1991) and interspecific cross between maize and gamagrass, *Tripsacum dactyloides* (Shavrukov et al., 2006; Shavrukov and Sokolov, 2015).

The mechanisms of Na^+ transport at the molecular, electrophysiological and whole plant level have been summarised by Munns and Tester (2008) and Ismail and Horie (2017). In contrast to Na^+, mechanisms of Cl^- transport in plants are poorly understood (Teakle and Tyerman, 2010).

4.4 TRAITS AND GENOTYPING

Many of the traits that explain plant adaptation to drought (such as phenology, root size and depth, hydraulic conductivity and the storage of reserves) are associated with plant development. The effects of drought and salt stresses on photosynthesis are either stomatal (stomatal closure which can be measured by a porometer) or 'non-stomatal' (damage to thylakoid membranes, which can be measured by chlorophyll fluorescence).

With the knowledge of adaptations and the mechanisms behind them, quantitative traits can be identified or hypothesized. These physiologically complex traits are easily measured, such as leaf shape or time to flowering, using a ruler or visual records, respectively. Other measurements can be complex because they need expensive equipment such as a mass spectrometer used for carbon isotope discrimination and finally for water use efficiency. In contrast, physiologically simple traits like leaf rolling could be difficult to quantify.

In case of easily measured traits and without a requirement of expensive equipment and tools, a large number of genotypes can be screened. If there is genetic variation, then a QTL analysis can be performed by either a genome-wide association study (GWAS) or bi-crossings of the contrasting parents and analyses of hybrids. Both methods required extensive mapping with molecular markers and screening for the trait (the phenotype) a large number of used genotypes (cultivars, breeding lines or landraces) or progenies in the F_2 or F_3 segregating populations, recombinant inbred lines or advanced backcrossed breeding lines. The QTL should be validated in different genetic background and environment, and fine mapping can be attempted with the aid of a model species, and a candidate gene may be revealed. If not, the original linked marker can be used for breeding purposes.

4.4.1 Drought

When considering traits that are useful in plant breeding in drought-prone environments, the term 'drought tolerance' is the most suitable (Langridge and Reynolds, 2015). The reaction of plants will depend on several factors, such as intensity and duration of drought as well as a stage of plant ontogenesis on the event of drought. It is important that plants can be tolerant and not die but also if plants can adapt and continue to develop growth and produce seeds. The term 'desiccation tolerance' is only applicable to plants that can undergo and survive a prolonged dry season, many months or many years.

'Drought resistance' and 'drought avoidance' are nebulous terms. It is more effective to ask, given a fixed and limiting water supply, what is the best growth rate or maximum biomass that can be produced. This focuses on the term 'water use efficiency' (WUE), that is the efficiency of the use of scarce water (Passioura and Angus, 2010).

Traits for grain production under drought have been comprehensively reviewed and discussed by Richards et al. (2002, 2010), Reynolds et al. (2007), Reynolds and Tuberosa (2008), and Langridge and Reynolds (2015, 2021). Richards et al. (2010) review the key 'best-bet' traits for water-limited environments. These are summarised in Table 4.1.

Measurable adaptive characters can be considered in: whole plant growth and development, root and canopy architecture and longevity, and biochemical properties. Some of these traits have QTLs that are reliable and useful for breeding, some have candidate genes, however most have only phenotype descriptions (see Table 4.1). Traits involving whole plant growth and shoot development are reviewed by Richards et al. (2010). These are seedling establishment, shoot vigour, tiller regulation, phenology and early flowering, and floret sterility.

Traits involving root vigour and architecture have attracted less interest, largely because they are so difficult to measure. Root system vigour describes the variation in rate of root growth that results in the capture of greater volumes of soil water and nutrients. Deeper roots can access more water, maintain high stomatal conductance and hence photosynthesis, and are indicated by cooler canopies (Lopes and Reynolds, 2010). Root systems research has been hampered by the difficulty and time-consuming methods of measurement, and so molecular markers would provide a new impetus for their selection and use in breeding.

Traits involving leaves are maintenance of green leaf area (longevity), leaf rolling to reduce evaporation and glaucousness (leaf wax properties to reflect light).

TABLE 4.1
Summary of the Most Important Traits and the Selection Method for Improving Yield of Temperate cereals in Water-Limited Environments

Trait	Markers or Genomic Regions Identified	Most Efficient Selection Method
Seedling establishment	Yes	Phenotype and marker
Shoot vigor[a]	Yes	Phenotype
Root vigour	Yes	Phenotype
Root depth	No	Phenotype
Transpiration efficiency (CID)[b]	Yes	Phenotype
Stem carbohydrate remobilization (WSC)	Yes	Phenotype
Tillering[a]	Yes	Phenotype or marker
Glaucousness	Yes	Phenotype
Leaf rolling	No	Phenotype
Flowering time	Yes	Phenotype

Source: Adapted from Richards et al. (2010).

[a] These may have a positive on yield, if soil water (although limited) is available throughout the season, but a negative effect if drought occurs during the phase of grain filling.

[b] Carbon isotope discrimination: ^{12}C:^{13}C discrimination (low Δ or $\delta^{13}C$).

Biochemically based properties, which are complex but for which there are markers (Richards et al., 2010) or candidate genes, are transpiration efficiency, stem carbohydrate storage and remobilisation, and osmotic adjustment and compatible solutes.

4.4.2 Salt-Specific Traits

Traits for saline soils were summarised by Colmer et al. (2005) and listed in Table 4.2. The first traits listed are salt-specific, the last four in common with drought.

In the list in Table 4.2, Cl^- exclusion as a trait is not mentioned, because Colmer et al. (2005) was focused only on wheat and barley, for which species there is no evidence of Cl^- toxicity, and no correlation between genotypic differences in Cl^- uptake and salinity tolerance. Genes important for the regulation of Cl^- transport in saline soil are reviewed by Teakle and Tyerman (2010). No QTLs for these genes have been used in plant breeding.

A decrease in leaf chlorophyll content is an indicator of tolerance, and an easily measurable phenotype, but there is a problem of assigning cause. A decrease in chlorophyll could be due to lower tolerance of tissue Na^+ or Cl^- concentrations, or premature senescence caused by signals from the roots.

4.4.3 Traits in Common with Drought

Osmotic adjustment occurs readily because of uptake of Na^+ and Cl^-, and in species like wheat and barley the uptake is regulated so that it does not become toxic. Turgor can be perfectly maintained

TABLE 4.2
Key Traits for Salt Tolerance in Wheat and Barley

Trait	Measurement	Plant Stage	Comment	Amenable to Large-Scale Evaluation?
Na^+ 'exclusion'	Leaf blade Na^+ concentration	At defined leaf age and stage of plant development	The most useful trait; requires only small tissue samples	Yes
Cl^- exclusion	Leaf blade concentration	As above	Leaf blade Cl^- does not correlate with salt tolerance in many species	Yes
K^+/Na^+ discrimination	Leaf blade Na^+ and K^+	As above	Leaf Na^+ alone is sufficient	Yes
Sheath retention of ions	Na^+ and/or Cl^- concentration in sheath and blade	As above	Not widely studied	Yes
Tissue tolerance	Leaf injury, coupled with high Na^+ and/or Cl^- accumulation	Seedling stage	Laborious to phenotype; uncertain of method	No
Ion partitioning into different-aged leaves	Leaf blade Na^+ concentration of different leaves	Pre- and post-stem elongation	Requires small samples of several leaves.	Yes
Compatible solutes like proline	Concentration of specific solutes	Seedling stage	A high-throughput analytical system	Solutes – Yes

Source: These can be measured without destroying the plant, so that the seed can be collected for breeding. Adapted from Colmer et al. (2005).

in moderate salinity with osmotic adjustment (Boyer et al., 2008). If Na^+ and Cl^- are sequestered in the vacuole of a cell to maintain osmotic adjustment, organic solutes that are compatible with metabolic activity even at high concentrations (hence 'compatible solutes') accumulate in the cytosol and organelles to balance the osmotic pressure of the ions in the vacuole (Flowers et al., 1977). The compounds that accumulate most commonly are sucrose, proline and glycine betaine, although other molecules can accumulate to high concentrations in certain species.

In many halophytes, proline or glycine betaine occur at sufficiently high concentrations in leaves (over 40 mM on a tissue water basis) to contribute to the osmotic pressure (over 0.1 MPa) in the cell as a whole (Flowers et al., 1977). In glycophytes, the concentrations of compatible solutes that accumulate are not so high, on the order of 10 mM, but if partitioned exclusively to the cytoplasm, they could generate a significant osmotic pressure and function as an osmolyte. At low concentrations, these solutes presumably have another role, perhaps in stabilizing the tertiary structure of proteins, and function as osmoprotectants (Rhodes et al., 2002).

4.5 CANDIDATE GENES AND MOLECULAR MARKERS

For Na^+ exclusion, and maintenance of high K^+/Na^+ ratios in leaves, *HKT* (high-affinity K^+ transporter) genes are important in the regulation of Na^+ and K^+ in higher plants, and thereby in mediating salinity tolerance (Munns and Tester, 2008; Horie et al., 2009; Pardo, 2010). *HKT* genes are important for cellular Na^+ and K^+ homeostasis. *HKT* members of sub-family 1 have a much higher selectivity for Na^+ over K^+ than do members of sub-family 2 (Pardo, 2010; Hauser and Horie, 2010). The transporters encoded by these *HKT1*-type genes are expressed in cells lining the xylem vessels and can remove Na^+ from the xylem and result in lower Na^+ in leaves and a higher K/Na^+ ratio which is important to maintain ion homeostasis of the cytoplasm (Munns and Tester, 2008).

Major QTLs in wheat and barley have been found for five Na^+ excluding genes – *Nax1* and *Nax2* in durum wheat, *Nax3* and *Nax4* in barley, and *Kna1* in bread wheat (Table 4.3).

In durum wheat, *Nax1* and *Nax2* enhance removal of Na^+ from the xylem, leading to low Na^+ concentrations in leaves (James et al., 2006). *Nax1* removes Na^+ from the xylem in roots and the lower parts of leaves, the leaf sheaths, while *Nax2* removes Na^+ from the xylem only in the roots (James et al., 2006). *Nax1* has a unique phenotype of a high sheath/blade ratio of Na^+ concentration, and a candidate gene, *TmHKT1;4-A*, was identified (Huang et al., 2006) and confirmed (Huang et al., 2008). The ability to remove Na^+ from the xylem while it flows along the leaf sheath into the leaf blade might be particularly beneficial when the soil is flooded and insufficient oxygen reaches the roots to energise ion transport activity. The candidate gene for *Nax2* was identified as *TmHKT1;5-A* (Byrt et al., 2007). Both *Nax1* and *Nax2* were derived from the ancient einkorn wheat, *T. monococcum* (diploid A genome), and had been crossed into modern durum wheat by conventional breeding (James et al., 2006).

Nax2 has the same phenotype as *Kna1*, the QTL for Na^+ exclusion and enhanced K/Na^+ selectivity in bread wheat, *Triticum aestivum* (Dvořak et al., 1994). *Nax2* on the A genome (*TmHKT1;5-A*) was shown to be homoeologous to *Kna1* on the D genome (*TaHKT1;5-D*) (Byrt et al., 2007).

Modern dwarf rice is less tolerant of salinity than wheat, but the landraces Pokkali and Nona Bokra are relatively tolerant. A QTL named *SKC1* (Shoot K Content) derived from Nona Bokra was mapped to chromosome 1 and shown to remove Na^+ from the xylem in roots (Ren et al., 2005). Independent studies gave the name '*Saltol*' QTL to the same chromosome region, and confirmed that the candidate gene is an orthologue of *TmHKT1;5-A* namely *OsHKT1;5* (Ismail and Horie, 2017).

QTLs for Na^+ exclusion in leaves of barley revealed other major genes. One of them named *Nax4* (Rivandi et al., 2011) was mapped in chromosome 1HL, with the favourable allele originating from the Algerian landrace, Sahara 3771. The *Nax4* locus contains a *SOS3* homologue, which is a regulator of the antiporter *SOS1* that effluxes Na^+ from root cells (Rivandi et al., 2011). In contrast, a very different gene was associated with the identified *Nax3* and mapped in chromosome 7HS, which originated from wild species *H. vulgare* ssp. *spontaneum*, CPI-71284-48 (Shavrukov et al., 2010). Further study identified *HVP10* as *HvPPase* encoding a vacuolar H^+-pyrophosphatase (V-H^+-PPase)

TABLE 4.3
Genes Mapped from QTLs with Confirmed Na^+ Exclusion and Salinity Tolerance

Locus/QTL	Gene Name	Species and Localization	Function	References
		Genes Controlling Na^+ Exclusion from Leaves		
Nax1	*TmHKT1;4-A*	Durum wheat, chromosome 2AL	Removes Na^+ from xylem in roots and leaf sheaths	Huang et al. (2006)
Nax2	*TmHKT1;5-A*	Durum wheat, chromosome 5AL	Removes Na^+ from xylem in roots	Byrt et al. (2007)
SKC1 and Saltol	*OsHKT1;5*	Rice, chromosome 1	Removes Na^+ from xylem in roots; orthologue of *Nax2*	Ren et al. (2005); Ismail and Horie (2017)
Kna1	*TaHKT1;5-D*	Bread wheat, chromosome 4DL	Removes Na^+ from xylem in roots; homeolog of *Nax2*	Dvořak et al. (1994); Byrt et al. (2007)
Nax3	*HVP10=HvPPase*	Barley, chromosome 7HS	Vacuolar H^+-pyrophosphatase (V-H^+-PPase)	Shavrukov et al. (2010, 2013)
	HVP1=HvPPase	Barley, chromosome 7HL	V-H^+-PPase responsive to osmotic stress	Shavrukov (2014)
Nax4	*SOS3*	Barley, chromosome 1HL	Na^+ exclusion in leaves. Orthologue of *SOS3*	Rivandi et al. (2011)
		Genes Controlling Tolerance to Combined Salinity and Drought		
Gh_D08G0921	*CIPK21*, CBL-interacting protein kinase 21	Cotton, chromosome D08	Upregulated under multiple abiotic stress conditions. Orthologue of At5g57630 in *Arabidopsis*	Abdelraheem et al. (2021)
Gh_A01G0392	*TTL3*, tetratricopeptide-repeat thioredoxin-like 3	Cotton, chromosome A01	Osmotic tolerance. Orthologue of At2g42580 in *Arabidopsis*	Abdelraheem et al. (2021)
Zinc finger with A20/AN1 domain	*HvSAP12* stress-associated proteins (SAP)	Barley, chromosome 2H	Salinity tolerance	Baidyussen et al. (2020)
Zinc finger with A20/AN1 and AN1/C2H2 domains	*HvSAP8* (= *HvSAP5*) and *HvSAP16*	Barley chromosome 7H and 2H	Tolerance to drought and salinity	Baidyussen et al. (2020, 2021)
Os08g41990	Aminotransferase	Rice, chromosome 8	Salinity tolerance in drought-selected breeding lines	Nounjan et al. (2020)

Source: Gene names and their origin are described in the text.

Note: *Nax1*, *Nax2*, *SKC1/Saltol* and *Kna1* are expressed in xylem parenchyma cells lining the xylem vessels, and reduce the amount of Na^+ arriving in the leaves.

which is under the control of a proton pump from cytosol to vacuole. A high gradient of protons across the vacuole is required to assist the Na^+/H^+ antiporters and relocate toxic Na^+ into vacuole (Shavrukov et al., 2013; Shavrukov, 2014). Interestingly, it was described that another gene, *HVP1*, originated from the same wild barley accession and mapped in long arm of chromosome 7HL, has the same function as for the proton pump, but expressed in response to osmotic stress or osmotic component of salinity and possibly acting in similar way for osmotic adjustment under drought (Shavrukov, 2014).

Efflux of Na^+ across the plasma membrane to the apoplast and ultimately to the soil solution is by Na^+/H^+ antiporters (Blumwald et al., 2000), which are well known but these have not been mapped.

4.6 CASE STUDIES ON COMBINED STRESSES OF DROUGHT AND SALINITY

We could then speculate on breeding approaches. Can we breed for drought tolerance and assume it will help when the soil becomes saline, or can we breed for the control of Na^+ uptake from saline soil like in barley and assume it will help in drought?

Here, we firstly discuss the methodology appropriate to answer these questions, and follow with five case studies showing the promise of breeding for one stress that would enhance its adaptation to the other stress. We point out that in most cases, the drought stress was imposed by reduced soil watering and maintaining it at a deficit for some time, not by letting the soil dry, nor by mimicking drought with plants in solution culture and the addition of osmotica such as polyethylene glycol. In arid and semi-arid environments, the abiotic stresses of drought and salinity often occur concomitantly and, moreover, soil drying causes a corresponding higher NaCl concentration. However, in experiments with controlled conditions, this is not so trivial to replicate drought and salinity, and even more complicated to replicate combined or multiple stresses acting together and simultaneously. For drought, watering was withheld and volumetric water content was decreased to 40%–20% of field capacity (Baidyussen et al., 2021), weight of pots with plants and soil was reduced to 60% of initial weight (Chevilly et al., 2021) or without control of soil water capacity (Gökçe et al., 2022). PEG-induced dehydration is very different and only simulating dehydration and, therefore, results can be used with caution for the comparison (Abdelraheem et al., 2021). For salinity, plants grown in pots with soil or hydroponics were used and a range of 150–220 mM NaCl was applied either with increments or at once (Abdelraheem et al., 2021; Baidyussen et al., 2020; Chevilly et al., 2021; Gökçe et al., 2022; Nounjan et al., 2020). For combined stress only one study was found where plants grown in soil had limited irrigation with 150 mM NaCl with 70% of soil saturation capacity (Gökçe et al., 2022). The following cases represent examples of the comparison between drought, salinity and both stresses together to identify QTLs and candidate genes for plant tolerance to the abiotic stresses.

4.6.1 Water (Osmotic) Potential and Na^+ Ion Accumulation with Na^+/K^+ Ratio Were Distinct for Drought and Salinity Stresses, but Not Overlapped: Case of Melon Study

Four genotypes (tolerant and sensitive to abiotic stresses) in two cultivars, Galia and Piel de Sapo (types cvs. *reticulatus* and *inodorus*, respectively), were subjected either to 220 mM NaCl for 6 days or drought for 9 days (Chevilly et al., 2021). Very strong and distinct traits were identified, specific for each stress, with clear differences between tolerant and sensitive genotypes. For salinity, Na^+ ion accumulation and Na^+/K^+ ratio were the most recognisable for salinity while water (osmotic) potential showed biggest differences under drought. Authors indicated these characteristics as the most specific for their application in melon breeding programmes to improve tolerance in each of the two individual stresses but not for both stresses together because no common overlapped traits were identified for both drought and salinity in this study (Chevilly et al., 2021).

4.6.2 In Addition to Specific Mechanisms, Photosynthesis and Antioxidants Were Common for Drought and Salinity: Case of Onion Study

In the experiments with three types of stresses (salinity, drought and combined salinity+drought), 32 breeding lines and commercial cultivars of onion (*Allium cepa* L.), a group of accessions was identified as salt-sensitive while another group was sensitive to drought (Gökçe et al., 2022). Five onion accessions (K41, K52, U12, U47 and U49) showed highest bulb weight under all conditions of stresses and it was not surprising that they all were the top-best tolerant genotypes under salinity, drought and combined stresses. These attributes were associated with better stability of photosynthesis, less damage of chlorophyll content, improved gaseous exchange and more effective antioxidant production, which were common characteristics for both salinity and drought stresses. However, there were onion accessions that had very contrasted reactions to the different abiotic stresses. Breeding lines K18 and K51 were identified as salt-tolerant but more sensitive to drought and, vice versa, genotypes K35 and K59 were failed to group as salt-tolerant but showed much better tolerance to drought (Gökçe et al., 2022). This finding indicated some specific mechanisms such as ability to manage with toxic ions within cells under NaCl stress and with osmolytes production under drought. Therefore, it can be hypothesized that the selection of genotypes based on specific criteria for salinity or drought will result in the improvement of the corresponding tolerance (that is, selection for NaCl improves salinity tolerance and selection for drought improves tolerance to drought). However, selection of genotypes for common characteristics (photosynthesis and oxidative stress) in condition of a single stress (either salinity or drought) is very likely accompanied by the improvement in both stresses. The proof of this hypothesis is supported by the study in onion (Gökçe et al., 2022).

4.6.3 QTL and Candidate Genes Identification Using GWAS under PEG-Induced Dehydration and Salinity Treatment: Case of Cotton Study

In upland cotton (*Gossypium hirsutum* L.), 550 RILs from crosses between 11 parental lines were studied in young plants grown in pots with soil and irrigated with either 10% of polyethylene glycol (PEG-8000) or 200 mM NaCl (50 mM NaCl×4 increments increased) (Abdelraheem et al., 2021). Plant height and shoot dry weight were evaluated as tolerance to PEG-dehydration and salinity after 2-weeks of the treatment compared to non-treated controls. GWAS was based on 473K SNP markers analysis. Among very diverse phenotypic distribution, authors identified one best and one worst RILs, MG18 and MG470, respectively, with corresponding common tolerance and sensitivity to both dehydration and NaCl stress. From 43 QTLs identified in total, nine QTLs were common for both dehydration and salinity, and authors listed 53 most suitable candidate genes and some of them grouped in six genetic clusters, which were overlapped with QTLs identified in other studies. One of the identified candidate gene was well-known *CIPK21* (*Gh_D08G0921*)-encoded CBL-interacting protein kinase 21 at the position 19.31 Mb of chromosome D08, and involved in response to multiple abiotic stresses. In chromosome A01, gene *TTL3* (*Gh_A01G0392*) encoding tetratricopeptide-repeat thioredoxin-like 3 protein showed strong response of cotton plants to osmotic stress, which occurred under both salinity and drought. The identified genes were induced by abiotic stresses and responded to hormonal regulation (Abdelraheem et al., 2021).

4.6.4 Molecular Markers for Zinc Finger Transcription Factors with A20/AN1 and AN1/C2H2 Domains Associated with Salinity and Drought Tolerance, and Yield Improvement: Case of Barley Study

A group of zinc finger genes encoding stress-associated proteins (SAP) were identified in barley segregating population (Baidyussen et al., 2020). Five genes were identified as highly expressed in experiments with hydroponics with 150 mM NaCl treatment applied in gradual increments. One of the genes with A20/AN1 domain, *HvSAP12*, was used for the development of SNP marker

KATU-B30, showing significant association with improved tolerance to both salinity and drought. Additionally, strong association was found between KATU-B30 SNP marker and grain yield in plants of 50 breeding lines grown in field trails with low–moderate salinity and drought (Baidyussen et al., 2020). Two other genes, *HvSAP8* (= *HvSAP5*) and *HvSAP16*, with A20/AN1 and AN1/C2H2 domains, respectively, showed high to moderate expressions in response to salt stress (Baidyussen et al., 2020), and they were used in further experiments with drought. Six selected breeding lines from two segregating populations with identified alleles of SNP markers HvSAP8 and HvSAP16 similar to their parents were reported to have increased expression in plants grown in pots with soil under drought. Higher yield and yield components (1,000 seed weight or number of tillers) were found strongly correlated with favourable alleles of SNP markers, HvSAP8 or HvSAP16, respectively, in breeding lines grown under drought (Baidyussen et al., 2021). In these experiments, breeding lines selected for *HvSAP* genes under salinity stress showed improved drought tolerance with higher grain yield and yield components (Baidyussen et al., 2020, 2021). Figure 4.1 illustrates experiments with barley plants grown in hydroponics in the presence of NaCl in growth solution

FIGURE 4.1 Experiments with plants in hydroponics (A and B), with work of PhD student Akmaral Baidyussen (B); in pots with soil and with NaCl treatment (C) and under drought (D). Field trials with barley plants with low–moderate salinity and drought (E and F), Akmola region, Kazakhstan, 2020–2021.

(Figure 4.1A and B), in pots with soil under drought (Figure 4.1C and D), and in field with moderate salinity and dry environment (Figure 4.1E and F). These photographs illustrate how the experimental setups influence plant development under the different environmental conditions, and the possibility of differences in salt tolerance occurring under such different growth conditions.

4.6.5 Selection for Drought Tolerance QTL (and Genes) Resulted in the Improvement of Salt Tolerance: Case of Rice Study

Strong evidence was reported that the selection of rice genotypes for drought tolerance can result in improved tolerance to NaCl (Nounjan et al., 2020). Two breeding lines of Jasmine Thai rice (*Oryza sativa* L.), CSSL8-103 and CSSL8-106, originated from the cross between popular but sensitive to drought and salinity cv. Khao Dawk Mali 105 (KDML105) and drought-tolerant line DH103. During several rounds of marker-assisted backcrossing, authors reported that a genetic segment in chromosome 8 with drought tolerance QTL has been transferred from DH103 parent to CSSL8-103 and CSSL8-106 breeding lines, with confirmed improved plant growth and yield production under drought. Young plants (21-days-old) of CSSL8-103 and CSSL8-106 lines and their drought-tolerant parent DH103 were exposed to salt stress (150 mM NaCl, 6 days) and showed improved tolerance to salinity compared to their sensitive parent KDML105. The possible mechanisms for tissue tolerance and a combination with osmotic adjustment rather than sodium exclusion were hypothesised by authors for the improved salt tolerance. It was supported by aminotransferase candidate gene (*Os08g41990*) identified in drought tolerance QTL, and involved in chlorophyll biosynthesis and stability, among two more candidate genes. Under NaCl stress, the expression profiles of *Os08g41990* candidate gene were unchanged in the breeding lines and their drought-tolerant parent DH103 together with superior salt-tolerant cv. Pokkali compared to downregulated expression in salt-sensitive parent KDML105 (Nounjan et al., 2020). In conclusion, aminotransferase and other candidate genes controlling chlorophyll and stable photosynthesis can share both pathways resulting in the successful independent breeding for tolerance to drought and salinity in rice.

4.7 CONCLUSIONS

Multiple stresses acting together are not the same as a simple 'sum' of the stresses. Drought and salinity have specific traits and plants may react differently for each of the stresses. However, osmotic adjustment represents a common element of both stresses and, therefore, reactions of plants occur in similar directions. Gene up- and downregulations also have differences or commonalities between the two stresses, where some genes are specific, for example for ion translocation and water transport, but other genes are in common for osmosic regulation. There is nothing unusual in such reaction of plants which are facing unpredictable combination of stresses, time, magnitude and duration of stress events. Therefore, better understanding of physiology and genetics of plant tolerance mechanism to single or multiple stresses is very important for both academic understanding of the plant biology and for crop breeding. This will result in significant improvement of yield and quality of agricultural production using genotypes more tolerant to multiple stresses (Langridge and Reynolds, 2021; Ashraf and Munns, 2022).

REFERENCES

Abdelraheem, A., Thyssen, G. N., Fang, D. D., Jenkins, J. N., Carty, J. C., Wedegaertner, T., et al. 2021. GWAS reveals consistent QTL for drought and salt tolerance in a MAGIC population of 550 lines derived from intermating of 11 Upland cotton (*Gossypium hirsutum*) parents. *Molecular Genetics and Genomics* 296:119–129. doi:10.1007/s00438-020-01733-2.

Ahmed, I. M., Nadira, U. A., Bibi, N., Zhang, G. P., and Wu, F. B. 2015. Tolerance to combined stress of drought and salinity in barley. In *Combined Stresses in Plants*. Ed. R. Mahalingam, 93–122. Switzerland: Springer. doi:10.1007/978-3-319-07899-1.

Asch, F., Dingkuhn, M., Dörffling, K., and Miezan, K. 2000. Leaf K/Na ratio predicts salinity induced yield loss in irrigated rice. *Euphytica* 113:109–118. doi:10.1023/A:1003981313160.

Ashraf, M., and Munns, R. 2022. Evolution of approaches to increase the salt tolerance of crops. *Critical Reviews in Plant Sciences* 41:128–160. doi:10.1080/07352689.2022.2065136.

Baidyussen, A., Aldammas, M., Kurishbayev, A., Myrzabaeva, M., Zhubatkanov, A., Sereda, G., et al. 2020. Identification, gene expression and genetic polymorphism of Zinc finger A20/AN1 stress-associated genes, *HvSAP*, in salt stressed barley from Kazakhstan. *BMC Plant Biology* 20:156. doi:10.1186/s12870-020-02332-4.

Baidyussen, A., Jatayev, S., Khassanova, G., Amantayev, B., Sereda, G., Sereda, S., et al. 2021. Expression of specific alleles of Zinc-finger transcription factors, HvSAP8 and HvSAP16, and corresponding SNP markers, are associated with drought tolerance in barley populations. *International Journal of Molecular Science* 22:12156. doi:10.3390/ijms222212156.

Blumwald, E., Aharon, G. S., and Apse, M. P. 2000. Sodium transport in plant cells. *Biochimica et Biophysica Acta – Biomembranes* 1465:140–151. doi:10.1016/S0005-2736(00)00135-8.

Boyer, J. S., James, R. A., Munns, R., Condon, T., and Passioura, J. 2008. Osmotic adjustment leads to anomalously low estimates of relative water content in wheat and barley. *Functional Plant Biology* 35:1172–1182. doi:10.1071/FP08157.

Byrt, C. S., Platten, J. D., Spielmeyer, W., James, R. A., Lagudah, E. S., Dennis, E. S., et al. 2007. HKT1;5-like cation transporters linked to Na+ exclusion loci in wheat, Nax2 and Kna1. *Plant Physiology* 143:1918–1928 doi:10.1104/pp.106.093476.

Chevilly, S., Dolz-Edo, L., Martínez-Sánchez, G., Morcillo, L., Vilagrosa, A., López-Nicolás, J. M., et al. 2021. Distinctive traits for drought and salt stress tolerance in melon (*Cucumis melo* L.). *Frontiers in Plant Science* 12:777060. doi:10.3389/fpls.2021.777060.

Colmer, T. D., Flowers, T. J., Munns, R. 2006. Use of wild relatives to improve salt tolerance in wheat. *Journal of Experimental Botany* 57:1059–1078. doi:10.1093/jxb/erj124.

Colmer, T. D., Munns, R., and Flowers, T. J. 2005. Improving salt tolerance of wheat and barley: future prospects. *Australian Journal of Experimental Agriculture* 45:1425–1443. doi:10.1071/EA04162.

Cuin, T. A., Parsons, D., and Shabala, S. 2010. Wheat cultivars can be screened for NaCl salinity tolerance by measuring leaf chlorophyll content and shoot sap potassium. *Functional Plant Biology* 37:656–664. doi:10.1071/FP09229.

Cuin, T. A., Tian, Y., Betts, S. A., Chalmandrier, Ŕ., and Shabala, S. 2009. Ionic relations and osmotic adjustment in durum and bread wheat under saline conditions. *Functional Plant Biology* 36:1110–1119 doi:10.1071/FP09051.

Dvořak, J., Noaman, M. M., Goyal, S., and Gorham, J. 1994. Enhancement of the salt tolerance of *Triticum turgidum* L. by the *Kna1* locus transferred from the *Triticum aestivum* L. chromosome 4D by homoeologous recombination. *Theoretical and Applied Genetics* 87:872–877 doi:10.1007/BF00221141.

Flowers, T. J., Troke, P. F., and Yeo, A. R. 1977. The mechanism of salt tolerance in halophytes. *Annual Review in Plant Physiology* 28:89–121. doi:10.1146/annurev.pp.28.060177.000513.

Garthwaite, A. J., Von Bothmer, R., and Colmer, T. D. 2005. Salt tolerance in wild *Hordeum* species is associated with restricted entry of Na^+ and Cl^- into the shoots. *Journal of Experimental Botany* 56(419):2365–2378. doi:10.1093/jxb/eri229.

Gökçe, Z. N. Ö., Gökçe, A. F., Junaid, M. D., and Chaudhry, U. K. 2022. Morphological, physiological, and biochemical responses of onion (*Allium cepa L.*) breeding lines to single and combined salt and drought stresses. *Euphytica* 218:29. doi:10.1007/s10681-022-02980-7.

Hauser, F., and Horie, T. 2010. A conserved primary salt tolerance mechanism mediated by HKT transporters: A mechanism for sodium exclusion and maintenance of high K^+/Na^+ ratio in leaves during salinity stress. *Plant, Cell and Environment* 33:552–565. doi:10.1111/j.1365-3040.2009.02056.x.

Hopmans, J. W., Qureshi, A. S., Kisekka, I., Munns, R., Grattan, S. R., et al. 2021. Critical knowledge gaps and research priorities in global soil salinity. *Advances in Agronomy* 169:1–191 doi:10.1016/bs.agron.2021.03.001.

Horie, T., Hauser, F., and Schroeder, J. I. 2009. HKT transporter-mediated salinity resistance mechanisms in *Arabidopsis* and monocot crop plants. *Trends in Plant Science* 14:660–668. doi:10.1016/j.tplants.2009.08.009.

Huang, S., Spielmeyer, W., Lagudah, E. S., James, R. A., Platten, J. D., Dennis, E. S., et al. 2006. A sodium transporter (HKT7) is a candidate for *Nax1*, a gene for salt tolerance in durum wheat. *Plant Physiology* 142:1718–1727 doi:10.1104/pp.106.088864.

Huang, S., Spielmeyer, W., Lagudah, E. S., and Munns, R. 2008. Comparative mapping of *HKT* genes in wheat, barley and rice, key determinants of Na+ transport and salt tolerance. *Journal of Experimental Botany* 59: 927–937. doi:10.1093/jxb/ern033.

Ismail, A. M., and Horie, T. 2017. Genomics, physiology, and molecular breeding approaches for improving salt tolerance. *Annual Review of Plant Biology* 68:405–434 doi:10.1146/annurev-arplant-042916-040936.

James, R. A., Davenport, R., and Munns, R. 2006. Physiological characterisation of two genes for Na+ exclusion in durum wheat: *Nax1* and *Nax2*. *Plant Physiology* 142:1537–1547 doi:10.1104/pp.106.086538.

James, R. A., Rivelli, A. R., Munns, R. and von Caemmerer, S. 2002. Factors affecting CO_2 assimilation, leaf injury and growth in salt-stressed durum wheat. *Functional Plant Biology* 29:1393–1403 doi:10.1071/FP02069.

Krishnamurthy, L., Serraj, R., Rai, K. N., Hash, C. T., and Dakheel, A. J. 2007. Identification of pearl millet [*Pennisetum glaucum* (L.) R. Br.] lines tolerant to soil salinity. *Euphytica* 158:179–188. doi:10.1007/s10681-007-9441-3.

Langridge, P., and Reynolds, M. P. 2015. Genomic tools to assist breeding for drought tolerance. *Current Opinion in Biotechnology* 32:130–135. doi:10.1016/j.copbio.2014.11.027.

Langridge, P., and Reynolds M. 2021. Breeding for drought and heat tolerance in wheat. *Theoretical and Applied Genetics* 134:1753–1769 doi:10.1007/s00122-021-03795-1.

Lopes, M. S., and Reynolds, M. P. 2010. Partitioning of assimilates to deeper roots is associated with cooler canopies and increased yield under drought in wheat. *Functional Plant Biology* 37:147–156. doi:10.1071/FP09121.

Mujeeb-Kazi, A., Munns, R., Rasheed, A., Ogbonnaya, F. C., Ali, N., Hollington, P., et al. 2019. Breeding strategies for structuring salinity tolerance in wheat. *Advances in Agronomy* 155:121–187. doi:10.1016/bs.agron.2019.01.005.

Munns, R. 1988. Why measure osmotic adjustment? *Australian Journal of Plant Physiology* 15:717–726 doi:10.1071/PP9880717.

Munns, R. 2002. Comparative physiology of salt and water stress. *Plant Cell and Environment* 25:239–250. doi:10.1046/j.0016-8025.2001.00808.x.

Munns, R. 2011. Plant adaptations to salt and water stress: differences and commonalities. *Advances in Botanical Research* 57:1–32. doi:10.1016/B978-0-12-387692-8.00001-1.

Munns, R. and James, R. A. 2003. Screening methods for salinity tolerance: a case study with tetraploid wheat. *Plant and Soil* 253:201–218. doi:10.1023/A:1024553303144.

Munns, R., Passioura, J. B., Colmer, T. D., and Byrt, C. S. 2020. Osmotic adjustment and energy limitations to plant growth in saline soil. *New Phytologist* 225:1091–1096 doi:10.1111/nph.15862.

Munns, R., and Tester, M. 2008. Mechanisms of salinity tolerance. *Annual Review of Plant Biology* 59:651–681. doi:10.1146/annurev.arplant.59.032607.092911.

Nounjan, N., Mahakham, W., Siangliw, J. L., Toojinda, T., and Theerakulpisut, P. 2020. Chlorophyll retention and high photosynthetic performance contribute to salinity tolerance in rice carrying drought tolerance quantitative trait loci (QTLs). *Agriculture* 10:620. doi:10.3390/agriculture10120620.

Pardo, J. M. 2010. Biotechnology of water and salinity stress tolerance. *Current Opinion in Biotechnology* 21:185–196. doi:10.1016/j.copbio.2010.02.005.

Passioura, J. B., and Angus, J. F. 2010. Improving productivity of crops in water-limited environments. *Advances in Agronomy* 106:37–75. doi:10.1016/S0065-2113(10)06002-5.

Rahnama, A., Munns, R., Posting, K., and Watt, M. 2011. A screening method to identify genetic variation in root growth response to a salinity gradient. *Journal of Experimental Botany* 62:69–77. doi:10.1093/jxb/erq359.

Ren, Z. H., Gao, J. P., Li, L. G., Cai, X. L., Huang, W., Chao, D. Y., et al. 2005. A rice quantitative trait locus for salt tolerance encodes a sodium transporter. *Nature Genetics* 37:1141–1146. doi:10.1038/ng1643.

Rengasamy, P. 2006. World salinization with emphasis on Australia. *Journal of Experimental Botany* 57:1017–1023. doi:10.1093/jxb/erj108.

Reynolds, M., Dreccer, F., and Trethowan, R. 2007. Drought-adaptive traits derived from wheat wild relatives and landraces. *Journal of Experimental Botany* 58:177–186. doi:10.1093/jxb/erl250.

Reynolds, M., and Tuberosa, R. 2008. Translational research impacting on crop productivity in drought-prone environments. *Current Opinion in Plant Biology* 11:171–179. doi:10.1016/j.pbi.2008.02.005.

Rhodes, D., Nadolska-Orczyk, A. and Rich, P. J. 2002. Salinity, osmolytes and compatible solutes. In *Salinity: Environment–Plants–Molecules*. Eds. A. Lauchli and U. Luttge, 181–204. Dordrecht: Kluwer. doi:10.1007/0-306-48155-3_9.

Richards, R. A., Rebetzke, G. J., Condon, A. G., and Van Herwaarden, A. F. 2002. Breeding opportunities for increasing the efficiency of water use and crop yield in temperate cereals. *Crop Science* 42:111–121. doi:10.2135/cropsci2002.1110.

Richards, R. A., Rebetzke, G. J., Watt, M., Condon, A.G., Spielmeyer, W., and Dolferus, R. 2010. Breeding for improved water productivity in temperate cereals: phenotyping, quantitative trait loci, markers and the selection environment. *Functional Plant Biology* 37:85–97 doi:10.1071/FP09219.

Rivandi, J., Miyazaki, J., Hrmova, M., Pallotta, M., Tester, M., and Collins, N. C. 2011. A *SOS3* homologue maps to HvNax4, a barley locus controlling an environmentally sensitive Na+ exclusion trait. *Journal of Experimental Botany* 62:1201–1216 doi:10.1093/jxb/erq346.

Schachtman, D. P., Munns, R., and Whitecross, M.I. 1991. Variation in sodium exclusion and salt tolerance in *Triticum tauschii. Crop Science* 31:992–997.

Shamaya, N. J., Shavrukov, Y., Langridge, P., Roy, S. J., and Tester, M. 2017. Genetics of Na^+ exclusion and salinity tolerance in Afghani durum wheat landraces. *BMC Plant Biology* 17:209. doi:10.1186/s12870-017-1164-6.

Shavrukov, Y. 2014. Vacuolar H+-PPase (HVP) genes in barley: Chromosome location, sequence and gene expression relating to Na+ exclusion and salinity tolerance. In: *Barley: Physical Properties, Genetic Factors and Environmental Impacts on Growth*. Ed. K. Hasunuma, 125–142. New York: NOVA Science Publisher.

Shavrukov, Y., Bovill, J., Afzal, I., Hayes, J. E., Roy, S. J., Tester, M., et al. 2013. *HVP10* encoding V-PPase is a prime candidate for the barley HvNax3 sodium exclusion gene: evidence from fine mapping and expression analysis. *Planta* 237:1111–1122 doi:10.1007/s00425-012-1827-3.

Shavrukov, Y., Gupta, N. K., Miyazaki. J., Baho. M. N., Chalmers, K. J., Tester, M., et al. 2010. *HvNax3* – a locus controlling shoot sodium exclusion derived from wild barley (*Hordeum vulgare* ssp. *spontaneum*). *Functional and Integrative Genomics* 10:277–291 doi:10.1007/s10142-009-0153-8.

Shavrukov, Y., Langridge, P., and Tester, M. 2009. Salinity tolerance and sodium exclusion in genus *Triticum. Breeding Science* 59 (5), 671–678. doi:10.1270/jsbbs.59.671.

Shavrukov, Y., Shamaya, N., Baho, M., Edwards, J., Ramsey, C., Nevo, E., et al. 2011. Salinity tolerance and Na^+ exclusion in wheat: Variability, genetics, mapping populations and QTL analysis. *Czech Journal of Genetics and Plant Breeding* 47:S85–S93. doi:10.17221/3260-cjgpb.

Shavrukov, Y., and Sokolov, V. 2015. Maize-gamagrass interspecific hybrid, *Zea mays×Tripsacum dactyloides*, shows better salinity tolerance and higher Na^+ exclusion than maize and sorghum. *International Journal of Latest Research in Science and Technology* 4:128–133.

Shavrukov, Y., Sokolov, V., Langridge, P., and Tester M. 2006. Interspecific hybrid, Zea mays L.×Tripsacum dactyloides L., a new fodder crop with large silage biomass production under abiotic stresses. In: *Groundbreaking Stuff. Proceedings of the 13th Australian Agronomy Conference. Australian Society of Agronomy*. Eds. N. Turner, and T. Acuna, Perth, Australia. Available at: http://www.regional.org.au/au/asa/2006/poster/environment/4578_shavrukovy.htm#TopOfPage.

Teakle, N. L., and Tyerman, S. D. 2010. Mechanisms of Cl^- transport contributing to salt tolerance. *Plant, Cell and Environment* 33:566–589. doi:10.1111/j.1365-3040.2009.02060.x.

Ul Haq, T., Gorham, J., Akhtar, J., Akhtar, N., and Steele K. A. 2010. Dynamic quantitative trait loci for salt stress components on chromosome 1 of rice. *Functional Plant Biology* 37:634–645. doi:10.1071/FP09247.

5 Multiple Stresses Underground in Soil (Salinity and Sodicity)

Neha Bhargavi, Oishika Sah, Saiyyeda Firdous, and Karthika Rajendran
VIT University

CONTENTS

DOI: 10.1201/9781003300564-5

5.1 INTRODUCTION

The existence of plants is inextricably linked to the properties of soil. Soil provides nutrients, water, and oxygen to plants, allowing them to grow and develop. Human activities and other natural processes (weathering, wind, and rain) contribute to the formation of salt-affected soils. By accumulating free soluble salts, soil salinization accelerates soil degradation. All soils and natural streams contain soluble salts; the amount of soluble salts in the root zone determines whether the soil is "normal" or "salt-affected," hence influencing crop growth. It's one of the most serious types of soil degradation since it reduces agricultural yield significantly (McWilliam, 1986; Rengasamy, 2006; Munns and Tester, 2008).

Stress is defined as any factor that impedes plant growth and productivity (Ashraf, 1994). Salt-affected soils not only cause osmotic and ion-specific stresses in plants, but also limit the bioavailability of important nutrients and pose agricultural production challenges. Crop cultivation in salt-affected soils is additionally hampered by water-related issues, such as diminished root growth due to mechanical impedance in drying soils, insufficient porosity, and waterlogging (Rengasamy, 2006). The interaction of hypoxia and saline conditions has a considerable negative effect on plant growth, and many salt-affected soils are also waterlogged, at least on a regular basis (Barrett-Lennard, 2003).

As the human population on Earth continues to grow, scientists work relentlessly to develop innovative methods to feed the planet. By 2050, global food production must be doubled in order to meet the demands of all humankind (Bhalla, 2006). Because high-quality land has already been used for agriculture over the last century, degraded land should be exploited for food production (Tilman et al., 2002; Abrol, 2004). Various management strategies, such as crop rotation, crop substitution, and salt leaching, can improve agricultural production in salt-affected soils. However, it is a costly and laborious task. The development of new cultivars that are tolerant to saline environments and provide a high yield could contribute to food security.

In spite of the fact that traditional breeding has created numerous cultivars over the past century, little progress has been made in generating cultivars with enhanced tolerance to salt-affected environments. It is mostly due to a lack of genetic diversity in the gene pool of many cultivated crops. In addition, the transfer of genes from wild to domesticated species is difficult due to reproductive barriers (Ashraf and Akram, 2009; Turan et al., 2012). In recent decades of plant genomics research, molecular marker technologies such as single nucleotide polymorphism (SNP) markers have enabled high-throughput genotyping at the DNA level in order to find out quantitative trait loci (QTL) linked to gene(s) for salinity tolerance. Alternately, it is feasible to modify the expression of genes already present in the same crop through genetic engineering. This technique also permits the transfer of stress-responsive genes from one crop to another. It has been established that overexpression of antiporter genes promotes plant salt tolerance and that it is a viable strategy for regulating the uptake of toxic ions and, thus, enhancing plant salt tolerance.

Although many reviews and chapters have been published on this subject, we discussed various stresses and response mechanisms in plants, as well as reclamation procedures in salt-affected soil, and highlight the use of breeding techniques and biotechnological tools to develop new crop cultivars with improved salinity tolerance in the following sections.

5.2 CAUSES, CLASSIFICATION, AND DISTRIBUTION OF SALT-AFFECTED SOILS

Weathering of parent rocks rich in soluble salts, coastal salt marshlands, and inland deserts formed by wind and rain are some of the natural processes that lead to soil salinization. However, these processes are not considered important for agriculture (Kumar and Sharma, 2020). On the other hand, secondary salinization, which can be linked to either improperly managed irrigation practices or the clearing of forest areas, is the leading cause of widespread contamination of agricultural fields throughout the world (Rao et al., 2012). In dry and semi-arid areas, salt-affected soil is widespread

TABLE 5.1
Classification of Salt-Affected Soils (U.S. Salinity Laboratory Staff. 1954)

Salt-Affected Soil	Soil pH	Electrical Conductivity (EC_e) (dS/m^{-1})	Sodium Adsorption Ratio (SAR)	Exchangeable Sodium Percentage (ESP)	Soil Physical Condition
Saline	<8.5	>4	<13	<15	Flocculated
Sodic	>8.5	<4	>13	>15	Dispersed
Saline–Sodic	>8.5	>4	>13	>15	Flocculated

(Gupta and Abrol, 1990; Ashraf, 1994). This is due to a lack of rainfall combined with an abundance of evaporation, which results in a limited quantity of salt leaching. It has been observed that as the level of dryness in the area increases, so does the severity of salt accumulation. Salt-affected soils can also be found in humid places susceptible to seawater intrusions, such as deltaic regions and low-lying areas. In 1954, the U.S. Salinity Laboratory (USSL) Staff classified salt-affected soils into three broad categories (Table 5.1): (a) saline soils; (b) sodic soils; and (c) saline–sodic soils, which are discussed in detail below.

5.2.1 Saline Soils

Saline soils have an electrical conductivity of the saturation extract (EC_e) of >4 deci Siemens per meter ($dS\ m^{-1}$; about 40 mM NaCl) at 25°C, a sodium adsorption ratio (SAR) of <13, an exchangeable sodium percentage of <15% (ESP), and pH <8.5. It contains high amounts of soluble salts, including chlorides (Cl^-) and sulfates (SO_4^{2-}) of sodium (Na^+), calcium (Ca^{2+}), and magnesium (Mg^{2+}). However, sodium chloride (NaCl) is the most frequent soluble salt in saline soils. Rarely, nitrates (NO_3^-) are found with Cl^- and SO_4^{2-} (Osman, 2018). Due to the presence of excessive salts and low concentrations of Na^+ at exchange sites, these soils are frequently flocculated. Their permeability is comparable to or greater than that of sodic and saline–sodic soils (Keren, 2005). They have favorable soil structure and tillage attributes for crop cultivation. Substantial quantities of gypsum and other calcium compounds with low solubility may be present in these soils. The surface of this soil is covered with a white salt crust.

5.2.2 Sodic Soils/Alkaline Soils

At 25°C, EC_e of sodic soil is <4 dS/m^{-1}, SAR is >13, ESP is >15%, and pH is >8.5. Carbonates (CO_3^{2-}) and bicarbonates (HCO_3^-) of Na^+, Ca^+, and Mg^{2+} predominate in sodic soil. The vast majority of Na^+ in the soil solution is exchangeable, and only trace amounts of soluble salts are present. Gypsum is almost absent in these soils. The physical condition of these soils is poor, dispersion of clay is very common, and soil structure is unstable. The surface of these soils is occasionally black due to the dispersion of organic materials and humus (Levy and Shainberg, 2005). When they are wet, they have a tendency to be sticky, but when they are dry, they are quite rigid (Bernstein, 1975).

5.2.3 Saline–Sodic Soils

At 25°C, saline–sodic soil has an EC_e of >4 dS/m^{-1}, a SAR of >13, and an ESP of >15%. These soils have both soluble and exchangeable Na^+. Soil is flocculated, and the pH is less than 8.5 as long as excess salts are present. Because of the high-salt concentrations, saline–sodic soils have a permeability that is closer to normal than non-saline–sodic soils. When such soils are leached, the soluble salt content decreases, and the soil reaction becomes strongly alkaline when exchangeable Na^+ is hydrolyzed (Bernstein, 1975; Sparks, 2003).

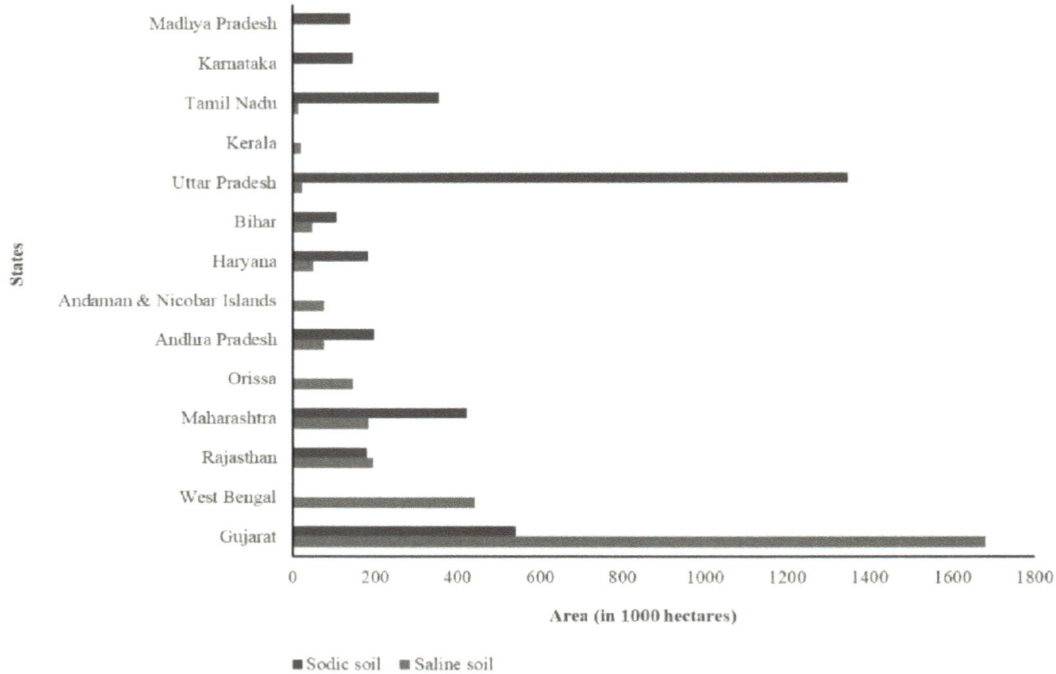

FIGURE 5.1 Distribution of saline soils and sodic soil in different states of India (Source: ICAR, 2017).

Salt-affected soils can be found in more than 75 countries across Asia, Africa, and America. There are 932.2 million hectares (Mha) of soils in total: 351 Mha of saline soils and 581 Mha of sodic soils (Osman, 2018). Salt accumulation affects 6.73 Mha in India, of which 3.77 Mha are sodic soils and 2.96 Mha are saline soils (ICAR, 2017). In 15 Indian states, salt-affected soil is present. Gujarat has the highest percentage of saline soils (1.6 Mha), while Uttar Pradesh has the highest percentage of sodic soils (1.3 Mha). Figure 5.1 illustrates the distribution of saline and sodic soils in India.

5.3 STRESS RESPONSES AND COPING MECHANISMS OF PLANTS IN SALT-AFFECTED SOILS

In saline soils, the presence of salts in a solution causes osmotic and ionic effects that reduce plant growth and have a significant impact on plant productivity. Despite the fact that they are frequently difficult to distinguish, these two effects are not mutually exclusive. Unlike ionic stress, osmotic effect has an instantaneous impact on plant growth and function (Tester and Davenport, 2003; Munns and Tester, 2008). Plant growth in sodic soil is hampered by poor soil physical structure, in addition to osmotic and ionic effects (Läuchli and Grattan, 2007). In saline–sodic soils, salinity predominates over sodicity. Even while saline, sodic, and saline–sodic environments have comparable units, there is no clear boundary between their stress factors: in the field, plants are regularly exposed to other stress factors such as temperature and nutrient deficits, and they experience numerous stresses (Läuchli and Epstein, 1990; Läuchli and Grattan, 2007). This section provides a summary of multiple stresses, responses, and mechanisms of plants that allow them to withstand the effects of salt-affected soils.

5.3.1 Osmotic Stress

The vast majority of studies have demonstrated that osmotic stress from the external medium has an immediate effect on plant growth in saline soils. According to the osmotic theory, soil salt exerts its effects on plants via osmotic pressure. High-osmotic-pressure soils often have a low water potential; nonetheless, a low water potential gradient developed between the plant and soil, resulting in insufficient driving power for root water uptake. This consequence is analogous to "physiological drought." When the EC_e is 4 dS/m or greater, which is comparable to 40 mM NaCl, and the osmotic pressure is 0.2 MPa, crop production is reduced in most crops (Bernstein, 1975; Munns and Tester, 2008).

Osmotic stress, which is frequent in plants under water stress, causes the symptoms. Osmotic stress inhibits the production of new leaves and slows leaf expansion. The leaf is small and green in color. Close inspection may reveal leaf succulence, thickness, and waxy cuticles that contribute to the bluish-green hue (Bernstein, 1975). An increase in soil salt concentration causes temporary water loss in leaf cells. Despite a decrease in cell elongation rates, cells recover their entire volume and turgor due to osmotic adjustment. Reduced cell elongation and division inhibit leaf growth and reduce its final size. Because the surface area is reduced more than depth, leaves get smaller and thicker. Overall, plant growth and dry matter production are reduced. The number of tillers that are produced by monocotyledon species and the number of branches that are produced by dicotyledon species has both decreased. Since root growth compensates for a lack of soil moisture by increasing the soil volume available to roots, the ratio of shoots to roots tends to decrease with time. This is hypothesized to preserve soil moisture and avoid a rise in soil salt concentration (Munns and Tester, 2008).

The effects of osmotic stress on plant growth are poorly understood. Under salt stress, hormonal signals may influence plant growth. Among plant hormones, abscisic acid (ABA), gibberellins, and jasmonates may play a crucial role. Changes in water relations and local ABA synthesis triggered by salinity have an immediate effect on stomatal conductance. It is also important to note that ABA's effect on the root–shoot ratio and the transfer of soil conditions to the leaves is especially a crucial mechanism in saline soils (Munns and Tester, 2008). DELLA proteins also mediate the growth-promoting actions of gibberellins and other plant hormones (Tester and Davenport, 2003). Some cell wall characteristics may have changed, although their precise nature is unknown. There is still a paucity of information on the long-distance and local signals that influence these cell wall properties and growth rates (Munns and Tester, 2008). Additionally, there are complex interactions between photosynthesis and growth rate. Although a reduction in photosynthesis leads to a loss in growth under salinity stress, fast changes in leaf expansion rates and an increase in stored carbohydrate, which signals underutilized assimilation, are also significant factors (Munns et al., 2000; Passioura and Munns, 2000; Munns and Tester, 2008).

Due to osmotic stress, plants produce more reactive oxygen species (ROS) when their growth is inhibited. Therefore, the synthesis of antioxidative enzymes such as catalase (CAT), superoxide dismutase (SOD), ascorbate peroxidase (APX), and glutathione-*S*-transferase (GST) increases in the cell. These enzymes can help to reduce the effects of oxidative stress, which can cause cell damage. SOD can scavenge superoxide radicals generated by chloroplasts, mitochondria, and microbodies. APX and CAT can neutralize hydrogen peroxide, a by-product of oxidation (Sairam and Tyagi, 2004). However, the use of these enzymes can have a negative impact on normal functioning of plants. Consequently, the metabolic processes of plants could be affected, including the breakdown of key biomolecules such as protein, DNA, lipid, and chlorophyll (Munns and Tester, 2008).

5.3.2 Ionic Stress

Solubilized salts, on the other hand, form ions such as Na^+, Ca^{2+}, Mg^{2+}, Cl^-, SO_4^{2-}, HCO_3^-, and K^+, which can harm plants due to their ion specificity (Bernstein, 1975). These ions are formed when salts dissolve in water. Even, at low salt concentrations, these ions can cause harmful effects. Na^+ and Cl^- are the most common ions causing ionic damage to plants in saline soils. It has been demonstrated that Cl^- is more damaging to woody perennial plants, such as, grapes (Downton, 1977) and citrus (Cole, 1985). This may be because Na^+ is so well retained in woody roots and stems that only a small amount reaches the leaves. In fact, after passing through the lamina, Cl^- becomes extremely harmful to plants (Munns and Tester, 2008).

The most common ion effect is ion or nutrient poisoning. Ions enter the plant system via transpiration and accumulate in the leaf blade: since Na^+ enters the leaves via the xylem stream and remains in the leaf apoplast when water evaporates, excessive concentrations of Na^+ in the leaf apoplast (up to several millimeters) can cause toxic damage or cell water loss, making the plant more susceptible to water stress. It leads to leaf burn symptoms (Öertli, 1968; Bernstein, 1975). Ionic stress induces chlorosis and necrosis at the leaf margins, leading to premature leaf senescence (Tester and Davenport, 2003). If plants shed older leaves quicker than they create new ones, they will be unable to meet their carbohydrate requirements, resulting in a slower growth rate. Senescence began when a tree leaf contained 0.5% Cl^- or 0.2% Na^+ by dry weight. The accumulation of Na^+ disrupts plant cell metabolism. More than 50 enzymes require a low Na^+: K^+ ratio in the cytoplasm to function properly and it is known that high Na^+ accumulation in the cytoplasm disrupts vital physiological and biochemical processes within the cell. K^+ is required for tRNA binding to ribosomes, whereas Na^+ suppresses protein synthesis (Tester and Davenport, 2003). In addition, Na^+ interferes with the function of microtubules, microfibrils, spherosomes, and ribosomes (Mansour et al., 1993).

Controlling Na^+ movement throughout the plant would be essential for ion exclusion. The strategies used by plants to reduce the amount of damage caused by high Na^+ levels vary by crop, and managing Na^+ necessitates the coordination of a number of different processes. Plants can, for example, reduce the amount of Na^+ taken into the plant, decrease the amount of Na^+ loaded into the xylem or increase the amount of Na^+ retrieved before it reaches the shoot, increase the amount of Na^+ recirculated out of the shoot via the phloem, allocate more Na^+ to specific parts of the shoot, such as old leaves, control transpiration, and secrete Na^+ onto the surface of the leaf (Tester and Davenport, 2003). In *Arabidopsis*, the HKT gene family is involved in Na^+ exclusion by transporting Na^+ from xylem vessels to xylem parenchyma cells and protecting leaves from salt stress (Davenport et al., 2007, Sunarpi et al., 2005).

Through vacuolar compartmentation, however, tolerance to high cytoplasmic Na^+ could be achieved. It is known as the tissue tolerance mechanism, and it permits plants to remain green even while they accumulate a large amount of Na^+ in their leaf tissue. The failure of salt-sensitive plants to regulate Na^+ entry into cells inhibits subsequent cellular processes, whereas salt-resistant phenotypes are the result of tissue tolerance to Na^+ via vacuolar compartmentation (Munns and James, 2003; Tester and Davenport, 2003). Numerous transporters and ion channels occur on the membrane of tonoplasts, including the NHX1, which is a Na^+/H^+ antiporter. In saline conditions, it selectively transports Na^+ and K^+ into the vacuole and is expressed in roots and leaves. It is present in various plant species, including *Arabidopsis* and barley, where it promotes high Na^+ accumulation and, thus, increased salinity tolerance. Similarly, in *Arabidopsis* and tobacco, a proton transporter situated on the tonoplast membrane (AVP1) enhances Na^+ sequestration in leaf vacuoles using energy derived from pyrophosphate breakdown to pump protons into the vacuole. Overexpressing NHX1 or AVP1 increases Na^+ accumulation and tolerance in *Arabidopsis*. This shows that better sequestration may increase tissue tolerance by lowering cytosolic Na^+ (Garbarino and Dupont, 1989; Apse et al., 1999; Blumwald et al., 2000; Gaxiola et al., 2001; Jha et al., 2010).

In addition, the salt overload-sensitive (SOS) signal pathway controls the major transport systems involved in ion homeostasis in plants grown in saline environments (Hasegawa et al., 2000). When Na^+ is elevated near roots, the SOS3 calcineurin B-like protein (CBL4) detects increases in cytosolic free Ca^{2+}. Subsequently, cytosolic Ca^{2+} elevations facilitate CBL4/SOS3 dimerization and interaction with CIPK24 (SOS2), a CBL-interacting protein kinase (Halfter et al., 2000). CBL4/CIPK24 (SOS3/SOS2) complex activates membrane-bound Na^+/H^+ antiporter SOS1 via phosphorylating and activating a myristoyl fatty acid chain covalently linked to the CBL4/SOS3 complex (Ishitani et al., 2000; Qiu et al., 2002; Shi et al., 2002).

Furthermore, due to the accumulation of sodium ions in the vacuole, the osmotic potential of the cytoplasm must be brought into balance with that of the vacuole. This is accomplished by producing organic solutes that do not hinder biological reactions in the cytoplasm of the cell. They only prevent reactions caused by an increase in the concentration of inorganic ions.

Secondary metabolites, such as quaternary ammonium compounds (glycinebetaine, proline betaine, B-alaninebetaine, and choline-*O*-Sulfate), polyhydric alcohols (d-mannitol and d-sorbitol), mono- and disaccharides (sucrose, glucose, and fructose), and tertiary sulfonium compounds, such as 3-dimethylsuphonio propionate, are examples of "osmoprotectants" in higher plants (Hu et al., 2000; Läuchli and Grattan, 2011).

5.3.3 Secondary Stresses

In saline conditions, mineral nutrients such as K^+, Ca^{2+}, and NO_3^- are often poorly absorbed. As a result, soil salinity can cause further damage to plants besides producing osmotic and ion-specific stresses (Greenway and Munns, 1980). Na^+ can compete with K^+ for binding sites on K^+ transporters in plants, leading to a deficiency of K^+ (Tester and Davenport, 2003). Ca^{2+} deficiency, which is induced by soil salinity, further reduces plant growth and yield. Similarly, saline conditions decrease NO_3^- absorption within 1 minute of salt stress being applied (Klobus et al., 1988).

In contrast, the presence of sodicity will not result only in osmotic and ionic effects. Poor soil physical properties can result in crusting, limited infiltration, and poor aeration, all of which are detrimental to plant and microbial growth (Tester and Davenport, 2003; Läuchli and Grattan, 2007). It makes tilling and crop cultivation difficult, limits seed germination and root development, and is prone to wind and water erosion (Oster et al., 1999). They have nutritional deficiencies. A surplus of exchangeable Na^+ elevates the pH of the soil. Sodic soils, with their extremely alkaline pH, reduce the amount of P, Fe, Mn, and Zn available to plants, despite the fact that soil pH has a minimal direct influence on plant growth. The formation of insoluble calcium and magnesium carbonates as a result of interactions with soluble sodium carbonate decreases the concentration of calcium and magnesium in the soil solution, causing a calcium deficit in plants. On the other hand, the ions of sodium (Na), molybdenum (Mo), and boron (B) are toxic to plants in sodic soils (B), even in minimal concentrations. The toxic levels of CO_3^{2-} and HCO_3^- toxicities can also harm or kill plants. The toxicity of Na^+ can be reduced if the soil has enough calcium (Osman, 2018).

The reactions of plants growing in saline–sodic soil are governed by the concentrations of ions rather than their proportions on the exchange complex, and all nutrient levels are adequate. Saline–sodic soils are ideal for plant growth. The total soluble salts are low at similar ESPs in non-saline–sodic soil because increases in the exchangeable Na^+ are balanced by decreases in the exchangeable Ca^{2+} and Mg^{2+} in the soil solution. Due to the unfavorable physical condition of non-saline–sodic soils, even crops that are able to endure high levels of exchangeable Na^+ from a nutritional standpoint usually fail. They typically have a decreased rate of fertility (Bernstein, 1975).

5.4 PREVENTIVE MEASURES AND SOIL AMENDMENTS

5.4.1 Saline Soils

Saline soils cannot be reclaimed using chemical amendments, conditioners, or fertilizers. Main methods to reclaim saline soils in order to remove salts from the plant root zone include leaching with water and drainage. Besides, scraping, flushing, leaching, drainage, management of irrigation water, fertilizer management, mulching, selection of tolerant crops, and planting deep-rooted perennials can be utilized. They are discussed as follows.

5.4.1.1 Scraping

The accumulated salts on the surface can be removed mechanically. This is the simplest and most cost-effective approach for removing salty soils from a small area.

5.4.1.2 Flushing

Surface salts are washed away by running water (Zaman et al., 2018). However, this is not a recommended method. This is especially useful in crusted soils with little permeability.

5.4.1.3 Leaching

The only realistic approach to remove too many salts from the soil is to use good irrigation water, or rain to leach them away. When high-quality water is combined with saline water and used for irrigation, the impact of salinity is greatly minimized. If there are drainage facilities, this strategy is beneficial because draining the salt-rich effluents lowers the water table and removes the salts. Leaching requirement (LR) is defined as "the percentage of irrigation water that must be leached through the root zone in order to maintain a particular level of soil salinity." It is determined by the amount of salt in the irrigation water as well as the amount of salt that may be present in the root zone (Osman, 2018).

5.4.1.4 Drainage

The drainage capacity of the soil also contributes to salinity control. The LR can occur only when the proper amount of water flows through the root zone. Salt leaching, as well as the construction of lateral and main drainage channels 60 cm deep and 45 cm wide, could aid in soil reclamation. Subsurface drainage is an excellent method for lowering the water table, removing surplus salts, and preventing ions such as Cl^-, SO_4^{2-}, and others from becoming more saline. Sometimes, *Phytophthora* root rot and other fungal infestations can be promoted by high moisture in saline and sodic soils. Improved water management, such as better irrigation or drainage, may avoid fungal infections linked with wetter soil conditions even if soil salinity isn't reduced (Bernstein, 1975).

5.4.1.5 Management of Irrigation

Plant growth can be enhanced by employing alternate furrow irrigation rather than flood irrigation. It is also more efficient to use a drip or sprinkler irrigation systems than traditional flood irrigation because it uses less water.

5.4.1.6 Fertilizer Management

The low nitrogen content of these soils can be remedied by providing 20%–25% of the necessary nitrogen level. The addition of organic manures such as FYM, compost, and so on helps to mitigate the negative effects of salinity caused by the release of organic acids during decomposition. Green manuring can also help to reduce the effects of salinity.

5.4.1.7 Mulching

In locations where water is scarce, organic mulch on the surface decreases the rate of surface evaporation and the rate at which the soil accumulates salt.

5.4.1.8 Selection of Salt-Tolerant Crops

Depending on the salt content of the soil and the ability of the crop to withstand salinity, a particular crop may be chosen. It has been stated by Galvani (2007), which is based on the Saint George Reclamation Consortium, Italy in 2005, that there are four categories of crops, which can tolerate varying amounts of EC_e. These categories are as follows: moderately sensible (0–4 mS/cm), sensible (4–6 mS/cm), tolerant (6–8 mS/cm), and very tolerant (8–12 mS/cm). Carrot, cucumber, watermelon, bean, strawberry, apple, pear, plum, radish, and celery are moderately sensible crops. Sensible crops are apricot, onion, lettuce, melon, potato, peach, rice, soybean, sorghum, grape, and pumpkin. Tolerant crops include oats, cabbage, wheat, sunflower, olive tree, tomato, rice, and spinach. Asparagus, beetroot, barley, and rye are crops with a high level of tolerance.

5.4.1.9 Planting Deep-Rooted Perennials

For the management of dryland salinities in Australia, Canada (Manitoba, Saskatchewan, and Alberta), the United States (Montana, North and South Dakota), South Africa, Iran, Afghanistan, Thailand, and India, planting perennials with deep roots that would draw water from greater depths of the soil is the best option (Osman, 2018).

5.4.2 Sodic Soils

Sodic soil management also involves leaching via irrigation and drainage, but before leaching, excess exchangeable Na^+ must be replaced with a sufficient cation (Ca^{2+}) to reduce the ESP to the appropriate level while growing suitable crops. The replacement requires soil amendments that either contains calcium or generates calcium ions after application. To break up the shallow, dense, sodic clay pans or natric layers that exist within 0.4 m of the soil's surface, deep ploughing and subsoiling have been used (Abdelgawad et al., 2004). Saline–sodic soils are associated with sodic soils because they have similar characteristics and management strategies. Chemical reclamation, deep tillage and subsoiling, periodic irrigation, tolerant crop selection, organic amendments, and phytoremediation have all been successfully applied in sodic and saline–sodic soils (Qadir et al., 2007). However, physicochemical and geological properties of the site must be considered while constructing a successful reclamation strategy. These classic reclamation methods are briefly described here.

5.4.2.1 Chemical Reclamation

By releasing soluble calcium or dissolving calcium to replace exchangeable sodium, adding divalent cations can quickly repair sodic soil. $CaSO_4$, calcium chloride ($CaCl_2.2H_2O$), sulfuric acid (H_2SO_4), limestone, and calcium carbonate ($CaCO_3$) are some of the chemicals used in chemical reclamation (Oster, 1993). Gypsum and calcium chloride are optimal for soils with a pH of 8.2 or above, whereas limestone and $CaCO_3$ are effective for soils with a pH of 7. Gypsum is a low-cost, high-solubility fertilizer that promotes EC_e and soil permeability. Calcium chloride, a highly soluble salt, has the same impact in sodic soil as gypsum (Oster and Frenkel, 1980). When pure sulfuric acid with a purity of 95% is applied to calcium carbonate-containing soil, calcium sulfate is formed. The hydrolysis process produces sulfuric acid, which combines with calcium carbonate in sodic soil to render sodium soluble when water is present (Miyamoto et al., 1975). Sulfur and pyrite are two other chemical amendments that can help restore sodic soils (Gangwar et al., 2020).

5.4.2.2 Deep Tillage and Subsoiling

The benefits of ploughing in the restoration of sodic soil include breaking up soil particles, increasing soil porosity, and encouraging root growth. As a result, microbial diversity, water storage capacity, and rhizosphere microorganisms have all increased, resulting in better root development and a delayed onset of plant senescence (Qin et al., 2008; Liang et al., 2010). One to two meters of deep ploughing can be utilized to break up concrete-like structures and enhance soil quality. Deep tillage exposes the sodic subsoil, which causes logging and infiltration issues. Both ridges and planting 0.25 m above furrows can contribute to soil aeration. Wet earth sinks and compacts quicker; therefore, deep ploughing alone will not solve the sodic soil problems. Ghafoor et al. (1985) found that the application of gypsum ($CaSO_42H_2O$) and tilling significantly increased wheat productivity.

5.4.2.3 Periodic Irrigation

In areas where water is not a limiting factor, sodic soil can be recovered. It can be accomplished by continuously diluting the soil with high-salt, divalent cation-rich water. High EC in water limits soil dispersion and increases soil flocculation (Murtaza et al., 2009). The exchangeable Na^+ is displaced by Ca^{2+} ions in the water, resulting in a decrease in sodicity. For good soil recovery, water depth should be 9–10 times the soil depth (Levy et al., 1998).

5.4.2.4 Use of Organic Matter

The integration of organic matter into sodic and saline–sodic soils is widely regarded as beneficial, albeit a time-consuming process. In fact, the processes help break down calcium compounds in sodic soil. It involves composting, crop rotation, and the use of green manures.

It is possible for roots to form macropores on the surface of moist soil, which can modify root zone characteristics by removing air from micropores and increasing water infiltration (Qadir and

Oster, 2004). Tisdall (1991) discovered that the in situ synthesis of polysaccharides and infectious hyphae at the root–soil interface boosts aggregate strength. It is believed that agroforestry frameworks improve the organic structure of sodic soil (Singh and Singh, 1997). Frequently, organic supplements are blended with gypsum (Abdel-Fattah, 2012). In conjunction with chemical amendments, the application of 10 t/ha^{-1} FYM enhanced soil physical properties such as bulk density, porosity, void ratio, water permeability, and hydraulic conductivity, and increased rice and wheat yields in sodic soil (Hussain et al., 2001). Rice straw, wheat straw, rice husk, and chopped salt grass can enhance the physical properties of saline–sodic soil (Osman, 2018). Press mud, a by-product of sugarcane juice including fiber, sugar, wax, ash, and soil, can also be used to improve the physical, chemical, and biological aspects of both sodic and saline–sodic soils (Barry et al., 2001).

5.4.2.5 Selection of Crops for Sodic Soils

Rice and dhaincha are tolerant to exchangeable Na^+, while wheat and bajra are moderately tolerant and mash (*Phaseolus mungo*) and lentil are sensitive (Osman, 2018). Since they are tolerant of high boron concentrations, sorghum, cotton, celery, and asparagus can all be cultivated on sodic soils. Extremely tolerant crops for saline–sodic soil conditions include beardless wild rye and tall wheatgrass, Altai wild rye, hybrid wheatgrass, slender wheatgrass, and Russian wild rye; highly tolerant crops include tall fescue, western wheatgrass, and crested wheatgrass; and moderately tolerant crops include creeping foxtail and meadow brome (USDA-NRCS, 2010).

5.4.2.6 Phytoremediation

Phytoremediation is a plant-based technique for remediating both sodic and saline–sodic soils. These plant roots can accelerate the dissolving rate of calcite, resulting in higher Ca^{2+} levels in the soil solution than Na^+. Phytoremediation could be more effective than gypsum treatment in reducing ESP. Essentially, this reaction is brought about by the coordinated actions of three different mechanisms in the root zone. First, CO_2 is converted to H_2CO_3, which then combines with $CaCO_3$; second, H_2CO_3 dissociates into H^+ and HCO_3^-, followed by an H^+ reaction with $CaCO_3$; and third, $CaCO_3$ is broken down, yielding Ca^{2+} and CO_3^{2-} (Qadir et al., 2007). For phytoremediation, the genotypes with high resistance to salinity, sodicity, and hypoxia are preferred. *Sesbania*, alfalfa, and Bermuda grass are excellent examples of plants used for phytoremediation of sodic and saline–sodic soils (Ahmad et al., 1990; Ilyas et al., 1993; Kelley, 1937).

5.5 PLANT-BASED TECHNIQUES

The management techniques may provide permanent solutions for salt-affected farms. However, it is expensive and the benefits are slow to occur. Alternatively, developing tolerant cultivars would be a cost-effective solution to these issues. Using traditional, marker-assisted selection, and genetic engineering techniques, considerable efforts have been made to develop salinity-tolerant cultivars, whereas sodicity-tolerant cultivars have received very little attention. This section illustrates some of the advancements made over the years.

5.5.1 Conventional Plant Breeding

The key to developing new crop varieties is to look for natural variation in the desired trait (Koornneef et al., 2004). For this purpose, both inter- and intra-species genetic diversity may be used. Research studies have utilized different physiological parameters such as photosynthesis and stomatal conductance in wheat (Mujeeb-Kazi et al., 2019) and rice (Srivastava and Sharma, 2022); chlorophyll content in cotton (Farooq et al., 2018); morphological parameters including shoot length and shoot dry in wheat weight in wheat (Al-Ashkar et al., 2020); pollen viability and grain yield in rice (Mohammadinezhad et al., 2010); number of bolls/plant and boll weight in cotton (Zafar et al., 2020); and grain yield and yield-contributing components in chickpea (Khamesi et al., 2020) to identify the novel source of genetic variations under the saline condition (EC_e >4 and PH <8). However, the gene pool of many

TABLE 5.2
List of Genotypes Identified for Tolerance to Soil Sodicity through Screening Morpho-Physiological Parameters

Crops	Parameters	Promising Genotypes	References
Wheat	Grain yield	MW#293	Genc et al. (2019)
	Grain yield	KRL 261, Perenjori, and KRL 283	Verma et al. (2014)
	Grain yield and biomass	KRL 3-4, KRL 99, KRL 210, and Kharchia 65	Sharma et al. (2018)
	K/Na ratio	Baxter, Hume, and a line HSF1-255	Dang et al. (2016)
Groundnut	Plant height, number of branches/plant, pod yield/plant, seed yield, hundred-seed weight, pod yield/ha, seed yield/ha	Bunting and Keriz	HajHussein et al. (2018)
Barley	K/Na ratio	Baronesse and Grout	Dang et al. (2016)
Barnyard millet	Seedling traits and yield-contributing traits	MDU-1 and BAR-119	Subramanian et al. (2020)
Lentil	Seed yield, uptake of Na+, H_2O_2 production	PDL-1, PSL-9, ILWL-15, ILWL- 192, and ILWL- 20	Singh et al. (2018)

crops lack genetic diversity for salinity tolerance (Turan et al., 2012). Table 5.2 lists the screening criteria for sodicity tolerance (EC_e >4 and pH >8) as well as potential genetic variation sources for increasing tolerance to sodic conditions. Nonetheless, relatively few efforts have been made to screen for sodicity tolerance. It could be owing to screening issues under controlled environmental conditions (Genc et al., 2019).

Salinity tolerance is a polygenic trait governed by multiple genes, each of which has a small effect. Because it is based on quantitative data such as mean, variance, and covariance, plant breeders can only see gains over time. In addition, salinity tolerance is frequently influenced by genotype–environment interactions and has low heritability (Ashraf, 1994; Gregorio et al., 2002). It seems that selection under extremely saline circumstances accelerates the evolution of salt-tolerant genotypes (Ashraf, 1994). The majority of breeding programs have used indirect selection criteria based on genotypes that have low leaf Na^+ concentrations; nevertheless, selection in combinations of the three primary components of salinity tolerance could be an efficient strategy to generate salinity-tolerant cultivars (Rajendran, 2012; Rajendran et al., 2009, 2015).

In self-pollinated crops, breeding techniques such as pedigree breeding and backcross breeding have been effective in developing salinity-tolerant cultivars. On the other hand, recurrent selection and mass selection were proven to be effective in alfalfa and other cross-pollinated crops (Noble et al., 1984). Conventional breeding has been successful in releasing salt-tolerant cultivars in many countries. There have been significant advances in rice production in Sri Lanka with Pokkali; Bangladesh with BRRI Dhan 47 and BINA Dhan 8; India with Kala Rata 1–24, Non Bokra, Bhura Rata; Thailand with FL530; Japan with Mantaro rice and Kanto 51; and the United States with American Rice. Sea Rice 86 (SR86) was the first rice in China to be developed to withstand saline-alkaline soils. More than 103 rice varieties with better salt tolerance have been developed by the International Rice Research Institute (IRRI), such as Pamodar, Jhona349, and IR9884-54-3 (Fernando, 1949; Ghose and Butany, 1959; Sun et al., 2017; Chen et al., 2017; Qin et al., 2020). Salinity-tolerant cultivars, on the other hand, do not yield well in sodic and saline–sodic soils (Bui, 2013). In particular, breeding for saline–sodic soil has been a difficult task for decades.

5.5.2 Genomics and Other Biotechnological Tools

With an understanding of genetic architecture, modern agriculture places greater emphasis on increasing crop tolerance to soil salinity. Plant genome mapping and molecular biology advances will help us better understand the role of stress tolerance genes in crop productivity (Sairam and Tyagi, 2004). QTL mapping is a common method for determining the approximate chromosomal location of a QTL associated with a gene of interest. Many crops have been tested for salt tolerance using QTL. Lindsay et al. (2004) discovered a QTL for Na^+ exclusion on chromosome 2A in durum wheat (38%), while Dubcovsky et al. (1996) discovered a QTL for K^+/Na^+ discrimination on chromosome 4D in bread wheat. Lin et al. (2004) reported shoot Na^+ concentration on chromosome 7 and shoot K^+ concentration on chromosome 1 in rice, then Thomson et al. (2010) discovered Na^+/K^+ discrimination (*Saltol*) on chromosome 1 in rice (2010). The low-cost sequencing also allows us to conduct genome-wide association studies (GWAS) for SNP analysis and uncover allelic variation in important genes for K^+/Na^+ ratio and K^+ and Na^+ concentration in barley (Hazzouri et al., 2018; Dissanayake et al., 2021).

Recent years have seen marked advancements in candidate gene mapping, map-based cloning, and particularly genetic engineering. The strategy of increasing salt tolerance in several crops was made possible by genetic transformation, which allowed scientists to alter the expression of a gene (or genes) in a single crop or to transfer them to a genetic background distinct from the crop's original background. Genetic engineering focused on genes that could affect the accumulation of Na^+, such as transporters responsible for Na^+ exclusion (*HKT* and *SOS1*), compartmentalization inside vacuoles (*NHX1*, *NHX5*, *AVP1*, and *AVP2*), osmolytes (*ostB*, *TPS1*, and *P5CS*), antioxidative enzymes (*Glyl* and *Gly2*), late embryogenesis abundant (LEA) proteins (*HVA1*), molecular chaperons (*RcHSP17.8* and *T30*), and transcription factors for gene regulation (*AtDREB1A*, *MtCBF4*, and *ZmbZI72*) (Turan et al., 2012). Salt tolerance genes might be targeted using the CRISPR-Cas9 technology, which was utilized to confer salinity tolerance. Wang et al. (2018) recently developed a simplified single transcriptional unit (SSTU) CRISPR system to target eight genes in the LEA family, and generated a higher proportion of heterozygous and chimeric mutants induced by Cpf1 despite the lower editing efficiency of Cpf1 compared to FnCpf1. Due to the accuracy of gene and base editing techniques, plant genetic engineering will almost probably lead to the alteration of genes involved in plant responses to salt stress and the development of tolerant crop plants with enhanced yields.

5.6 CONCLUSIONS

Mismanaged agricultural irrigation practices are predicted to increase the area of salt-affected soils over time. There are a variety of soil reclamation techniques that can reverse soil conditions to a limited extent, but they are not taken seriously and their benefits take time to reflect on the farm. The effective implementation of land reclamation procedures will allow for sustainable land usage and prevent its abandonment. On the other hand, the development of tolerant cultivars has been proposed as a viable solution to address this problem. In the germplasm, promising sources of salinity- and sodicity-tolerant genotypes for various crops have been identified. This information can be utilized by breeders for hybridization and the identification of salinity-tolerant QTLs/gene(s). Furthermore, cutting-edge techniques such as Crispr-CAS and gene pyramiding, as well as other approaches such as epigenetic control of gene expression and small/micro-RNAs, could be used to develop commercial salt-tolerant cultivars.

REFERENCES

Abdel-Fattah, M. K. 2012. Role of gypsum and compost in reclaiming saline-sodic soils. *IOSR Journal of Agriculture and Veterinary Science* 1(3):30–38. doi: 10.9790/2380-0133038.

Abdelgawad, A., Arslan, A., Awad, F., and Kadouri, F. 2004. Deep plowing management practice for increasing yield and water use efficiency of vetch, cotton, wheat and intensified corn using saline and non-saline irrigation water. In *Proceedings of the 55th IEC Meeting of the International Commission on Irrigation and Drainage (ICID)*, 67–78. Moscow, Russia.

Abrol, Y. P. 2004. Wild, A. Soils, land and food: managing the land during the twenty-first century. *Annals of Botany* 93(6):785–786. doi: 10.1093/aob/mch104.

Ahmad, N., Qureshi, R. H., and Qadir, M. 1990. Amelioration of a calcareous saline-sodic soil by gypsum and forage plants. *Land Degradation & Development* 2:277–284. doi: 10.1002/ldr.3400020404.

Al-Ashkar, I., Alderfasi, A., Romdhane, B. W., Seleiman, M. F, El-Said, R. A., and Al-Doss, A. 2020. Morphological and genetic diversity within salt tolerance detection in eighteen wheat genotypes. *Plants* 9(3):287. doi: 10.3390/plants9030287.

Apse M. P., Aharon G. S., Snedden W. A., and Blumwald, E. 1999. Salt tolerance conferred by overexpression of a vacuolar Na^+/H^+ antiport in *Arabidopsis*. *Science* 285:1256–1258. doi: 10.1126/science.285.5431.1256.

Ashraf, M. 1994. Breeding for salinity tolerance in plants. *Critical Reviews in Plant Sciences* 13(1):17–42. doi: 10.1080/07352689409701906.

Ashraf, M., and Akram, N. A. 2009. Improving salinity tolerance of plants through conventional breeding and genetic engineering: an analytical comparison. *Biotechnology Advances* 27: 744–752. doi: 10.1016/j.biotechadv.2009.05.026.

Barrett-Lennard, E. G. 2003. The interaction between waterlogging and salinity in higher plants: causes, consequences and implications. *Plant and Soil* 253(1):35–54. doi: 10.1023/A:1024574622669.

Barry, G. A., Rayment, G. E., Jeffery A. J., and Price, A. M. 2001. Changes in cane soil properties from application of sugar mill by-products. In *Proceeding Conference of the Australian Society of Sugarcane Technology*. ed. D.M. Hogarth, 185–191. PK Editorial Services Pty Ltd, Brisbane.

Bernstein, L. 1975. Effects of salinity and sodicity on plant growth. *Annual Review of Phytopathology* 13(1): 295–312. doi: 10.1146/annurev.py.13.090175.001455.

Bhalla, P. L. 2006. Genetic engineering of wheat- current challenges and opportunities. *Trends in Biotechnology* 24(7):305–311. doi: 10.1016/j.tibtech.2006.04.008.

Blumwald, E., Aharon, G. S., and Apse, M. P. 2000. Sodium transport in plant cells. *Biochimica et Biophysica Acta (BBA)-Biomembranes* 1465(1–2):140–151. doi: 10.1016/S0005-2736(00)00135-8.

Bui, E. 2013. Possible role of soil alkalinity in plant breeding for salt-tolerance. *Biology Letters* 9(5): 20130566. doi: 10.1098/rsbl.2013.0566.

Chen, R., Cheng, Y., Han, S., Van Handel, B., Dong, L., Li, X., and Xie, X. 2017. Whole genome sequencing and comparative transcriptome analysis of a novel seawater adapted, salt-resistant rice cultivar – sea rice 86. *BMC Genomics* 18:655. doi: 10.1186/s12864-017-4037-3.

Cole, P. J. 1985. Chloride toxicity in citrus. *Irrigation Science* 6(1):63–71. doi: 10.1007/BF00272476.

Dang, Y. P., Jack T. C., and Ram C. D. 2016. Genetic diversity in barley and wheat for tolerance to soil constraints. *Agronomy* 6(4):55. doi: 10.3390/agronomy6040055.

Davenport, R. J., Munoz-Mayor, A., Jha, D., Essah, P. A., Rus, A., Tester, M. 2007. The Na+transporter AtHKT1 controls xylem retrieval of Na^+ in *Arabidopsis*. *Plant Cell & Environment* 30:497–507. doi: 10.1111/j.1365-3040.2007.01637.x.

Dissanayake, R., Cogan, N. O. I., Smith, K. F., and Kaur, S. 2021. Application of genomics to understand salt tolerance in lentil. *Genes* 12(3):332. doi: 10.3390/genes12030332.

Downton, W. J. S. 1977. Chloride accumulation in different species of grapevine. *Scientia Horticulturae* 7(3):249–253. doi: 10.1016/0304-4238(77)90021-8.

Dubcovsky, J., María, G. S., Epstein, E., Luo, M. C., and Dvořák, J. 1996. Mapping of the K^+/Na^+ discrimination locus *Kna1* in wheat. *Theoretical and Applied Genetics* 92(3):448–454. doi: 10.1007/BF00223692.

Farooq, A. M., Shakeel, A., Muhammad A. R., and Saleem, F. M. 2018. Genetic variability studies for salinity tolerance in *Gossypium hirsutum*. *International Journal of Agriculture and Biology* 20(12):2871–1278. doi: 10.1080/15440478.2020.1838996.

Fernando, L. H. 1949. The performance of salt resistant paddy, Pokkali in Ceylon. *Journal of Tropical Agriculture* 105:124–126.

Galvani, A. 2007. The challenge of the food sufficiency through salt tolerant crops. *Reviews in Environmental Science and Bio/Technology* 6:3–16. doi: 10.1007/s11157-006-0010-3.

Gangwar, P., Singh, R., Trivedi, M., and Tiwari, R. K. 2020. Sodic soil: management and reclamation strategies. In *Environmental Concerns and Sustainable Development*, 175–190. Singapore: Springer. doi: 10.1007/978-981-13-6358-0_8.

Garbarino, J., and DuPont, F. M. 1989. Rapid induction of Na^+/H^+ exchange activity in barley root tonoplast. *Plant Physiology* 89(1):1–4. doi: 10.1104/pp.89.1.1.

Gaxiola, R. A., Li, J., Undurraga, S., Dang, L. M., Allen, G. J., Alper, S. L., and Fink, G. R. 2001. Drought- and salt-tolerant plants result from overexpression of the AVP1 H^+-pump. *Proceedings of the National Academy of Sciences* 98(20):11444–11449. doi: 10.1073/pnas.191389398.

Genc, Y., Taylor, J., Lyons, G., Li, Y., Cheong, J., Appelbee, M., Oldach, K., and Sutton, T. 2019. Bread wheat with high salinity and sodicity tolerance. *Frontiers in Plant Science* 10:1280. doi: 10.3389/fpls.2019.01280.

Ghafoor, A., Muhammad, S., and Ahmad, N. 1985. Reclamation of Khurrianwala saline-sodic soil. *Bull. (J.) Pakistan Council of Research in Water Resources* 15(1):23–28

Ghose, R. L., and Butany, W. T. 1959. Botanical improvement of varieties—general characters of Indian varieties and the application of genetics to rice Improvement. *Proceedings / Indian Academy of Sciences* 49(5):287–302. doi: 10.1007/bf03052051.

Greenway, H., and Munns, R. 1980. Mechanisms of salt tolerance in non-halophytes. *Annual review of Plant Physiology and Plant Molecular Biology* 31(1): 149–190.

Gregorio, G. B., Senadhira, D., Mendoza, R. D., Manigbas, N. L., Roxas, J. P., and Guerta, C. Q. 2002. Progress in breeding for salinity tolerance and associated abiotic stresses in Rice. *Field Crops Research* 76(2–3):91–101. doi: 10.1016/s0378-4290(02)00031-x.

Gupta, R. K., and Abrol, I. P. 1990. Salt-affected soils: their reclamation and management for crop production. *Advances in Soil Science* 11:223–288. doi: 10.1007/978-1-4612-3322-0_7.

HajHussein, O., Assar, A. H. B., Fraah, A. D. M., and Al Sir, A. 2018. Variability heritability and genetic advance of some groundnut genotypes (*Arachis hypogaea L.*) under saline sodic soil. *Annals of Review and Research* 1:1–5 doi: 10.1980/ARR.2018.01.555554.

Halfter, U., Ishitani, M., and Zhu, J. K. 2000. The *Arabidopsis* SOS2 protein kinase physically interacts with and is activated by the calcium-binding protein SOS3. *Proceedings of the National Academy of Sciences* 97:3735–3740. doi: 10.1073/pnas.97.7.3735.

Hasegawa, P., Bressan, R.A., Zhu, J.K., and Bohnert, H.J. 2000. Plant cellular and molecular responses to high salinity. *Annual review of Plant Physiology and Plant Molecular Biology* 51: 463–499.

Hazzouri, K. M., Khraiwesh, B., Amiri, K. M., Pauli, D., Blake, T., Shahid, M., Mullah, S. K., Nelson, D., Mansour, A. L., Salehi-Ashtiani, K., Purugganan, M., and Masmoudi, K. 2018. Mapping of *HKT1;5* gene in barley using GWAS approach and its implication in salt tolerance mechanism. *Frontiers in Plant Science* 9:156. doi: 10.3389/fpls.2018.00156.

Hu, Y., Schnyder, H., and Schmidhalter, U. 2000. Carbohydrate deposition and partitioning in elongating leaves of wheat under saline soil conditions. *Functional Plant Biology* 27(4):363–370.doi: 10.1071/PP99147.

Hussain, N., Hassan, G., Arshadullah, M., and Mujeeb, F. 2001. Evaluation of amendments for the improvement of physical properties of sodic soil. *International Journal of Agriculture and Biology* 3(3): 319–322.

ICAR. 2017. Technological interventions for enhancing farm productivity in salt-affected vertisols. ICAR-CSSRI/Bharuch/Technical Manual/2017/07. https://krishi.icar.gov.in/jspui/bitstream/123456789/10700/1/Technological%20Intervention%20in%20salt%20affected%20Vertisols.pdf (accessed June 6, 2022).

Ilyas, M., Miller, R. W., and Qureshi, R. H. 1993. Hydraulic conductivity of saline-sodic soil after gypsum application and cropping. *Soil Science Society of America Journal* 57(6):1580–1585.

Ishitani, M., Liu, J. P., Halfter, U., Kim, C. S., Shi, W. M., and Zhu, J. K. 2000. SOS3 function in plant salt tolerance requires N-myristoylation and calcium binding. *Plant Cell* 12:1667–77. doi: 10.1105/tpc.12.9.1667.

Jha, D., Shirley, N., Tester, M., and Roy, S. J. 2010. Variation in salinity tolerance and shoot sodium accumulation in *Arabidopsis* ecotypes is linked to differences in the natural expression levels of transporters involved in sodium transport. *Plant, Cell & Environment* 33(5):793–804. doi: 10.1111/j.1365-3040.2009.02105.x.

Kelley, W. P. 1937. The reclamation of alkali soils. *Bulletin (California Agricultural Experiment Station)* 617:1–40.

Keren, R. 2005. Salt-affected soils, reclamation. In *Encyclopedia of Soils in the Environment*, ed. D. Hillel, 454–461. Oxford: Elsevier. doi: 10.1016/B0-12-348530-4/00503-8.

Khamesi, F., Amini, A., and Ehsanzadeh, P. 2020. Chickpea response to saline water: Concurrence of ion homeostasis sustainment and antioxidative defense measures. *South African Journal of Botany* 133:245–252. doi: 10.1016/j.sajb.2020.08.001.

Klobus, G., Ward, M. R., and Huffaker, R. C. 1988. Characteristics of injury and recovery of net No_3^- transport of barley seedlings from treatment of NaCl. *Plant Physiology* 87: 878–882. doi: 10.1104/pp.87.4.878.

Koornneef, M., Alonso-Blanco, C., and Vreugdenhil, D. 2004. Naturally occurring genetic variation in *Arabidopsis thaliana*. *Annual Review of Plant Biology* 55:141–172. doi: 10.1146/annurev.arplant.55.031903.141605.

Kumar, P., and Sharma, P. K. 2020. Soil salinity and food security in India. *Frontiers in Sustainable Food Systems* 4:533781. doi: 10.3389/fsufs.2020.533781.

Läuchli, A., and Epstein, E. 1990. Plant responses to saline and sodic conditions. In *Agricultural Salinity Assessment and Management*, ed. K. K. Tanji, 113–137. New York: American Society of Civil Engineers.

Läuchli, A., and Grattan, S. R. 2007. Plant growth and development under salinity stress. In *Advances in Molecular Breeding toward Drought and Salt Tolerant Crops*, eds. M. A. Jenks, P. M. Hasegawa and S. M. Jain, 1–32. Dordrecht, The Netherlands: Springer.

Läuchli, A., and Grattan, S. R. 2011. Plant responses to saline and sodic conditions. In *Agricultural Salinity Assessment and Management*, eds. W. W. Wallender, and K. K. Tanji, 169–205. Reston, VA: American Society of Civil Engineers. doi: 10.1061/9780784411698.ch06.

Levy, G. J., and Shainberg, I. 2005. Sodic soils. In *Encyclopedia of Soils in the Environment*, eds. D. Hillel, J. L. Hatfield, D. S. Powlson, C. Rosenzweig, K. M. Scow, M. J. Singer and D. L. Sparks, 504–513. Amsterdam, The Netherlands: Academic Press. doi: 10.1016/B0-12-348530-4/00218-6.

Levy, G. J., Shainberg, L., and Miller, W. P. 1998. Physical properties of sodic soils. In *Sodic Soils: Distribution, Properties, Management and Environmental Consequences*, eds. M. E. Sumner and R. Naidu, 74–94. Oxford: Oxford University Press.

Liang, J. F., Qi, Q. Z., Jia, X. H., Gong, S. J., and Huang, Y. F. 2010. Effects of different tillage management on soil properties and corn growth. *International Journal of Ecology and Environmental Sciences* 19(4):945–950.

Lin, H. X., Zhu, M. Z., Yano, M., Gao, J. P., Liang, Z. W., Su, W. A., Hu, X. H., Ren, Z. H., and Chao D. Y. 2004. QTLs for Na^+ and K^+ uptake of the shoots and roots controlling rice salt tolerance. *Theoretical and Applied Genetics* 108(2):253–260. doi: 10.1007/s00122-003-1421-y.

Lindsay, M. P., Lagudah, E. S., Hare, R. A., and Munns, R. 2004. A locus for sodium exclusion (Nax1), a trait for salt tolerance, mapped in durum wheat. *Functional Plant Biology* 31(11):1105–1114. doi: 10.1071/FP04111.

Mansour, M. M. F., Lee-Stadelmann, O. Y., and Stadelmann, E. J. 1993. Solute potential and cytoplasmic viscosity in *Triticum aestivum* and *Hordeum vulgare* under salt stress. A comparison of salt resistant and salt sensitive lines and cultivars. *Journal of Plant Physiology* 142(5):623–628. doi: 10.1016/S0176-1617(11)80409-8.

McWilliam, J. R. 1986. The national and international importance of drought and salinity effects on agricultural production. *Functional Plant Biology* 13(1):1–13. doi: 10.1071/PP9860001.

Miyamoto, S., Prather, R. J., and Stroehlein, J. L. 1975. Sulfuric acid and leaching requirements for reclaiming sodium-affected calcareous soils. *Plant Soil* 43:573–585. doi: 10.1007/BF01928520.

Mohammadinezhad, G., Singh, R.K., Arzani, A., Rezaei, A.M., Sabouri, H., and Gregorio, G.B. 2010. Evaluation of salinity tolerance in rice genotypes. *International Journal of Plant Production* 4(3): 199–208.

Mujeeb-Kazi, A., Munns, R., Rasheed, A., Ogbonnaya, F. C., Ali, N., Hollington, P., Dundas, I., Saeed, N., Wang, R., Rengasamy, P., Saddiq, M. S., Díaz De León, J. L., Ashraf, M., and Rajaram, S. 2019. Breeding strategies for structuring salinity tolerance in wheat. *Advances in Agronomy* 155:121–187 doi: 10.1016/bs.agron.2019.01.005.

Munns, R., Guo, J., Passioura, J. B., and Cramer, G. R. 2000. Leaf water status controls day-time but not daily rates of leaf expansion in salt-treated barley. *Functional Plant Biology* 27(10):949–957. doi: 10.1071/PP99193.

Munns, R., and James, R.A. 2003. Screening methods for salinity tolerance: a case study with tetraploid wheat. *Plant Soil* 253:201–18. doi: 10.1023/A:1024553303144.

Munns, R., and Tester, M. 2008. Mechanisms of salinity tolerance. *Annual Review of Plant Biology* 59:651–681. doi: 10.1146/annurev.arplant.59.032607.092911.

Murtaza, G., Ghafoor, A., Owens, G., Qadir, M., and Kahlon, U. Z. 2009. Environmental and economic benefits of saline-sodic soil reclamation using low-quality water and soil amendments in conjunction with a rice-wheat cropping system. *Journal of Agronomy and Crop Science* 195:124–136. doi: 10.1111/j.1439-037X.2008.00350.x.

Noble, C. L., Halloran, G. M., and West, D. W. 1984. Identification and selection for salt tolerance in Lucerne (*Medicago sativa L.*). *Australian Journal of Agricultural Research* 35(2):239. doi: 10.1071/ar9840239.

Öertli, J. J. 1968. Extracellular salt accumulation a possible mechanism of salt injury in plants. *Agrochimica* 12:461–469.

Osman, K. T. 2018.Saline and sodic soils. In *Management of Soil Problems*, ed. K. T. Osman, 255–298. Cham: Springer. doi: 10.1007/978-3-319-75527-4_10.

Oster, J. D. 1993. Sodic soil reclamation. In *Towards the Rational Use of High Salinity Tolerant Plants*, ed. H. A. Leith and A. L. Masoom, 485–490. Amsterdam, The Netherlands: Kluwer Academic Publishers. doi: 10.1007/978-981-13-6358-0_8.

Oster, J. D., and Frenkel, H. 1980. The chemistry of reclamation of sodic soils with gypsum and lime. *Soil Science Society of American Journal* 44:41–45. doi: 10.2136/sssaj1980.03615995004400010010x.

Oster. J. D., Shainberg, I., and Abrol, I. P. 1999. Reclamation of salt-affected soils. In *Agricultural Drainage*, eds. R. W. Skaggs and J. Van Schilfgaarde, 659–691. Madison: ASA-CSSA-SSSA. doi: 10.2134/agronmonogr38.c19.

Passioura, J. B., and Munns, R. 2000. Rapid environmental changes that affect leaf water status induce transient surges or pauses in leaf expansion rate. *Functional Plant Biology* 27(10):941–948. doi: 10.1071/PP99207.

Qadir, M., and Oster, J. D. 2004. Crop and irrigation management strategies for saline-sodic soils and waters aimed at environmentally sustainable agriculture. *Science of the Total Environment* 329:1–19. doi: 10.1016/j.scitotenv.2003.10.012.

Qadir, M., Oster, J. D., Schubert, S., Noble, A. D., and Sahrawat, K. L. 2007. Phytoremediation of sodic and saline-sodic soils. *Advances in Agronomy* 96:197–247. doi: 10.1016/S0065-2113(07)96006-X.

Qin, H., Li, Y., and Huang, R. 2020. Advances and challenges in the breeding of salt-tolerant rice. *International Journal of Molecular Sciences* 21(21):8385. doi: 10.3390/ijms21218385.

Qin, H. L., Gao, W. S., Ma, Y. C., Ma, L., and Yin, C. M. 2008. Effects of subsoiling on soil moisture under no-tillage 2 years later. *Scientia Agricultura Sinica* 41(1):78–85. doi: 10.1016/S1671-2927(08)60026-7.

Qiu, Q. S., Guo, Y., Dietrich, M. A., Schumaker, K. S., and Zhu, J. K. 2002. Regulation of SOS1, a plasma membrane Na^+/H^+ exchanger in *Arabidopsis thaliana*, by SOS2 and SOS3. *Proceedings of the National Academy of Science USA* 99:8436–8441 doi: 10.1073/pnas.122224699.

Rajendran, K. 2012. *Components of Salinity Tolerance in Wheat*. PhD diss., The University of Adelaide, Australia.

Rajendran, K., Somanagouda P., and Kumar, S. 2015. Phenotyping for problem soils. In *Phenomics in Crop Plants: Trends, Options and Limitations*, ed. J. Kumar, A. Pratap and S. Kumar, 129–146. New Delhi: Springer. doi: 10.1007/978-81-322-2226-2.

Rajendran, K., Tester, M., and Roy, S. J. 2009. Quantifying the three main components of salinity tolerance in cereals. *Plant, Cell and Environment* 32(3):237–249. doi: 10.1111/j.1365-3040.2008.01916.x.

Rao, G., Khandelwal, K. M., Arora, S., and Sharma. D. K. 2012. Salinity ingress in coastal Gujarat: appraisal of control measures. *Journal of Soil Salinity and Water Quality* 4(2):102–113.

Rengasamy, P. 2006. World Salinization with emphasis on Australia. *Journal of Experimental Botany* 5:1017–1023. doi: 10.1093/jxb/erj108.

Sairam, R. K., and Tyagi, A. 2004. Physiology and molecular biology of salinity stress tolerance in plants. *Current science* 86(3): 407–421.

Sharma, S. K., Kulshreshtha, N., Kumar, A., Yaduvanshi, N. P. S., Singh, M., Prasad, K. R. K., and Basak, N. 2018. Waterlogging effects on elemental composition of wheat genotypes in sodic soils. *Journal of Plant Nutrition* 41(10):1252–1262. doi: 10.1080/01904167.2018.1434541.

Shi, H. Z., Quintero, F. J., Pardo, J. M., and Zhu, J. K. 2002. The putative plasma membrane Na^+/H^+ antiporter SOS1 controls long-distance Na^+ transport in plants. *Plant Cell* 14:465–77. doi: 10.1105/tpc.010371.

Singh, D., Singh, C. K., Singh, Y. P., Singh, V., Singh, R., Tomar, R. S., Sanwal, S. K., Karwa, S., Mishra, V. K., Sarkar, S. K., Pal, M., Kumar, A., Yadav, R. K., and Sharma, P. C. 2018. Evaluation of cultivated and wild genotypes of lens species under alkalinity stress and their molecular collocation using microsatellite markers. *PLoS One* 13(8). doi: 10.1371/journal.pone.0199933.

Singh, G., and Singh, N. T. 1997. Effect of land use practices on organic carbon dynamics of sodic soils. In *Soil Properties and their Management for Carbon Sequestration*, eds. R. Lal, J. Kimble and R. Follett, 89–105. Lincoln, NE: USDA-Natural Resources Conservation Service, National Soil Survey Centre.

Sparks, D. L. 2003. The chemistry of saline and sodic soil. In *Environmental Soil Chemistry*, ed. D.L. Sparks, 285–300. San Diego, CA: Academic Press. doi: 10.1016/B978-0-12-656446-4.X5000-2.

Srivastava, S., and Sharma, P. K. 2022. Morpho-physiological and biochemical tolerance mechanisms in two varieties of *Oryza sativa* to salinity. *Russian Journal of Plant Physiology* 69(2). doi: 10.1134/s1021443722020194.

Subramanian, A., Raj, Nirmal, R., and Jeyaprakash, P. 2020. In vitro and in vivo screening of barnyard millet (*Echinochloa frumentacea*) germplasm for salinity tolerance. *Plant Archives* 20(2):7389–7397

Sun, M. F., Yan, G. H., Wang, A. M., Zhu, G. Y., Tang, H. S., He, C. X., Ren, Z. L., Liu, K., Zhang, G. Y., and Shi. W. 2017. Research progress on the breeding of salt-tolerant rice varieties. *Barley and Cereal Sciences* 34(4): 1–9.

Sunarpi, H. T., Motoda, J., Kubo, M., Yang, H., Yoda, K., Horie, R., Chan, W. Y., Leung, H. Y., Hattori, K., Konomi, M., and Osumi, M. 2005. Enhanced salt tolerance mediated by AtHKT1 transporter-induced Na^+ unloading from xylem vessels to xylem parenchyma cells. *The Plant Journal* 44:928–38. doi: 10.1111/j.1365-313X.2005.02595.x.

Tester, M., and Davenport, R. 2003. Na^+ tolerance and Na^+ transport in higher plants. *Annals of Botany* 91(5): 503–527. doi: 10.1093/aob/mcg058.

Thomson, M. J., de Ocampo, M., Egdane, J., Rahman, M. A., Sajise, A. G., Adorada, D. L., Tumimbang-Raiz, E., Blumwald, E., Seraj, Z. I., Singh, R. K., Glenn, B. G., and Abdelbagi. M. I. 2010. Characterizing the *saltol* quantitative trait locus for salinity tolerance in rice. *Rice* 3:148–160. doi: 10.1007/s12284-010-9053-8.

Tilman, D., Cassman, K. G., Matson, P. A., Naylor, R., and Polasky, S. 2002. Agricultural sustainability and intensive production practices. *Nature* 418(6898):671–677. doi: 10.1038/nature01014.

Tisdall, J. M. 1991. Fungal hyphae and structural stability of soil. *Australian Journal of Soil Research* 29:729–743. doi: 10.1071/SR9910729.

Turan, S., Cornish, K., and Kumar, S. 2012. Salinity tolerance in plants: breeding and genetic engineering. *Australian Journal of Crop Science* 6(9):1337–1348.

USDA-NRCS. 2010. Plants for saline to sodic soil conditions. Technical Note: Plant Materials 9A, USDA-NRCS, Salt Lake City, Utah.

U.S. Salinity Laboratory Staff. 1954. Diagnosis and improvement of saline and alkali soils. In L.A. Richards (Ed.) USDA Handbook 60. pp 100–130. Washington, DC: U.S. Gov. Printing Office.

Verma, P. N., Singh, B. N., Singh, G., Singh, M. K., and Setter, T. L. 2014. Genetic diversity analysis for yield and other agronomic traits in bread wheat under water logged sodic soil condition. *Journal of Wheat Research* 6(1):51–58.

Wang, M., Mao, Y., Lu, Y., Wang, Z., Tao, X., and Zhu, J. K. 2018. Multiplex gene editing in rice with 1608 simplified CRISPR-Cpf1 and CRISPR-Cas9 systems. *Journal of Integrative Plant Biology* 60:626–631. doi: 10.1111/jipb.12667.

Zafar, M. M., Razzaq, A., Farooq, M. A., Rehman, A., Firdous, H., Shakeel, A., Mo, H., Ren, M., Ashraf, M., and Youlu, Y. 2020. Genetic variation studies of ionic and within boll yield components in cotton (*Gossypium hirsutum* L.) under Salt Stress. *Journal of Natural Fibers* 19(8):1–20. doi: 10.1080/15440478.2020.1838996.

Zaman, M., Shahid, S. A., and Heng, L. 2018. *Guideline for Salinity Assessment, Mitigation and Adaptation Using Nuclear and Related Techniques*. Cham, Switzerland: Springer Nature. doi: 10.1007/978-3-319-96190-3.

6 Wild Barley Relatives – Potential Donors of Salinity Tolerance for Cereal Crops

Stanislav Isayenkov
Institute of Food Biotechnology and Genomics
National Academy of Science of Ukraine

CONTENTS

6.1 INTRODUCTION

High salinity is one of the biggest threats to modern agriculture and crop productivity, leading to an annual estimated economic loss of over US$27 billion (Qadir et al., 2014). Saline soils cover about 380–995 Mha of the Earth's land surface (>6% of the planet's total land area) (Szabolcs, 1994; IAEA, 1995; FAO, 2016). The area of salinized soils is reported to be increasing at a rate of 10% per year and is an issue in more than 100 countries worldwide (Rengasamy 2006; Shrivastava and Kumar, 2015). Around 4.03 billion people live in the 13 countries most affected by soil salinity and represent about 52% of the world's population (Liu et al., 2020a). According to recent predictions, around half of all arable land in the world will be salinized by 2050 (Butcher et al., 2016). In order to meet the food demands of the growing world population a substantial increase in irrigation has to be done (Liu et al., 2020a). However, freshwater resources are mainly concentrated in the northern part of the world and the level of water consumption increased by more than 600% during the last century, with nearly 70% of these resources consumed for irrigation (Koyro et al., 2011). Furthermore, the important consequence of irrigation techniques' application is additional land salinization. According to a recent estimation, the irrigation and use of fertilizers practice will add from 3 to 6 tons of salt per hectare of farmland every year (Liu et al., 2020a). These problems are enhanced by the worldwide shortage of freshwater, caused by a number of reasons. First, climate change-driven rise in sea levels leads to the intrusion of groundwater by seawater. Second, the continuous and unsustainable use of groundwater significantly reduced the level of underground water tables. Around 20% of water resources used for agricultural irrigation are delivered from groundwater (IPCC, 2013). For example, the water tables of Texas, Oklahoma, and Kansas decreased by more than 30 m due to high agricultural use (Whitmee et al., 2015). There are several possible solutions to combat the problem of soil salinization and freshwater shortage hampering crop productivity.

DOI: 10.1201/9781003300564-6

One of them is increasing freshwater supply by desalinization of seawater; however, the current desalinization technology demands high input of energy. The second way is to increase the salt tolerance of our crops. High levels of soil salinity result in impaired plant growth and development. Unfortunately, the majority of the world's crops species and varieties belong to glycophytes, which are sensitive to salt stress. According to a recent estimation, the two world's major staple crops rice (*Oryza sativa*) and wheat (*Triticum aestivum*) comprise together 39% of all caloric intake by humans (Liu et al., 2020a). Growing these crops on saline soils would lead to estimated yield losses of ~70% for rice and ~40% for wheat (Shabala and Munns, 2017). The development of salt-tolerant crops is the most efficient approach to overcome the worldwide rising issue of soil salinity. Therefore, evaluation of salt-tolerance mechanisms and identification and isolation of relevant genotypes or genes in plant species exhibiting high salt tolerance are very important steps toward this goal.

6.2 GENERAL MECHANISMS OF SALT TOLERANCE IN PLANTS

Soil salinity has a negative impact on plant growth and development through various mechanisms, such as osmotic stress due to loss of cellular water content, cytotoxicity due to excessive uptake of Na^+ and Cl^- ions, oxidative stress due to the generation of reactive oxygen species (ROS), and nutritional imbalance (Isayenkov and Maathuis, 2019). The mechanism of salt tolerance is complex and involves maintaining cellular ion homeostasis, osmotic adjustments, and ROS scavenging (Isayenkov, 2012; Isayenkov and Maathuis, 2019). According to the growth performance in saline environments, plants are classified as salt-tolerant (halophytes) and salt-sensitive (glycophytes) (Flowers and Colmer, 2008, 2015; Shabala, 2013; Shabala et al., 2014; Mishra and Tanna, 2017; Isayenkov and Maathuis, 2019). Halophytic plants are capable of tolerating high Na^+ and Cl^- concentrations in their shoots when compared with the glycophytic plants. The salt tolerance of halophytes and other plants is facilitated by two major mechanisms – salt exclusion or inclusion. Thus, plants can be grouped into two types – includers and excluders. The representatives of the first group are capable of accumulating high NaCl concentrations in their tissues equal to or higher than the water salinity level. The glycophytic plants are mostly regarded as salt excluders due to their capacity to prevent the accumulation of salts in photosynthetic tissues. The majority of halophytic plants belong to includers and can transport Na^+ actively from root to shoot (Flowers et al., 1977; Läuchli, 1984: Isayenkov, 2019). Interestingly, the cultivated common barley (*Hordeum vulgare* L.) exhibits a halophytic includer phenotype (Mian et al., 2011). The includer-type plants employ progressive mechanisms of intracellular NaCl compartmentation and effectively regulate cytoplasmic homeostasis and synthesis of compatible solutes. However, in comparison to glycophytic plants, the increased salt tolerance of halophytes is achieved predominantly by specific anatomy or greater robustness of employed mechanisms rather than qualitative differences (Flowers and Colmer, 2008; Isayenkov and Maathuis, 2019). Although the main cytotoxic ions (Na^+ and Cl^-) are important contributors to the adjustment of cellular osmotic potential, they have to be compartmentalized into vacuoles or endosomal compartments to avoid toxic concentrations of these ions in the cytosol (Flowers and Colmer, 2015; Isayenkov and Maathuis, 2019). The process of vacuolar and endosomal ion sequestration is mediated by ion exchangers and the H^+ pumps localized to the tonoplast or endosomal membranes (Isayenkov and Maathuis, 2019, Isayenkov, 2019). The accumulation of typical organic osmolytes, such as sugars, sugar alcohols, and proline, occurred in the cytoplasm of halophytic plant species to balance the osmotic potential of Na^+ and Cl^-; however, the osmotic adjustments are achieved more efficiently and consumed less energy by using the ions than of organic solutes (Flowers and Colmer, 2008; Slama et al., 2015; Zhao et al., 2020). It is believed that the mechanism of osmotic adjustment by ions is prevalent for many halophytic species (Flowers and Colmer, 2015; Zhao et al., 2020).

It is worth mentioning that majority of crops are monocots. Monocotyledonous halophytes contain lower Na^+ concentrations and exhibit higher selectivity for K^+ uptake as compared to dicotyledonous halophytes (Flowers and Colmer, 2008; 2015). Thus, the majority of halophytic monocot species exhibit a strong excluder phenotype (Colmer et al., 2005a; 2005b; Garthwaite et al., 2005;

Wang et al., 2016; Zhang et al., 2020a, 2020b, 2020c). During evolution, many plant species have evolved unique ways and adaptation strategies to survive in saline environments. Some halophytic plants, mainly dicot species, have developed special structures and organs, such as epidermal bladder cells to accumulate an excess of Na^+ in their vacuoles, or hydathodes, which actively secrete salt and reduce the concentration of toxic ions in the cells (Shabala et al., 2014; Yuan et al., 2016; Mishra and Tanna, 2017; Isayenkov, 2019). Although the existence of leaf microhairs (modified trichomes) for salt secretion was reported for the wild relative of rice *Oryza coarctata* (Flowers et al., 1990; Zhao et al., 2020), the majority of halophytic monocotyledon species do not exhibit such specialized organs but have evolved other strategies to survive under saline conditions. For example, no strong evidence for the existence of specific secretory organs or structures and no succulence were observed in the genus of *Hordeum* (Wang et al., 2016). The wild relatives of the cereal crops with halophyte phenotype deserve particular attention due to significant similarities in anatomy, physiology, and genome structure with major agricultural cereals. Furthermore, the process of crop domestication has led to the narrowing and truncation of genetic variation. Therefore, closely related species of crops are valuable genetic resources for modern agriculture and breeding (Dai et al., 2018; Isayenkov, 2019).

Cultivated plant species and their wild relatives could differ significantly in the strength of their salt tolerance. Several wild species within the Triticeae tribe, to which the major crops wheat (*Triticum aestivum*) and barley (*Hordeum vulgare*) belong, exhibit the ability to survive on saline soils (Isayenkov, 2019). However, the majority of these halophytic species from the Triticeae tribe belong to facultative halophytes, which do not require salt for their growth and development (Polunin, 1960). Furthermore, no succulence was observed in the Triticeae tribe.

6.3 WILD RELATIVES OF CULTIVATED BARLEY

Cultivated barley belongs to one of the oldest domesticated cereal crops. It was reported that this cereal was domesticated from its wild relative *Hordeum vulgare* ssp. *spontaneum* approximately around 10,000 years ago in the Fertile Crescent (Badr et al., 2000; Jakob et al., 2014). However, more recent data suggest that the origin of our cultivated barley is more complex and derived from multiple sources of wild population (Poets et al., 2015). In comparison with wheat and other cultivated species of Triticeae, *Hordeum vulgare* is considered a salt-tolerant cereal capable of growing on salt concentrations as high as 150 mM NaCl (Greenway, 1962). However, this level of salinity leads to the reduction of biomass up to 55%–58%. The genus *Hordeum* is comprised of 33 species distributed all over the world, sharing four distinct genomes: H, I, Xa, and Xu (Table 6.1; Brassac and Blattner, 2015). Many wild barley species exhibit superior salt tolerance compared to cultivated barley (*Hordeum vulgare*) (Mano and Takeda, 1998; Garthwaite et al., 2005; Isayenkov, 2019). However, salt-tolerance mechanisms were mainly investigated in *H. vulgare* ssp. *vulgare* and *H. vulgare* ssp. *spontaneum* (Gorham et al., 1990; Nevo et al., 1993; Forster et al., 1994; Fatehi et al., 2012; Shen et al., 2016; Ebrahim et al., 2020; Liu et al., 2020b; Capasso et al., 2021). Despite suggestions that the progenitor of cultivated barley, *H. vulgare* ssp. *spontaneum*, could be a good source of salt tolerance (Nevo et al., 1993). Recent studies demonstrated some limitations in salt tolerance of *H. vulgare* ssp. *spontaneum* in comparison with cultivars of *H. vulgare* ssp. *vulgare* and some wild species of the genus *Hordeum* (Forster et al., 1994; Ellis et al., 2000), whereas *H. vulgare* ssp. *spontaneum* can be considered as a promising germplasm for drought- and heat-tolerance traits (Isayenkov, 2019).

According to an ecogeographical study of the genus (Bothmer et al., 1995), the wild barley species are widely distributed in arid and temperate areas of the Northern Hemisphere, South America, and South Africa (Bothmer 1992; Bothmer et al., 1995; Blattner, 2006; Blattner et al., 2010). Furthermore, more than half of the wild *Hordeum* species are capable of surviving and reproducing in various saline environments ranging from steppe saline grasslands colonized by *H. brevisubulatum* (Li et al., 2012) to Salt Lake shores inhabited by *H. patagonicum* (Bothmer et al., 1995; Garthwaite et al., 2005), to salt-affected dryland habitats populated by wild barley species such as *H. bogdanii*, *H. secalinum*, and *H. depressum* (Bothmer and Jacobsen, 1985).

6.3.1 Physiological Adjustments of Wild Barley Species to Salinity Stress

As mentioned earlier the cultivated barley (*H. vulgare*) is considered to be salt-tolerant, with most known barley varieties capable of growing in the presence of up to 50 mM NaCl (Mian et al., 2011). *H. vulgare* exhibits a high capacity to translocate Na^+ to the shoot rather than to accumulate or compartmentalize this cation in leaf tissues exhibiting salt-including behavior (Figure 6.1). Despite some advantages to having a salt-includer phenotype, this strategy might work to a certain extent until the tissue concentration of toxic ions will expand over the physiological threshold. In contrast with *H. vulgare* the eight wild *Hordeum* species – *H. bogdanii*, *H. intercedens*, *H. jubatum*, *H. lechleri*, *H. marinum*, *H. murinum*, *H. patagonicum*, and *H. secalinum* studied under salinity stress – had better capacities to exclude Na^+ and Cl^-, and to maintain a higher concentration of K^+ in the leaf tissues. A decrease in the tissue water content under high salinity was shown to be a notable actor in the studied wild barley species in contrast with *H. vulgare* (Garthwaite et al., 2005). Taking together, these studies have shown that *H marinum*, *H. intercedens*, and *H. secalinum*, exhibit the highest degree of salt tolerance in comparison with other species, demonstrating the smallest reduction in shoot and root dry masse and relative growth rate under salinity treatment (Garthwaite et al., 2005). The further analysis of physiological adaptation mechanisms of *H. marinum* to high salinity revealed higher water saturation deficit and osmotic potential in comparison with that of cultivated barley due to higher accumulation of proline, glycine betaine, and dehydrins (Maršálová et al., 2016; Ferchichi et al., 2018). It was suggested that the salt tolerance of *H. marinum* correlates with superior Na^+/K^+ homeostasis, energy-saving strategy, and osmotic adjustments in tissues via recruitment of inorganic ions such as Na^+ and K^+ as "cheap" osmolites to compensate osmotic imbalance (Huang et al., 2018; Isayenkov et al., 2020; Dell'Aversana et al., 2021). Despite stress-induced K^+ leakage from the root tissues *H. marinum* is capable of retaining large amounts of K^+ and prevent Na^+ accumulation in the shoot's tissues (Isayenkov et al., 2020; Dell'Aversana et al., 2021). The ^{13}C feeding experiments using *H. vulgare* and *H. marinum* plants subjected to salinity or osmotic stress demonstrated ~two-fold higher efficiency of ^{13}C uptake by *H. marinum* shoots in comparison with *H. vulagre*. On the contrary, the ^{13}C accumulation was barely detectable in salt-stressed roots of *H. vulagare* due to the inhibited photosynthetic ^{13}C fixation by the shoots (Isayenkov et al., 2020). Furthermore, the ^{13}C assimilation in *H. vulgare* shoots was significantly decreased under osmotic stress and almost negligible under salinity stress. The efficiency of ^{13}C assimilation was slightly decreased under osmotic stress for *H. marinum* plants, while it did not lead to significant changes under salinity stress. Thus, *H. marinum* plants demonstrate the ability to maintain a substantial photosynthetic activity rate under salinity stress (Isayenkov et al., 2020). Furthermore, the NMR analysis revealed morphological changes in the *H. marinum* roots associated with adaptation to high salinity. The salt-treated roots remained in their primordial state and the root surface area was significantly increased facilitating enhanced metabolite uptake and active ion efflux and improved root–rhizosphere interaction (Isayenkov et al., 2020). Thus, according to the recent observations, *H. marinum* plants are capable of controlling Na^+ and Cl^- concentrations in leaves at high salinity, the roots appear to play the central role in salinity adaptation, ensured by the development of thinner and more lignified roots creating an apoplastic barrier to prevent water and solute losses and to reduce ionic flow along with fine-tuning of membrane transport for effective management of restriction of ion entry and sequestration, accumulation of osmolytes, and minimization of energy costs (Figure 6.1) (Isayenkov et al., 2020).

Several studies suggest that mechanisms of physiological adaptation of *H. brevisubulatum* to high salinity rely on low Na^+ accumulation and stable K^+ content in tissues (Wang et al., 2016; Zhang et al., 2020a, 2020b). The application of micro-electrode ion flux estimation technique (MIFE) helped to uncover the physiological mechanism of salt tolerance for *H. brevisubulatum* under long-term salt stress (Wang et al., 2016). It was revealed that rapid Na^+ accumulation occurred in the shoot to respond to the initial salt shock. The next step of this mechanism is to trigger Na^+ efflux from the shoot tissues and activate K^+ influx for maintaining a stable K^+/Na^+ ratio in plant tissues (Wang et al., 2016). Thus, salt tolerance of *H. brevisubulatum* is based on the coordinated regulation of Na^+ and K^+ during salinity stress.

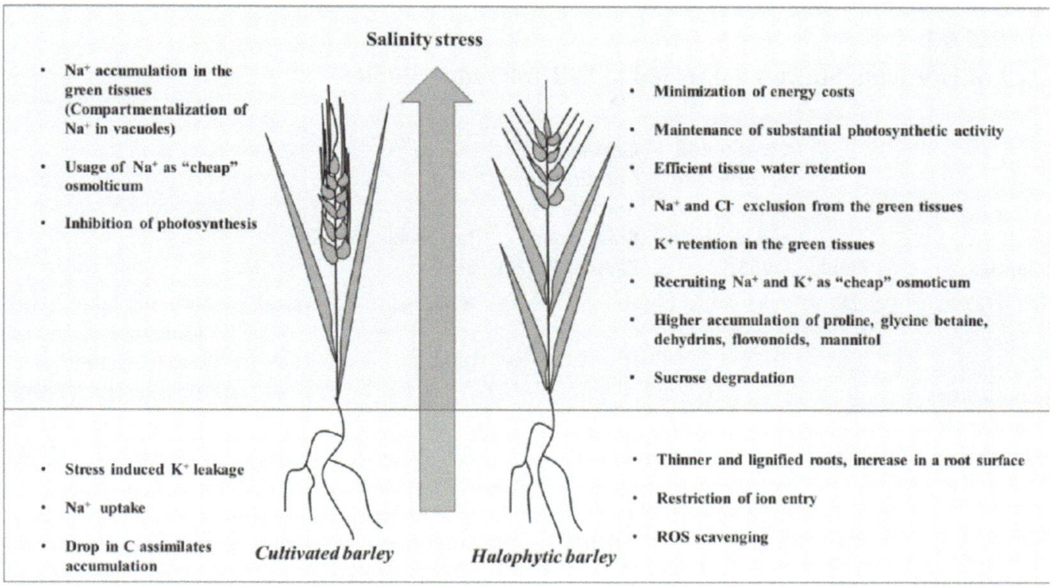

FIGURE 6.1 Differences in physiological adjustments between cultivated barley (*H. vulgare*) and halophytic barley species under salinity.

The main conclusion that could be drawn from the described studies is that in contrast to cultivated barley exhibiting Na^+ includer phenotype, the majority of wild barley species belong to the Na^+ excluders (Figure 6.1). Furthermore, halophytic barley species demonstrate high efficiency in regulation and tissue distribution of Na^+, K^+, and, possibly Cl^-. This regulation of inorganic ion fluxes is based on the coordinated work of membrane transporters and regulatory proteins (Figure 6.1). Furthermore, some of the halophytic barley species are capable of conducting substantial biosynthetic activities, efficiently accumulating inorganic and organic osmolites, intensifying the processes of ROS detoxification.

6.4 GENOMIC DETERMINANTS OF SALT TOLERANCE IN BARLEY

The comprehensive study of salt tolerance of different wild barley species suggested that *Hordeum* species containing the X and H genomes are enriched with traits associated with salt tolerance (Garthwaite et al., 2005). However, the recent revisions of genomic background of *Hordeum* species suggest completely different types of genomes for the wild barely species studied by Garthwaite et al. (2005) (Table 6.1; Blattner, 2006; Brassac and Blattner, 2015). Thus, we can assume that the original paper by Garthwaite et al. (2005) suggests enhanced salt tolerance for the wild barley species with Xa and I genomes. Furthermore, to support this idea, the typical salt-tolerant perennial barley species *H. brevisubulatum* (Trin.) Link has I type genome construction as well. On the other hand, some populations of *H. murinum* L. ssp. *glaucum* (Steud.) Tzvel with Xu genome can grow and reproduce on the Salt Lake shores. Perhaps, the salt tolerance of wild barley species rather depends on species-specific evolutional adaptation to harsh environments than the genome type.

Only a few scientific studies have addressed the potential of wild barley species for the adaptation of cultivated crops to salinity stress. For example, introgression of the halophytic barley *H. chilense* increased the salinity tolerance of durum wheat (Forster et al., 1990). *H. brevisubulatum* is a perennial *Hordeum* halophytic species, *and* indigenous to Asiatic salinized meadow grassland (Tao et al., 2015). Furthermore, this *Hordeum* species is considered a saline-tolerant grass for soil improvement in North China (Li et al., 2012). This species is used as a common forage material in the quality improvement and breeding of salt- and alkaline-tolerant vegetation,

TABLE 6.1
List of *Hordeum* Species Subjected to Salt-Tolerance Studies

Species	Ploidy	Genomic Constitution According to Bothmer and and Jacobsen (1985)	Genomic Constitution According to Brassac and Blattner (2015)	Taxonomic Section	Salt Tolerance	References
H. vulgare ssp. *vulgare*	2X	I	H	Hordeum	Control	Mano and Takeda (1998); Garthwaite et al. (2005)
H. vulgare L. ssp. *spontaneum* (C. Koch.) Thell.	2X		H	Hordeum		Nevo et al. (1993); Bothmer et al. (1995); Mano and Takeda (1998)
H. bogdanii Will.	2X	H	I	Stenostachys Nevski Series Sibirica Nevski	Tolerant	Garthwaite et al. (2005)
H. brevisubulatum (Trin.) Link	2X, 4X, 6X		I, II, III	Stenostachys Nevski Series Sibirica Nevski	Tolerant	Mano and Takeda (1998); Lu et al. (2005); Li et al. (2012); Wang et al. (2016); Zhang et al. (2020a, 2020b)
H. marinum Huds. ssp. *Gussoneanum* (Parl.) Thell.	2X, 4X	XX	Xa, XaXa	Trichostachys Dum	Tolerant	Mano and Takeda (1998); Garthwaite et al. (2005) Islam et al. (2007) Maršálová et al., 2016
H. marinum Huds. ssp. *marinum*	2X,		Xa	Trichostachys Dum	Tolerant	Mano and Takeda (1998); Huang et al. (2018); Huang et al. (2019); Hmidi et al. (2019); Isayenkov et al. (2020)
H. murinum ssp. *glaucum* (Steud.) Tzvel	2X	Y	Xu	Trichostachys Dum	Tolerant	Mano and Takeda (1998); Garthwaite et al. (2005)
H. murinum L ssp. *murinum*	4X		XuXu	Trichostachys Dum	Tolerant	Garthwaite et al. (2005)
H. murinum L ssp. *leporinum* (Link) Arc.	4X, 6X		XuXu, XuXuXu	Trichostachys Dum	Tolerant	Garthwaite et al. (2005)
H. intercedens Nevski	2X	H	I	Critesion (Raf.) Blattner	Tolerant	Garthwaite et al. (2005)
H. patagonicum (Haum.) Covas	2x	H	I	Critesion (Raf.) Blattner	Tolerant	von Bothmer et al. (1995); Garthwaite et al. (2005)
H. chilense Roem. & Schult	2X		I	Critesion (Raf.) Blattner	Tolerant	Forster et al. (1990)
H. jubatum L.	4X	HH	II	Critesion (Raf.) Blattner	Tolerant	Garthwaite et al. (2005)
H. lechleri L.	6X	HHH	III	Critesion (Raf.) Blattner	Tolerant	Garthwaite et al. (2005)
H. secalinum Schreb	4X	XH	IXa	Nodosa (Nevski) Blattner	Tolerant	Garthwaite et al. (2005)

as exemplified by the *H. brevisubulatum* cultivar "Junxu No.1," created in China in 2003 (Li et al., 2004). The genetic information about this unique halophytic barley species is still obscure. However, recently chloroplast (cp) genome of *H. brevisubulatum* was sequenced and analyzed indicating that this barley species is closely related to *H. bogdanii* and its cp genome was relatively conserved (Su et al., 2020; Cui et al., 2021). The detailed study of *H. brevisubulatum* revealed that the high salt tolerance of this species occurred due to low Na^+ accumulation and stable K^+ content in comparison with *H. vulgare* (Wang et al., 2016; Zhang et al., 2020a, 2020b). Furthermore, several key genes involved in the salt tolerance of *H. brevisubulatum* were cloned and characterized (Table 6.2).

TABLE 6.2
Functional Characterization of Barley Genes (Traits) Related to Salt Tolerance

Species	Genes/Crosses	Expression Systems/ Transgenic Plant/ Introgressions	Functions	References
Hordeum brevisubulatum	Sodium/hydrogen exchanger HbNHX1	*Nicotiana tabacum*	Vacuolar sequestration of Na^+/K^+	Lu et al. (2005)
	Protein kinase HbCIPK2	*Arabidopsis thaliana*	Positive regulator of salt and osmotic stress tolerance	Li et al. (2012)
	Potassium channel HbHAK1	*A. thaliana*	Facilitates K^+ retention and contributes to salt tolerance	Zhang et al. (2020a)
	Calcineurin B-like protein HbCBL1	*A. thaliana*; yeast	HbCBL1 combined with HbCIPK2 could activate HbVGKC1-channel to uptake K^+	Zhang et al. (2020b)
	Calcineurin B-like protein HbCBL4	*A. thaliana*; yeast	HbCBL4/10–HbCIPK2 complex modulated HbSOS1L-pump to exclude Na^+	Zhang et al. (2020b)
	Calcineurin B-like protein HbCBL10	*A. thaliana*; yeast	HbCBL4/10–HbCIPK2 complex modulated HbSOS1L-pump to exclude Na^+	Zhang et al. (2020b)
	Sodium/hydrogen exchanger HbSOS1L	*A. thaliana*; yeast	HbCBL4/10–HbCIPK2 complex modulated HbSOS1L-pump to exclude Na^+	Zhang et al. (2020b)
	Potassium channel *HbVGKC1*	Xenopus oocytes; yeast	HbCBL1 combined with HbCIPK2 could activate HbVGKC1-channel to uptake K^+	Zhang et al. (2020b)
	Multiprotein bridging factor 1 HbMBF1a	*A. thaliana*	HbMBF1a is a positive regulator of salt and ABA responses	Zhang et al. (2020c)
Hordeum marinum	*T. aestivum–H. marinum*	*Triticum aestivum*	Improved salt tolerance	Islam et al. (2007); Alamri et al. (2013)
	Cation transporter HmHKT2;1	*Xenopus* oocytes	Contribute to *H. marinum* adaptation to salinity	Hmidi et al. (2019)
	Cation transporter *HmHKT1;5*	*Oriza sativa* cv. Nipponbare; *Xenopus* oocytes	HmHKT1;5 is involved in Na^+ uptake but not in Na^+ translocation from roots to shoots	Huang et al. (2019)
Hordeum chilense	*T. turgidum–H. chilense*	*Triticum turgidum ssp. durum*	Improved salt tolerance	Forster et al. (1990)

For example, functional characterization of HbHAK1 from *H. brevisubulatum* revealed the participation of this potassium channel transport protein in potassium retention and significant contribution to salt tolerance (Zhang et al., 2020a). Interestingly, the HbHAK1 demonstrated better K^+ selectivity and transport activity under extremely low-K^+ conditions compared to its homologue of cultivated barley (Zhang et al., 2020a). Furthermore, cloning and functional characterization of CBL-interacting protein kinase HbCIPK2 from *H. brevisubulatum* revealed participation in salt and osmotic stress tolerance via prevention of K^+ loss in root and accumulation of fewer Na^+ in the shoot tissues (Li et al., 2012; Wang et al., 2016). This suggests that HbCIPK2 is involved in maintaining optional K^+/Na^+ homeostasis and protection of root cells from death (Li et al., 2012).

The subsequent work on mechanisms involved in maintaining an adequate K^+/Na^+ ratio during long-term adaptation to salinity shed the light on the important role of HbCIPK2-interacting proteins including upstream Ca^{2+} sensors, HbCBL1, HbCBL4, and HbCBL10 as well as downstream phosphorylated targets, including voltage-gated K^+ channel HbVGKC1 and SOS1-like transporter HbSOS1L. These studies suggest that a key role in mediating crosstalk between the sodium response and the potassium uptake signaling pathways belongs to HbCIPK2 (Table 6.2) (Zhang et al., 2015).

The typical Mediterranean halophyte – annual seaside barley (*Hordeum marinum*) is considered to be one of the major genetic sources of salinity tolerance (Garthwaite et al., 2005). The *H. marinum* comprises two subspecies – sea-side barley *H. marinum* subsp. *marinum* and Mediterranean barley *H. marinum* ssp. *gussoneanum*. While *H. marinum* ssp. *marinum* is typical diploid, the different populations of *H. marinum* subsp. *gussoneanum* could be diploids or tetraploids (Table 6.1; Brassac and Blattner, 2015; Blattner, 2018). These two Mediterranean wild barley grasses represent typical vegetation for saline coastal or inland meadows and marshes, but their distribution in other areas is rather limited (Lombardi and Lupi, 2006). It was shown that amphidiploid wheat hybrids, generated by crossing with *H. marinum*, exhibit improved salt tolerance (Islam et al., 2007; Alamri et al., 2013), which is likely linked to the increased level of proteins involved in energy metabolism (Maršálová et al., 2016). In addition, analysis of antioxidant enzymes in seaside barley revealed that this group of enzymes have significantly higher activity in *H. marinum* plants grown at high salinity (Seckin et al., 2010). Application of transcriptional profiling suggests that the salt-tolerance strategy of *H. marinum* comprises low energy consumption, utilization of inorganic ions (Na^+ and K^+) as cheap osmotic agents, and changes in the activity of the HmHKT1;5 and HmHKT2;1 transporter (Huang et al., 2018; Huang et al., 2019; Hmidi et al., 2019). The recent complex study of the physiological and transcriptional response of *H. marinum* to high salinity and osmotic stress revealed the ability to support photosynthetic activity and carbon fixation under salinity stress as well as maintain the optional physiological ratio of Na^+/K^+ (Isayenkov et al., 2020).

6.5 MULTI-OMICS OF WILD BARLEY SPECIES AS A TOOL FOR THE IDENTIFICATION OF KEY TOLERANCE GENES

The salt-tolerance mechanisms of obligate halophytes rely on specific structural and developmental properties, which include succulence, leaf shedding, presence of salt-excluding organs such as salt glands, salt hairs, or even Kranz-like leaf architecture (Shabala et al., 2014; Yong et al., 2022). Many halophytic plant species acquired salt tolerance via specific and efficient management of Na^+/Cl^- transport, including inhibition of Na^+/Cl^- uptake, vacuolar sequestration, efficient tissue redistribution via the employment of various PM and tonoplast ion channels and transporters (Flowers and Colmer, 2015; Isayenkov and Maathuis, 2019). Although nearly all plant species share similar cellular mechanisms of salt tolerance, in halophytic plants these mechanisms are manifested at higher levels.

As was mentioned earlier, the wild barley species belong to the facultative halophytes and can grow and reproduce in non-saline conditions as well as under high soil salinity. These species do not have specialized organs for salt excretion or special anatomical modifications and have to rely on the basic mechanisms of plant salt tolerance. However, the general management of gene network, regulation, and physiological adjustments to high salinity are highly efficient. Furthermore, the species-specific salt-tolerance mechanisms were developed during barley evolution. Therefore, the comparative studies of salt-tolerance mechanisms in various wild species are very perspective, making halophytic barleys a valuable resource for the identification and isolation of genes involved in salt tolerance. Undoubtedly, facultative halophytic plants deserve more scientific attention to be explored for this purpose due to their similarity in anatomy, physiology, and basic principles of salt tolerance with the majority of plant species, including cereals.

Traditional agricultural practices and breeding programs led to the production of highly productive crop cultivars, including barley. However, a wide range of common barley cultivars usually have limited genetic variability in comparison with wild relatives, which might serve as a valuable source of genes for genetic improvement (Varshney et al., 2007; Dawson et al., 2015; Isayenkov, 2019; Liu et al., 2020a; 2020b). Many wild barley species exhibit the promising potential to improve drought and salt tolerance and waterlogging (Garthwaite et al., 2005; Alamri et al., 2013; Isayenkov, 2019). Numerous studies were dedicated to evaluating salt stress response by transcript profiling, including RNA-seq, metabolomics and ionomics, and other techniques. However, nearly all of them were focused on the evaluation of transcriptional or metabolic response just for one species – *H. vulgare*, with a dominant majority of these studies based on comparative analysis of different salt-sensitive or salt-tolerant cultivars of *H. vulgare* ssp. *vulgare*, or ecogeographical accessions of *H. vulgare* ssp. *spontaneum* (Colmer et al., 2006; Bahieldin et al., 2015; Capasso et al., 2021; Ouertani et al., 2021).

Only two distinct wild barley species (*H. marinum* and *H. brevisubulatum*) were subjected to transcriptional and metabolomic profiling (Maršálová et al., 2016; Huang et al., 2018; Isayenkov et al., 2020; Zhang et al., 2020a), but genome sequences of these species are still needed to be deciphered. However, analysis of some transcriptomic data of salt-treated *H. brevisubulatum* plants facilitated the identification of salt-responsive genes such as 13 members of HAK/KUP/KT family, multiprotein bridging factor 1 (MBF1), HbVGKC1, and HbCBL1/2/10 members (Table 6.3; Zhang et al., 2020a, 2020b, 2020c). Additionally, Huang et al. (2018) identified some key genes expressed under salinity, investigating *H. vulgare* ssp *vulgare* cv. Golden promise, Tibetan wild barley (*H. vulgare* ssp. *spontaneum*), and *H. marinum*. By application of RNA-seq analysis, it was found that advanced maintaining of Na^+/K^+ homeostasis in *H. marinum* may rely on the regulation of ion transporters including SOS1 and HKT1;1. Furthermore, comparative analysis of metabolite and transcript profiles among three different genotypes suggested that *H. marinum* applied energy-saving strategies such as larger-scale gene downregulation, efficient utilization of Na^+ and K^+ as a source of "cheap" osmotica, enhanced level of glycolysis and the TCA cycle, and accumulation of compatible solutes in shoots (Table 6.3). The induction of genes involved in the process ROS detoxification is far more advanced in *H. marinum* plants (Huang et al., 2018). A further study of the adaptation mechanisms of *H. marinum* plants to high salinity and osmotic stress revealed other significant differences. Applications of different omics approaches, such as ionomics, metabolomics, and RNA-seq revealed significant differences in the adaptation of *H. marinum* to high salinity and drought (Table 6.3; Isayenkov et al., 2020). *H. marinum* plants exhibited better growth capacity, water retention, and shoot development under salinity stress in contrast to osmotic stress, with genes involved in photosynthesis or chlorophyll metabolism strongly repressed under osmotic stress conditions. The observed upregulation of genes involved in the TCA cycle and, in part, glycolysis indicates increased energy metabolism during salinity and osmotic stress. Overexpression of SWEET proteins indicates increased sucrose transport to the roots, and its utilization for energy retrieval (Isayenkov et al., 2020). RNA-seq analysis revealed upregulation of genes encoding K^+ transporter SKOR, an ammonium transporter, a cation/H^+ antiporter, and Mg^{2+} transporter in salt-treated roots. Despite the proposed roles of SOS1, HKT1;1, HKT1;5, and HKT2;2 in establishing ionic balance in *H. marinum* under salinity (Huang et al., 2018; 2019; Hmidi et al., 2019), none of the corresponding genes were found to be differentially expressed in this study (Isayenkov et al., 2020). Instead, other transporters, including K^+ transporter SKOR, an ammonium transporter, a

TABLE 6.3
Barley Genes Potentially Be Involved in Salt Tolerance (Transcriptomic Analysis of Salt-Stress-Responsive Genes)

Barley Species	Gene	Tissue	Potential Function	References
H. brevisubulatum	Putative potassium channel AKT1	Roots and shoots	Voltage-gated potassium channel activity (GO0005249)	Zhang et al. (2020b)
	Putative potassium channel AKT2	Roots and shoots	Voltage-gated potassium channel activity (GO0005249)	Zhang et al. (2020b)
	Putative potassium channel KAT1	Roots and shoots	Voltage-gated potassium channel activity (GO0005249)	Zhang et al. (2020b)
	Potassium channel HbVGKC1	Roots and shoots	K^+ transport	
H. marinum	Cation transporter HKT1;1	Roots and shoots	Transports Na^+ to the phloem sap	Huang et al. (2018)
	Sodium/hydrogen exchanger NHX7 (SOS1)	Roots and shoots	Na^+ efflux from the cell	Huang et al. (2018)
	Potassium transporter 13 (Barley_newGene_20569)	Roots and shoots	K^+ transport	Huang et al. (2018)
	Potassium transporter 23 (gene:MLOC_58287)	Roots	K^+ transport	Huang et al. (2018)
	Potassium transporter 7 (gene:MLOC_16185)	Shoots	K^+ transport	Huang et al. (2018)
	Sodium/hydrogen exchanger NHX6	Shoots	Endosomal sequestration of Na^+/K^+	Huang et al. (2018)
	Vacuolar cation/proton exchanger 1a (CAX 1a)	Shoots	Vacuolar sequestration of Ca^{2+} or other cations	Huang et al. (2018)
	Cation/H(+) antiporter 16	Roots and shoots	Na^+/K^+ transport	Huang et al. (2018); Isayenkov et al. (2020)
	Cation/H(+) antiporter 19	Shoots	Na^+/K^+ transport	Huang et al. (2018)
	Potassium channel SKOR	Roots	K^+ transport to xylem	Isayenkov et al. (2020)
	Anion channel SLAH2	Roots	Cl^- transport	Isayenkov et al. (2020)
	Anion channel SLAH3	Roots	Cl^- transport	Isayenkov et al. (2020)
	Anion transporter NRT1/ PTR FAMILY 7.3,	Shoots	Anion transport	Isayenkov et al. (2020)
	Boron transporter BORT4	Roots	Boron transport/anion transport	Isayenkov et al. (2020)
	Ammonium transporter 2	Roots	Ammonium transport	Isayenkov et al. (2020)
	Sucrose transporter SWEET12	Roots	Sucrose export	Isayenkov et al. (2020)
	Sucrose transporter SWEET13a	Roots	Sucrose export	Isayenkov et al. (2020)
	Sucrose transporter SWEET14b	Roots	Sucrose export	Isayenkov et al. (2020)
	Glutathione S-transferase	Shoots	Glutathione metabolism	Huang et al. (2018)
	Gamma-glutamyl phosphate reductases	Shoots	Proline metabolism	Huang et al. (2018)
	Pyrroline-5-carboxilate reductases	Shoots	Proline metabolism	Huang et al. (2018)

(*Continued*)

TABLE 6.3 (*Continued*)
Barley Genes Potentially Be Involved in Salt Tolerance (Transcriptomic Analysis of Salt-Stress-Responsive Genes)

Barley Species	Gene	Tissue	Potential Function	References
	Prolyl 4-hydroxylase	Shoots	Proline metabolism	Isayenkov et al. (2020)
	Tyrosine decarboxylase	Shoots	Tyrosine metabolism	Isayenkov et al. (2020)
	Tryptophan synthase	Shoots	Tryptophan metabolism	Isayenkov et al. (2020)
	Anthranilate phosphoribosyltransferase	Shoots	Phenylalanine Metabolism	Isayenkov et al. (2020)
	Anthranilate synthase	Shoots	Phenylalanine metabolism	Isayenkov et al. (2020)
	Ascorbate oxidase	Roots	Ascorbate degradation	Isayenkov et al. (2020)
	Isovaleryl-CoA-dehydrogenase	Roots	Production of diterpenoids	Isayenkov et al. (2020)
	Copalyl-diP synthase	Roots	Production of diterpenoids	Isayenkov et al. (2020)
	Indole-3-glycerol phosphate synthase	Roots	Tryptophan-independent IAA biosynthetic pathway	Isayenkov et al. (2020)

cation/H^+ antiporter, Mg^{2+} transporter, along with S-type anion channels SLAH2 and SLAH3, and several boron transporters were strongly upregulated in salt-treated roots (Isayenkov et al., 2020). Taking together, despite the obscure understanding of adaptation mechanisms to high salinity, the accumulated data suggest substantial differences in the salt-tolerance mechanisms between wild barley species and cultivars of *H. vulgare*.

It has been also found that the decrease in expression of GLR3.4 and a cyclic nucleotide-gated channel potentially could minimize Na^+ uptake by the roots (Isayenkov et al., 2020). Increased expression of several S and P transporter genes, as well as genes for NRT1/PTR protein family, indicates activation in some anion uptake to compensate for the accumulation of Cl^-. Besides transport protein, strong transcriptional upregulation of amino acid-degrading enzymes in the roots were observed. Another interesting fact was the specific accumulation of flavonoids in salinity stressed roots of *H. marinum* plants, whereas their biosynthesis was repressed under osmotic stress. Overexpression of chalcone synthase was observed during salinity stress as well (Isayenkov et al., 2020). However, apart from a couple of multi-omics studies conducted on *H. marinum*, the outstanding features of this interesting species remain largely unexplored. It would be a good opportunity for further scientific projects to conduct full genome sequencing of *H. brevisubulatum* and *H. marinum* accessions.

Taking into account the importance of maintenance of ionic homeostasis in halophytic barley relatives, further study directions have to be focused on the functional study of membrane transporters from wild barley species (HKT, SKOR, SLAH or NRT/PTR and others), transcription factors, and regulatory proteins with potential applications in the domesticated cereals. The wide range of halophytic barley species has great potential for novel gene discovery and adaptation mechanisms. Application of multi-omics approaches to evaluate the mechanisms of salt and drought tolerance on the physiological, biochemical, and transcriptional levels will significantly help to move forward our understanding of tolerance mechanisms and the feasibility to transfer these traits to our modern crops. Thus, the genomic resources from different halophytic barley species, including partially studied *H. marinum* or *H. brevisubulatum*, could be explored more in deep as a useful resource to design new breeding strategies or create biotechnological plants to counteract environmental stress.

6.6 CONCLUSIONS AND OUTLOOKS

Contrasting many traditional cereal crops, halophytic barley relatives can grow and reproduce under ambient salinity. Such capacities to tolerate high salt concentration or osmotic imbalance are mediated by proper modulation of gene arrays playing key roles in supporting salt-tolerance mechanisms. Although despite the substantial amount of multi-omics research focused on salt or drought tolerance of domesticated barley cultivars or its wild counterpart *H. vulgare* ssp. *spontaneum*, investigation of other members of the *Hordeum* genus with more pronounced halophytic properties attracted much less attention. Application of high-throughput transcriptome sequencing techniques such as massively parallel cDNA sequencing (RNA-seq) or expressed exome capture sequencing (EecSeq) have to be increasingly applied to understand the responses of plant species of Triticeae tribe to abiotic stresses at the level of gene expression. Furthermore, with the continuous rise of next-generation sequencing (NGS) affordability, the whole-genome sequencing for a wide range of halophytic barley species represents the future important and very promising target. Analysis of genome structure, SNPs, and gene copy numbers will provide valuable information for further advances in molecular breeding programs, gene engineering and genome editing, and synthetic biology. The complex application of high-throughput sequencing techniques in combination with advanced metabolomics, proteomics, and ionomics will promote uncovering species-specific as well as general properties of salt tolerance in halophytic barley species. Future breeding programs will significantly benefit from the implementation of NGS and high-throughput phenotyping tools. The fast development of novel research techniques including NMR analysis, single-cell sequencing, and microscopic ion tracing will provide a big advantage to study mechanisms of wild barley adaptation to high salinity as well as other environmental stresses. Therefore, exploration of the salt-tolerance potential of different halophytic barley species by application of modern advanced research techniques will significantly move the progress in the development of crops with superior tolerance and productivity. Increased knowledge around salt tolerance in halophytic wild relatives is crucial for sustainable agriculture in salinity-affected areas to reduce the negative impacts of global climate change.

REFERENCES

Alamri, S.A., Barrett-Lennard, E.G., Teakle, N.L., and Colmer, T.D. 2013. Improvement of salt and waterlogging tolerance in wheat: comparative physiology of *Hordeum marinum-Triticum aestivum* amphiploids with their *H. marinum* and wheat parents. *Functional Plant Biology* 40(11): 1168–1178. doi: 10.1071/FP12385.

Badr, A. Sch, K.M.R., El Rabey, H., Effgen, S., Ibrahim, H.H., Pozzi, C.M., et al. 2000 On the Origin and Domestication History of Barley (*Hordeum vulgare*). *Molecular Biology and Evolution* 17: 499–510. doi: 10.1093/oxfordjournals.molbev.a026330.

Bahieldin, A., Atef, A., Sabir, J.S., Gadalla, N.O., Edris, S., Alzohairy, A.M., et al. 2015. RNA-Seq analysis of the wild barley (*H. spontaneum*) leaf transcriptome under salt stress. *Comptes Rendus Biologies* 338: 285–297. doi: 10.1016/j.crvi.2015.03.010.

Blattner, F.R. 2006. Multiple intercontinental dispersals shaped the distribution area of *Hordeum* (Poaceae). *New Phytologist* 169: 603–614. doi: 10.1111/j.1469-8137.2005.01610.x.

Blattner, F. 2018. Taxonomy of the genus *Hordeum* and Barley (*Hordeum vulgare*). In *The Barley Genome. Compendium of Plant Genomes*, ed. N. Stein and G. Muehlbauer, 11–23. Cham: Springer. doi: 10.1007/978-3-319-92528-8_2.

Blattner, F.R., Pleines, T., and Jakob S.S. 2010. Rapid radiation in the barley genus *Hordeum* (Poaceae) during the Pleistocene in the Americas. In *Evolution in action,* ed. M. Glaubrecht, 17–33. Berlin, Heidelberg: Springer. doi: 10.1007/978-3-642-12425-9_2C.

Bothmer, R. 1992. The wild species of *Hordeum*: relationships and potential use for improvement of cultivated barley. In *Barley: Genetics, Biochemistry, Molecular Biology and Biotechnology*, ed. P.R. Shewry, 3–18. Oxon, UK: CAB International.

Bothmer, R., and Jacobsen, N. 1985. Origin, taxonomy, and related species. In *Barley,* ed. D.C. Rasmusson. Vol. 26, 19–56. Madison, WI: American Society of Agronomy, Crop Science Society of America, Soil Science Society of America; ASA Agronomy Monograph.

Bothmer, R., Jacobsen, N., Baden, C., Jørgensen, R.B., and Linde-Laursen, I. 1995. An ecogeographical study of the genus *Hordeum*. In *Origin, Taxonomy, and Related Species*, ed. D.C. Rasmusson, 2nd edn., 19–56 Rome: International Plant Genetic Resources Institute.
Brassac, J., and Blattner, F.R. 2015. Species-level phylogeny and polyploid relationships in *Hordeum* (Poaceae) inferred by next-generation sequencing and in silico cloning of multiple nuclear loci. *Systematic Biology* 64: 792–808. doi: 10.1093/sysbio/syv035.
Butcher, K., Wick, A.F., DeSutter, T., Chatterjee, A., and Harmon, J. 2016. Soil salinity: a threat to global food security. *Agronomy Journal* 108: 2189–2200. doi: 10.2134/agronj2016.06.0368.
Capasso, G., Santini, G., Petraretti, M., Esposito, S., and Landi, S. 2021. Wild and traditional barley genomic resources as a tool for abiotic stress tolerance and biotic relations. *Agriculture* 11: 1102. doi: 10.3390/agriculture11111102.
Colmer, T.D., Flowers, T.J., and Munns, R. 2006. Use of wild relatives to improve salt tolerance in wheat. *Journal of Experimental Botany* 57: 1059–1078. doi:10.1093/jxb/erj124.
Colmer, T.D., Garthwaite, A.J., Islam, A.K.M.R., Islam, S., Malik, A.I, and von Bothmer, R. 2005b. Salinity and waterlogging tolerance in wild *Hordeum* species: physiological basis and prospects for use in cereal improvement. In *Plant Nutrition for Food Security, Human Health and Environmental Protection*, ed. C.J. Li et al., 8–9. Beijing: Tsinghua University Press.
Colmer, T.D., Munns, R., and Flowers, T.J. 2005a Improving salt tolerance of wheat and barley: future prospects. *Australian Journal of Experimental Agriculture* 45: 1425–1443. doi: 10.1071/EA04162.
Cui, G., Wang, C., Wei, X., Wang, H., Wang, X., Zhu, X., et al. 2021. Complete chloroplast genome of *Hordeum brevisubulatum*: genome organization, synonymous codon usage, phylogenetic relationships, and comparative structure analysis. *PLoS One* 16(12): e0261196. doi: 10.1371/journal.pone.0261196. eCollection 2021.
Dai, X., Long, C., Xu, J., Guo, Q., Zhang, W., Zhang, Z., et al. 2018. Are dominant plant species more susceptible to leaf-mining insects? A case study at Saihanwula Nature Reserve, China. *Ecology and Evolution* 8(15): 7633–7648. doi: 10.1002/ece3.4284.
Dawson, I., Russell, J., Powell, W., Steffenson, B., Thomas, W.T.B., and Waugh, R. 2015. Barley: a translational model for adaptation to climate change. *New Phytologist* 206: 913–931. doi: 10.1111/nph.13266.
Dell'Aversana, E., Hessini, K., Ferchichi, S., Fusco, G.M., Woodrow, P., and Ciarmiello, L.F. et al. 2021. Salinity duration differently modulates physiological parameters and metabolites profile in roots of two contrasting barley genotypes. *Plants* 10(2): 307. doi: 10.3390/plants10020307.
Ebrahim, F, Arzani, A, Rahimmalek, M, Sun, D, and Peng, J. 2020. Salinity tolerance of wild barley *Hordeum vulgare ssp. spontaneum*. *Plant Breeding* 139: 304–316. doi: 10.1111/pbr.12770.
Ellis, R.P., Forster, B.P., Robinson, D., Handley, L.L., Gordon, D.C., and Russell, J.R., et al. 2000. Wild barley: a source of genes for crop improvement in the 21st century? *Journal of Experimental Botany* 51: 9–17. doi: 10.1093/jexbot/51.342.9.
Fatehi, F., Hosseinzadeh, A., Alizadeh, H., and Brimavandi, T. 2012. The proteome response of *Hordeum spontaneum* to salinity stress. *Cereal Research Communications* 41: 78–87. doi: 10.1556/CRC.2012.0017.
Ferchichi, S., Hessini, K., Dell'Aversana, E., D'Amelia, L., Woodrow, P., and Ciarmiello, L.F., et al. 2018. *Hordeum* vulgare and *Hordeum maritimum* respond to extended salinity stress displaying differenterent temporal accumulation pattern of metabolites. *Functional Plant Biology* 45: 1096–1109. doi: 10.1071/FP18046.
Flowers, T.J., and Colmer, T.D. 2008. Salinity tolerance in halophytes. *New Phytologist* 179: 945–963. doi: 10.1146/annurev.pp.28.060177.000513.
Flowers, T.J, and Colmer, T.D. 2015. Plant salt tolerance: adaptations in halophytes. *Annual Botany* 115: 327–331. doi: 10.1093/aob/mcu267.
Flowers, T.J., Flowers, S.A., Hajibacheri, M.A., and Yeo, A.R. 1990. Salt tolerance in the halophytic wild rice, *Porteresia coarctata* Tateoka. *New Phytologist* 114: 675–684. doi:10.1111/j.1469-8137.1990.tb00439.x
Flowers, T.J., Troke, P.F., and Yeo, A.R. 1977. The mechanism of salt tolerance in halophytes. *Annual Review of Plant Physiology and Plant Molecular Biology* 28: 89–121. doi: 10.1146/annurev.pp.28.060177.000513.
Food and Agriculture Organization. 2016. *The State of Food and Agriculture: Climate Change, Agriculture and Food Security*. Roma, Italy: FAO, p. 173.
Forster, B.P., Pakniyat, H., Macaulay, M., Matheson, W., Philips, M.S., Thomas. W.T.B., et al. 1994. Variation in the leaf sodium content of the *Hordeum vulgare* (barley) cultivar Maythorpe and its derived mutant cv. Golden promise. *Heredity* 73: 249–253 doi: 10.1038/hdy.1994.130.
Forster, B.P., Phillips, M.S., Miller, T.E., Baird, E., and Powell, W. 1990. Chromosome location of genes controlling tolerance to salt (NaCl) and vigour in *Hordeum vulgare* and *H. chilense*. *Heredity* 65: 99–107 doi: 10.1038/hdy.1990.75.

Garthwaite, A.J., Bothmer, R., and Colmer, T.D. 2005. Salt tolerance in wild *Hordeum* species is associated with restricted entry of Na^+ and Cl^- into the shoots. *Journal of Experimental Botany* 56: 2365–2378. doi: 10.1093/jxb/eri229.

Gorham, J., Bristol, A., Young, E.M., Wyn Jones, R.G., and Kashour, G. 1990. Salt tolerance in the Triticeae: K/Na discrimination in barley. *Journal of Experimental Botany* 41: 1095–1101. doi:10.1093/jxb/41.9.1095

Greenway, H. 1962. Plant response to saline substrates. I. Growth and ion uptake of several varieties of *Hordeum* during and after sodium chloride treatment. *Australian Journal of Biological Science* 15: 16–38 doi: 10.1071/bi9650763.

Hmidi, D., Messedi, D., Corratgï-Faillie, C., Marhuenda, T.O., Fizames, C.C., Zorrig, W., et al. 2019. Investigation of Na^+ and K^+ transport in halophytes: Functional analysis of the HmHKT2;1 transporter from *Hordeum maritimum* and expression under saline conditions. *Plant Cell Physiology* 60: 2423–2435. doi: 10.1093/pcp/pcz136.

Huang, L., Kuang, L., Li, X., Wu, L., Wu, D., and Zhang, G. 2018. Metabolomic and transcriptomic analyses reveal the reasons why *Hordeum marinum* has higher salt tolerance than *Hordeum vulgare*. *Environmental and Experimental Botany* 156: 48–61. doi: 10.1016/j.envexpbot.2018.08.019.

Huang, L., Kuang, L., Wu, L., Wu, D., and Zhang, G. 2019. Comparisons in functions of HKT1;5 transporters between *Hordeum marinum* and *Hordeum vulgare* in responses to salt stress. *Plant Growth and Regulation* 89: 309–319. doi: 10.1007/s10725-019-00538-7.

IAEA, 1995. *Management Strategies to Utilize Salt-Affected Soils, IAEATECDOC-814*. Vienna: International Atomic Energy Agency.

IPCC 2013. *Climate Change 2014: Synthesis Report. Contribution of Working Groups I, II and III to the Fifth Assessment Report of the Intergovernmental Panel on Climate Change R. K. P. a. L. A. M. e. Core Writing Team*. Geneva, Switzerland: IPCC, p. 151.

Isayenkov, S.V. 2012. Physiological and molecular aspects of salt stress in plants. *Cytology and Genetics* 46, 302–318. doi:10.3103/S0095452712050040.

Isayenkov, S.V. 2019. Genetic sources for the development of salt tolerance in crops. *Plant Growth and Regulation* 89: 1–17. doi: 10.1007/s10725-019-00519-w.

Isayenkov, S.V., Hilo, A., Rizzo, P., Tandron Moya, Y.A., Rolletschek, H., et al. 2020. Adaptation strategies of halophytic barley *Hordeum marinum* ssp. *marinum* to high salinity and osmotic stress. *International Journal of Molecular Science* 21: 9019. doi: 10.3390/ijms21239019.

Isayenkov, S.V., and Maathuis, F.J.M., 2019. Plant salinity stress: many unanswered questions remain. *Frontiers in Plant Science* 10: 80. doi: 10.3389/fpls.2019.00080.

Islam, S., Malik, A.I., Islam, A.K., and Colmer, T. D. 2007. Salt tolerance in a *Hordeum marium-Triticum aestivum amphiploid*, and its parents. *Journal of Experimental Botany* 58(5): 1219–1229. doi: 10.1093/jxb/erl293.

Jakob, S.S., Rödder, D., Engler, J.O., Shaaf, S., Özkan, H., Blattner, F.R., Kilian, B. 2014. Evolutionary history of wild barley (*Hordeum vulgare* sub sp. *spontaneum*) analyzed using multilocus sequence data and paleodistribution modeling. *Genome Biology and Evolution* 6: 685–702. doi: 10.1093/gbe/evu047.

Koyro, H.-W., Khan, M. A., and Lieth, H. 2011. Halophytic crops: a resource for the future to reduce the water crisis? *Emirates Journal of Food and Agriculture* 23(1): 1–16. doi: 10.9755/ejfa.v23i1.5308.

Läuchli, A. 1984. Salt exclusion: an adaptation of legumes for crops and pastures under saline conditions. In *Salinity Tolerance in Plants: Strategies for Crop Improvement*. ed. R.C. Staples, 171–187. New York, NY: Wiley.

Li, R., Zhang, J., Wu, G., Wang, H., Chen, J., and Wei, J. 2012. HbCIPK2, a novel CBL-interacting protein kinase from halophyte *Hordeum brevisubulatum*, confers salt and osmotic stress tolerance. *Plant, Cell and Environmens* 35: 1582–1600 doi: 10.1111/j.1365-3040.2012.02511.x.

Li, Y. F., Shen, J. L., Li, X. W., Zhang, Y. L., Cheng, X. R., and Yang, B. M. 2004. Studies on breeding of *Hordeum brevisubulatum* Junxu No. 1. *Jilin Animal Husbandry and Veterinarian* 8: 31–33.

Liu, M., Pan, T., Allakhverdiev, S. I., Yu, M., and Shabala, S. 2020a. Crop Halophytism: an environmentally sustainable solution for global food security. *Trends in Plant Science* 25(7): 630–634. doi: 10.1016/j.tplants.2020.04.008.

Liu, M., Li, Y., Ma, Y., Zhao, Q., Stiller, J., Feng, Q., et al. 2020b. The draft genome of a wild barley genotype reveals its enrichment in genes related to biotic and abiotic stresses compared to cultivated barley. *Plant Biotechnology Journal* 18: 443–456. doi: 10.1111/pbi.13210.

Lombardi, T, and Lupi, B. 2006. Salt-tolerance in wild *Hordeum* (Poaceae) species: differences between *H. maritimum* with and *H. hystrix* Roth. *Atti della Società toscana di scienze naturali, residente in Pisa. Memorie* B(113): 31–35.

Lu, SY., Jing, YX., Shen, SH., Zhao, HY., Ma, LQ., Zhou, XJ., et al. 2005. Antiporter gene from *Hordeum brevisubulatum* (Trin.) link and its overexpression in transgenic tobaccos. *Journal of Integrative Plant Biology* 47: 343–349 doi: 10.1111/j.1744-7909.2005.00027.x.

Mano, Y., and Takeda, K. 1998. Genetic resources of salt tolerance in wild *Hordeum* species. *Euphytica* 103: 137–141. doi: 10.1023/A:1018302910661.

Maršálová, L., Vítámvás, P., Hynek, R., Prášil, I.T., and Kosová, K. 2016. Proteomic response of *Hordeum vulgare* cv.Tadmor and *Hordeum marinum* to salinity stress: similarities and differences between a glycophyte and a halophyte. *Frontiers in Plant Science* 7: 1154. doi: 10.3389/fpls.2016.01154.

Mian, A., Oomen, R.J., Isayenkov, S., Sentenac, H., Maathuis, F.J. and Véry, A.-A. 2011. Over-expression of an Na^+- and K^+-permeable HKT transporter in barley improves salt tolerance. *The Plant Journal* 68: 468–479. doi: 10.1111/j.1365-313X.2011.04701.x.

Mishra, A., and Tanna, B. 2017. Halophytes: potential resources for salt stress tolerance genes and promoters. *Frontiers in Plant Science* 8: 829. doi: 10.3389/fpls.2017.00829.

Nevo, E., Krugman, T. and Beiles, A. 1993. Genetic resources for salt tolerance in the wild progenitors of wheat (*Triticum dicoccoides*) and barley (*Hordeum spontaneum*) in Israel. *Plant Breeding* 110: 338–341. doi: 10.1111/j.1439-0523.1993.tb00599.x.

Ouertani, N, R., Arasappan, D., Abid, G., Ben Chikha, M., Jardak, R., Mahmoudi, H., et al. 2021. Transcriptomic analysis of salt-stress-responsive genes in barley roots and leaves. *International Journal of Molecular Sciences* 22: 8155. doi: 10.3390/ijms22158155.

Poets, A.M., Fang, Z., Clegg, M.T., and Morrell, P.L. 2015. Barley landraces are characterized by geographically heterogeneous genomic origins. *Genome Biology* 16: 1. doi: 10.1186/s13059-015-0712-3.

Polunin, N. 1960. *Introduction to Plant Geography*. New York, NY: McGraw-Hill Book Company, Inc. doi: 10.5962/bhl.title.10242.

Qadir, M., Quillerou, E., Nangia, V., Murtaza, G., Singh, M., Thomas, R. J., et al. 2014. Economics of salt-induced land degradation and restoration. *Natural Resources Forum* 38: 282–295. doi: 10.1111/1477-8947.12054.

Rengasamy, P. 2006. World salinization with emphasis on Australia. *Journal of Experimental Botany* 7: 1017–1023. doi: 10.1093/jxb/erj108.

Seckin, B., Turkan, I., Sekmen, A. H., and Ozfidan, C. 2010. The role of antioxidant defense systems at differential salt tolerance of *Hordeum marinum* Huds. (sea barley grass) and *Hordeum vulgare* L. (cultivated barley). *Environmental and Experimental Botany* 69: 76–85. doi: 10.1016/j.envexpbot.2010.02.013.

Shabala, S. 2013. Learning from halophytes: physiological basis and strategies to improve abiotic stress tolerance in crops. *Annals of Botany* 112: 1209–1221. doi: 10.1093/aob/mct205.

Shabala, S., Bose, J., and Hedrich R. 2014. Salt bladders: do they matter? *Trends Plant Science* 19: 687–691. doi: 10.1016/j.tplants.2014.09.001.

Shabala, S., and Munns, R. 2017. Salinity stress: physiological constraints and adaptive mechanisms. In *Plant Stress Physiology*, 2nd edn., ed. S. Shabala, 24–63, Wallingford, UK: CABI. doi: 10.1079/9781780647296.0024.

Shen, Q., Fu, L., Dai, F., Jiang L., Zhang G., et al. 2016. Multi-omics analysis reveals molecular mechanisms of shoot adaption to salt stress in Tibetan wild barley. *BMC Genomics* 17: 889. doi: 10.1186/s12864-016-3242-9.

Shrivastava, P., and Kumar, R. 2015. Soil salinity: a serious environmental issue and plant growth promoting bacteria as one of the tools for its alleviation. *Saudi Journal of Biological Sciences* 22: 123–131. doi: 10.1016/j.sjbs.2014.12.001.

Slama, I., Abdelly, C., Bouchereau, A., Flowers, T., and Savouré, A. 2015. Diversity, distribution and roles of osmoprotective compounds accumulated in halophytes under abiotic stress. *Annals of Botany* 115: 433–447 doi: 10.1093/aob/mcu239.

Su, X., Zhao, J., and Wang, Z. 2020. The complete chloroplast genome of *Hordeum brevisubulatum*. *Mitochondrial DNA Part B* 5(3): 3006–3007. doi: 10.1080/23802359.2020.1797552.

Szabolcs, I. 1994. *Handbook of Plant and Crop Stress*, ed. M. Pessarakli, 3–11. New York: Marcel Dekker.

Tao, Y., Li, F., Gao, F.Q., Sun, Q.Z. 2015. Microbial characteristics of *Hordeum brevisubulatum* silage and screening for lactic acid bacteria with high fermentation performance. *Acta Prataculturae Sinica* 24(12): 6673.

Varshney, R.K., Langridge, P., and Graner, A. 2007. Application of genomics to molecular breeding of wheat and barley. *Advances in Genetics* 58: 121–155.

Wang, C. M., Xia, Z. R., Wu, G. Q., et al. 2016. The coordinated regulation of Na^+ and K^+ in *Hordeum brevisubulatum* responding to time of salt stress. *Plant Science* 252: 358–366.

Whitmee, S., Haines, A., Beyrer, C., Boltz, F., Capon, A.G., de Souza Dias, B.F., et al. 2015. Safeguarding human health in the Anthropocene epoch: report of the rockefeller foundation-lancet commission on planetary health. *The Lancet* 386: 1973–2028. doi: 10.1016/S0140-6736(15)60901-1.

Yuan, F., Leng, B., and Wang, B. 2016. Progress in studying salt secretion from the salt glands in recretohalophytes: how do plants secrete salt? *Frontiers in Plant Science* 7: 977. doi: 10.3389/fpls.2016.00977.

Yong, MT., Solis, C.A., Amatoury, S., Sellamuthu, G., Rajakani, R., Mak, M., et al. 2022. Proto Kranz-like leaf traits and cellular ionic regulation are associated with salinity tolerance in a halophytic wild rice. *Stress Biology* 2: 8. doi: 10.1007/s44154-021-00016-z.

Zhang, C., Ge, R.C., Zhang, J.W., Chen, Y.J., Wang, H.Z., et al. 2015. Identification and expression analysis of a novel HbCIPK2-interacting ferredoxin from halophyte *H. brevisubulatum. PLoS ONE* 10(12): e0144132. doi: 10.1371/journal.pone.0144132.

Zhang, H., Feng, H., Zhang, J., Ge, R., Zhang, L., Wang, Y., et al. 2020a. Emerging crosstalk between two signaling pathways coordinates K^+ and Na^+ homeostasis in the halophyte *Hordeum brevisubulatum. Journal of Experimental Botany* 71: 4345–4358 doi: 10.1093/jxb/eraa191.

Zhang, H., Xiao, W., Yu, W., Jiang, Y., and Li, R. 2020b. Halophytic *Hordeum brevisubulatum* HbHAK1 facilitates potassium retention and contributes to salt tolerance. *International Journal of Molecular Sciences* 21: 5292. doi: 10.3390/ijms21155292.

Zhang, L., Wang, Y., Zhang, Q., Jiang, Y., Zhang, H., and Li, R. 2020. Overexpression of HbMBF1a, encoding multiprotein bridging factor 1 from the halophyte *Hordeum brevisubulatum,* confers salinity tolerance and ABA insensitivity to transgenic *Arabidopsis thaliana. Plant Molecular Biology* 102(1–2): 1–17. doi: 10.1007/s11103-019-00926-7.

Zhao, C., Zhang, H., Song, C., Zhu, J.K., and Shabala, S. 2020. Mechanisms of plant responses and adaptation to soil salinity. *Innovation* 1: 100017. doi: 10.1016/j.xinn.2020.100017.

7 Rehabilitation and Management of Multiple Stresses in Saline and Sodic Soils for Agriculture Sustainability

Ratneswar Poddar, Arup Sen, Arindam Sarkar, and Sanmay Kumar Patra
Bidhan Chandra Krishi Viswavidyalaya

Akbar Hossain
Bangladesh Wheat and Maize Research Institute

CONTENTS

DOI: 10.1201/9781003300564-7

7.1 INTRODUCTION

For optimal growth and development, crops require sufficient or balanced amounts of available water and nutrients in the rhizospherezone (Ondrasek et al., 2011). However, the sustenance of healthy soil properties, requiring ready supply of essential inputs to plants, is severely impaired by either natural or anthropogenic-mediated soil degradation, mainly due to soil salinity and sodicity. The occurrence of thissoil is the characteristic of the regions,where the evapotranspiration greatly exceeds the total precipitation. Soil salinity refers to the presence of excess amounts of different water-soluble salts such as sodium (Na^+), potassium (K^+), calcium (Ca^{2+}), magnesium (Mg^{2+}), and chloride (Cl^-) ions with electrical conductivity of saturation extract (ECe)>4 dS/m^{-1} within soil system to dissipate its productivity (Gangwar et al.,2020). Likewise, the accumulation of excess Na^+ in the soil exchangeable complex with exchangeable sodium percentage (ESP) >15, which negatively affects the soil and crop productivity, is called sodic or alkali soil (Sharma et al., 2016). Soil salinity and sodicity are one of the important abiotic environmental stresses, which resulted in not only the degradation of land area but also declinein the productivity and quality of crops (Qadir et al., 2007; Shahbaz and Ashraf, 2013). The dense soil conditions with nutritional imbalance encountered in sodic soils significantly affect the hydraulic conductivity and permeability of both water and air, because of the deterioration of soil physical conditions (Singh et al., 2012). The problems of soil salinity and sodicity are not specific/restricted only to arid and semiarid regions, saline and sodic soil can also develop anywhere in humid and subhumid regions with faulty agricultural practices, and coastal areas. The concentration of mineral salts accumulates in the lower layer and come to the surface by capillary action during dry seasons as a result of high evaporation under shallow water table condition, prolonged use of brackish water for irrigation, and faulty irrigation practices in agriculture. However, other factors such as heavy textured soil, inadequate drainage, scanty rainfall, weathering of rocks, excessive application of soil amendments and soluble fertilizers, the capillary rise of saline ground water, prolonged use of salt-containing irrigation water, and tidal seawater ingression, specifically in coastal regions also play key roles in soil salinization and sodification processes. High salt concentrations of different compositions adversely affect the soil's physical, chemical, and biological properties, influencing the crop growth and marketable yield (Hafez et al., 2021). Soil salinity due to the excessive water-soluble salt concentrations increases the soil solution osmotic potential, causing less water and nutrient uptake by plants; whereas sodicity is mainly associated with poor soil physical structure with reduced permeability, infiltration and hydraulic conductivity, and poor air–water relations as a result of high dispersion and swellingof clays in the presence of sodium (Wong et al., 2010). Globally, an estimated area of 1,128 million ha is largely affected by salinity and sodicity stresses (Mandal et al., 2018), which in addition to reducing net cultivable area, has serious implications for agricultural productivity, it's quality, and livelihood security of the people. The over-increasing human and livestock population growth throughout the world is creating a huge challenge to feed them, and it is predicted that by 2050, the estimated demand for food and feed requirements of the world's population will become double (Foley et al., 2011). Therefore, a thorough understanding of the land degradation through soil salinization and sodification, its caustic effects on the soil–water–plant continuum, and the possible planning for effective remediation and management strategies to restore such soils for ensuring the food and nutritional security of the world population is the necessity of the hour.

7.2 VARIOUS EFFECTS UNDER SALINITY AND SODICITY STRESS

7.2.1 Effect on Plant Growth

Plants are negatively affected by salt stress in three ways: osmotic imbalance, ionic toxicity, and a decrease in nutrient absorption (Selvakumar et al., 2014). Because of the osmotic imbalance caused by the excessive salinity, it is more difficult for plants to take up water from the soil through their roots. The initial impact of salt exposure is a decrease in turgidity, followed by cell dehydration, and, eventually, cell death. The salt stress could decrease the chlorophyll content and photosynthetic efficiency and transportation of the assimilate/hormones that are linked to the developing tissues of the plants (Yang et al., 2020). Plants growing in the salty environment showed ionic toxicity when Na^+ is absorbed replacing K^+ in the solution medium, causing an alternation of protein structure (Win and Oo, 2017). Many recent studies have found that salt stress has a negative impact on plant growth and development, as well as seed germination, seedling growth, and enzyme function (Seckin et al., 2009). High salinity has been shown to cause the generation and build-up of reactive oxygen species (ROS) in plant cells (Chawla et al., 2013).The effects of alkalinity stress on agricultural plants are strikingly comparable to those of salt stress (Xu et al., 2013), though carbonate salts induce stress that is sometimes greater than that caused by NaCl and Na_2SO_4 salinization (Shi and Sheng, 2005). Apart from essential element stresses, alkalinity stress also causes osmotic, ion-induced damage, or high pH consequences, which are the major constraints to plant growth (Lynch and Clair, 2004). Most alkalinity-stressed plants show stunted growth owing to inadequate nutrient absorption, as well as leaf chlorosis due to high and low Na^+ and Fe^{2+} uptake, respectively (Singh et al., 2018). High amounts of Na^+ interfere with stomatal closure worsens the problem of plant water loss (Msimbira and Smith, 2020). This is a regular occurrence in saline circumstances, but it may also happen under alkalinity conditions. Due to high concentrations of HCO_3^- in the soil solution, alkalinity stress inhibits root development, albeit this varies by crop species. The reduction of root respiration by HCO_3^- is linked to the suppression of root development (Msimbira and Smith, 2020).

7.2.2 Effect on Major Plant Nutrients Dynamics

The solubility and bioavailability of most nutrient elements in the soil aremostly controlled by soil pH, thus salt-affected soils (saline, saline–sodic, and sodic soils) are expected to hamper the availability of plant nutrients, including micronutrient elements. Soil salinity and sodicity are reported to affect plant growth directly through their effect on the concentration distribution of bioavailable plant nutrients in soils and their uptake by plants along with water (Litalien and Zeeb, 2020; Stavi et al., 2021), and indirectly, through the degradation of soil physicochemical environment (Driessen et al., 2001).Plants exposed to salt-affected environments experience two types of stress viz. the osmotic stress and nutrient stress. The osmotic stress is due to the low osmotic potential of water in saline soils,which adversely affects water absorption by plants. The nutrient stress is attributable to both nutrient toxicity (Na, Cl, B) and nutrient deficiency (N, Ca, K, P, Fe, Zn), causing a nutritional imbalance in the soil. As a result of the increased osmotic pressure of the soil solution due to increased salt concentration, the plants have to expendextra energy to pump water and nutrient ions from the soil and transport it further upwards, after excluding maximum Na^+ and $Cl^-$from the root interface (Choudhary and Yaduvanshi, 2016). The dynamics of plant-available nutrients in the soil are governed by soil pH. However, the persistence of high pH conditions in the soils is not conducive to the solubilization of nutrients and their optimum availability to plants. Also, the toxicity effects of certain cationic and anionic elements in saline and sodic soil hampers plant nutrient adsorption due to competition between ions and reduced movement of nutrients in the soil as a result of the low osmotic potential of the solution (Jekabsone et al., 2022). Greater salt concentration and high pH also suppress microbial activity in the soil, thus affecting sequestration of organic carbon and nutrient dynamics and their concentrations in these soils. In addition, the very slow rate of soil weathering taking place in arid regions and without any percolating waterallows various salts and minerals to accumulate in the soil solution, contributing to high levels of certain toxic elements in the soil's rhizosphere, affecting the soil–plant–animal system.

7.2.2.1 Effect on SOC Dynamics

Soil is the largest terrestrial carbon sink and contains two-thirds of the world's terrestrial carbon. However, soil organic carbon (SOC) stocks in saline and sodic soilsarelow due to organic matter contents as a result of reduced plant biomass production and decreased decomposition rates on account of lower microbial activity. In addition, the hazards such as soil erosion, dispersion, and leaching losses can also contribute lower C concentrations in soils. The microbial community structure, fungal population growth, and enzymic activity as an active indicator of SOC dynamics were found to be more sensitive to salinity stress (Gros et al., 2003). Low osmotic potential in soil solution, especially in saline soil, water is pulled out of the plant cells to the soil solution (i.e., plasmolysis or reverse osmosis), thereby causing root cells to shrink and collapse. The microbial carbon utilization efficiency decreases due to declining vegetation health, caused by an increase in soil salinity and sodicity (Rietz and Haynes, 2003). The poor soil physicochemical properties of the salt-affected soils affect the SOC. The reduced SOC mineralization was noticed with increasing soil salinity (Tripathi et al., 2006; Ghollarata and Raiesi, 2007; Yuan et al., 2007).

7.2.2.2 Nitrogen Dynamics

Nitrogen is the most important crop nutrient for crop vegetative growth and isrequired in the highest amount among all macronutrients. However, salt-affected soil isreported to be deficient in N. Nitrogen in the soil is generally present in plant-unavailable organic form (98%) and only a fraction (2%) is present in plant-available inorganic form (Weil and Brady, 2017). Though the organic form of N is in plant-unavailable form,microbe-mediated mineralization of organic N into inorganic N results inthe available form of N in soil. Also, processes relating to N dynamics in soil, such as oxidation, ammonification, nitrification, and N-fixation are carried out by several microorganisms. However, many studies have indicated the negative impact of salt-induced stress on soil microbes, reducing their numbers, activity, and biodiversity, and restricting those N-mineralization processes. Decreased urease activity at higher levels of salinity and/or sodicity level results in lower urea hydrolysis, attributable to the toxic effects of high pH and $CaCO_3$ contents on the urease and other N-transforming bacteria (Choudhary and Yaduvanshi, 2016). In a study on the coastal saline soil, Tan and Thanh (2021) found that nitrogen content in saline soil was significantly decreased with 10% increase in salt concentration, due to reduced microbial growth, which is incapable for degradation of organic matter. Salinity is also reported to reduce the growth of nodules in legume crops, reducing the efficiency to fix atmospheric N in soil (Fageria et al., 2011).

7.2.2.3 Phosphorus Dynamics

Phosphorus is the second most important nutrient element after nitrogen. However, total P content in saline and sodic soils is relatively low and most of the phosphorus compounds present in soil are highly insoluble. While added P-containing fertilizers also have a tendency to become fixed in soil rendering them unavailable for crop uptake. In saline soil,the availability of P is greatly reduced due to the antagonistic effect between $H_2PO_4^{-1}$ and Cl^- and $SO4^{2-}$ as the latter two are present in high concentrations. Common ion effects from Cl^- and $SO4^{2-}$ also affect the dissolution of phosphorus-carrying minerals (Weil and Brady, 2017). An excessive amount of Ca^{2+} present in these soils reacts with P from added soluble fertilizer and tends to form calcium-phosphate compounds that become increasingly insoluble over time. Salinity-induced P deficiency has negative effects on root growth and proliferation, thus reducing the root surface area in contact with soil P and adversely affecting the plant P uptake; besides, inhibiting the photosynthetic activity through decreasing stomatal conductance, chlorophyll content, and CO_2 assimilation(Dey et al., 2021).

7.2.2.4 Potassium Dynamics

After N and P,potassium (K) is the third most important nutrient element limiting crop growth and productivity. However, its concentration in the soil is influenced primarily by soilcation exchange properties and mineral weathering, rather than by microbiological processes. Only 1%–2% of the total soil K is in readily available form and the remaining 98% is in soil mineral lattices, occurring relatively in unavailable forms (Weil and Brady, 2017). Generally, K availability in salt-affected soil is low. Common ion effects due to high Na^+,Ca^{2+}, and Mg^{2+} concentration in the soil solution may reduce K uptake by the plant because cations tend to compete against one another for root K absorption (Wakeel et al., 2011; Wakeel, 2013). In plants, Na^+ competes with K^+ in the process of transport across the cell membrane during uptake, making it difficult for plants to obtain the needed K^+ from the saline–sodic or sodic soils (Bidalia et al., 2019).

7.2.3 Effect on Soil Microbes

Soil salinity has a detrimental influence on microbial diversity, community structure, and abundance (Wen-wen et al., 2019). Soil salinizationwas found to reduce soil respiration, enzyme activity, soil microbial biomass, and bacterial growth rate, and thus play a pivotal role in regulating ecosystem function (Qi and Yang, 2017). Microorganisms found in naturally saline environments are thought to have developed a technique for coping with high salt concentrations and several adaptations for keeping their population active while living with such harsh conditions. From a genetic standpoint, these species exhibit an under-or overexpression of specific genes and metabolites that enable them to cope with osmotic stress (Dion and Nautiyal, 2008). Salinity influences the global microbial composition in saline soil more than any other severe chemical factor, such as temperature or pH, according to a recent meta-analysis of soil microbial communities (Ma and Gong, 2013). They revealed that approximately 50% of archaeal diversity and less than 25% of overall bacterial diversity comprise saline soil habitats. Increased salinity of the soil had an impact on root exudation and microbial breakdown of organic materials (Ondrasek et al., 2010).The main determinants of phytomicrobiome composition are environmental variables,with soil pH having a significant impact on microbial community structure (Zhalnina et al., 2015). According to Cho et al. (2016), the soil pH, a critical factor of the environment, has a significant impact on the diversity and richness of bacterial lifeforms in soil. However, for appropriate physiological processes, all living cells have optimum pH requirements. Because the availability of nutrients is adequate,the pH range of 5.5–6.5 is ideal for plant growth. The bonds that hold the DNA helix strands together are disrupted by high pH, and lipid hydrolysis occurs more readily as the environment becomes more basic (Shen et al., 2019). To thrive in high pH environments, most bacteria adapt their surrounding medium to near neutrality. Microbes need Na^+ to maintain their intracellular pH because it allows H^+/Na^+ antiporters to exchange into the culture media (Satyanarayana et al., 2005). Furthermore, through the proton motive force, H^+ concentration gradients across membranes play a crucial role in creating ATP during cellular respiration (Celiker and Gore, 2013). It is unclear how PGPM can utilize sodium-dependent ATP synthases as a method for alkali tolerance.

7.3 RECLAMATION AND MANAGEMENT STRATEGY OF SALINE AND SODIC SOILS

The reclamation of salt-affected soils involves physical, chemical, hydro-technical, and biological methods for sustaining the soil and crop productivity, and enhancing the income of the farmers. The physical methods include deep ploughing, sub-soiling, sand-filling, profile inversion, and salt scrapping. The hydro-technical methods include salt leaching, provision of artificial surface and subsurface drainage technology, etc. The chemical methods include the application of gypsum, calcium chloride, calcite, phospho-gypsum, and iron pyrites. The biological methods accommodate green manuring, the addition of FYM and compost, incorporation of crop residues, press mud cake, municipal and industrial solid wastes, microbial consortium, bio-saline agriculture, and use of salt-tolerant crops and varieties. The most common reclamation strategy for saline soils is to flush the salts out of the root zone with good-quality irrigation water. The other technological interventions involve continuous ponding, intermittent drainage, and sprinkling to tackle the problems of soil salinity (Arora and Sharma, 2017; Kumar and Sharma, 2020).

7.3.1 Irrigation Practices

7.3.1.1 Continuous Ponding

Continuous ponding or flooding involves the application of large volumes of standing water until enough salts havebeen removed from the crop root zone. The desired soil salinity to be accomplished depends on the salinity tolerance of the crop. The soil type and the initial EC value will determine the requirement of the actual amount of clean water for salt disposal.

7.3.1.2 Advanced Irrigation Management

To minimize aquifer recharge with excess salts and thus limit soil salinization,injudicious flood and furrow irrigation should be negated wherever possible. Similarly, in non-flooding irrigation methods that apply routine, frequent high amounts of irrigation water to leach salts from the rhizosphere increase the risk of groundwater salinization (Devkota et al., 2015). This risk is particularly prominent in dry lands, where the risk of soil salinization is high even when good-quality water is used for irrigation (Cuevas et al., 2019). Conversely, water-saving irrigation methods, such as microirrigation systems, and specifically drip irrigation systems, can substantially decrease the risk of groundwater recharge,while effectively leaching salts from the dripper-formed "wet bulbs,"from which the plant roots extract water (Stavi, 2020).

7.3.1.3 Intermittent Ponding or Sprinkler Irrigation

Intermittent ponding or sprinkling reclamation methods can be used to save freshwater while reclaiming soils. In this approach,a series of smaller amounts of water is applied as ponded- or sprinklerirrigation and allowed to drain after each application. The wetting and drying cycles efficiently leach salts from soil micropores using one-third to two-thirds less water than continuous ponding. The quantity of water needed will depend upon initial and desired salt levels, water quality, application methods, soil texture, and crop chosen. Finer soils will require more leaching water than coarser soils because of their capacity to retain more water.

7.3.2 Drainage

Adequate drainage is needed to flush out the surface accumulated salt crusts in heavy textured saline soils with sufficient amounts of good-quality water, where the flushed salts from the soil surface layers enter the drainage system and are eliminated from the target land (Shahid et al., 2018). The breaking-up of the impervious or hardpan layers of the soil profile with deep ploughing or sub-soiling can help mitigate saline conditions. The soils with a shallow water table depth require a subsurface drainage system to leach the excess soluble salts through drainage water. The adoption of subsurface drainage technology in saline soils increasedthe yields of paddy,wheat, and cotton by about 45%, 111%, and 215%, respectively, and farmers' income three-fold (Sharma et al., 2014). Besides, it significantly increased cropping intensity and socio-economic benefits in terms of on-farm employment and 128 man-days/ha/annum additional employment generation (Singh, 2009; Sharma et al., 2011).

7.3.3 Salt-Tolerant Crops and Varieties

In areas where salt leaching with water is not feasible or economical, the cultivation of salt-tolerant crops and varieties is another way to address the problem of soil salinization. This technique is viable and cost-effective and suits well to the small and marginal farmers who are unable to bear the high costs of chemical amendment-based reclamation technologies. Numerous salt-tolerant varieties of rice, wheat, mustard, and other crops, grasses, shrubs, fruit trees, and medicinal and aromatic plants have been developed/identified for commercial cultivation in salt-affected soils (Singh, 2009; Sharma et al., 2011). In general, perennial plants, especially some grass forages, possess the highest tolerance to salts, while legumes are typically the most sensitive to salts. The plant's tolerance to salts is not constant and can differ throughout the growing season or under periods of stress. Crops such as sugar beets, alfalfa, and barley are all sensitive to salt toxicity during emergence but become more resistant at maturity. The cultivation of salt-tolerant multipurpose trees, bushes, grasses, fruit plants, medicinal and aromatic plants, etc. is another opportunity of managing salt-affected soils.

7.3.4 Organic Amendments

The application of organic supplements, such as livestock-derived organic amendments, composts, and waste materials, such as municipal solid wastes, increases the SOC sequestration and improves the soil's macro-aggregation processes and structure formation. Bacteria and fungi release various glue-like substances through their metabolic processes that contribute to the adhesion of soil particles, creating soil aggregates. These processes improve soil permeability, increase soil porosity and aeration, and enhance the salts leaching from the rhizosphere, and simultaneously decrease evaporation rates, reducing salt accumulation in the surface soil layer (Lakhdar et al., 2009; Diacono and Montemurro, 2015). In sodic lands, the application of organic amendments combined with gypsum was reported to better alleviate soil sodicity than the application of gypsum alone (Gonçalo et al., 2020). The low cost of organic amendments is widely used for alleviation of soil salinity and sodicity. In addition to these "regular" organic additives, specific, designated organic soil conditioners are more effective in sodic soils when combined with gypsum (Cuevas et al., 2019).

7.3.5 Fertilizing and Manuring

The use of chemical fertilizer may exacerbate rhizosphere salinization. This is not only wasteful in terms of fertilizing efficiency but also increases the salinity of aquifers (Finlayson et al., 2010). Recent advances in precision fertilizer systems may increase fertilizer useefficiency while reducing the risk of salinization and contamination of both underground and surface waterbodies. The fertigation technologies, in which dissolved fertilizers in several splits according to the crop requirements are distributed through drip irrigation systems, were shown to optimize fertilizer useefficiency and reduce agriculture's environmental foot print (Azad et al., 2020).

7.3.6 Phytoremediation

Phytoremediation or bio-drainage of salt-affected soils is the process of removing excess salts from the soil by growing different types of plants with deep root systems, high biomass production, and greater evapotranspirative demand. The growing of salt-tolerant trees, shrubs, and grasses is a cost-effective and environmental-friendly way of restoring salt-affected soils to productive ones (Qadir et al., 2007). Reclamation and rehabilitation of sodic and saline-waterlogged soils through afforestation and agroforestry practices arewell established (Singh et al., 2012; Dagar, 2014). Some of the promising species for saline and sodic soil reclamation include *Prosopis juliflora*, *Acacia nilotica*, *Casuarina equisetifolia*, *Tamarix articulate*, *Eucalyptus tereticornis*, and *Leptochloa fusca*; for waterlogged saline soils include *Prosopis juliflora*, *Tamarix articulata*, *Casuarina glauca*, *Terminalia arjuna*, *Pongamia pinnata*, *Syzygium cuminii*, *Bambusa arundinacea*, *Acacia farnesiana*, *Acacia nilotica*, *Acacia tortilis*, *Parkinsonia aculeata*; and plant species like *Eucalyptus tereticornis*, *Populus deltoids*, and *Tectona grandis* are effective for reclaiming sodic soils (Dagar et al., 2016; Patra and Banik, 2018). Promising agroforestry models, fruit-based agroforestry models, silvi-pastoral models, etc. along with appropriate planting and management techniques have been developed specifically for saline/sodic/saline–sodic conditions. The medicinal and aromatic plants such as isabgol (*Planta goovata*), aloe (*Aloe barbadensis*), kalmegh (*Andrographis paniculata*), *Matricaria chamomilla*, *Vetiveria zizanioides*, *Cymbopogon martini*, and *Cymbopogon flexuosus* have been reported to be highly effective in restoring saline soils (Kumar and Sharma, 2020). Although bio-drainage is an emerging technology around the world for land reclamation; however, its feasibility needs to be explored scientifically region-wise. The bio-drainage as partial substitution or in conjunction with conventional surface and subsurface drainage could be the most viable option at the present state of knowledge to remediate effectively the sodic and saline-waterlogged lands.

7.3.7 Bioremediation

The bioremediation approach, which involves plant–microbial interaction, is used for enhancing the productivity of salt-affected soils (Arora et al., 2014). The microorganisms have the capability of rapid adjustment toward environmental changes and deterioration and can play an important role in the maintenance and sustainability of any ecosystem. Microorganisms possess some unique properties such as salt stress tolerance, genetic diversity, synthesis of compatible solutes, production of plant growth-promoting hormones, biocontrol potential, and their interaction with crop plants. If these traits are suitably exploited, microorganisms can be a weapon in alleviating salt effects injury on the crops (Shrivastava and Kumar, 2015). A low-cost microbial bio-formulation "CSR-BIO,"a consortium of *Bacillus pumilus*, *Bacillus thuringiensis*, and *Trichoderma harzianum*, is rapidly becoming popular with the farmers (Damodaran et al., 2013a). This bioformulation acts as a soil conditioner and nutrient mobilizer and has been found to increase the productivity of the high-value crops such as banana, vegetables, and gladiolus in sodic and normal soils by about 22%–43%.

7.3.7.1 Plant Growth-Promoting Rhizobacteria for Salinity Management

Physiological impacts of soil salinity reduce crop yield significantly; nevertheless, plant growth-promoting rhizobacteria (PGPR) has been found to boost crop productivity by minimizing the physiological harm caused by excessive ion concentrations in the soil (Ilangumaran and Smith, 2017; Dodd and Pérez-Alfocea, 2012; Munns and Tester, 2008). This response is due to the varied methods that these helpful bacteria use to regulate the adverse effects of excessive salt concentrations at different levels (Egamberdieva et al., 2019). Salt tolerance, generation of plant growth-promoting hormones, genetic variety, synthesis of suitable solutes, and favorable interactions with crop plants should all be considered while developing a prospective bio-inoculant (Ahmed et al., 2019). Beneficial soil microbes can help plant growth and production through a variety of methods under salinity stress, including nutrient mobilization, hormone secretion, and disease suppression (Table 7.1).

7.3.7.2 PGPR in Enhancing Tolerance to Alkalinity

Haloalkaliphilic PGPM have the capacity to reduce both salinity and alkalinity stress in plants and increase growth (Siddikee et al., 2011). Alkaliphilic bacteria are those that can keep their intrinsic pH below 9, while the environmental pH is 9–11. A cytoplasmic membrane proton transport mechanism is used to achieve alkalinity tolerance (Torbaghanet al., 2017). The Bacillaceae family of rhizobacteria is one of the best-studied in this regard, with a diverse variety of host plants; the PGPR group includes 15 genera, including *Alkalibacillus*, *Bacillus*, and *Haloalkalibacillus*, among others (Torbaghan et al., 2017). The application of alkalinity-tolerant rhizobacteria capable of producing indole acetic acid (IAA) and 1-aminocyclopropane-1-carboxylate (ACC) reduces ethylene synthesis while boosting K levels. PGPM also alleviates sodicity stress in plants via altering hormones,enzymes,photosynthesis,and the release of organic acids and secondary metabolites (Dixit et al., 2020). Furthermore, rhizobacteria are important in the

TABLE 7.1
Example of Some PGPR Involved in Salinity Tolerance to Crop Plant

PGPR	Plant Species	Mechanism	Effect	Reference
Bradyrhizobium japonicum	*Glycine max*	Nodulation and maintaining ion homeostasis	Improved growth and N assimilation	Ilangumaran et al. (2021)
Pseudomonas mendocina	*Lactuca sativa*	Production of exopolysaccharide	Improved soil aggregation	Kohler et al. (2006)
Bacillus aquimaris	*Triticumaestivum*	Production of proline	Improved shoot and root growth	Upadhyay and Singh (2015)
Azospirillum brasilense	*Zea mays*	Production of trehalose	Increased root and shoot biomass	Rodríguez-Salazar et al. (2009)
Pseudomonas pseudoalcaligenes and Bacillus pumilis	*Oryza sativa*	Accumulation of osmoprotectants	Increased shoot biomass	Jha et al. (2011)
Brachybacterium sp., Brevibacterium sp.	*Arachis hypogaea*	IAA production and ion homeostasis	Increased uptake of nutrients and higher production	Shukla et al. (2012)
Pseudomonas putida	*Gossypium hirsutum*	Alteration in phytohormone production	Increased nutrient content and productivity	Yao et al. (2010)
Bacillus spp.	*Triticum aestivum*	Production of exopolysaccharide	Increased biomass production and soil aggregation	Ashraf et al. (2004)
Streptomyces sp.	*Triticum aestivum*	Production of IAA and siderophore	Higher biomass production, shoot length, and nutrient uptake	Sadeghi et al. (2012)
Pseudomonas fluorescens and Enterobacter aerogenes	*Zea mays*	ACC deaminaseproduction	Higher biomass production, NPK uptake, and modulation of K^+/Na^+ balance	Nadeem et al. (2009)
Bacillus pumilis	*Zea mays*	ABA production and accumulation in leaves	Higher water uptake and osmotic potential	Yasmin et al. (2017)
Pseudomonas psimiae	*Glycine max*	Production of volatile organic compounds and higher phosphatase activity	Improved seed germination	Vaishnav et al. (2016)
Bacillus amyloliquefaciens	*Zea mays*	Production of polyamine	Na^+ toxicity reduction	Chen et al. (2017)

cycling of essential nutrients such as N and C, ensuring long-term nutrient storage in the soil. Because legumes have an intrinsic tendency to acidify the soil, they are also recognized as a way of alkalinity control and have been used in numerous regions of the world. Combining legumes and their symbionts to address alkalinity issues is a long-term solution for reclaiming unproductive soils for agricultural use. AM fungi and rhizobia are among the oldest and best-documented plant symbioses, and our study of them has led to new ways to help plants cope with a variety of stresses (Kumar et al., 2009). Under alkalinity stress, faba bean performed well when inoculated with rhizobia and mycorrhizae (Abd-Allaet al., 2014). Examples of some PGPR involved in reducing alkalinity stress on cropplant are available in Table 7.2.

TABLE 7.2
Example of Some PGPR Involved in Reducing Alkalinity Stress to Crop Plant

PGPR	Plant Species	Effect	Reference
Bacillus clausii	*Triticum aestivum*	Increased biomass production and crop yield	Torbaghanet al. (2017)
Bacillus sp.	*Zea mays*	Improved sugar content and photosynthetic pigment	Dixit et al. (2020)
Burkholderia spp.	*Glycine max*	Increased biomass production and N/P ratio	Fernández et al. (2007)
Bacillus subtilis	*Vicia faba*	Higher germination and IAA production	Yousef (2018)
Bacillus simplex	*Triticum aestivum*	Reduction in soil pH and higher root P concentration	Hansen et al. (2020)
Mesorhizobium ciceri	*Cicer arietinum*	Higher nodulation and improved crop yield	Müller et al. (2016)
Bacillus thurigiensis	*Musa spp*	Higher brunch weight	Damodaran et al. (2013b)
Azotobacter chroococcum, Pseudomonas koreensis	*Zea mays*	The higher value of leaf area index, chlorophyll, and sugar content	Nehela et al. (2021)

7.3.8 Recycling of Wasteful Drainage Waters

The reuse of saline water and amelioration of salt-affected soil is becoming an increasingly vital tool to improve crop production. The cyclic strategy involves the use of saline–sodic drainage water and non-saline irrigation water in crop rotations that include both moderately salt-sensitive and salt-tolerant crops. Typically, the non-saline water is also used before planting and during the initial growth stages of the salt-tolerant crop, while saline water is usually used after seedling establishment (Qadir and Oster, 2004; Patra, 2017). The reuse of drainage water by blending, which consists of mixing good and poor quality water, is an alternative option for land irrigation. However, blending may be more practical for situations where the drainage water or shallow groundwater is not too saline for the crops and insufficient quantities of freshwater are available to meet crop water requirements. The sequential strategy provides the means of minimizing the volume of drainage water produced by irrigated agriculture. The blended use of bad and good-quality waterfor irrigation can mitigate the water scarcity and effective use of saline soils for crop production.

7.3.9 Conjunctive Use of Waters

The strategy of conjunctive use of brackish canal water and fresh groundwater makes a deterrent against the soil salinity emergence. Summer rice production with groundwater irrigation in the coastal basin becomes costlier due to the high investment forlifting water from a deep aquifer. The repeated use of the abundantly available canal water for irrigation develops soil salinity,deteriorates the soil health,nutrient imbalance, and depressesyield. The results showed that irrigation with the cheapest canal water and costly groundwater at 1:1 proportion increased rice yield by 77% and BCR by 56% (Table 7.3). This low-cost indirect drainage technology is an economically viable option, which helps gradual removal of bad quality canal water, the excessive accumulation of which otherwise prove to be hazardous to the soil and crop environment. Besides,this strategy is found useful to curtail the excessive demand forlimited fresh groundwater resources by partially using the brackish canal water for irrigation.

TABLE 7.3
Conjunctive Use of Poor Canal Water (CW) and Fresh Groundwater (GW) on Summer Ricein Coastal Saline Soils of West Bengal, India

CW:GW	Grain Yield (t/ha)	Straw Yield (t/ha)	Benefit–Cost Ratio
4:0	3.0	3.4	1.07:1
1:1	5.3	5.5	1.67:1
0:4	5.6	5.9	1.60:1

Source: Patra (2017).

7.3.10 Rainwater Harvesting and Recycling

The rainfall occurs during a limited period ofa year and if not conserved,this fresh water may not be available for the rest of the year. This natural resourceis to be harnessed through proper soil and water conservation measures for mitigating irrigation water demand during the dry spells in the coastal region. It is an integrated approach torainwater management,dealing with on-farm harvest and storage of excess rainwater during monsoon,and recycling the same for irrigation for dry season crops. This is an efficient method to utilize a large quantity of water, which otherwise goes waste to the sea. This approach preserves the best source of fresh water, helps in deciding a multi-cropping system, and maximizes the water use efficiency. This conservation technique improves the groundwater quality,reduces the groundwatersalinity, and develops the groundwater potential (aquifer yield) of the area for subsequent crop irrigation.

7.3.11 Alternative Land Use Planning

Improved land management practices with special care to efficient drainage in monsoon season (June–September) and judicious water management in post-monsoon months are imperative to obtain high yields of crops in low-lying saline-waterlogged lands. Raised bed-pond system of land configuration could help to dispose of the excess rainfall-runoff water and prevent waterlogging in the rainy season and, simultaneously, it also stores a significant amount of fresh water that could be used for both protective or supplemental irrigation to dry season crops and pisciculture. High-value vegetable crops were grown on different parcels of raised bed land during *kharif*, rabi, and summer seasons. The excavated pond was used to harvest monsoon rain for supplemental irrigation in the dry season crops in the raised bed and the storagepond water was used for pisciculture, duckery, etc.

7.3.12 Mulching

It is a protective measure that inhibits salt accumulation on the soil surface because it reduces evaporation by 50%–80%. Plastic cover and organic mulches are both effective in controlling evaporation. Plastic mulches may require lower management and labor cost than organic mulches,but these are less effective than organic mulches as regards to reducing salt build-up. However, it can intercept precipitation, which reduces the potential leaching effect from rainfall incidence. Plastic mulches also deteriorate over time and its disposal is a problem (Aragues et al., 2014). Organic mulches,on the other hand,may decompose rapidly and need to be replaced more often. These enrich the soil ecosystem by promoting microflora and macroflora populations, especially the earthworm population (Jodaugiene et al., 2010). Zero or minimum tillage with minimum soil disturbances can also reduce evaporation considerably from the saline soils.

7.4 CONCLUSION

Soil salinization and alkalization are considered the most critical environmental stress, which can negatively affect plant growth, induce soil degradation, lower agricultural productivity, and worsen farmers' income, particularly in arid and semiarid regions of the world. The problems are associated with high total salts (saline soils), excess exchangeable sodium (sodic soils), or both (saline–sodic soils). Salt-affected soils can severely reduce land value and soil productivity. For restoration of saline soils,provision of effective surface and subsurface drainage for flushing out or leaching of excess soluble salts is necessary. In sodic and saline–sodic soils, chemical remediation measures, especially using the gypsum-based soil amendments followed by salt leaching, could be effective in restoring the affected lands. Further, the practices of conservation agriculture, such as reduced or minimum tillage, crop residue management, appropriate manuring/composting, precision fertilizer schemes, and proper crop and varietal selection can considerably reduce the risk of soil salinity and sodicity. Specifically, the adoption of advanced forestry and agroforestry systems has tremendous potential in restoring the saline and sodic soils.

REFERENCES

Abd-Alla, M. H., El-Enany, A. W. E., Nafady, N. A., Khalaf, D. M., and Morsy, F. M. 2014. Synergistic interaction of rhizobium leguminosarumbv. Viciae and arbuscularmycorrhizal fungi as a plant growth promoting biofertilizers for faba bean (*Viciafaba* L.) in alkaline soil. *Microbiological Research*169:49–58. https://doi.org/10.1016/j.micres.2013.07.007.

Ahmed, S., Roy Choudhury, A., Chatterjee, P., Samaddar, S., Kim, K., Jeon, S., and Sa, T. 2019. The role of plant growth-promoting rhizobacteria to modulate proline biosynthesis in plants for salt stress alleviation. In: Sayyed, R., Arora, N., and Reddy, M. (eds.).*Plant Growth Promoting Rhizobacteria for Sustainable Stress Management. Microorganisms for Sustainability*, vol.12.Singapore:Springer. https://doi.org/10.1007/978-981-13-6536-2_1.

Aragues, R., Medina, E. T., and Claveria, I. 2014. Effectiveness of inorganic and organic mulching for soil salinity and sodicity control in a grapevine orchard drip-irrigated with moderately saline waters. *Spanish Journal of Agricultural Research*12(2):501–508. http://dx.doi.org/10.5424/sjar/2014122-5466.

Arora, S., and Sharma, V. 2017. Reclamation and management of salt-affected soils for safeguarding agricultural productivity. *Journal of Safe Agriculture*1:1–10.

Arora, S., Vanza, M., Mehta, R., Bhuva, C., and Patel, P. 2014. Halophilic microbes for bio-remediation of salt-affected soils. *African Journal of Microbiology Research*8:3070–3078.https://doi.org/10.5897/AJMR2014.6960.

Ashraf, M., Hasnain, S., Berge, O., and Mahmood, T. 2004. Inoculating wheat seedlings with exopolysaccharide-producing bacteria restricts sodium uptake and stimulates plant growth under salt stress. *Biology and Fertility of Soils*40:157–162.

Azad, N., Behmanesh, J., Rezaverdinejad, V., Abbasi, F., and Navabian, M. 2020. An analysis of optimal fertigation implications in different soils on reducing environmental impacts of agricultural nitrate leaching. *Scientific Reports*10:7797. https://doi.org/10.1038/s41598-020-64856-x.

Bidalia, A., Vikram, K., Yamal, G., and Rao, K.S. 2019. Effect of salinity on soil nutrients and plant health. In: Akhtar M. S. (ed.) *Salt Stress, Microbes, and Plant Interactions: Causes and Solution*.Singapore: Springer, pp. 273–297.

Celiker, H., and Gore, J. 2013. Cellular cooperation: insights from microbes. *Trends in Cell Biology*23:9–15. https://doi.org/10.1016/j.tcb.2012.08.010.

Chawla, S., Jain, S., and Jain, V. 2013. Salinity induced oxidative stress and antioxidant system in salt-tolerant and salt-sensitive cultivars of rice (*Oryza sativa* L.). *Journal of Plant Biochemistry and Biotechnology*22(1): 27–34.

Chen, L., Liu, Y., Wu, G., Zhang, N., Shen, Q., and Zhang, R. 2017. Beneficial Rhizobacterium Bacillus amyloliquefaciens SQR9 induces plant salt tolerance through spermidine production. *Molecular Plant-Microbe Interactions*30:423–432.

Cho, S., Kim, M., and Lee, Y. 2016. Effect of pH on soil bacterial diversity. *Journal of Ecology and Environment*40:10. https://doi.org/10.1186/s41610-016-0004-1.

Choudhary, O.P., and Yaduvanshi, N. P. S. 2016. Nutrient management in salt-affected soils. *Indian Journal of Fertilisers*12(12):20–35.

Cuevas, J., Daliakopoulos, I. N., del Moral, F., Hueso, J. J., and Tsanis, I. K.2019. A review of soil-improving cropping systems for soil salinization. *Agronomy*9:295. https://doi.org/10.3390/agronomy9060295.

Dagar, J. C. 2014. Greening salty and waterlogged lands through agroforestry systems for livelihood security and better environment. In:Dagar, J. C., Singh, A. K., and Arunachalam, A. (eds.). *Agroforestry Systems in India: Livelihood Security & Ecosystem Services, Advances in Agroforestry 10*. Berlin: Springer, pp:273–332. https://doi.org/10.1007/978-81-322-1662-9_9.

Dagar, J. C., Khajanchi, L., Jeet, R., Mukesh, K., Chaudhary, S. K., Yadav, R. K., Ahamad, S., Singh, G., and Kaur, A. 2016. Eucalyptus geometry in agroforestry on waterlogged saline soils influences plant and soil traits in North-West India. *Agriculture Ecosystem and Environment*233:33–42.https://doi.org/10.1016/j.agee.2016.08.025.

Damodaran, T., Mishra, V. K., Sharma, D. K., Jha, S. K., Verma, C. L., Rai, R. B., Kannan, R., Nayak, A. K., and Dhama, K. 2013b. Management of sub-soil sodicity for sustainable banana production in sodic soil – An approach. *International Journal of Current Research*5(7):1930–1934.

Damodaran, T., Rai, R. B., Jha, S. K., Sharma, D. K., Mishra, V. K., Dhama, K., Singh, A. K., and Sah, V. 2013a. Impact of social factors in adopon of CSRBIO- A cost effective, eco-friendly bio-growth enhancer for sustainable crop production. *South Asian Journal of Experimental Biology*3(4):158–165.

Devkota, M., Martius, C., Gupta, R. K., Devkota, K. P., McDonald, A. J., and Lamers, J. P. A. 2015. Managing soil salinity with permanent bed planting in irrigated production systems in Central Asia. *Agriculture Ecosystem and Environment*202:90–97.https://doi.org/10.1016/j.agee.2014.12.006.

Dey, G., Banerjee, P., Sharma, R.K., Maity, J.P., Etesami, H., Shaw, A.K., Huang, Y.-H., Huang, H.-B., and Chen, C.-Y. 2021. Management of phosphorus in salinity-stressed agriculture for sustainable crop production by salt-tolerant phosphate-solubilizing bacteria—a review. Agronomy11:1552. https://doi.org/10.3390/ agronomy11081552.

Diacono, M., and Montemurro, F. 2015. Effectiveness of organic wastes as fertilizers and amendments in salt-affected soils. *Agriculture*5:221–230. https://doi.org/10.3390/agriculture5020221.

Dion, P., and Nautiyal, C. S. 2008. *Microbiology of Extreme Soils*. Berlin, Heidelberg: Springerhttps://doi.org/10.1007/978-3-540-74231-9.

Dixit, V. K., Misra, S., Mishra, S. K., Tewari, S. K., Joshi, N., and Chauhan, P. S. 2020. Characterization of plant growth-promoting alkalotolerantAlcaligenes and Bacillus strains for mitigating the alkaline stress in *Zea mays*. *Antonie Van Leeuwenhoek*113:889–905.https://doi.org/10.1007/s10482-020-01399-1.

Dodd, I. C., and Pérez-Alfocea, F. 2012. Microbial amelioration of crop salinity stress. *Journal of Experimental Botany*63:3415–3428.

Driessen, P., Deckers, J., Spaargaren, O., and Nachtergaele, F. 2001. *Lecture Notes on the Major Soils of the World*. Rome:Food and Agriculture Organization. Available at: http://www.fao.org/3/y1899e/y1899e.pdf.

Egamberdieva, D., Wirth, S., Bellingrath-Kimura, S. D., Mishra, J., and Arora, N. K. 2019. Salt-tolerant plant growth promoting rhizobacteria for enhancing crop productivity of saline soils. *Frontiers in Microbiology*10:2791.

Fageria, N. K., Gheyi, H. R., and Moreira, A. 2011. Nutrient bioavailability in salt affected soils. *Journal of Plant Nutrition*34(7):945–962. https://doi.org/10.1080/01904167.2011.555578.

Fernández, L. A., Zalba, P., Gómez, M. A., and Sagardoy, M. A. 2007. Phosphate solubilization activity of bacterial strains in soil and their effect on soybean growth under greenhouse conditions. *Biology and Fertility of Soils*43:805–809.https://doi.org/10.1007/s00374-007-0172-3.

Finlayson, J., Bathgate, A., Nordblom, T., Theiveyanathan, T., Farquharson, B., Crosbie, R., Mitchell, D., and Hoque, Z. 2010. Balancing land use to manage river volume and salinity: economic and hydrological consequences for the little river catchment in central west, New South Wales, Australia. *Agricultural System*103:161–170.https://doi.org/10.1016/j.agsy.2009.12.003.

Foley, J. A., Ramankutty, N., Brauman, K. A., Cassidy, E. S., Gerber, J. S., Johnston, M., et al. 2011. Solution for a cultivated planet. *Nature*478(7369):337–342.

Gangwar, P., Singh, R., Trivedi, M., and Tiwari, R. K. 2020. Sodic soil: management and reclamation strategies. In: Shukla, V., and Kumar, N. (eds.). *Environmental Concerns and Sustainable Development*,pp. 175–190. Singapore:Springer Nature. https://doi.org/10.1007/978-981-13-6358-0_8.

Ghollarata, M., and Raiesi, F. 2007. The adverse effects of soil salinization on the growth of *Trifoliumalexandrinum* L. and associated microbial and biochemical properties in a soil from Iran. *Soil Biology and Biochemistry*39(7): 1699–1702.

Gonçalo, F. F., Dias, N. D., Suddarth, S. R. P., Ferreira, J. F. S., Ray, G., Anderson, R. G., et al. 2020. Reclaiming tropical saline-sodic soils with gypsum and cow manure. *Water*12:57. https://doi.org/10.3390/w12010057.

Gros, R., Poly, F., JocteurMonrozier, L. and Faivre, P., 2003. Plant and soil microbial community responses to solid waste leachates diffusion on grassland. *Plant and Soil*255(2):445–455.

Hafez, M., Popov, A. I. and Rashad, M.2021. The response of saline-sodic soils to reclamation using biological and organic amendments under arid regions of Egypt. In *E3S Web of Conferences 247:01047, EDP Sciences*. https://doi.org/10.1051/e3sconf/202124701047

Hansen, V., Bonnichsen, L., Nunes, I., Sexlinger, K., Lopez, S., Van Der Bom, F., et al. 2020. Seed inoculation with *Penicilliumbilaiae* and Bacillus simplex affects the nutrient status of winter wheat. *Biology and Fertility of Soils*56:97–109.

Ilangumaran, G., Schwinghamer, T. D., and Smith, D. L. 2021. Rhizobacteria from root nodules of an indigenous legume enhance salinity stress tolerance in soybean. *Frontiers in Sustainable Food Systems*4:617978. https://doi.org/10.3389/fsufs.2020.617978.

Ilangumaran, G., and Smith, D. L. 2017. Plant growth promoting rhizobacteria in amelioration of salinity stress: a systems biology perspective. *Frontiers in Plant Science*8:1768.

Jekabsone, A., Andersone-Ozola, U., Karlsons, A., Romanovs, M., and Ievinsh, G. 2022. Effect of salinity on growth, ion accumulation and mineral nutrition of different accessions of a crop wild relative legume species, *Trifoliumfragiferum. Plants*11:797. https://doi.org/10.3390/ plants11060797.

Jha, Y., Subramanian, R. B., and Patel, S. 2011. Combination of endophytic and rhizospheric plant growth promoting rhizobacteria in oryza sativa shows higher accumulation of osmoprotectant against saline stress. *Acta PhysiologiaePlantarum*33:797–802.

Jodaugiene, D., Pupaliene, R., Sinkeviciene, A., Marcinkeviciene, A., Zebrauskaite, K., Baltaduonyte, M., and Cepuliene, R. 2010. The influence of organic mulches on soil biological properties. *Zemdirbyste Agriculture*97(2):33–40.

Kohler, J., Caravaca, F., Carrasco, L., and Rolda, A. 2006. Contribution of pseudomonas mendocina and glomusintraradices to aggregate stabilization and promotion of biological fertility in rhizosphere soil of lettuce plants under field conditions. *Soil Use and Management*22:298–304.

Kumar, A., Sharma, S., and Mishra, S. 2009. Effect of alkalinity on growth performance of Jatrophacurcas inoculated with PGPR and AM fungi. *Journal of Phytology*1:177–184.

Kumar, P., and Sharma, P. K. 2020. Soil salinity and food security in India. *Frontiers in Sustainable Food Systems*4:533781. https://doi.org/10.3389/fsufs.2020.533781.

Lakhdar, A., Rabhi, M., Ghnaya, T., Montemurro, F., Jedidi, N., and Abdelly, C. 2009. Effectiveness of compost use in salt-affected soil. *Journal of Hazardous Materials*171:29–37. https://doi.org/10.1016/j.jhazmat.2009.05.132.

Litalien, A., and Zeeb, B. 2020. Curing the earth: a review of anthropogenic soilsalinization and plant-based strategies for sustainable mitigation. *Science of Total Environment*698:134235. https://doi.org/10.1016/j.scitotenv.2019.134235.

Lynch, J. P., and Clair, S. B. S. 2004. Mineral stress: the missing link in understanding how global climate change will affect plants in real world soils. *Field Crops Research*90:101–115.https://doi.org/10.1016/j.fcr.2004.07.008.

Ma, B., and Gong, J. 2013. A meta-analysis of the publicly available bacterial and archaeal sequence diversity in saline soils. *World Journal of Microbiology and Biotechnology*29:2325–2334. https://doi.org/10.1007/s11274-013-1399-9.

Mandal, S., Raju, R., Kumar, A., Kumar, P., and Sharma, P. 2018. Current status of research, technology response and policy needs of salt-affected soils in India – A review. *Journal of Indian Society of Coastal Agricultural Research*36(2): 40–53.

Msimbira, L. A., and Smith, D. L. 2020. The roles of plant growth promoting microbes in enhancing plant tolerance to acidity and alkalinity stresses. *Frontiers in Sustainable Food Systems*4:106. https://doi.org/10.3389/fsufs.2020.00106.

Müller, D. B., Vogel, C., Bai, Y., and Vorholt, J. A. 2016. The plant microbiota: systems-level insights and perspectives. *Annual Review of Genetics*50:211–234.https://doi.org/10.1146/annurev-genet-120215-034952.

Munns, R., and Tester, M. 2008. Mechanisms of salinity tolerance. *Annual Review of Plant Biology*59:651–681.

Nadeem, S. M., Zahir, Z. A., Naveed, M., and Arshad, M. 2009. Rhizobacteria containing ACC-deaminase confer salt tolerance in maize grown on salt-affected fields. *Canadian Journal of. Microbiology*55:1302–1309.

Nehela, Y., Mazrou, Y. S. A., Alshaal, T., Rady, A. M. S., El-Sherif, A. M. A., Omara, A. E. D., Abd El-Monem, A. M., and Hafez, E. M. 2021. The integrated amendment of sodic-saline soils using biochar and plant growth-promoting rhizobacteria enhances maize (*Zea mays* L.) resilience to water salinity. *Plants*10:1960. https://doi.org/10.3390/plants10091960.

Ondrasek, G., Rengel, Z., Romic, D., and Savic, R. 2010. Environmental salinisation processes in agro-ecosystem of neretva river estuary. *Növénytermelés*59:223–226.

Ondrasek, G., Rengel, Z., and Veres, S. 2011. Soil salinisation and salt stress in crop production. In: Shanker, A., and Venkateswarlu, B. (ed.). *Abiotic Stress in Plants - Mechanisms and Adaptations*,pp. 171–190. InTech Open. ISBN: 978-953-307-394-1.http://dx.doi.org/10.5772/22248.

Patra, S. K. 2017. Saline-waterlogged lands of West Bengal – priorities and management. In:Patra, S.K., Saha, S., Bhattacharyya, K., Biswas, B. and Poddar, R. (eds.).*Crop and Water Productivity in Agriculture*, pp. 21–32.ICAR-BCKV. Kolkata: Renu Publishers.

Patra, S. K., and Banik, M. 2018. Bioremediation of waterlogging and soil salinity for sustainability of agriculture: problems and prospects. *International Journal of Multidisciplinary Research and Development*5(1):144–152.

Qadir, M., and Oster, J. D. 2004. Crop and irrigation management strategies for saline-sodic soils and waters aimed at environmentally sustainable agriculture. *Science of Total Environment*323:1–19

Qadir, M., Oster, J. D., Schubert, S., Noble, A. D., and Sahrawat, K. L. 2007. Phytoremediation of sodic and saline-sodic soils. *Advances in Agronomy*96:197–247.https://doi.org/10.1016/s0065-2113(07)96006-x.

Qi, L., and Yang, J. 2017. Microbial community composition regulates SOC decomposition response to forest conversion in a Chinese temperate forest. *Ecological Research*32:163–172.

Rietz, D. N., and Haynes, R. J. 2003. Effects of irrigation-induced salinity and sodicity on soil microbial activity. *Soil Biology and Biochemistry*35(6):845–854.

Rodríguez-Salazar, J., Suárez, R., Caballero-Mellado, J., and Iturriaga, G. 2009. Trehalose accumulation in Azospirillumbrasilense improves drought tolerance and biomass in maize plants. *FEMS Microbiology Letters*296:52–59.

Sadeghi, A., Karimi, E., Dahaji, P. A., Javid, M. G., Dalvand, Y., and Askari, H. 2012. Plant growth promoting activity of an auxin and siderophore producing isolate of Streptomyces under saline soil conditions. *World Journal of Microbiology and Biotechnology*28:1503–1509.

Satyanarayana, T., Raghukumar, C., and Shivaji, S. 2005. Extremophilic microbes: diversity and perspectives. *Current Science*89(1):78–90.

Seckin, B., Sekmen, A. H., and Türkan, I. 2009. An enhancing effect of exogenous mannitol on the anti- oxidant enzyme activities in roots of wheat under salt stress. *Journal of Plant Growth Regulation*28(1):12.

Selvakumar, G., Kim, K., Hu, S., and Sa, T. 2014. Effect of salinity on plants and the role of arbuscular mycorrhizal fungi and plant growth-promoting rhizobacteria in alleviation of salt stress. In: Ahmad P. and Wani M. R. (eds.) *Physiological Mechanisms and Adaptation Strategies in Plants Under Changing Environment*, pp 115–144. Singapore: Springer.

Shahbaz, M., and Ashraf, M. 2013. Improving salinity tolerance in cereals. *Critical Review in Plant Science*32:237–249.

Shahid, S. A., Zaman, M., and Heng, L. 2018. Salinity and sodicity adaptation and mitigation options. In:Zaman, M., Shahid, S. A., and Heng, L. (eds.). *Guideline for Salinity Assessment, Mitigation and Adaptation Using Nuclear and Related Techniques*, pp. 55–89. Vienna: International Atomic Energy Agency. https://doi.org/10.1007/978-3-319-96190-3.

Sharma, D. K., Chaudhari, S. K., and Singh, A. 2014. *CSSRI Vision 2050*.Karnal, India: Central Soil Salinity Research Institute.

Sharma, D. K., Dey, P., Gupta, S. K. and Sharma, P. C. 2011. *CSSRI Vision 2030*. Karnal:Central Soil Salinity Research Institute. Available online at: http://krishikosh.egranth.ac.in/handle/1/50338 (accessed December 26, 2019).

Sharma, D. K., Singh, A., Sharma, P. C., Dagar, J. C., and Chaudhari, S. K. 2016. Sustainable management of sodic soils for crop production: opportunities and challenges. *Journal of Soil Salinity and Water Quality*8(2):109–130.

Shen, C., Shi, Y., Fan, K., He, J.-S., Adams, J. M., Ge, Y., et al. 2019. Soil pH dominates elevational diversity pattern for bacteria in high elevation alkaline soils on the Tibetan Plateau. *FEMS Microbiology Ecology*95:fiz003.

Shi, D., and Sheng, Y2005. Effect of various salt–alkaline mixed stress conditions on sunflower seedlings and analysis of their stress factors. *Environmental and Experimental Botany*54:8–21.

Shrivastava, P., and Kumar, R. 2015. Soil salinity: a serious environmental issue and plant growth promoting bacteria as one of the tools for its alleviation. *Saudi Journal of Biological Sciences*22:123–131.https://doi.org/10.1016/j.sjbs.2014.12.001.

Shukla, P. S., Agarwal, P. K., and Jha, B. 2012. Improved salinity tolerance of *Arachishypogaea* (L.) by the Interaction of halotolerant plant-growth-promoting rhizobacteria. *Journal of Plant Growth Regulation*31:195–206.

Siddikee, M. A., Glick, B. R., Chauhan, P. S., Jong Yim, W., and Sa, T. 2011. Enhancement of growth and salt tolerance of red pepper seedlings (*Capsicum annuum* L.) by regulating stress ethylene synthesis with halotolerant bacteria containing 1-aminocyclopropane-1-carboxylic acid deaminase activity. *Plant Physiology and Biochemistry*49:427–434.https://doi.org/10.1016/j.plaphy.2011.01.015.

Singh, D., Singh, C. K., Singh, Y. P., Singh, V., Singh, R., Tomar, R. S. S., et al. 2018. Evaluation of cultivated and wild genotypes of lens species under alkalinity stress and their molecular collocation using microsatellite markers. *PLoS One*13:e0199933. https://doi.org/10.1371/journal.pone.0199933.

Singh, G. 2009. Salinity-related desertification and management strategies: indian experience. *Land Degradation and Development*20:367–385.https://doi.org/10.1002/ ldr.933.

Singh, K., Pandey, V. C., Singh, B., and Singh, R. R. 2012. Ecological restoration of degraded sodic lands through afforestation and cropping. *Ecological Engineering*43:70–80.https://doi.org/10.1016/j.ecoleng.2012.02.029.

Stavi, I. 2020. On-site use of plant litter and yard waste as mulch in gardening and landscaping systems. *Sustainability*12:7521. https://doi.org/10.3390/su12187521.

Stavi, I., Thevs, N., and Priori, S. 2021. Soil salinity and sodicity in drylands: a review of causes, effects, monitoring, and restoration measures. *Frontiers in Environmental Science* 9:712831. http://dx.doi.org/10.3389/fenvs.2021.712831.

Tan, L. V., and Thanh, T. 2021. The effects of salinity on changes in characteristics of soils collected in a saline region of the Mekong Delta, Vietnam. *Open Chemistry*19:471–480.https://doi.org/10.1515/chem-2021-0037.

Torbaghan, M. E., Lakzian, A., Astaraei, A. R., Fotovat, A., and Besharati, H. 2017. Salt and alkali stresses reduction in wheat by plant growth promoting haloalkaliphilic bacteria. *Journal of Soil Science and Plant Nutrition*17:1058–1087.https://doi.org/10.4067/s0718-95162017000400016.

Tripathi, S., Kumari, S., Chakraborty, A., Gupta, A., Chakrabarti, K., and Bandyapadhyay, B. K. 2006. Microbial biomass and its activities in salt-affected coastal soils. *Biology and Fertility of Soils*42(3):273–277.

Upadhyay, S. K., and Singh, D.P. 2015. Effect of salt-tolerant plant growth-promoting rhizobacteria on wheat plants and soil health in a saline environment. *Plant Biology*17:288–293.

Vaishnav, A., Kumari, S., Jain, S., Varma, A., Tuteja, N., and Choudhary, D. K. 2016. PGPR-mediated expression of salt tolerance gene in soybean through volatiles under sodium nitroprusside. *Journal of Basic Microbiology*56:1274–1288.

Wakeel, A. 2013. Potassium–sodium interactions in soil and plant under saline-sodic conditions. *Journal of Plant Nutrition and Soil Science*176(3):344–354.

Wakeel, A., Farooq, M., Qadir, M., and Schubert, S. 2011. Potassium substitution by sodium in plants. *Critical reviews in plant sciences*30(4):401–413.

Weil, R.R., and Brady, N.C. 2017. Soils of dry regions: alkalinity, salinity, and sodicity. *The Nature and Properties of Soils*, 15th ed., pp. 438–481. Upper Saddle River, NJ: Pearson.

Wen-wen, Z., Chong, W., Rui, X., and Li-jie, W. 2019. Effects of salinity on the soil microbial community and soil fertility. *Journal of Integrative Agriculture*18(6):1360–1368.https://doi.org/10.1016/S2095-3119(18)62077-5.

Win, K. T., and Oo, A. Z. 2017. Salt-stress-induced changes in protein profiles in two blackgram (*VignaMungo L.*) varieties differing salinity tolerance. *Advances in Plants and Agriculture Research*7(1):206–210. https://doi.org/10.15406/apar.2017.07.00239.

Wong, V. N. L., Greene, R. S. B., Dalal, R. C., and Murphy, B. W., 2010. Soil carbon dynamics in saline and sodic soils: a review. *Soil Use and Management*26(1):2–11. https://doi.org/10.1111/j.1475-2743.2009.00251.x.

Xu, W., Jia, L., Shi, W., Baluška, F., Kronzucker, H. J., Liang, J., et al. 2013. The tomato 14-3-3 protein TFT4 modulates H+ efflux, basipetal auxin transport, and the PKS5-J3 pathway in the root growth response to alkaline stress. *Plant Physiology*163:1817–1828.https://doi.org/ 10.1104/pp.113.224758.

Yang, Z., Li, J.-L., Liu, L.-N., Xie, Q., and Sui, N. 2020. Photosynthetic regulation under salt stress and salt-tolerance mechanism of sweet sorghum. *Frontiers in Plant Science*10:1722. https://doi.org/10.3389/fpls.2019.01722.

Yao, L., Wu, Z., Zheng, Y., Kaleem, I., and Li, C. 2010. Growth promotion and protection against salt stress by Pseudomonas putida Rs-198 on cotton. *European Journal of Soil Biology*46:49–54.

Yasmin, H., Nosheen, A., Naz, R., Bano, A., and Keyani, R. 2017. L-tryptophan-assisted PGPR-mediated induction of drought tolerance in maize (*Zea mays* L.). *Journal of Plant Interactions*12:567–578.

Yousef, N. M. 2018. Capability of plant growth-promoting rhizobacteria (PGPR) for producing indole acetic acid (IAA) under extreme conditions. *European Journal of Biological Research*8:174–182.

Yuan, B. C., Li, Z. Z., Liu, H., Gao, M., and Zhang, Y. Y. 2007. Microbial biomass and activity in salt affected soils under arid conditions. *Applied Soil Ecology*35(2):319–328.

Zhalnina, K., Dias, R., De Quadros, P. D., Davis-Richardson, A., Camargo, F. A., Clark, I. M., et al. 2015. Soil pH determines microbial diversity and composition in the park grass experiment. *Microbial Ecology*69:395–406. https://doi.org/10.1007/s00248-014-0530-2.

8 Reactive Oxygen Species and Alternative Oxidase in Multiple Stress Tolerance

Crystal Sweetman and Kathleen L. Soole
Flinders University

CONTENTS

8.1 INTRODUCTION

Plants have adapted numerous coping mechanisms to withstand abiotic stresses. When categorised broadly, many of these mechanisms overlap between species and across different types of stress. These "general responses" could provide avenues for improving crop multi-stress tolerance (Pandey et al., 2015). Manipulation of some of these mechanisms in transgenic plants can alter resilience to a certain stressor. In some cases, the plants exhibit improved resilience to multiple, separately applied stressors. For example, overexpression of an antioxidant enzyme in rice led to increased tolerance to individual drought, salt and cold stresses (Zhang et al., 2013). However, concurrent exposure of a

DOI: 10.1201/9781003300564-8

plant to more than one stress is rarely tested. That is, few studies have explored whether manipulating general stress mechanisms can elicit tolerance to multiple simultaneous stresses.

One reason for this gap is the increased complexity of experiments with multiple simultaneous stresses. For example, an imbalance in the severity of stresses, or the presence of a dominant stressor, may result in a response largely resembling that of the more dominant/severe stress. On the other hand, a combination of stressors may be recognised by the plant as a novel stress, rather than a variation on one type of stress. Stress responses can also differ with the sequence of stress events, plant age, tolerance level (i.e., different cultivars or species) and stress history (an individual may have already adapted to a particular stress).

Common biochemical responses of plants exposed to different stresses can be divided into broad categories, including modification of the cuticle, altered membrane fatty acid composition, up-regulation of molecular chaperones, augmentation of reactive species scavengers and accumulation of compatible solutes (He et al., 2018). As an example of the latter, the osmotic effects incurred by drought, salinity and cold temperature stresses typically lead to the accumulation of osmoprotectants (Chinnusamy et al., 2004).

Common physiological responses include the adjustment of stomatal aperture, transport processes and various photosynthetic parameters. For example, both drought and salinity incur a water deficit that leads to stomatal closure, consequently restricting CO_2 availability and this may limit photosynthesis. Meanwhile, heat and salinity both affect transport and compartmentation of ions but via different mechanisms (Pandey et al., 2015).

Many abiotic stresses including drought, high light, heat, cold, salinity, waterlogging, metal toxicity, nutrient imbalance and ozone exposure also invoke diverse secondary stresses, but each are unified in their elicitation of oxidative stress (Sachdev et al., 2021). Therefore, mechanisms that minimise oxidative stress are promising targets for engineering plant multi-stress resistance (He et al., 2018).

8.2 OXIDATIVE STRESS AND REACTIVE OXYGEN SPECIES

Oxidative stress occurs when reactive oxygen species (ROS), including singlet oxygen ($^1O_2$), the superoxide anion (O_2^-), hydrogen peroxide (H_2O_2) and the hydroxyl radical ($^{\bullet}OH$), accumulate to a level that overwhelms ROS detoxification systems (Thannickal and Fanburg, 2000). Excessive ROS can lead to cellular damage due to their ability to react, in the presence of heavy metals, with almost all cellular components. ROS accumulation in the apoplast, for example during exposure to ozone, can lead to the formation of lesions. Cellular and organellar membranes exposed to ROS will form lipid peroxidation products. Lipid-derived radicals are also highly reactive and further promote oxidative stress. DNA adducts, DNA strand breakage, oxidation of purines, protein modifications (e.g., sulfonylation, carbonylation, glutathionylation and *S*-nitrosylation) and formation of protein–DNA cross links can all result from reactions between ROS, proteins and DNA (Van Breusegem and Dat, 2006). Some of these products, such as protein carbonylation products and the lipid peroxidation product malondialdehyde, can be used as biochemical indicators of oxidative stress. Ultimately the cumulative effects of ROS-related damage can lead to cell lysis and death.

Under normal cellular conditions ROS accumulate at basal levels and are beneficial to the cell, often acting as signals for developmental or regulatory processes (Foyer and Noctor, 2003; Mittler et al., 2011). More recently, it has been proposed that the stress-induced overaccumulation of ROS and ROS-related damage (e.g., enzyme inactivation) are also beneficial to the cell; participating in signal networks to promote cellular and physiological responses and improve tolerance (Foyer et al., 2017; Mittler, 2017). Thereby, ROS accumulation is seen as a positive cellular response that promotes cell survival or programmed cell death as necessary, rather than a destructive path that leads directly to damage and necrosis. Consistent with these roles, ROS interact with hormones involved in both developmental and stress processes.

ROS levels are dictated by the balance between generation rate and the capacity for their processing, or "scavenging". Therefore, increased accumulation of ROS may be caused by enhanced production and/or impaired scavenging systems, both of which can occur via numerous pathways in different cellular locations depending on the type of damage or stress.

8.2.1 Generation of ROS

ROS are generated at several cellular and tissue locations via enzymatic and non-enzymatic mechanisms. Some are inherent products of integral processes, such as photosynthesis and cellular respiration. Ubiquitous to eukaryotes, mitochondrial respiration results in electron leakage to O_2 (see Section 5.2) and the resultant O_2^- is rapidly converted to H_2O_2. In plant photosynthetic tissues, ROS are generated at higher rates due to electron leakage at the photosynthetic apparatus in the chloroplast during light exposure (see Section 3.1), as well as in the peroxisomes during the metabolism of glycolate from photorespiration, which generates H_2O_2 (Foyer and Noctor, 2003).

Other notable sites of ROS generation within the cell include the ER, plasma membrane, apoplast and the cell wall, each with important functions, particularly as developmental and environmental cues (Karkonen and Kuchitsu, 2015; Yoboue et al., 2018). NADPH oxidase utilises cytosolic NADPH to generate superoxide in the apoplast, which is rapidly converted to H_2O_2 either spontaneously or catalysed by superoxide dismutase. These ROS are important for transmitting information between cells. ROS signals are auto-propagating and have been seen to transmit through the plant at a rate of 8 cm per minute (Miller et al., 2009). This process occurs as a "ROS-induced ROS release (RIRR)", by which ROS released at one location leads to an increase in the rate of generation and release of ROS at other locations, both intra- and intercellular. RIRR was initially described in animal cells, whereby accumulation of mitochondrial ROS could trigger a collapse in the membrane potential and the release of ROS, signalling further ROS release at other organelles. Regular liberation of mitochondrial ROS could act as a safety valve to prevent harmful local accumulation of ROS, as well as notify nearby organelles of normal, or perturbed mitochondrial function. In plants, RIRR has been more closely associated with cell-to-cell signalling, but some studies suggest that organelle-derived ROS could trigger ROS generation at the plasma membrane by NADPH oxidase activity (Zandalinas and Mittler, 2018).

8.2.2 ROS Processing

ROS levels can be down-regulated by either capping their generation or enhancing ROS-processing activities (or both). A suite of enzymes and antioxidant compounds make up the ROS-processing systems, which differ between cellular compartments and tissues but each are critical to the regulation of ROS levels in plants.

Non-enzymatic antioxidants include (1) ascorbate, or vitamin C, which is a highly abundant organic acid that can donate electrons to a wide range of molecules including hydrogen peroxide, hydroxyl radicals, superoxide and tocopheroxyl radical (to regenerate tocopherol), either spontaneously or catalysed by enzymes; (2) GSH, or reduced glutathione, which can also donate electrons to a range of molecules including hydrogen peroxide, hydroxyl radicals, superoxide and singlet oxygen and is located in almost all cellular compartments and has a very diverse set of functions, not only in ROS processing, (3) tocopherols, which are lipophilic antioxidant compounds essential for processing ROS and the prevention of lipid peroxidation in the thylakoid membrane, in the chloroplast, (4) xanthophylls, B-carotene, zeaxanthin and other carotenoids, which accumulate in and around the photosynthetic apparatus and provide photoprotection to the photosynthetic apparatus: an essential role, because damage to the photosynthetic protein complexes can cause release of their metal cofactors, which will rapidly amplify ROS production and cellular damage, and (5) flavonoids, including flavanols, flavones, isoflavones and anthocyanins, which are specialised secondary metabolites that can also process ROS to some degree (Das and Roychoudhury, 2014).

ROS processing can be facilitated by a series of enzymes, including (1) superoxide dismutases (SODs), which are a set of metalloproteins that convert superoxide to hydrogen peroxide, ensuring that superoxide is not available to generate hydroxyl radicals; (2) catalase, which detoxifies hydrogen peroxide to water almost exclusively in the peroxisome; (3) ascorbate peroxidase, which makes part of the ascorbate–glutathione antioxidant system, is present in almost all cell compartments and catalyses the reduction of hydrogen peroxide to water, using ascorbate as a reducing agent; (4) glutathione peroxidase, which also converts hydrogen peroxide to water but uses GSH as an electron donor and is present in almost all cellular compartments; (5) monodehydroascorbate reductase (MDHAR), which can regenerate ascorbate from its partially oxidised form, monodehydroascorbate, using electrons from NADPH; (6) dehydroascorbate reductase (DHAR), which also regenerates ascorbate but from the fully oxidised form, dehydroascorbate, utilising reduced glutathione GSH, as an electron donor and maintaining a reduced ascorbate pool in the cell and apoplast; and (7) glutathione reductase, which is required to regenerate GSH from oxidised glutathione, using NADPH as a reducing agent (Das and Roychoudhury, 2014).

8.3 STRESS-SPECIFIC MECHANISMS OF ROS ACCUMULATION

ROS accumulate when metabolism is perturbed, for example in response to an abiotic stress. All abiotic stress conditions have the potential to induce ROS accumulation (Pandey et al., 2015; Sachdev et al., 2021), and while the mechanism of ROS accumulation depends on the type of stress, there are some common features. Some stresses and their ROS accumulation mechanisms are outlined below.

8.3.1 High Light

Light energy, in the form of photons, is required for the "light reactions" of photosynthesis. Photons are absorbed by chlorophyll and utilised by photosystem II (PSII) to energise electrons, which originate from the oxidation of water. These electrons are then passed to plastoquinone, cytochrome b6f and photosystem I (PSI). Another photon is harnessed by PSI, which energises the electron further, sufficient to generate NADPH from NADP. Meanwhile, a proton is pumped across the membrane by action of the cytochrome b6f, generating an electrochemical gradient that is used to drive ATP synthesis via ATP synthase. During these processes, ROS generation can occur.

As light intensity increases, photon availability can surpass the capacity of the photosystems, and energy absorbed by chlorophyll cannot be fully utilised by the PS reaction centres. Thus, limitation of electron transport occurs at the electron acceptor side of PSII and the plastoquinone pool becomes over-reduced. Back transfer of electrons may then result in the formation of singlet chlorophyll, which reacts with another chlorophyll to form highly reactive triplet chlorophyll, which further reacts with O_2 to form singlet oxygen ($^1O_2$). Superoxide can also be formed via electron leakage at the PSII electron acceptor side, while hydrogen peroxide is formed via incomplete water oxidation at the PSII electron donor side (Pospisil, 2016).

PSII is susceptible to damage, necessitating high turnover of components of this enzyme complex to maintain function. Excess light and the accumulation of singlet oxygen can damage the photosystems, or more specifically, damage the repair systems of PSII (Takahashi and Badger, 2011). Damaged photosynthetic machinery can perpetuate ROS production and overwhelm the ROS-processing systems, further affecting photosynthesis and plant growth. This mechanism of impaired photosynthesis is known as photoinhibition.

8.3.2 Drought

Drought-limited photosynthesis is a common cause of ROS production during stress (Laxa et al., 2019). However, water limitation is first detected by the roots. ROS generation at the root tip is involved in the release of abscisic acid, which is transmitted as a stress signal to the leaves.

Here, abscisic acid leads to stomatal closure, which in turn limits CO_2 accessibility for the dark reactions of photosynthesis. Without the Calvin–Benson cycle, reducing equivalents from the light reactions (NADPH) accumulate, while $NADP^+$ becomes limiting. This stalls photosynthetic electron transport, leading to overreduction of PSII. The use of O_2 as an alternative electron acceptor leads to generation of superoxide, O_2^-, as a by-product. With an extended duration and/or severity of drought, chloroplast structure and function deteriorate and this further impacts photosynthesis.

Limited CO_2 availability during drought stress also leads to the oxygenation of RuBisCO and photorespiration, thus generation of H_2O_2 in the peroxisome. The highly reduced NADPH pool stimulates the malate shuttle to transport reducing equivalents from the chloroplast to the cytosol and mitochondrion. Meanwhile, excess glycine generated in the peroxisome during photorespiration is also imported into the mitochondria, where it is metabolised to serine and generates NADH (Douce et al., 2001). The influx of reducing equivalents from chloroplast and peroxisome into the mitochondria can be oxidised via the mitochondrial electron transport chain (mETC). However, ROS are also generated here, particularly when the mETC becomes over-reduced due to excessive electron influx or limitation by respiratory complexes or other respiratory components (i.e., adenylate control, see Section 5.2).

8.3.3 Heat

One of the primary effects of heat stress is an increase in membrane fluidity. As a result, membrane-bound processes such as photosynthesis and mitochondrial respiration are greatly affected (Suzuki and Mittler, 2006). Due to the involvement of high-energy electron transfer in these processes, and local availability of O_2 (especially if CO_2 becomes limiting), impairment of photosynthetic and mETC components leads to overreduction, electron leakage and the generation of ROS. During high temperatures, singlet oxygen can be produced in the dark via lipid peroxidation, which is catalysed by lipoxygenase. Meanwhile, incomplete water oxidation can lead to hydrogen peroxide formation, which can then form a hydroxyl radical in the presence of manganese (Pospisil, 2016). In addition, the plasma membrane NADPH oxidase, RBOH, generates ROS signals that are important for tolerance to heat stress (Larkindale et al., 2005).

In mammals, the production of superoxide in response to heat stress has been proposed to occur via reduction of O_2 by mitochondrial semi-quinones in the mETC; thermal denaturation of FAD-containing redox proteins, thus exposing the FAD to autooxidation by NADH; a decrease in SOD protein and enzyme activity; and an increase in iron release from ferritin at higher temperatures (Slimen et al., 2014). Similar mechanisms may occur in plants but more research is required in this area.

8.3.4 Cold

ROS also accumulate in low temperature stress (O'Kane et al., 1996). One of the first responses of a plant to chilling stress is increased membrane rigidity (Ding et al., 2019). This may affect functionality of membrane-bound enzymes such as mitochondrial respiration and photosynthesis protein complexes. Furthermore, cold treatments have been seen to invoke thylakoid loosening and degradation of chloroplast ultrastructures, thus impairing photosynthetic machinery (Zhang et al., 2020). Stomatal closure is another common response to chilling stress (Ding et al., 2019), which may lead to CO_2 limitation. Slowing of photosynthesis but continued light exposure leads to ROS generation, especially at PSII (as described in Section 3.1).

Decreased activity of ROS-processing enzymes at temperatures below their thermal optima may also impair the plant's ability to manage ROS, although overexpression of antioxidant enzymes such as APX, Cu/Zn-SOD and GRX have been shown to improve tolerance to cold stress and reduce ROS levels during cold stress (Dreyer and Dietz, 2018; You and Chan, 2015). In addition, cold stress leads to a decrease in the rate of mitochondrial respiration through the classical mETC, resulting in ROS accumulation (Prasad et al., 1994).

In the root, programmed cell death of columella stem cells is utilised as a protective mechanism against cold stress (Hong et al., 2017). Programmed cell death typically involves a ROS burst; therefore, root vasculature is also likely an important site for ROS metabolism during cold stress.

8.3.5 Salinity

The response of plants to high concentrations of sodium chloride in the growth medium varies between species and cultivars and over time, but can generally be classified in two ways: (1) avoidance, or exclusion, whereby Na^+ and Cl^- are actively sequestered or excreted, at great energetic cost to the plant or (2) tolerance, whereby the plant adapts to life with high concentrations of Na^+ and Cl^- (Munns and Tester, 2008).

Photosynthesis during salinity stress is greatly affected by ionic imbalance, stomatal conductance and oxidative damage to photosynthetic apparatus. Stomatal closure is a typical response to excessive salt exposure, due to osmotic stress and water uptake limitations. As CO_2 becomes limiting for the Calvin cycle, (1) the presence of light will exceed that which can be utilised by PSI and II, causing the production of ROS (see Section 3.1) and (2) the rate of photorespiration will increase, also increasing ROS (Bose et al., 2014).

Salinity stress can affect thylakoid stacking in chloroplasts, thus decreasing the activity of PSII, which congregates in these stacks, and Na^+ and/or Cl^- may affect the surface charge of thylakoids, which could affect PSI activity and the electrochemical gradient used for ATP synthesis. Salt stress and accumulation of Na^+ and Cl^- also leads to a loss of K^+ in the cytosol and chloroplast, which can further decrease photosynthesis rates, while saturation of the Cl^-binding sites in PSI has also been proposed to lead to ROS production during salinity stress (Bose et al., 2017).

Mitochondrial respiration can also be responsible for ROS generation during salinity stress. This relates to overreduction of the mitochondrial electron transport chain, as overexpression of either alternative oxidase or uncoupling protein led to the improvement of salinity tolerance and a reduction in ROS levels in plants (Begcy et al., 2011; Smith et al., 2009). These enzymes embed within the mitochondrial inner membrane and can release feedback inhibition on electron transfer activities, which could prevent overreduction of the chain (see Section 5.2).

During salinity stress, plants can secrete spermidine, which can be oxidised by polyamine oxidase to generate H_2O_2 in the apoplast. NADPH oxidase is also enlisted to generate H_2O_2 in the apoplast in response to salinity stress, potentially triggering the activity of antioxidant enzymes, peroxidase and catalase, within the cell (Liu et al., 2021).

8.3.6 Nutrient Deficiency

Oxidative stress is a common effect of macronutrient and micronutrient deficiencies, although the mechanisms and regulation can differ. Shared effects include (1) activation of ROS production at the apoplast via NADPH oxidases (RBOH proteins) and/or type III peroxidases, which are particularly important for root and root hair growth (e.g., nitrogen, phosphorous, potassium and magnesium); (2) down-regulation of photosynthetic and respiratory electron transport due to the lack of enzyme cofactors (e.g., nitrogen, phosphorous, sulphur, iron, manganese, copper); (3) overreduction of photosynthetic electron transport components and chloroplast NADPH pools due to either stomatal closure (e.g., nitrogen, potassium and zinc deficiency) or decreased efficiency of the Calvin cycle and RuBisCO (e.g., magnesium and sulphur deficiency); and (4) impaired antioxidant defence systems (e.g. sulphur, iron, zinc, copper and manganese) (Tewari et al., 2021; Hajiboland, 2012).

The most common mechanism for oxidative stress in response to micronutrient deficiency is via impaired ROS-processing capacities. Iron, zinc, copper and manganese are essential for functionality of different superoxide dismutase isoforms. Typically, Fe-SOD is localised to the chloroplast, Mn-SOD to mitochondria and Cu/Zn-SOD to the chloroplast and cytosol. Deficiencies of any of these micronutrients will impair the antioxidant response of particular cellular compartments during stress, although a lack of Cu/Zn-SOD may be complemented by up-regulation of Fe-SOD. Zinc is also important for synthesis of glutathione reductase and ascorbate peroxidase, even though it is not a cofactor for these, while iron is also necessary for the haem groups of catalase and peroxidases (Hajiboland, 2012).

Nutrient deficiencies can cause limitations to mitochondrial respiration. Iron is necessary for Fe-S and haem groups within mitochondrial electron transport chain components and copper is a cofactor for respiratory Complex IV. Therefore, deficiencies in these minerals can also lead to interruption of cellular respiration and potentially ROS production within mitochondria.

8.3.7 Hypoxia and Recovery

Hypoxia can occur during flooding events or heavy rainfall and waterlogging, leading to localised O_2 limitation, particularly for roots but in some instances an entire plant may become submerged. To survive this kind of stress, the plant may utilise biochemical, morphological and physiological adaptations to maintain ATP production in the absence of O_2 or increase the supply of O_2 to essential tissues (Pucciariello and Perata, 2021).

Paradoxically, hypoxia or even anoxia triggers an oxidative burst despite the low availability of O_2. The oxidative burst is generated by NADPH oxidase activity at the plasma membrane and can trigger further responses to the stress, for example, programmed cell death of root cortical cells. The latter is a critical step in the formation of aerenchyma, which can improve air circulation through the submerged roots and therefore this ROS response is positive for plant tolerance (Yamauchi et al., 2017). Emergence of adventitious roots is another physiological response to hypoxia.

Mitochondrial respiration may also be a source of ROS during hypoxia. This pathway requires O_2 directly, for the process of ATP formation. Under limiting O_2, the inhibition of mitochondrial respiration could cause overreduction of the electron transport chain and ROS generation.

The removal of a plant from hypoxia, for example, after a flood subsides, elicits a new stress. The sudden return of O_2 (and light, if it had been blocked during submergence) leads to ROS production. Again, the formation of ROS via RBOH appears to be imperative to plant survival, while ROS overaccumulation is detrimental. Therefore, the presence of ROS-generating, ROS-processing and ROS-prevention pathways are essential to maintaining homeostatic control (Pucciariello and Perata, 2021).

8.3.8 Extended Darkness

Senescence is a common response to extended darkness in plants. It is triggered by ROS that may originate primarily from the mitochondria and peroxisome, but upon the loss of photosynthetic capacity, will be amplified in the chloroplast. It occurs in response to sucrose starvation signals and can be delayed by application of ascorbate. Numerous ROS-responsive transcription factors are also involved, suggesting that it is a tightly controlled process. However, once the cellular membranes begin to deteriorate, ROS are auto-propagated through lipid peroxidation (Rogers and Munne-Bosch, 2016; Rosenwasser et al., 2011).

8.3.9 Ozone

Ozone enters the leaf via stomata and reaches the apoplast, where much of the reactive molecule may be reduced via the apo-plastic pool of ascorbate. Some may be rapidly converted to ROS that could either diffuse across the membrane or cause membrane lipid peroxidation. The signalling mechanism for acute ozone exposure is not known, though the cell response is analogous to a hypersensitive response. There is an initial oxidative burst most likely driven by NADPH oxidases and programmed cell death may be initiated (Baier et al., 2005). The photosynthetic electron transport system can become over-reduced due to ozone-induced stomatal closure and impaired Calvin cycle.

Mitochondrial respiration increases as a mechanism to move excess reducing equivalents from the chloroplast, and mitochondrial ROS production increases as a result (Gandin et al., 2021).

8.3.10 Metal Toxicity

Most heavy metal exposure occurs in the root. The plant deploys avoidance tactics to prevent uptake, chelate metal ions by phytochelatins and metallothionein proteins or sequester them at the cell wall, trichome or inside the vacuole. However, upon entry to the cell, ROS production from free metals such as iron, copper, zinc, lead, cadmium, chromium, nickel, arsenic and mercury can lead to significant oxidative stress, invoking activity of ROS-processing enzymes (Dutta et al., 2018).

Redox-active metals such as iron and copper can catalyse the Haber–Weiss/Fenton reactions, converting H_2O_2 to the more reactive OH^-. Redox-inactive metals instead bind sulfhydryl groups on proteins. As a result, they may displace the correct metal ion, which may lead to ROS and they also inhibit enzyme activity, again leading to ROS, depending on the enzyme. For example, displacement of Zn from Cu/Zn-SOD may inactivate the enzyme, causing a shortage in the cell's ROS-scavenging capacity. Heavy metals can also consume chemical antioxidants such as glutathione, preventing them from ROS scavenging. The presence of metals inside the cell can also trigger ROS production via NADPH oxidase and this is likely a signalling response (Shahid et al., 2014).

8.4 OXIDATIVE STRESS AS A RESULT OF COMBINED STRESSES

Oxidative stress is a common response to abiotic stress. Each individual stress described in the previous section led to increased accumulation of ROS. Furthermore, while ROS accumulation may have been triggered via different mechanisms, the amplification of ROS often ultimately occurs via impaired photosynthesis. In a recent study, increasing plurality of abiotic stress combinations from 2 to 6 led to a consistent increase in ROS levels in *Arabidopsis* plants along with decreased plant growth and survival (Zandalinas et al., 2021). From this it can be hypothesised that many combinations of stresses will lead to oxidative stress (Figure 8.1). However, the mechanisms by which ROS are generated and the cellular locations may differ. For example, ROS generated during drought stress alone are mostly derived from chloroplast and mitochondria due to limitations on photosynthesis and the effects of photorespiration. However, drought combined with heat seems to shift the damaging effects of ROS to the cytosol (Koussevitzky et al., 2008).

The most common mechanism for the induction of oxidative stress highlighted throughout Section 8.3 was the damage and loss of photosynthetic components, leading to excessive ROS generation. Photoinhibition can be partially stymied by photorespiration, but excess reducing equivalents must be exported from the chloroplast and oxidised to enable metabolism to continue and prevent further amplification of ROS.

Particular combinations of abiotic stress have been postulated to worsen crop productivity, including **drought** plus high light, heat, UV light or nutrient stress, **salinity** plus heat, nutrient stress or ozone, **heat** plus high light or ozone, **chilling** plus high light (Mittler and Blumwald, 2010). Such combination stresses are likely to generate high levels of chloroplast ROS due to photosynthetic damage, and there are studies to show this. Some examples are given below.

Example 1: High light intensity simultaneous to cold or drought can overwhelm the photosynthetic capacity of the plant, exacerbating ROS generation and leading to greater cell damage. Furthermore, overexpressing antioxidant enzymes such as Cu/Zn-SOD, ascorbate peroxidase, glutathione reductase and glutathione peroxidase can improve tolerance to combined temperature and high light stresses, while mutants deficient in antioxidant compounds such as ascorbate or glutathione are less able to survive such conditions (Choi et al., 2002; Suzuki and Mittler, 2006).

Example 2: Combined heat and drought stress shows greater oxidative stress (measured as the increase in lipid peroxidation product, malondialdehyde) compared to either stress alone, in a

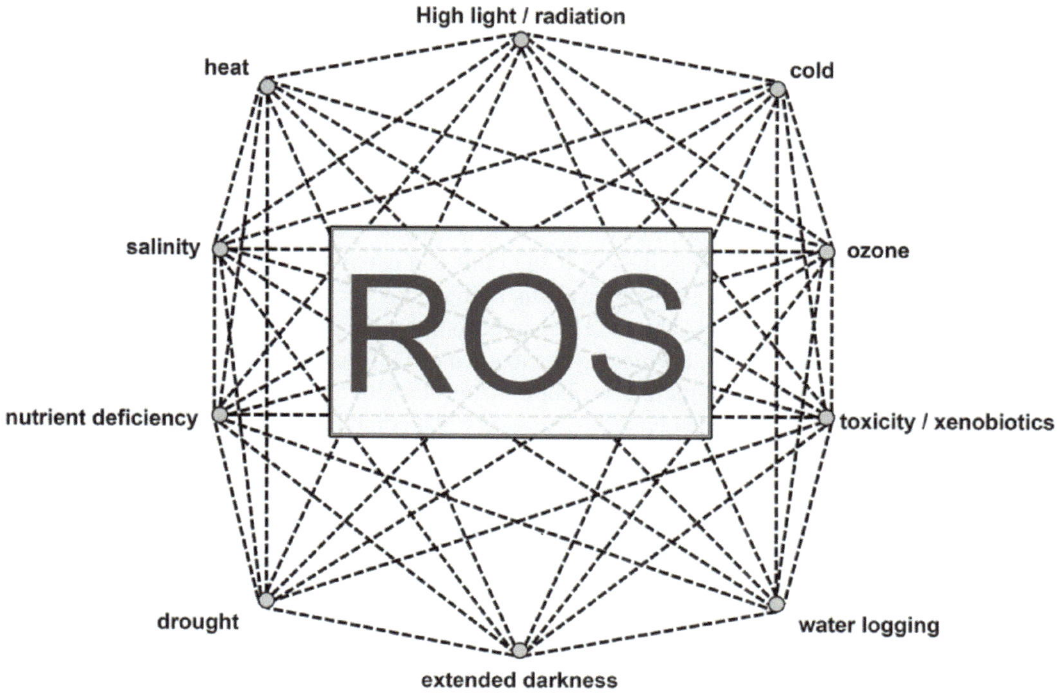

FIGURE 8.1 Graphical representation of abiotic stresses and their elicitation of ROS accumulation. Nodes represent individual stresses, each of which can cause increased ROS levels. Connections indicate combinations of stresses, which together, may maintain or further amplify ROS levels.

range of plants from monocots such as maize, to citrus, and even succulents such as purslane. This is likely because the physiological responses to drought and heat are at odds with each other. The stomatal response during the combined stress favours that of drought – closing to conserve water. This restricts the use of evaporative cooling. Meanwhile, dark respiration increases, favouring the heat response but utilising carbon despite impaired photosynthesis (Rizhsky et al., 2002; Pandey et al., 2015; Laxa et al., 2019).

Example 3: Nutrient deficiencies that lead to a metabolic or energetic deficit can debilitate high-cost defences associated with other stresses such as salinity stress, thus overcoming the ability of the plant to extrude or sequester Na^+ and Cl^- (Ahanger et al., 2019). Plants that are deficient in micronutrients such as iron, zinc, copper and manganese may limit the activity of ROS scavengers or other redox reactions (e.g., SODs), making the plant more susceptible to ROS generated upon exposure to other abiotic stresses such as high light, cold or drought. Meanwhile, the photoinhibitory effects of some macronutrient and micronutrient deficiencies also lead to light sensitivity (e.g., potassium, magnesium, iron and manganese) (Choudhury et al., 2017; Hajiboland, 2012).

Not all combination stresses will result in cumulative oxidative stress. In fact, some combinations may counter it. For example, in *Medicago truncatula* the increase in ROS levels observed upon drought or ozone stress was abolished when the two stresses were applied in combination (Iyer et al., 2013). This was likely due to early stomatal closure, a phenomenon seen on multiple occasions in the presence of ozone or high CO_2 (Laxa et al., 2019), although increased levels of ascorbate and glutathione were also observed in the combined treatment plants. Likewise, combinations of salinity and heat stress can (but not always; Suzuki et al., 2016) improve tolerance relative to salinity stress alone, exemplified in tomato and peanut plants with increased stomatal opening, increased compatible osmolytes and decreased levels of ROS and lipid peroxidation products (Colmenero-Flores and Rosales, 2014).

8.5 STRESS, MITOCHONDRIA AND ALTERNATIVE RESPIRATION

Removing excess reducing equivalents from the chloroplast appears to be a common objective in the cellular response to many of the individual and combined stresses discussed in previous sections. It is central to preventing further ROS production and enabling photosynthesis to continue. The plant mitochondrial electron transport chain (mETC) is equipped to utilise reducing equivalents from both within the mitochondrial matrix and other cellular compartments (Van Aken, 2021). The shuttling of reducing equivalents into the mitochondria (e.g., via the malate valve) (Selinski and Scheibe, 2019) increases the input of electrons into the mETC and this can occur safely, as long as there is a clear path for the electrons to traverse and exit the chain, ultimately reducing O_2 to water. However, during abiotic stress some components of the mETC may become limiting, leading to overreduction, electron leakage and ROS formation. Fortunately, there are mechanisms for minimising ROS production at the plant mETC, of which the main contributor is the alternative pathway (AP) of respiration. An increase in the expression of mitochondrial AP genes is a common molecular response to abiotic stresses in plants. The presence of these enzymes can also be critical to plant stress responses.

8.5.1 THE MITOCHONDRIAL ELECTRON TRANSPORT CHAIN OF PLANTS

In plants, AP enzymes co-exist with the "classical" components of the mETC, that is, Complexes I, II, III and IV, as well as electron carriers' ubiquinone and cytochrome c. A schematic depicting the mETC of plants in the presence or absence of AP enzymes is given in Figure 8.2.

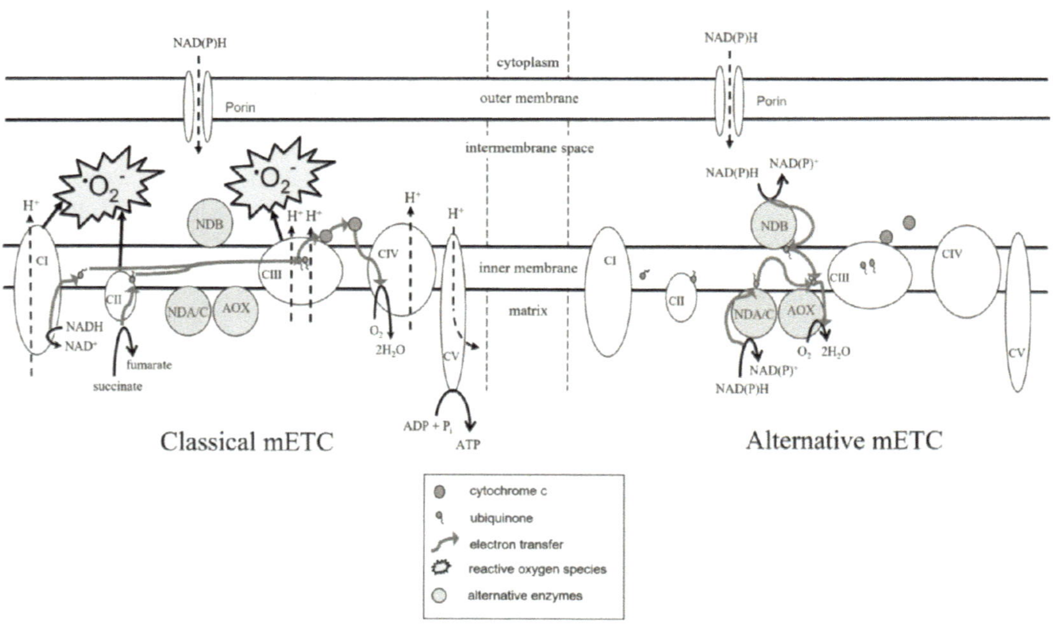

FIGURE 8.2 Mitochondrial electron transport chain components and electron movements. CI–CIV, Complex I–Complex IV of the classical mETC. CV, Complex V (ATP synthase). Solid arrows indicate electron movement from NAD(P)H or succinate, through the mETC complexes or alternative respiration enzymes. Energy from electron transport through the classical mETC is harnessed for proton pumping at Complexes I, III and IV, generating a proton gradient that drives ATP synthesis via ATP synthase. ROS generation can also occur at Complexes I, II and III. Electron transport through the alternative mETC does not lead to proton pumping, nor ROS synthesis. This schematic shows two extreme scenarios of electron transport, either entirely through the classical mETC or entirely through the alternative mETC; however, in reality electrons may transfer through combinations of classical and alternative mETC components.

The classical mETC complexes utilise energy from electron transfer events to pump protons from the mitochondrial matrix to the intermembrane space. This forms an electrochemical gradient, or "proton motive force" that drives ATP synthesis. Therefore, more efficient proton pumping will lead to more efficient ATP synthesis (setting aside the potential for proton leak back into the matrix, e.g., via uncoupling protein).

Electrons are donated to the mETC from NADH or succinate (via $FADH_2$) through the activities of Complex I and Complex II, respectively; therefore, these complexes can be considered as electron entry points. They both donate their electrons to ubiquinone, which thus becomes ubiquinol. NADH oxidation by Complex I leads to proton pumping, while $FADH_2$ oxidation by Complex II does not (i.e., electron flux through Complex II results in less ATP production than electron flux through Complex I). Other entry points for electrons into the mETC include the set of type II NAD(P)H dehydrogenase (ND) enzymes, which can oxidise NADH and/or NADPH from within either the mitochondrial matrix or the intermembrane space, depending on the isozyme. NDs donate their electrons to ubiquinone but do not contribute to proton pumping. They make up part of the mitochondrial AP of respiration (Rasmusson et al., 2004).

Further along the chain, electrons are passed from ubiquinol to Complex III, and are then transferred, via cytochrome c, to Complex IV. Electrons exit the classical mETC via Complex IV, also known as the terminal oxidase or "cytochrome c oxidase", which donates the electrons to O_2, forming water. Both Complex III and IV utilise energy from their electron transfer events to pump protons. Another exit point for electrons of the mETC is via the alternative oxidase (AOX), which also donates the electrons to O_2 but does not participate in proton pumping. This also makes up part of the mitochondrial AP of respiration (Vanlerberghe and McIntosh, 1997).

8.5.2 Alternative Respiration in Plant Mitochondria

There are many confirmed and proposed roles for alternative respiration in plant mitochondria, including thermogenesis (Miller et al., 2011), flowering and pollen formation (Chai et al., 2010, 2014), carbon–nitrogen balance (Gandin et al., 2014), stress signalling (Van Aken et al., 2009), protection of the classical mitochondrial electron transport chain (Dahal and Vanlerberghe, 2017) and metabolic flexibility during stress and stress recovery (Oh et al., 2022). Overall, AP enzymes can play a role in balancing cell energy state, metabolic flux and ROS levels (Vanlerberghe, 2013; Vanlerberghe et al., 2016, 2020). For the purposes of this chapter the focus remains heavily on the roles of the AP in oxidative stress, ROS mitigation and balancing redox equivalents.

AP enzymes bypass the energy-conserving reactions of the classical mETC complexes, therefore NADH oxidation becomes uncoupled from ATP synthesis (see Figure 8.2), and the pathway is sometimes referred to as "energy wasteful". This also means that ATP feedback limitation on the mETC can be circumvented. ATP feedback limitation occurs as ADP concentration (and ADP:ATP) becomes too low to act as a substrate for mitochondrial ATP synthase, thus the proton motive force cannot be utilised to generate ATP. The electrochemical gradient builds up across the inner mitochondrial membrane, thus limiting mETC flux. This is known as adenylate control. It can also lead to stalling of the TCA cycle due to an imbalance in NAD:NADH and FAD:$FADH_2$ in the mitochondrial matrix. Engagement of the AP enzymes enables an increase in the rate of NAD(P)H and $FADH_2$ oxidation without generating ATP, thus recycling $NAD(P)^+$ and FAD^+ needed for the TCA cycle or for oxidising excess NAD(P)H in the cytosol (Vanlerberghe et al., 2020).

The classical mETC complexes are also sites of mitochondrial ROS generation. ROS generation is thought to occur via electron leakage at Complexes I, II and III, especially when flux through the chain becomes restricted. As the activity of NDs and AOX can relieve mETC flux restrictions and bypass the sites of ROS generation, they can dampen the production of mitochondrial ROS (Cvetkovska and Vanlerberghe, 2012). Many studies have demonstrated the ROS-mitigating effects of AOX overexpression during stress or perturbations, from plant cell cultures to whole plant systems (see Table 8.1; Maxwell et al., 1999; Panda et al., 2013).

TABLE 8.1
Summary of Altered Tolerance to Individual Stresses in Plants with Modified AOX Expression

Species	Target Gene or Construct	Stress	Tolerance	Reference
		Overexpression Lines		
Nicotiana tabacum	*NtAOX1a*	Moderate drought	Increased	Dahal et al. (2014)
		Extreme drought	Increased	Dahal and Vanlerberghe (2017)
		Long-term drought	Increased	Dahal and Vanlerberghe (2018)
		Hypoxia	Increased	Jayawardhane et al. (2020)
	NtAOX1a	Cold	Increased	Zhang et al. (2009)
Arabidopsis thaliana	*AtAOX1a*	Cold	Increased	Fiorani et al. (2005)
		Salinity	Increased	Smith et al. (2009)
		Aluminium toxicity	Increased	Liu et al. (2014)
		Hypoxia	Increased	Vishwakarma et al. (2018)
Brassica napus	*BnaAOX1b*	Salinity	Increased	Yang et al. (2019)
		Osmotic	Increased	Yang et al. (2019)
Oryza sativa	*OsAOX1a*	Heat	Increased	Murakami and Toriyama (2008)
		Knockdown or Knockout Lines		
Nicotiana tabacum	*NtAOX1a* – RNAi	Extreme drought	Decreased	Wang and Vanlerberghe (2013)
		Drought	Decreased	Dahal et al. (2014)
		Extreme drought	Decreased	Dahal and Vanlerberghe (2017)
		Long-term drought	Decreased	Dahal and Vanlerberghe (2018)
		Hypoxia	Decreased	Jayawardhane et al. (2020)
	NtAOX1a – AS	Cold	Decreased	Zhang et al. (2009)
Arabidopsis thaliana	*AtAOX1a* – AS	Cold	Decreased	Fiorani et al. (2005)
		Aluminium exposure	Decreased	Liu et al. (2014)
		Hypoxia	Decreased	Vishwakarma et al. (2018)
		Salinity	Decreased	Manbir et al. (2022)
	AtAOX1a – TDNA	High light	Decreased	Zhang et al. (2010)
	AtAOX1a – TDNA	Cold	No change	Watanabe et al. (2008)
		High light	No change	Giraud et al. (2008)
		Drought	No change	Giraud et al. (2008)
		Nitrogen depletion	No change	Watanabe et al. (2010)
		Cadmium exposure	No change	Keunen et al. (2015)

Note: Tolerance measures vary between studies. Limited to stable plant transgenic or mutant lines where a change was observed in germination, growth, morphology or photosynthesis in the transgenic or mutant line(s) relative to their respective control lines in response to stress. Studies using the same plant lines are pooled into the same cell in the "Gene" column. RNAi, RNA interference; AS, antisense; TDNA, insertion mutant.

8.5.3 Alternative Pathway Genes

Alternative oxidase and type II NAD(P)H dehydrogenase enzymes are encoded by multigene families. *AOX* genes can be divided into two groups, *AOX1* and *AOX2*, which can each be divided into two subgroups: *AOX1a-c/e*, *AOX1d*, *AOX2a-c* and *AOX2d*. Not all of these groups are represented in all plants, though all species to date contain at least one *AOX1* gene. Generally, plants contain multiple *AOX1* isoforms and a single *AOX2* isoform. Exceptions include monocots, which typically contain only *AOX1* genes, and legumes, which typically contain one or two *AOX1* genes and an expanded *AOX2* gene set. The most typically stress-responsive *AOX* genes are *AOX1a*, *AOX1d* and *AOX2d*, though there are exceptions (Costa et al., 2014).

NDs are inner mitochondrial membrane enzymes that oxidise NADH and/or NADPH and are classified as either internal-facing (NDI) or external-facing (NDE). NDI proteins can oxidise NADH generated in the mitochondrial matrix, while NDE proteins can oxidise NAD(P)H from the intermembrane space or cytosol (NAD(P)H can pass through the outer mitochondrial membrane) (Rasmusson et al., 2004). NDI proteins are encoded by *NDA* or *NDC* genes, while NDE proteins are encoded by *NDB* genes.

This chapter will mainly focus on the ROS-mitigating effects of AOX, which has been more thoroughly studied than the NDs.

8.5.4 Regulation of Alternative Pathway Components

Transcription of *AOX1a* is responsive to a myriad of stresses (Feng et al., 2013). However, *AOX1a* may not only be involved in carrying out a stress response, but also as a regulator of stress responses (Van Aken et al., 2009). Up-regulation is often time-dependent, with an early peak in *AOX* transcripts between 1 and 24 hours, but potentially becoming important again during prolonged stress. The response of *AOX* genes can also be tissue-specific. For example, in response to salinity stress AOX transcripts increased in shoots and roots of *Arabidopsis* and barley plants, while rice AOX transcripts showed a consistent decrease in roots despite increasing in the shoots (Smith et al., 2009; Wanniarachchi et al., 2018). This suggests that the roles of AOX during salinity stress may differ between shoots and roots and across plant species.

As for *AOX*, some *ND* genes are stress-responsive, including the *Arabidopsis AtNDB2*, which is often co-expressed with *AtAOX1A*, particularly in response to stress. This observation was also recently demonstrated for *AOX1A* and *NDB2* orthologs in cereals and legumes (Clifton et al., 2005; Wanniarachchi et al., 2018; Sweetman et al., 2020). Activities of different ND isoforms have mostly been characterised *in vitro*. Isoforms can be differentially regulated by substrate (NADH or NADPH) and cellular conditions such as pH and calcium levels (Geisler et al., 2007). Some studies in *Arabidopsis* have also determined the effect of altered *ND* gene expression on plant growth and metabolism (Wallström et al., 2014a, 2014b). However, few have explored the involvement of NDs in stress tolerance (Sweetman et al., 2019) and generally ND proteins are under-researched, particularly in non-model species.

In addition to a dynamic transcriptional response, *AOX* is also highly regulated at the post-translational level. The enzyme is active as a non-covalently linked dimer and as long as the sulfhydryl groups are reduced, the enzyme is also subject to activation by specific organic acids such as pyruvate. AOX isoforms are differentially activated by a range of organic acids and they also harbour different activation kinetics with pyruvate (Møller et al., 2020; Selinski et al., 2018; Sweetman et al., 2022).

Measuring the maximal activity (capacity) of alternative respiration enzymes in purified mitochondria can be achieved using an oxygen electrode and combinations of mETC substrates and inhibitors, including the Complex IV inhibitor potassium cyanide and AOX inhibitor *n*-propyl gallate. There is a key distinction between AOX capacity and activity, or engagement of the enzyme. The latter represents what proportion of total electron flow is directed through AOX rather than

Complex IV. The engagement or in vivo activity of AOX can be measured in whole plant tissues using an oxygen isotope fractionation technique, as AOX and Complex IV differentially discriminate against the heavy isotope $^{18}O_2$. This technique has been used to demonstrate AOX activity and engagement in a variety of plant species and stresses (Del-Saz et al., 2018).

8.6 AOX AND STRESS TOLERANCE

AOX offers a bypass of ROS generation steps during mitochondrial respiration. As such, AOX activity may dampen local ROS production and, therefore, decrease ROS signalling, which seems counterproductive, considering the beneficial roles of ROS especially during the early stages of stress. For example, mitochondrial ROS may play a role in RIRR signalling, to enhance the stress response and acclimation (Zandalinas and Mittler, 2018). While AOX transcript levels are induced within hours of stress elicitation (Smith et al., 2009) and AOX activity increases in response to a range of stresses, both long- and short-term, the proportion of electrons directed to AOX shows variable responses (Del-Saz et al., 2018). Changes in AOX transcript and protein are generally not predictive of activity, which seems to be more determined by post-translational regulation (Guy and Vanlerberghe, 2005), which can occur by altering protein redox state and especially via local accumulation of organic acids such as pyruvate, glyoxylate and oxaloacetate. Therefore, varying AOX activity and engagement responses across species and stressors may be due to inherently different metabolic conditions and kinetic characteristics of AOX orthologs. These could play a major role in fine-tuning AOX activity such that ROS signalling can proceed when required, without risking ROS overaccumulation and oxidative stress.

Greater benefits of AOX may come into effect during longer-term stress, when ROS may begin to switch from advantageous to disadvantageous roles and overwhelm antioxidant responses. Indeed, AOX seems to be particularly important during long-term drought stress as indicated by (1) the pronounced increase in engagement of AOX from 10% to 40% upon extreme drought stress in soybean (Ribas-Carbo et al., 2005), (2) the loss of mitochondrial and photosynthetic activities and impaired survival during extended or extreme drought in plants lacking *AOX1a* (Giraud et al., 2008; Wang and Vanlerberghe, 2013; Dahal et al., 2014; Dahal and Vanlerberghe, 2017, 2018; Sweetman et al., 2019) and (3) increased tolerance to extended drought in plants overexpressing *AOX1a* (Dahal et al., 2014; Dahal and Vanlerberghe, 2017, 2018; Sweetman et al., 2019). During prolonged or extreme stress AOX activity may delay cell death by shielding the mitochondria against large-scale ROS production and oxidative stress.

Alternative oxidase has also demonstrated importance in the management of salinity stress (Smith et al., 2009; Yang et al., 2019). The defences against salinity stress are energetically costly, including ion transport processes for extrusion or sequestration of Na^+ and Cl^- as well as ROS processing. There is also often a decrease in photosynthetic rates as mentioned earlier. As such, an increase in efficiency of ATP synthesis seems key to combatting salinity stress. It is therefore counterintuitive that the expression and capacity of the "energy wasteful" mitochondrial alternative respiration enzymes increase during salinity stress (Smith et al., 2009) and that enhancement of AOX can improve tolerance to salinity (Table 8.1). The AP may be enlisted to decrease mitochondrial ROS generation and mitigate oxidative stress, and the importance of this role may, at times, outweigh the energetic cost of alternative respiration (Munns et al., 2020). As the engagement of AOX during salinity stress has shown variable responses in different experiments (Del-Saz et al., 2018), there may be other variables that influence AOX activity during salinity stress, for example, severity or duration of stress, the strategy used for managing salinity (i.e., extrusion versus tissue tolerance), cellular ROS levels and metabolic state.

Recently it was demonstrated that plants can up-regulate AOX activity during nutrient deficiency, appearing to sacrifice carbohydrate to improve nutrient uptake capabilities. Some plant roots can form beneficial associations with mycorrhiza in the soil, activating the mycorrhizal nutrient uptake pathway, which can also affect shoot AOX activity, most likely to facilitate an increase in photosynthesis for transport of assimilate to the fungus (Ortiz et al., 2020). AOX has also been proposed to impart the root cell with a level of metabolic flexibility, such that large amounts of organic acids malate and citrate can be secreted into the soil, to improve P availability. AOX may contribute by enabling regeneration of NAD^+ (without the barrier of adenylate control) for continued operation of the TCA cycle and continued malate and citrate biosynthesis (Ortiz et al., 2020). This may also prevent mitochondrial ROS accumulation, by ensuring that the mETC does not become over-reduced.

8.6.1 The Potential Roles of AOX during Combined Stresses

While many studies have confirmed an up-regulation of AOX transcripts and/or protein in response to various stresses, including combination stresses, there are currently no measurements of AOX engagement in response to combined abiotic stresses. Similarly, transgenic plants with altered AOX expression exhibit altered stress tolerance characteristics when exposed to multiple stresses individually (Table 8.1); however, few studies have utilised combined stress treatments with such plants (Table 8.2). One of these studies also demonstrated the importance of NDB2 in combined drought and increased light stress in *Arabidopsis* (Sweetman et al., 2019). AOX inhibitor treatments have also been used to demonstrate increased sensitivity of photosynthetic parameters during combined drought and high light treatments in wheat plants lacking AOX activity (Bartoli et al., 2005), consistent with findings from the transgenic and mutant plants lacking *AOX1a* (Table 8.2). Some hypotheses have also emerged around the role of AOX in mobilising reducing equivalents from the chloroplast during combined stresses such as nitrogen deficiency and high light (Watanabe et al., 2010).

As the AP can facilitate oxidation of excess NADPH from photo-inhibited photosynthesis and NADH from metabolism of the photorespiratory product, glycine, without limitation by adenylate control, there is almost certainly a role for this pathway in response to combined stresses that affect photosynthesis (Bykova et al., 2014; Sunil et al., 2019). Furthermore, with photoinhibition a common symptom of many individual and combined stresses, there is a clear benefit to ensuring that all plants have the resources to enlist the AP when necessary.

Future studies should measure the effects of single versus combined stress treatments on AOX activity and engagement, and test the tolerance of plants with modified levels of AOX for resilience to multiple simultaneous stresses. Although increased availability of *AOX1a* can improve

TABLE 8.2
Summary of Altered Tolerance to Combined Abiotic Stresses in Plants with Modified *AOX* Expression

Species	Target Gene or Construct	Stress	Tolerance	Reference
		Overexpression Lines		
Arabidopsis thaliana	*AtAOX1a*	Drought and increased light	Increased	Sweetman et al. (2019)
		Knockdown or Knockout Lines		
Nicotiana tabacum	*NtAOX1a* – RNA1	Extreme drought and high light	Decreased	Wang and Vanlerberghe (2013)
Arabidopsis thaliana	*AtAOX1a* – TDNA	Drought and increased light	Decreased	Giraud et al. (2008)
		Drought and increased light	Decreased	Sweetman et al. (2019)

stress tolerance, it is also becoming increasingly apparent that metabolic status can determine the activity and thereby the role of AOX and alternative respiration during stress (Selinski et al., 2018; Sweetman et al., 2022), therefore experiments with multiple stresses in both transgenic and non-transgenic plants should also seek to explore metabolic effects, especially for organic acids and preferably at the sub-cellular level.

8.7 CONCLUSION

The abiotic stress combinations encountered by a plant in the field are potentially limitless, and each combination may elicit different biochemical and physiological effects. This makes it difficult to breed or engineer plants that are broadly stress tolerant, such that they can withstand increasingly extreme swings in seasonal conditions each year. ROS accumulation is a common effect of abiotic stresses that may remain, or further amplify, in the presence of multiple stresses. While ROS are useful in eliciting cellular and tissue responses to stress, limiting ROS overaccumulation can improve stress tolerance. Plants with enhanced ROS-scavenging or ROS-prevention mechanisms, including components of the AP of respiration, have demonstrated improved tolerance to numerous independent stresses as well as combined drought and light stress. Therefore, this pathway, and others involved in ROS processing, offer promising targets for improving tolerance to a wide range of individual and combined stresses.

REFERENCES

Ahanger, M. A., Qin, C., Begum, N., Qi, M. D., Dong, X. X., El-Esawi, M., et al., 2019. Nitrogen availability prevents oxidative effects of salinity on wheat growth and photosynthesis by up-regulating the antioxidants and osmolytes metabolism, and secondary metabolite accumulation. *BMC Plant Biology* 19:479. doi:10.1186/s12870-019-2085-3.

Baier, M., Kandlbinder, A., Golldack, D., and Dietz, K. J. 2005. Oxidative stress and ozone: perception, signalling and response. *Plant Cell and Environment* 28:1012–1020 doi:10.1111/j.1365-3040.2005.01326.x.

Bartoli, C. G., Gomez, F., Gergoff, G., Guiamet, J. J., and Puntarulo, S. 2005. Up-regulation of the mitochondrial alternative oxidase pathway enhances photosynthetic electron transport under drought conditions. *Journal of Experimental Botany* 56:1269–1276. doi:10.1093/jxb/eri111.

Begcy, K., Mariano, E. D., Mattiello, L., Nunes, A. V., Mazzafera, P., Maia, I. G., and Menossi, M. 2011. An Arabidopsis mitochondrial uncoupling protein confers tolerance to drought and salt stress in transgenic tobacco plants. *PLoS One* 6:23776. doi:10.1371/journal.pone.0023776.

Bose, J., Munns, R., Shabala, S., Gilliham, M., Pogson, B., and Tyerman S. D. 2017. Chloroplast function and ion regulation in plants growing on saline soils: lessons from halophytes. *Journal of Experimental Botany* 68:3129–3143. doi:10.1093/jxb/erx142.

Bose, J., Rodrigo-Moreno, A., and Shabala, S. 2014. ROS homeostasis in halophytes in the context of salinity stress tolerance. *Journal of Experimental Botany* 65:1241–1257. doi:10.1093/jxb/ert430.

Bykova, N. V., Moller, I. M., Gardestrom, P., and Igamberdiev, A. U. 2014. The function of glycine decarboxylase complex is optimized to maintain high photorespiratory flux via buffering of its reaction products. *Mitochondrion* 19:357–364. doi:10.1016/j.mito.2014.01.001.

Chai, L, Wang, J. M., Fan, Z. Y., Liu, Z. B., Wen, G. Q., Li, X. F., and Yang, Y. 2014. Regulation of the flowering time of Arabidopsis thaliana by thylakoid ascorbate peroxidase. *African Journal of Biotechnology* 11:7151–7157. doi:10.5897/AJB11.3718.

Chai, T. T., Simmonds, D., Day, D. A., Colmer, T. D., and Finnegan, P. M. 2010. Photosynthetic performance and fertility are repressed in GmAox2b antisense soybean. *Plant Physiology* 152:1638–1649. doi:10.1104/pp.109.149294.

Chinnusamy, V., Schumaker, K. and Zhu, J. K. 2004. Molecular genetic perspectives on cross-talk and specificity in abiotic stress signalling in plants. *Journal of Experimental Botany* 55:225–236 doi:10.1093/jxb/erh005.

Choi, S. M., Jeong, S. W., Jeong, W. J., Kwon, S. Y., Chow, W. S., and Park, Y. I. 2002. Chloroplast Cu/Zn-superoxide dismutase is a highly sensitive site in cucumber leaves chilled in the light. *Planta* 216:315–324. doi:10.1007/s00425-002-0852-z.

Choudhury, F. K., Rivero, R. M., Blumwald, E., and Mittler, R. 2017. Reactive oxygen species, abiotic stress and stress combination. *Plant Journal* 90:856–867 doi:10.1111/tpj.13299.

Clifton, R, Lister, R., Parker, K. L., Sappl, P. G., Elhafez, D., Millar, A.H., et al., 2005. Stress-induced co-expression of alternative respiratory chain components in Arabidopsis thaliana. *Plant Molecular Biology* 58:193–212 doi:10.1007/s11103-005-5514-7.

Colmenero-Flores, J. M., and Rosales, M. A. 2014. Interaction between salt and heat stress: when two wrongs make a right. *Plant, Cell and Environment* 37:1042–1045. doi:10.1111/pce.12229.

Costa, J. H., McDonald, A. E., Arnholdt-Schmitt, B., and de Melo, D. F. 2014. A classification scheme for alternative oxidases reveals the taxonomic distribution and evolutionary history of the enzyme in angiosperms. *Mitochondrion* 19:172–183. doi:10.1016/j.mito.2014.04.007.

Cvetkovska, M., and Vanlerberghe, G. C. 2012. Alternative oxidase modulates leaf mitochondrial concentrations of superoxide and nitric oxide. *New Phytologist* 195:32–39. doi:10.1111/j.1469-8137.2012.04166.x.

Dahal, K., and Vanlerberghe, G. C. 2017. Alternative oxidase respiration maintains both mitochondrial and chloroplast function during drought. *New Phytologist* 213:560–571. doi:10.1111/nph.14169.

Dahal, K., and Vanlerberghe, G. C. 2018. Improved chloroplast energy balance during water deficit enhances plant growth: more crop per drop. *Journal of Experimental Botany* 69:1183–1197. doi:10.1093/jxb/erx474.

Dahal, K., Wang, J., Martyn, G. D., Rahimy, F. and Vanlerberghe, G. C. 2014. Mitochondrial alternative oxidase maintains respiration and preserves photosynthetic capacity during moderate drought in *Nicotiana tabacum*. *Plant Physiology* 166:1560–1574. doi:10.1104/pp.114.247866.

Das, K, and Roychoudhury, A. 2014. Reactive oxygen species (ROS) and response of antioxidants as ROS-scavengers during environmental stress in plants. *Frontiers in Environmental Science* 2:53. doi:10.3389/fenvs.2014.00053.

Del-Saz, N. F., Ribas-Carbo, M., McDonald, A. E., Lambers, H., Fernie, A. R., and Florez-Sarasa, I. 2018. An *in vivo* perspective of the role(s) of the alternative oxidase pathway. *Trends in Plant Science* 23:206–219. doi:10.1016/j.tplants.2017.11.006.

Ding, Y. L., Lv, J., Shi, Y. T., Gao, J. P., Hua, J., Song, C. P. et al. 2019. EGR2 phosphatase regulates OST1 kinase activity and freezing tolerance in Arabidopsis. *EMBO Journal* 38: e99819. doi:10.15252/embj.201899819.

Douce, R., Bourguignon, J., Neuburger, M., and Rebeille, F. 2001. The glycine decarboxylase system: a fascinating complex. *Trends in Plant Science* 6:167–176 doi:10.1016/s1360-1385(01)01892-1.

Dreyer, A., and Dietz, K. J. 2018. Reactive oxygen species and the redox-regulatory network in cold stress acclimation. *Antioxidants* 7:169. doi:10.3390/antiox7110169.

Dutta, S., Mitra, M., Agarwal, P., Mahapatra, K., De, S., Sett, U., and Roy, S. 2018. Oxidative and genotoxic damages in plants in response to heavy metal stress and maintenance of genome stability. *Plant Signaling and Behavior* 13:e1460048. doi:10.1080/15592324.2018.1460048.

Feng, H. Q., Guan, D. D., Sun, K., Wang, Y. F., Zhang, T. G., and Wang, R. F. 2013. Expression and signal regulation of the alternative oxidase genes under abiotic stresses. *Acta Biochimica et Biophysica Sinica* 45:985–994. doi:10.1093/abbs/gmt094.

Fiorani, F., Umbach, A. L., and Siedow, J. N. 2005. The alternative oxidase of plant mitochondria is involved in the acclimation of shoot growth at low temperature. A study of Arabidopsis *AOX1a* transgenic plants. *Plant Physiology* 139:1795–1805 doi:10.1104/pp.105.070789.

Foyer, C. H., and Noctor, G. 2003. Redox sensing and signalling associated with reactive oxygen in chloroplasts, peroxisomes and mitochondria. *Physiologia Plantarum* 119:355–364 doi:10.1034/j.1399-3054.2003.00223.x.

Foyer, C. H., Ruban, A. V., and Noctor, G. 2017. Viewing oxidative stress through the lens of oxidative signalling rather than damage. *Biohemical Journal* 474:877–883. doi:10.1042/bcj20160814.

Gandin, A., Denysyuk, M., and Cousins, A. B. 2014. Disruption of the mitochondrial alternative oxidase (AOX) and uncoupling protein (UCP) alters rates of foliar nitrate and carbon assimilation in *Arabidopsis thaliana*. *Journal of Experimental Botany* 65:3133–3142. doi:10.1093/jxb/eru158.

Gandin, A., Dizengremel, P., and Jolivet, Y. 2021. Integrative role of plant mitochondria facing oxidative stress: The case of ozone. *Plant Physiology and Biochemistry* 159:202–210 doi:10.1016/j.plaphy.2020.12.019.

Geisler, D. A., Broselid, C., Hederstedt, L., and Rasmusson, A. G. 2007. Ca^{2+}-binding and Ca^{2+}-independent respiratory NADH and NADPH dehydrogenases of *Arabidopsis thaliana*. *Journal of Biological Chemistry* 282:28455–28464. doi:10.1074/jbc.M704674200.

Giraud, E., Ho, L. H. M., Clifton, R., Carroll, A., Estavillo, G., Tan, Y. F., et al. 2008. The absence of alternative oxidase1a in Arabidopsis results in acute sensitivity to combined light and drought stress. *Plant Physiology* 147:595–610. doi:10.1104/pp.107.115121.

Guy, R. D., and Vanlerberghe, G. C. 2005. Partitioning of respiratory electrons in the dark in leaves of transgenic tobacco with modified levels of alternative oxidase. *Physiologia Plantarum* 125:171–180. doi:10.1111/j.1399-3054.2005.00557.x.

Hajiboland, R. 2012. Effect of micronutrient deficiencies on plants stress responses. In *Abiotic Stress Responses in Plants: Metabolism, Productivity and Sustainability*, eds. P. Ahmad and M. N. V. Prasad, 283–329. New York: Springer. doi:10.1007/978-1-4614-0634-1.

He, M., He, C. Q., and Ding, N. Z. 2018. Abiotic stresses: general defenses of land plants and chances for engineering multistress tolerance. *Frontiers in Plant Science* 9:1771. doi:10.3389/fpls.2018.01771.

Hong, J. H., Savina, M., Du, J., Devendran, A., Ramakanth, K. K., Tian, X., et al. 2017. A sacrifice-for-survival mechanism protects root stem cell niche from chilling stress. *Cell* 170:102–113. doi:10.1016/j.cell.2017.06.002.

Iyer, N. J., Tang, Y., and Mahalingam, R. 2013. Physiological, biochemical and molecular responses to a combination of drought and ozone in *Medicago truncatula*. *Plant Cell and Environment* 36:706–720 doi:10.1111/pce.12008.

Jayawardhane, J., Cochrane, D. W., Vyas, P., Bykova, N. V., Vanlerberghe, G. C., and Igamberdiev, A. U. 2020. Roles for plant mitochondrial alternative oxidase under normoxia, hypoxia, and reoxygenation conditions. *Frontiers in Plant Science* 11:566. doi:10.3389/fpls.2020.00566.

Karkonen, A., and Kuchitsu, K. 2015. Reactive oxygen species in cell wall metabolism and development in plants. *Phytochemistry* 112:22–32. doi:10.1016/j.phytochem.2014.09.016.

Keunen, E., Schellingen, K., Van Der Straeten, D., Remans, T., Colpaert, J., Vangronsveld, J., et al. 2015. ALTERNATIVE OXIDASE1a modulates the oxidative challenge during moderate Cd exposure in *Arabidopsis thaliana* leaves. *Journal of Experimental Botany* 66:2967–2977. doi:10.1093/jxb/erv035.

Koussevitzky, S., Suzuki, N., Huntington, S., Armijo, L., Sha, W., Cortes, D., et al. 2008. Ascorbate Peroxidase 1 plays a key role in the response of *Arabidopsis thaliana* to stress combination. *Journal of Biological Chemistry* 283:34197–34203. doi:10.1074/jbc.M806337200.

Larkindale, J., Hall, J. D., Knight, M. R., and Vierling, E. 2005. Heat stress phenotypes of arabidopsis mutants implicate multiple signaling pathways in the acquisition of thermotolerance. *Plant Physiology* 138:882–897 doi:10.1104/pp.105.062257.

Laxa, M., Liebthal, M., Telman, W., Chibani, K., and Dietz, K. J. 2019. The role of the plant antioxidant system in drought tolerance. *Antioxidants* 8:94. doi:10.3390/antiox8040094.

Liu, J. H., Fu, C. C., Li, G. J., Khan, M. N., and Wu, H. H. 2021. ROS homeostasis and plant salt tolerance: plant nanobiotechnology updates. *Sustainability* 13:3552. doi:10.3390/su13063552.

Liu, J., Li, Z., Wang, Y. Q., and Xing, D. 2014. Overexpression of ALTERNATIVE OXIDASE1a alleviates mitochondria-dependent programmed cell death induced by aluminium phytotoxicity in Arabidopsis. *Journal of Experimental Botany* 65:4465–4478. doi:10.1093/jxb/eru222.

Manbir, S. P., Kumari, A., and Gupta, K. J. 2022. Alternative oxidase plays a role in minimizing ROS and RNS produced under salinity stress in *Arabidopsis thaliana*. *Physiologia Plantarum* 174:e13649. doi:10.1111/ppl.13649.

Maxwell, D. P., Wang, Y., and McIntosh, L. 1999. The alternative oxidase lowers mitochondrial reactive oxygen production in plant cells. *Proceedings of the National Academy of Sciences of the USA* 96:8271–8276. doi:10.1073/pnas.96.14.8271.

Miller, G., Schlauch, K., Tam, R., Cortes, D., Torres, M. A., Shulaev, V. et al. 2009. The plant NADPH oxidase RBOHD mediates rapid systemic signaling in response to diverse stimuli. *Science Signaling* 2:ra45. doi:10.1126/scisignal.2000448.

Miller, R. E., Grant, N. M., Giles, L., Ribas-Carbo, M., Berry, J. A., Watling, J. R., et al. 2011. In the heat of the night - alternative pathway respiration drives thermogenesis in *Philodendron bipinnatifidum*. *New Phytologist* 189:1013–1026. doi:10.1111/j.1469-8137.2010.03547.x.

Mittler, R. 2017. ROS are good. *Trends in Plant Science* 22:11–19. doi:10.1016/j.tplants.2016.08.002.

Mittler, R., and Blumwald, E. 2010. Genetic engineering for modern agriculture: challenges and perspectives. In *Annual Review of Plant Biology*, vol. 61, eds. S. Merchant, W. R. Briggs, and D. Ort, 443–462. doi:10.1146/annurev-arplant-042809-112116.

Mittler, R., Vanderauwera, S., Suzuki, N., Miller, G., Tognetti, V. B., Vandepoele, K. et al. 2011. ROS signaling: the new wave? *Trends in Plant Science* 16:300–309. doi:10.1016/j.tplants.2011.03.007.

Møller, I. M., Igamberdiev, A. U., Bykova, N. V., Finkemeier, I., Rasmusson, A. G., and Schwarzlander, M. 2020. Matrix redox physiology governs the regulation of plant mitochondrial metabolism through post-translational protein modifications. *Plant Cell* 32:573–594 doi:10.1105/tpc.19.00535.

Munns, R., Day, D. A., Fricke, W., Watt, M., Arsova, B., Barkla, B. J., et al. 2020. Energy costs of salt tolerance in crop plants. *New Phytologist* 225:1072–1090 doi:10.1111/nph.15864.

Munns, R., and Tester, M. 2008. Mechanisms of salinity tolerance. *Annual Review of Plant Biology* 59: 651–681. doi:10.1146/annurev.arplant.59.032607.092911.

Murakami, Y., and Toriyama, K. 2008. Enhanced high temperature tolerance in transgenic rice seedlings with elevated levels of alternative oxidase, OsAOX1a. *Plant Biotechnology* 25:361–364 doi:10.5511/plantbiotechnology.25.361.

Oh, G. G. K., O'Leary, B. M., Signorelli, S., and Millar, A. H. 2022. Alternative oxidase (AOX) 1a and 1d limit proline-induced oxidative stress and aid salinity recovery in Arabidopsis. *Plant Physiology* 188: 1521–1536 doi:10.1093/plphys/kiab578.

O'Kane, D., Gill, V., Boyd, P., and Burdon, B. 1996. Chilling, oxidative stress and antioxidant responses in *Arabidopsis thaliana* callus. *Planta* 198:371–377 doi:10.1007/bf00620053.

Ortiz, J., Sanhueza, C., Romero-Munar, A., Hidalgo-Castellanos, J., Castro, C., Bascunan-Godoy, L., et al. 2020. In vivo metabolic regulation of alternative oxidase under nutrient deficiency-interaction with arbuscular mycorrhizal fungi and *Rhizobium bacteria. International Journal of Molecular Sciences* 21:4201. doi:10.3390/ijms21124201.

Panda, S. K., Sahoo, L., Katsuhara, M., and Matsumoto, H. 2013. Overexpression of alternative oxidase gene confers aluminum tolerance by altering the respiratory capacity and the response to oxidative stress in tobacco cells. *Molecular Biotechnology* 54:551–563. doi:10.1007/s12033-012-9595-7.

Pandey, P., Ramegowda, V., and Senthil-Kumar, M. 2015. Shared and unique responses of plants to multiple individual stresses and stress combinations: physiological and molecular mechanisms. *Frontiers in Plant Science* 6:723. doi:10.3389/fpls.2015.00723.

Pospisil, P. 2016. Production of reactive oxygen species by photosystem II as a response to light and temperature stress. *Frontiers in Plant Science* 7:1950. doi:10.3389/fpls.2016.01950.

Prasad, T. K., Anderson, M. D., and Stewart, C. R. 1994. Acclimation, hydrogen-peroxide, and abscisic-acid protect mitochondria against irreversible chilling injury in maize seedlings. *Plant Physiology* 105: 619–627 doi:10.1104/pp.105.2.619.

Pucciariello, C., and Perata, P. 2021. The oxidative paradox in low oxygen stress in plants. *Antioxidants* 10:332. doi:10.3390/antiox10020332.

Rasmusson, A. G., Soole, K. L., and Elthon, T. E. 2004. Alternative NAD(P)H dehydrogenases of plant mitochondria. *Annual Review of Plant Biology* 55:23–39. doi:10.1146/annurev.arplant.55.031903.141720.

Ribas-Carbo, M., Taylor, N. L., Giles, L., Busquets, S., Finnegan, P. M., and Day, D. A. 2005. Effects of water stress on respiration in soybean leaves. *Plant Physiology* 139:466–473. doi:10.1104/pp.105.065565.

Rizhsky, L., Liang, H. J. and Mittler, R. 2002. The combined effect of drought stress and heat shock on gene expression in tobacco. *Plant Physiology* 130:1143–1151. doi:10.1104/pp.006858.

Rogers, H., and Munne-Bosch, S. 2016. Production and scavenging of reactive oxygen species and redox signaling during leaf and flower senescence: similar but different. *Plant Physiology* 171:1560–1568. doi:10.1104/pp.16.00163.

Rosenwasser, S., Rot, I., Sollner, E., Meyer, A. J., Smith, Y., Leviatan, N., et al. 2011. Organelles contribute differentially to reactive oxygen species-related events during extended darkness. *Plant Physiology* 156:185–201. doi:10.1104/pp.110.169797.

Sachdev, S., Ansari, S. A., Ansari, M. I., Fujita, M., and Hasanuzzaman, M. 2021. Abiotic stress and reactive oxygen species: generation, signaling, and defense mechanisms. *Antioxidants* 10:277. doi:10.3390/antiox10020277.

Selinski, J., Hartmann, A., Deckers-Hebestreit, G., Day, D. A., Whelan, J., and Scheibe, R. 2018. Alternative oxidase isoforms are differentially activated by tricarboxylic acid cycle intermediates. *Plant Physiology* 176:1423–1432. doi:10.1104/pp.17.01331.

Selinski, J., and Scheibe, R. 2019. Malate valves: old shuttles with new perspectives. *Plant Biology* 21:21–30. doi:10.1111/plb.12869.

Shahid, M., Pourrut, B., Dumat, C., Nadeem, M., Aslam, M., and Pinelli, E. 2014. Heavy-metal-induced reactive oxygen species: phytotoxicity and physicochemical changes in plants. In *Reviews of Environmental Contamination and Toxicology*, vol 232, ed. D. M. Whitacre, 1–44. doi:10.1007/978-3-319-06746-9.

Slimen, I. B., Najar, T., Ghram, A., Dabbebi, H., Mrad, M. B., and Abdrabbah, M. 2014. Reactive oxygen species, heat stress and oxidative-induced mitochondrial damage. A review. *International Journal of Hyperthermia* 30:513–523 doi:10.3109/02656736.2014.971446.

Smith, C. A., Melino, V. J., Sweetman, C., and Soole, K. L. 2009. Manipulation of alternative oxidase can influence salt tolerance in *Arabidopsis thaliana*. *Physiologia Plantarum* 137:459–472. doi:10.1111/j.1399-3054.2009.01305.x.

Sunil, B., Saini, D., Bapatla, R. B., Aswani, V., and Raghavendra, A. S. 2019. Photorespiration is complemented by cyclic electron flow and the alternative oxidase pathway to optimize photosynthesis and protect against abiotic stress. *Photosynthesis Research* 139:67–79. doi:10.1007/s11120-018-0577-x.

Suzuki, N., Bassil, E., Hamilton, J. S., Inupakutika, M. A., Zandalinas, S. I., Tripathy, D., et al. 2016. ABA is required for plant acclimation to a combination of salt and heat stress. *PLoS One* 11:e0147625. doi:10.1371/journal.pone.0147625.

Suzuki, N., and Mittler, R. 2006. Reactive oxygen species and temperature stresses: A delicate balance between signaling and destruction. *Physiologia Plantarum* 126:45–51. doi:10.1111/j.0031-9317.2005.00582.x.

Sweetman, C., Miller, T. K., Booth, N. J., Shavrukov, Y., Jenkins, C. L. D., Soole, K. L., et al. 2020. Identification of alternative mitochondrial electron transport pathway components in chickpea indicates a differential response to salinity stress between cultivars. *International Journal of Molecular Sciences* 21:3844. doi:10.3390/ijms21113844.

Sweetman, C., Selinski, J., Miller, T. K., Whelan, J., and Day, D. A. 2022. Legume alternative oxidase isoforms show differential sensitivity to pyruvate activation. *Frontiers in Plant Science* 12:813691. doi:10.3389/fpls.2021.813691.

Sweetman, C., Waterman, C. D., Rainbird, B. M., Smith, P. M. C., Jenkins, C. L. D., Day, D. A., et al. 2019. AtNDB2 is the main external NADH dehydrogenase in mitochondria and is important for tolerance to environmental stress. *Plant Physiology* 181:774–788. doi:10.1104/pp.19.00877.

Takahashi, S., and Badger, M. R. 2011. Photoprotection in plants: a new light on photosystem II damage. *Trends in Plant Science* 16:53–60. doi:10.1016/j.tplants.2010.10.001.

Tewari, R. K., Yadav, N., Gupta, R., and Kumar, P. 2021. Oxidative stress under macronutrient deficiency in plants. *Journal of Soil Science and Plant Nutrition* 21:832–859 doi:10.1007/s42729-020-00405-9.

Thannickal, V. J., and Fanburg, B. L. 2000. Reactive oxygen species in cell signaling. *American Journal of Physiology-Lung Cellular and Molecular Physiology* 279:L1005–L1028. doi:10.1152/ajplung.2000.279.6.L1005.

Van Aken, O. 2021. Mitochondrial redox systems as central hubs in plant metabolism and signaling. *Plant Physiology* 186:36–52 doi:10.1093/plphys/kiab101.

Van Aken, O., Giraud, E., Clifton, R., and Whelan, J. 2009. Alternative oxidase: a target and regulator of stress responses. *Physiologia Plantarum* 137:354–361. doi:10.1111/j.1399-3054.2009.01240.x.

Van Breusegem, F., and Dat, J. F. 2006. Reactive oxygen species in plant cell death. *Plant Physiology* 141: 384–390. doi:101104/pp.106.078295.

Vanlerberghe, G. C. 2013. Alternative oxidase: A mitochondrial respiratory pathway to maintain metabolic and signaling homeostasis during abiotic and biotic stress in plants. *International Journal of Molecular Sciences* 14:6805–6847. doi:10.3390/ijms14046805.

Vanlerberghe, G. C., Dahal, K., Alber, N. A., and Chadee, A. 2020. Photosynthesis, respiration and growth: A carbon and energy balancing act for alternative oxidase. *Mitochondrion* 52:197–211 doi:10.1016/j.mito.2020.04.001.

Vanlerberghe, G. C., Martyn, G. D., and Dahal, K. 2016. Alternative oxidase: a respiratory electron transport chain pathway essential for maintaining photosynthetic performance during drought stress. *Physiologia Plantarum* 157:322–337. doi:10.1111/ppl.12451.

Vanlerberghe, G. C., and McIntosh, L. 1997. Alternative oxidase: from gene to function. *Annual Review of Plant Physiology and Plant Molecular Biology* 48:703–734. doi:10.1146/annurev.arplant.48.1.703.

Vishwakarma, A., Kumari, A., Mur, L. A. J., and Gupta, K. J. 2018. A discrete role for alternative oxidase under hypoxia to increase nitric oxide and drive energy production. *Free Radical Biology and Medicine* 122:40–51. doi:10.1016/j.freeradbiomed.2018.03.045.

Wallström, S. V., Florez-Sarasa, I., Araujo, W. L., Aidemark, M., Fernandez-Fernandez, M., Fernie, A. R., et al. 2014a. Suppression of the external mitochondrial NADPH dehydrogenase, NDB1, in *Arabidopsis thaliana* affects central metabolism and vegetative growth. *Molecular Plant* 7:356–368 doi:10.1093/mp/sst115.

Wallström, S. V., Florez-Sarasa, I., Araujo, W. L., Escobar, M. A., Geisler, D. A., Aidemark, M., et al. 2014b. Suppression of NDA-type alternative mitochondrial NAD(P)H dehydrogenases in Arabidopsis thaliana modifies growth and metabolism, but not high light stimulation of mitochondrial electron transport. *Plant and Cell Physiology* 55:881–896 doi:10.1093/pcp/pcu021.

Wang, J., and Vanlerberghe, G. C. 2013. A lack of mitochondrial alternative oxidase compromises capacity to recover from severe drought stress. *Physiologia Plantarum* 149:461–473. doi:10.1111/ppl.12059.

Wanniarachchi, V. R., Dametto, L., Sweetman, C., Shavrukov, Y., Day, D. A., Jenkins, C. L. D., et al. 2018. Alternative respiratory pathway component genes (AOX and ND) in rice and barley and their response to stress. *International Journal of Molecular Sciences* 19:915. doi:10.3390/ijms19030915.

Watanabe, C. K., Hachiya, T., Takahara, K., Kawai-Yamada, M., Uchimiya, H., Uesono, Y., et al. 2010. Effects of AOX1a deficiency on plant growth, gene expression of respiratory components and metabolic profile under low-nitrogen stress in *Arabidopsis thaliana. Plant and Cell Physiology* 51:810–822 doi:10.1093/pcp/pcq033.

Watanabe, C. K., Hachiya, T., Terashima, I., and Noguchi, K. 2008. The lack of alternative oxidase at low temperature leads to a disruption of the balance in carbon and nitrogen metabolism, and to an up-regulation of antioxidant defence systems in *Arabidopsis thaliana* leaves. *Plant, Cell & Environment* 31: 1190–1202. doi: 10.1111/j.1365-3040.2008.01834.x.

Yamauchi, T., Yoshioka, M., Fukazawa, A., Mori, H., Nishizawa, N. K., Tsutsumi, N., Yoshioka, H., and Nakazono, M. 2017. An NADPH oxidase RBOH functions in rice roots during lysigenous aerenchyma formation under oxygen-deficient conditions. *The Plant Cell* 29:775–790. doi:10.1105/tpc.16.00976.

Yang, H. L., Deng, L. B., Liu, H. F., Fan, S. H., Hua, W., and Liu, J. 2019. Overexpression of *BnaAox1b* confers tolerance to osmotic and salt stress in rapeseed. *G3: Genes Genomes Genet* 9:3501–3511. doi:10.1534/g3.119.400366.

Yoboue, E. D., Sitia, R., and Simmen, T. 2018. Redox crosstalk at endoplasmic reticulum (ER) membrane contact sites (MCS) uses toxic waste to deliver messages. *Cell Death and Disease* 9:331. doi:10.1038/s41419-017-0033-4.

You, J., and Chan, Z. L. 2015. ROS regulation during abiotic stress responses in crop plants. *Frontiers in Plant Science* 6:1092. doi:10.3389/fpls.2015.01092.

Zandalinas, S. I., and Mittler, R. 2018. ROS-induced ROS release in plant and animal cells. *Free Radical Biology and Medicine* 122:21–27. doi:10.1016/j.freeradbiomed.2017.11.028.

Zandalinas, S. I., Sengupta, S. Fritschi, F. B., Azad, R. K., Nechushtai, R., and Mittler, R. 2021. The impact of multifactorial stress combination on plant growth and survival. *New Phytologist* 230:1034–1048 doi:10.1111/nph.17232.

Zhang, D. W., Xu, F., Zhang, Z. W., Chen, Y. E., Du, J. B., Jia, S. D., et al. 2010. Effects of light on cyanide-resistant respiration and alternative oxidase function in Arabidopsis seedlings. *Plant Cell and Environment* 33:2121–2131. doi:10.1111/j.1365-3040.2010.02211.x.

Zhang, F. H., Lu, K., Gu, Y. Y., Zhang, L., Li, W. Y., and Li, Z. 2020. Effects of low-temperature stress and brassinolide application on the photosynthesis and leaf structure of tung tree seedlings. *Frontiers in Plant Science* 10:1767. doi:10.3389/fpls.2019.01767.

Zhang, Y., Xi, D. M., Wang, J., Zhu, D. F., and Guo, X. Q. 2009. Functional analysis reveals effects of tobacco alternative oxidase gene (*NtAOX1a*) on regulation of defence responses against abiotic and biotic stresses. *Bioscience Reports* 29:375–383. doi:10.1042/bsr20080133.

Zhang, Z. G., Zhang, Q., Wu, J. X., Zheng, X., Zheng, S., Sun, X. H., et al. 2013. Gene knockout study reveals that cytosolic ascorbate peroxidase 2(*OsAPX2*) plays a critical role in growth and reproduction in rice under drought, salt and cold stresses. *PLoS One* 8:e57472. doi:10.1371/journal.pone.0057472.

9 Molecular Understanding of Crosstalk Between Plant Growth Hormones and Plant Growth Regulators Under Multiple and Combined Abiotic Stress Tolerance

S. G. Harsha
Indian Institute of Horticultural Research
Indira Gandhi Krishi Vishwavidyalaya

H. S. Sheela, K. S. Shivashankara, R. H. Laxman, and M. D. Prathibha
Indian Institute of Horticultural Research

CONTENTS

9.1 INTRODUCTION

All living species' growth and development are regulated by interactions between their genetic makeup and their changing environment. Sessile plants face a series of growth-limiting conditions before completing their life cycles. Thus, they have developed a variety of complex signaling systems to respond to external and internal inputs, which regulate growth and development under stressful situations. Abiotic stressors (heat, drought, salinity, waterlogging, heavy metal toxicity, soil erosion, etc.) negatively influence plant growth, development, and yield under natural conditions, resulting in increased economic losses at the price of global food security (Hossain et al., 2020). The combined

DOI: 10.1201/9781003300564-9

impact of many abiotic stressors on the biological and economic output of a variety of crops is larger than the effects of each stress individually (Javeed et al., 2021). Plant growth regulators (PGRs), defined as natural biosynthetic molecules that control crop plant development, play an important role in abiotic stress mitigation (Takahashi et al., 2019). PGRs interact with complicated signaling networks to equilibrate responses to evolve environmental friendly strains and as a result, reverse the damages induced by stressful environmental circumstances (Iqbal et al., 2019; Kohli et al., 2013). Plants have evolved complicated ways to detect external environmental stimuli and can respond optimally to stress situations with the help of PGRs, which primarily govern plant defense responses through synergistic and antagonistic actions (called signaling crosstalk) (Berens et al., 2019). One of the most dynamic changes noticed under stress is perturbed phytohormone fluxes and its impact in signaling. Selective proteolytic pathways appear to be involved in the highly controlled response attributed to these main hormones in response to varied environmental stressors. Plants respond to growth, developmental, and environmental stimuli by modulating gene expression via phytohormone-mediated signaling pathways in general. PGRs crosstalk with numerous inorganic and organic chemicals such as nitrates, H_2O_2, H_2S, reactive oxygen species, and NO to balance plant growth and development under adverse settings (Xuan et al., 2020). There are differences in receiving numerous hormonal signals from the early phase of stress to the continuous phase of stress during the stress perception process, allowing plants to proceed to a well-organized development. PGRs such as abscisic acid (ABA), salicylic acid (SA), ethylene (ET), and jasmonates (JAs) have been linked to improving plant response to a variety of stressors. PGRs, such as cytokinins (CKs), gibberellins (GAs), auxin, and more recently discovered PGRs including strigolactones (SLs), nitric oxide (NO), and brassinosteroids (BRs) are important in plant growth and development in both normal and stressed environments. This book chapter attempts to illustrate the key aspects of the present knowledge on hormonal and PGR's role in several plant processes and crosstalk concerning abiotic stresses.

9.2 AUXIN

Auxins are a class of plant hormones that are naturally occurring and has primary role in cell division, elongation, and differentiation. It also has a vital role in embryonic development, root and stem tropisms, and apical dominance. The apical meristem of shoots, immature leaves, and seeds is where most auxin is produced. Auxin transport is mediated by carriers such as inflow and efflux carriers, and it plays a crucial role in controlling plant growth and development. The PIN family of auxin efflux carriers is the most studied auxin transporter and they are essential for many developmental processes (Bennett, 2015). Overexpression of putative auxin efflux carrier gene *PIN3t* is involved in the drought tolerance in the rice system, especially in root development (Ren et al., 2016; Zhang et al., 2012). Overexpression of rice *YUC* genes causes the plants to produce a large number of adventitious roots, which helps them adapt to severe environments (Yamamoto et al., 2007). Overexpression of *Yucca 6* gene (a gene involved in auxin biosynthesis pathway) showed enhanced tolerance to water stress in potato (Im Kim et al., 2013).

Auxins in combination with ethylene regulate root growth and its architecture under drought, salinity, flooding, and nutrient stress conditions and are well studied in *Arabidopsis*. Here, ethylene has a negative role in lateral root formation and a positive role in adventitious root formation through modulation of auxin transport was proposed (Strader and Bartel, 2011). Brassinolide treatment increases lateral root development in a transgenic *Arabidopsis* strain overexpressing cytokinin oxidase/dehydrogenase, revealing different hormonal pathways to similar morphologies (Vercruyssen et al., 2011). Auxin and strigolactones, a hormone that governs branching, converge for the regulation of root hair elongation, and ethylene works as an epistatic to strigolactones, according to the findings (Kapulnik and Koltau, 2014). Similar processes were controlled by distinct hormones through generally non-overlapping transcriptional responses in the superroot2 (SUR2) auxin overproducing mutant, revealing intricate connections between positively and negatively active transcription factors (Nemhauser et al., 2006). RAG1, a member of the small auxin up-regulated (SAUR) protein family, regulates hypocotyl elongation and fertility in *Arabidopsis*, indicating that auxin, GA, and maybe ABA interact via this protein. The following description goes on to show how auxin interacts with other hormones to influence developmental processes and stress responses.

9.3 GIBBERELLINS

Gibberellins are PGRs that influence a variety of developmental processes such as stem elongation, germination, blooming, sex expression, and enzyme induction, among others (Yamaguchi, 2008; Iqbal et al., 2011). Gibberellic acids are mostly produced in growth areas such as elongating stems, leaves, and root tips. Anthers and scutellar epithelium have high rates of synthesis (Hirano et al., 2008). Currently there are more than 100 gibberellins isolated and characterized, they are GA1, GA2, GA3, and so on. Among GAs, GA3 is the most commonly and widely studied. Several studies have reported that GA3 plays a positive role in mitigating abiotic stresses.

In tomato, external application of GA3 reduced the stomatal resistance and enhanced plant water use at low salinity (Maggio et al., 2010). Further, the adverse effect of salt stress in soybean was mitigated by the external application of GA3, which in turn increased the level of phytoestrogen content (daidzein and genistein) in the soybean leaves (Hamayun et al., 2010). GA regulates a wide range of physiological processes and stimulates development by inactivating DELLA proteins, which function as repressors. Abiotic stresses, particularly water and salt stress, reduce GA content, which aids in the stabilization of DELLA and inhibits growth. DELLA mutants are more drought and salinity-resistant. DELLA expression is regulated by both ABA and ethylene, by limiting DELLA-dependent growth. ABA and ethylene build in significant amounts in response to water stress and salinity stress, activating their signaling pathways. As a result, ABA and ethylene produce DELLA buildup and, as a result, growth inhibition. These examples demonstrate how stress-induced ABA, ethylene, and GA signals interact and crosstalk at the DELLA level to produce context-dependent growth phenotypes and stress resistance. According to study, the interplay of environmentally driven GA, ABA, ethylene, and cytokinin signals determines plant stress responses (Krouk et al., 2011).

Similar to GA, cytokinin levels drop in response to abiotic stress and nutrient deficiencies, resulting in a significant buildup of ABA. According to recent investigations, ABA-dependent responses help to explain stress-induced cytokinin depletion and growth inhibition. The exact mechanism by which cytokinin depletion affects DELLA proteins in stressed plants is unknown, however, given the high levels of ABA in stressed plants, an indirect role for DELLA in cytokinin homeostasis cannot be ruled out. Similarly, the high-affinity binding of JAZ and DELLA proteins causes growth inhibition in both biotic and abiotic stress situations, which is mediated cooperatively by JA and GA (Qin et al., 2011). It's worth noting that in water- and salt-stressed situations, ethylene enhances DELLA formation and stability while inhibiting GA synthesis. In rice, entrapped ethylene boosts GA sensitivity and upregulates various cellular activities, including stem and leaf development, when submerged. This example emphasizes the numerous ways in which hormonal interactions may trigger growth responses in response to various stressors.

9.4 CYTOKININS

Cytokinins (CKs) are a family of PGRs that are largely responsible for cell division in plant root and shoot systems. This hormone affects apical dominance and delays leaf senescence by stimulating cell proliferation, development, and differentiation. These hormones are found in all the plants, bacteria, mosses, and fungi. In the plant system, cytokinins are present in the rapidly dividing tissues such as root apices and growing shoots. The major function of cytokinin is to increase cell division by maintaining the protein production that is important for mitosis. Overexpression of cytokinin synthesis gene in a perennial grass species showed improved tolerance to heat stress (Xu et al., 2010). A subset of the cytokinin two-component signaling system plays a role in cold, salinity, and temperature stress response in *Arabidopsis* (Feng et al., 2019).

Isopentenyl transferase (IPT) is a major enzyme catalyzing cytokinin production, increasing sink capacity, and also mobilizing major nutrients to different parts of plant. In CK-deficient tobacco shoots and roots, altered sink/source relationships were discovered (Werner et al., 2008). Moisture stress-induced CK synthesis increases the N and C absorption in transgenic plants (Reguera et al., 2013). The C–N interaction governs CK production in response to soil N availability, while CK-regulated N absorption controls N levels in the plant (Kiba et al., 2011). In a C–N-dependent way, transgenic PSARK: IPT rice plants displayed CK-mediated enhanced N assimilation (Reguera et al., 2013). It was hypothesized that a reduction in root tip turgor would disrupt the synthesis and passage of CKs in the root tip, obstructing their passage as well as the accumulation of nutrients accessible in the soil during drought stress (Pavlů et al., 2018).

9.5 ETHYLENE

Ethylene is a major plant hormone involved in regulating plant responses across several stressful environments. Profound knowledge of enzymes, transcription factors, promoters, and inhibitors that act upon ethylene biosynthesis provides information about the possible interplay of ethylene and its associated compounds with other phytohormones and regulators during multiple stresses.

Plants accumulate ethylene under several abiotic stresses, and in turn, it acts upon ABA, auxin, and GA differentially to confer tolerance to these stresses. Crosstalk between ethylene signaling and GA biosynthesis affects root growth via EIN3-DELLA-dependent mechanism, thereby maintaining normal root growth under salinity stress (Achard et al., 2006). Ethylene regulates plants' responses to drought stress by acting on stomatal closure by inhibiting ABA-mediated stomatal movements. This stomatal regulatory function of ethylene occurs in two ways, either by inhibiting ABA-mediated stomatal closure by EIN3 and ETR1 (Tanaka et al., 2005) or by promoting NADPH oxidase-mediated ROS production and antioxidant flavanol synthesis by EIN2, which accumulates in guard cells under stress conditions, thereby negatively regulating stomatal closure by inhibiting ABA-mediated ROS production (Kazan, 2015; Watkins et al., 2014). Ethylene also induces programmed cell death under abiotic stress in a synergistic association with GA. Ethylene biosynthesis downstream transcriptional factors EIN3/EIL1 positively regulate root hair development in association with jasmonic acid by acting upon *JAZ1* gene and thus enhance drought tolerance in *Arabidopsis*. Under flooding stress, ethylene is rapidly accumulated in cells of submerged parts and interacts with ABA and GA to promote stem elongation. Under the flooded condition, ET-responsive ERF transcription factors reduce the inhibitory effect of ABA on GA signaling such that stem elongation can occur (Kazan, 2015). Ethylene-induced ERF transcription factors further accelerate the activation of SNORKELs (SK1 and SK2) and SUB1A, which promote stem growth under water submergence by activating GA responses required for stem elongation (Fukao et al., 2011; Hattori et al., 2009). Under complete submergence, SUB1A regulates the expression of SLENDER RICE1 (SLR1) and SLENDER RICE-LIKE 1 (SLRL1), which in turn suppress GA-mediated stem elongation, thereby preventing flowering inhibition and aids in the completion of the plant's life cycle (Peña-Castro et al., 2011). Ethylene-induced ERF gene SUBMERGENCE1A (SUB1A) increases BR levels by upregulating the transcription of BR biosynthetic genes, which then work together to suppress shoot elongation, keeping the plants in a quiescent state and enabling plant survival under submerged conditions in rice. In drought and ozone stress, ethylene interacts with ABA through stomatal opening (Wilkinson and Davies, 2010), while in floods and low-oxygen stress, it interacts with GA and calcineurin B-like interacting protein kinase (CIPK) through joint action with ERFs (Bailey-Serres and Voesenek, 2010).

Ethylene interacts with auxin to govern root growth during abiotic stressors like salt and drought. A common downstream transcription factor NAC2, which encourages lateral root expansion for improved survival under stressful conditions, has been shown to control ethylene and auxin production concurrently in numerous plants (Nuruzzaman et al., 2013; Shan et al., 2014). Ethylene and auxin also regulate normal root growth at the signaling level via RING finger E3 ubiquitin ligase

MHZ2 (SOR1), which remains inactive along with the auxin receptor and gets released and activated under the presence of ethylene and helps in the degradation of auxin inhibitor, thereby promoting gravitropic root growth (Chen et al., 2018). Under submergence stress in rice, ethylene also increases the expression of MHZ4 and MHZ5, which encode enzymes involved in ABA and carotenoid biosynthesis, respectively, and helps raise ABA content in roots by directing the flow of carotenoid biosynthesis into the ABA biosynthesis pathway (Ma et al., 2014; Yin et al., 2015; Zhou et al., 2020).

9.6 ABSCISIC ACID

ABA signaling includes several receptors, secondary messengers and subcellular compartments, protein kinases (SnRK2), and phosphatases (PP2C) (Xu et al., 2013). During abiotic stresses, two bZIP-TFs, ABI5 and AREB/ABFs, predominantly mediate ABA-mediated gene expression via the ABA-responsive element (ABRE) (Fujita et al., 2011). In rice, production of ABI5-Like1 (ABL1) is induced by salinity, drought, and osmotic stress and it binds to ABREs (G-box), which transcriptionally activates several stress-responsive genes, including ABRE-containing WRKY genes and ABRE-containing genes related to auxin metabolism (Yang et al., 2011). Thus, ABA maintains normal shoot growth under stressed conditions by interacting with auxin and promoting auxin-induced stress-related gene synthesis.

ABA also influences strigolactone biosynthesis as both hormones are synthesized by a common pool of precursors, that is, carotenoids (López-Ráez et al., 2010). NCED, a rate-limiting enzyme in ABA biosynthesis regulates strigolactone biosynthesis genes CCD7 and CCD8, thereby controlling axillary bud growth and root development and confers tolerance against abiotic stresses (Umehara et al., 2010; Koltai, 2011). ABA also confers tolerance against salt stress through a dynamic interplay with strigolactone by increasing the production of secondary messenger H_2O_2, which in turn stimulates the synthesis of strigolactone-related stress-responsive genes (Ren et al., 2018). Several secondary messengers such as mitogen-activated protein kinases (MAPKs), β-aminobutyric acid (BABA), and ROS are also involved in ABA signaling in response to abiotic stress. These compounds help in the interplay between ABA and other stress-related hormones such as SA, jasmonic acid, and ethylene under drought, heat, and biotic stress (Kumar et al., 2019; Yao et al., 2020). ABA also interacts with several gasotransmitters such as nitric oxide (NO), hydrogen sulfide (H_2S), and carbon monoxide (CO), which form an upstream network in ABA-mediated calcium signaling that governs stomatal closure under abiotic stress in association with ethylene and auxin (Bharath et al., 2021).

ABA alters the auxin distribution and regulates seed germination and also root development during abiotic stress by inducing the synthesis of auxin response factor2 (arf2), which is involved in auxin signaling (Wang et al., 2011). R2R3-type MYB transcription factor, MYB96, appears to regulate stress tolerance by integrating ABA and auxin signaling, which further results in the regulation of some auxin-responsive genes under drought stress (Seo et al., 2009) and two Cys2/His2 zinc-finger proteins, AZF1 and AZF2, are involved in suppressing ABA-repressive and auxin-inducible genes during salinity stress (Kodaira et al., 2011). Downstream transcription factor for ABA, MYB96 promotes cuticular wax production via upregulating the enzyme 3-ketoacyl-CoA synthase 2 (KCS2), which is involved in very-long-chain fatty acid biosynthesis, as well as the *RD22* gene, which inhibits lateral root development (Seo et al., 2009; Atkinson and Urwin, 2012). This MYB96 is also responsible for ABA–SA-mediated stomatal closure under drought by increasing the levels of SA by influencing the synthesis of SA-INDUCTION DEFICIENT 2 (SID2), the gene responsible for expressing isochorismate synthase, a key enzyme in SA biosynthetic pathway (Munemasa et al., 2015).

9.7 BRASSINOSTEROIDS

Brassinosteroids act together with other phytohormones to carry out certain reactions under heat and oxidative stress to confer tolerance such as H_2O_2 creation, respiratory burst oxidase homolog 1 (RBOH1) quality articulation, and NADPH oxidase action (Zhou et al., 2014). Xia et al.

(2014) reported that BR regulates stomatal movements and it induces stomatal closing by enhancing endogenous ABA level and regulation of H_2O_2-dependent redox poise, which in turn modulated the GSSG: GSH pool in the guard cells and induces stomatal closure. BR induces tolerance against many abiotic stresses by means of increasing the production of H_2O_2 that acts as a secondary messenger in signaling and activates diverse stimuli, enhancing the activities of a variety of stress-related molecules, such as dehydrins, transcription factors, heat shock proteins, and antioxidant enzymes (Zhu et al., 2013). Brassinosteroids also exert a positive effect on root development under drought by helping stomatal closure by triggering the burst of NO synthesis (Shi et al., 2015). Hydrogen sulfide (H_2S) acts as a second signaling messenger downstream of BR signaling and plays an irreplaceable role in BR-mediated stomatal movement and prevents oxidative damage to the photosynthetic system during chilling and drought stresses (Ma et al., 2020).

BR also interplays with auxin under stressed conditions and it has been found that it upregulates the expression of *AXR3/IAA17, AXR2/IAA7, SLR/IAA14*, and *IAA28*, which in turn activate several auxin-responsive genes that control root development under stress and among these AXR3/IAA17 plays a central link between these hormonal crosstalks (Kim et al., 2006). Under stressed conditions, BR perception increases cellular levels of H_2O_2, causing oxidative modification of BZR1 and BES1, the key transcription factors in BR signaling, promoting their interaction with PIF4 (phytochrome interacting factors 4) and ARF6 (auxin response factor 6), promoting root meristem development (Lv et al., 2018). Under drought, salt, and chilling stresses, brassinosteroids increase the production of ethylene and the expression level of AOX, which eliminates ROS accumulation and prevents photo-oxidation in the plant photosystem, and enhances tolerance against these stress conditions in *Cucumis sativus* L. (Wei et al., 2015). BRs are also involved in crosstalk with ethylene-mediated defense responses by its involvement in attenuating the ethylene-induced xylanase (Eix), an effective elicitor of plant defense responses (Bar et al., 2010).

Brassinosteroids interact with ABA and regulates ABA signaling by phosphorylating SnRK2.2, SnRK2.3, and SnRK2.6. BR regulates this with the help of BIN2 kinase (Belin et al., 2006), BRI1 co-receptor BAK1 (Shang et al., 2016), and downstream transcription factor BR-INSENSITIVEEMS-SUPPRESSOR 1 (BES1) (Lopez-Molina et al., 2002), which together upregulate expression of ABA-related stress genes through phosphorylation of the ABI5 transcription factor (Wang et al., 2018). BR confers tolerance against oxidative stress by decreasing the uptake of Cu and upregulating the expression of RsCOPT2 and RsCOPT1 genes, which are involved in spermidine and putrescine biosynthesis (Takahashi and Kakehi, 2010).

9.8 NITRIC OXIDE

Nitric oxide (NO) is a volatile gasotransmitter that plays a major role in gravitropism, seed germination, photosynthesis, root growth, pollen tube growth, floral transition, fruit ripening, mitochondrial functioning, senescence and seed dormancy, and stomatal movement (Mostofa et al., 2015). NO is a crucial signaling molecule that interacts with the main hormones and other endogenous chemicals in the signaling cascade. In this case, NO serves as a supplementary messenger for stress-related plant hormones. Recent research has revealed that NO plays a vital role in the modulation of the signaling link during control and stressful situations. It also plays a role in post-translational changes along with regulation of oxidative stress via the antioxidant defense system (Sharma et al., 2020). NO regulates tissue water content, enzymes, nutritional status, oxidative status, and primary and secondary metabolism (Dong et al., 2014; Wang X. et al., 2020). NO reduces the effects of stress on plants by upregulating the antioxidant defense system, which is mediated via SA and H_2O_2 (Singh et al., 2015).

ROS generation under stressful conditions activates antioxidants and transcription factors, resulting in ABA–NO crosstalk (Lu et al., 2009). Plant stress response is influenced by other signaling molecules namely cGMP, MAPK, and type 2C protein phosphatases, which are downstream of the NO–ABA interaction and antioxidant system (Mioto and Mercier, 2013). In mutant studies, the function of this crosstalk in salt stress (Kong et al., 2016), drought stress, heat stress (Zandalinas et al., 2016), and plant callus thermotolerance has been hypothesized. GAs have been discovered to influence hypocotyl formation in combination with DELLA protein degradation. ET–NO crosstalk stimulates MAPK pathways under cadmium stress in soybean and pea seedlings, resulting in polyamine synthesis (Rodríguez-Serrano et al., 2006). Treatment with cadmium and spermine in wheat seedlings increases NO generation in the roots, delaying root growth (Groppa et al., 2008). Crosstalk between NO and ET disrupts efficiency signaling, prompting the activation of a slew of genes involved in iron accumulation and transport (García et al., 2010).

9.9 STRIGOLACTONES

Strigolactones (SLs) are a class of phytohormones derived by carotenoids. These were first recognized as germination stimulants in parasitic weeds and as a plant–microbe symbiotic interaction (Akiyama et al., 2005; Xie et al., 2010). SLs also have important functional roles in plant development, leaf senescence, photomorphogenesis, flower development, and repression of shoot branching (Waters et al., 2017). Because of their physiological and molecular regulation systems against diverse abiotic stressors, SLs have got a lot of attention. In *Arabidopsis*, where SL-mediated stomatal closure was controlled through *PIN1*, *PIN4*, *PIN3*, or *PIN2* genes, but cytokinin negatively regulates the stomatal closure process (Nishiyama et al., 2011; Urquhart et al., 2015). Furthermore, SL-deficient mutants are vulnerable to water deficiency and salt stress, indicating that positive regulatory mechanisms in stress acclimation are involved (Cardinale et al., 2018). The combined activities of cytokinin, ABA, and SL in stomatal control and leaf senescence for plant adaptation to harsh situations show the complexity level involved in crosstalk mechanism (Czarnecki et al., 2013). Thus, these studies provide evidence for crosstalk between SL and other phytohormones in response to adverse environments.

9.10 SALICYLIC ACID

Salicylic acid is a phenolic compound that participates in several processes during plant development, such as seed germination, growth, development, stomatal regulation, photosynthesis, and fruit ripening (Pérez-Llorca et al., 2019). NO signaling is also known to control plant growth and stress responses in association with other key regulators. Under many abiotic stress conditions, stomatal closure is controlled through SA-induced protein kinases, cGMP (cyclic 30–50 guanosine monophosphate), and cADPR (cyclic ADPribose), and in NO accumulation in *Arabidopsis* guard cells (Hao et al., 2010). In *Arabidopsis*, SA aided the restoration of membrane potential and prevented salt stress-induced K+ loss via the GORK channel (Jayakannan et al., 2013). The SA improved salt stress tolerance by increasing the Chl, carotenoids, maintaining membrane integrity through accumulation of K^+ and soluble sugars in barley (El-Tayeb, 2005). The SA improved photosynthetic process through enhanced nitrogen and sulfur uptake. It was followed by improved antioxidant metabolism in mung bean cultivars under salt stress (Nazar et al., 2011).

9.11 CONCLUSIONS

Individual and combined abiotic stresses affect biological and commercial yields of multiple crops globally, necessitating the development of physiologically feasible stress management techniques. Hormones and PGRs, which primarily control plant defensive responses through synergistic and antagonistic events, have evolved multifaceted ways governed by various receptors to obtain external signals and then activate an ideal response counter to stress circumstances via hormones and PGRs, which primarily control plant defensive responses through synergistic and antagonistic events, a phenomenon known to be signaling crosstalk. Plant abiotic stress management is well-known for the interplay between phytohormones with stress. The major option under such conditions includes both traditional and recently found phytohormones, and the list is constantly increasing as new growth-stimulatory phytochemicals are identified. Hormone and PGR signaling is extremely complicated, and it has a widespread effect on plant growth and development. Understanding the complicated process of abiotic stress tolerance necessitates a thorough understanding of hormone responses at several periods and in response to diverse stressors. Furthermore, new PGRs such as nitric oxide, SA, strigolactones, and other signaling complexes play critical roles during stress and hormone response regulation, which could be studied more closely in the near future to better understand the cell-signaling crosstalk network. Thus, quite a few transcription factors and genes have played vital roles in plant growth and development in recent years, as well as responding to multiple abiotic stimuli regulated by PGRs, with the intention of developing stress-tolerant crops in the future.

REFERENCES

Achard, P., Cheng, H., De Grauwe, L., Decat, J., Schoutteten, H., Moritz, T., Van Der Straeten, D., Peng, J., and Harberd, N. P. 2006. Integration of plant responses to environmentally activated phytohormonal signals. *Science* 311(5757): 91–94.

Akiyama, K., Matsuzaki, K. I., and Hayashi, H. 2005. Plant sesquiterpenes induce hyphal branching in arbuscular mycorrhizal fungi. *Nature* 435(7043): 824–827.

Atkinson, N. J., and Urwin, P. E. 2012. The interaction of plant biotic and abiotic stresses: From genes to the field. *Journal of Experimental Botany* 63: 3523–3544.

Bailey-Serres, J., and Voesenek, L. A. C. J. 2010. Life in the balance: A signaling network controlling survival of flooding. *Current Opinion in Plant Biology* 13: 489–494.

Bar, M., Sharfman, M., Ron, M., and Avni, A. 2010. BAK1 is required for the attenuation of ethylene-inducing xylanase (Eix)-induced defense responses by the decoy receptor LeEix1. *The Plant Journal* 63(5): 791–800.

Belin, C., de Franco, P. O., Bourbousse, C., Chaignepain, S., Schmitter, J. M., Vavasseur, A., Giraudat, J., Barbier-Brygoo, H., and Thomine, S. 2006. Identification of features regulating OST1 kinase activity and OST1 function in guard cells. *Plant Physiology* 141(4): 1316–1327.

Bennett, T. 2015. PIN proteins and the evolution of plant development. *Trends in Plant Science* 20(8): 498–507.

Berens, M. L., Wolinska, K. W., Spaepen, S., Ziegler, J., Nobori, T., Nair, A., Krüler, V., Winkelmüller, T. M., Wang, Y., Mine, A., and Becker, D. 2019. Balancing trade-offs between biotic and abiotic stress responses through leaf age-dependent variation in stress hormone cross-talk. *Proceedings of the National Academy of Sciences* 116(6): 2364–2373.

Bharath, P., Gahir, S., and Raghavendra, A. S. 2021. Abscisic acid-induced stomatal closure: An important component of plant defense against abiotic and biotic stress. *Frontiers in Plant Science* 12: 324.

Cardinale, F., Korwin Krukowski, P., Schubert, A., and Visentin, I. 2018. Strigolactones: Mediators of osmotic stress responses with a potential for agrochemical manipulation of crop resilience. *Journal of Experimental Botany* 69(9): 2291–2303.

Chen, H., Ma, B., Zhou, Y., He, S. J., Tang, S. Y., Lu, X., Xie, Q., Chen, S. Y., and Zhang, J.S. 2018. E3 ubiquitin ligase SOR1 regulates ethylene response in rice root by modulating stability of Aux/IAA protein. *Proceedings of the National Academy of Sciences* 115(17), 4513–4518.

Czarnecki, O., Yang, J., Weston, D. J., Tuskan, G. A., and Chen, J. G. 2013. A dual role of strigolactones in phosphate acquisition and utilization in plants. *International Journal of Molecular Sciences* 14(4): 7681–7701.

Dong, Y. J., Jinc, S. S., Liu, S., Xu, L. L. and Kong, J., 2014. Effects of exogenous nitric oxide on growth of cotton seedlings under NaCl stress. *Journal of Soil Science and Plant Nutrition* 14(1): 1–13.

El-Tayeb, M. A. 2005. Response of barley grains to the interactive e. ect of salinity and salicylic acid. *Plant Growth Regulation* 45(3): 215–224.

Feng, Y., Liu, J., Zhai, L., Gan, Z., Zhang, G., Yang, S., Wang, Y., Wu, T., Zhang, X., Xu, X., and Han, Z. 2019. Natural variation in cytokinin maintenance improves salt tolerance in apple rootstocks. *Plant, Cell & Environment* 42(2): 424–436.

Fujita, Y., Fujita, M., Shinozaki, K., and Yamaguchi-Shinozaki, K., 2011. ABA-mediated transcriptional regulation in response to osmotic stress in plants. *Journal of Plant Research* 124(4): 509–525.

Fukao, T. et al. 2011. The submergence tolerance regulator SUB1A mediates crosstalk between submergence and drought tolerance in rice. *Plant Cell* 23: 412–427.

García, M. J., Lucena, C., Romera, F. J., Alcántara, E., and Pérez-Vicente, R. 2010. Ethylene and nitric oxide involvement in the up-regulation of key genes related to iron acquisition and homeostasis in Arabidopsis. *Journal of Experimental Botany* 61(14): 3885–3899.
Groppa, M. D., Rosales, E. P., Iannone, M. F., and Benavides, M. P. 2008. Nitric oxide, polyamines and Cd-induced phytotoxicity in wheat roots. *Phytochemistry* 69(14): 2609–2615.
Hamayun, M., Khan, S. A., Khan, A. L., Shinwari, Z. K., Hussain, J., Sohn, E. Y., Kang, S. M., Kim, Y. H., Khan, M. A., and Lee, I. J. 2010. Effect of salt stress on growth attributes and endogenous growth hormones of soybean cultivar Hwangkeumkong. *Pakistan Journal of Botany* 42(5): 3103–3112.
Hao, F., Zhao, S., Dong, H., Zhang, H., Sun, L., and Miao, C. 2010. Nia1 and Nia2 are involved in exogenous salicylic acid-induced nitric oxide generation and stomatal closure in Arabidopsis. *Journal of Integrative Plant Biology* 52(3): 298–307.
Hattori, Y. et al. 2009. The ethylene response factors SNORKEL1 and SNORKEL2 allow rice to adapt to deep water. *Nature* 460: 1026–1030
Hirano, K., Ueguchi-Tanaka, M., and Matsuoka, M. 2008. GID1-mediated gibberellin signaling in plants. *Trends in Plant Science* 13(4): 192–199.
Hossain, A., Farooq, M., EL Sabagh, A., Hasanuzzaman, M., Erman, M., and Islam, T. 2020. Morphological, physiobiochemical and molecular adaptability of legumes of fabaceae to drought stress, with special reference to *Medicago sativa* L. In Hasanuzzaman, M., Araújo, S., and Gill, S. S. *The plant family fabaceae* (pp. 289–317). Singapore: Springer Nature.
Im Kim, J., Baek, D., Park, H. C., Chun, H. J., Oh, D. H., Lee, M. K., Cha, J. Y., Kim, W. Y., Kim, M. C., Chung, W. S., and Bohnert, H. J. 2013. Overexpression of arabidopsis YUCCA6 in potato results in high-auxin developmental phenotypes and enhanced resistance to water deficit. *Molecular Plant* 6(2): 337–349.
Iqbal, M. A., Abdul, H., Siddiqui, M. H., Imtiaz, H., Tanveer, A., Saira, I., et al. 2019. A meta-analysis of the impact of foliar feeding of micronutrients on productivity and revenue generation of forage crops. *Planta Daninha* 37: e019189237. doi: 10.1590/s0100-83582019370100046.
Iqbal, N., Nazar, R., Khan, M. I. R., Masood, A., and Khan, N. A. 2011. Role of gibberellins in regulation of source–sink relations under optimal and limiting environmental conditions. *Current Science* 10(7): 998–1007.
Javeed, H. M. R., Ali, M., Skalicky, M., Nawaz, F., Qamar, R., Rehman, A. U., Faheem, M., Mubeen, M., Iqbal, M. M., Rahman, M. H. U., and Vachova, P. 2021. Lipoic acid combined with melatonin mitigates oxidative stress and promotes root formation and growth in salt-stressed canola seedlings (*Brassica napus* L.). *Molecules* 26(11): 3147.
Jayakannan, M., Bose, J., Babourina, O., Rengel, Z., and Shabala, S. 2013. Salicylic acid improves salinity tolerance in arabidopsis by restoring membrane potential and preventing salt-induced K^+ loss via a GORK channel. *Journal of Experimental Botany* 64(8): 2255–2268.
Kapulnik, Y., and Koltai, H. 2014. Strigolactone involvement in root development, response to abiotic stress, and interactions with the biotic soil environment. *Plant Physiology* 166(2): 560–569.
Kazan, K. 2015. Diverse roles of jasmonates and ethylene in abiotic stress tolerance. *Trends in Plant Science* 20(4): 219–229.
Kiba, T., Kudo, T., Kojima, M., and Sakakibara, H. 2011. Hormonal control of nitrogen acquisition: Roles of auxin, abscisic acid, and cytokinin. *Journal of Experimental Botany* 62(4): 1399–1409.
Kim, H., Park, P. J., Hwang, H. J., Lee, S. Y., Oh, M. H., and Kim, S. G. 2006. Brassinosteroid signals control expression of the AXR3/IAA17 gene in the cross-talk point with auxin in root development. *Bioscience, Biotechnology, and Biochemistry* 70(4): 768–773.
Kodaira, K. S., Qin, F., Tran, L.S. et al. 2011. *Arabidopsis* Cys2/His2 zinc-finger proteins AZF1 and AZF2 negatively regulate abscisic acid-repressive and auxin-inducible genes under abiotic stress conditions. *Plant Physiology* 157: 742–756.
Kohli, A., Sreenivasulu, N., Lakshmanan, P., and Kumar, P. P. 2013. The phytohormone crosstalk paradigm takes center stage in understanding how plants respond to abiotic stresses. *Plant Cell Reports* 32(7): 945–957.
Koltai, H. 2011. Strigolactones are regulators of root development. *New Phytologist* 190(3): 545–549.
Kong, J., Dong, Y., Song, Y., Bai, X., Tian, X., Xu, L., Liu, S., and He, Z. 2016. Role of exogenous nitric oxide in alleviating iron deficiency stress of peanut seedlings (*Arachis hypogaea* L.). *Journal of Plant Growth Regulation* 35(1): 31–43.
Krouk, G., Ruffel, S., Gutiérrez, R. A., Gojon, A., Crawford, N. M., Coruzzi, G. M., and Lacombe, B. 2011. A framework integrating plant growth with hormones and nutrients. *Trends in Plant Science* 16(4): 178–182.

Kumar, M., Kesawat, M. S., Ali, A., Lee, S. C., Gill, S. S., and Kim, H. U. 2019. Integration of abscisic acid signaling with other signaling pathways in plant stress responses and development. *Plants* 8(12): 592.

Lopez-Molina, L., Mongrand, S., McLachlin, D. T., Chait, B. T., and Chua, N. H. 2002. ABI5 acts downstream of ABI3 to execute an ABA-dependent growth arrest during germination. *The Plant Journal* 32(3): 317–328.

López-Ráez, J. A., Kohlen, W., Charnikhova, T., Mulder, P., Undas, A. K., Sergeant, M. J., Verstappen, F., Bugg, T. D., Thompson, A. J., Ruyter-Spira, C., and Bouwmeester, H. 2010. Does abscisic acid affect strigolactone biosynthesis? *New Phytologist* 187(2): 343–354.

Lu, S., Su, W., Li, H., and Guo, Z. 2009. Abscisic acid improves drought tolerance of triploid bermudagrass and involves H2O2-and NO-induced antioxidant enzyme activities. *Plant Physiology and Biochemistry* 47(2): 132–138.

Lv, W., Sui, L., Yan, X., Xie, H., Jiang, L., Geng, C., Li, Q., Yao, X., Kong, Y., and Cao, J. 2018. ROS-dependent Atg4 upregulation mediated autophagy plays an important role in Cd-induced proliferation and invasion in A549 cells. *Chemico-Biological Interactions* 279: 136–144.

Ma, B., Yin, C. C., He, S. J., Lu, X., Zhang, W. K., Lu, T. G., Chen, S. Y., and Zhang, J. S. 2014. Ethylene-induced inhibition of root growth requires abscisic acid function in rice (*Oryza sativa* L.) seedlings. *PLoS Genetics* 10(10): e1004701.

Ma, Y., Shao, L., Zhang, W., and Zheng, F. 2020. Hydrogen sulfide induced by hydrogen peroxide mediates brassinosteroid-induced stomatal closure of *Arabidopsis thaliana*. *Functional Plant Biology* 48(2): 195–205.

Maggio, A., Barbieri, G., Raimondi, G., and De Pascale, S. 2010. Contrasting effects of GA3 treatments on tomato plants exposed to increasing salinity. *Journal of Plant Growth Regulation* 29(1): 63–72.

Mioto, P. T., and Mercier, H. 2013. Abscisic acid and nitric oxide signaling in two different portions of detached leaves of *Guzmania monostachia* with CAM up-regulated by drought. *Journal of Plant Physiology* 170(11): 996–1002.

Mostofa, M. G., Fujita, M., and Tran, L. S. P. 2015. Nitric oxide mediates hydrogen peroxide-and salicylic acid-induced salt tolerance in rice (*Oryza sativa* L.) seedlings. *Plant Growth Regulation* 77(3): 265–277.

Munemasa, S., Hauser, F., Park, J., Waadt, R., Brandt, B., and Schroeder, J. I. 2015. Mechanisms of abscisic acid-mediated control of stomatal aperture. *Current Opinion in Plant Biology* 28: 154–162.

Nazar, R., Iqbal, N., Syeed, S., and Khan, N. A., 2011. Salicylic acid alleviates decreases in photosynthesis under salt stress by enhancing nitrogen and sulfur assimilation and antioxidant metabolism differentially in two mungbean cultivars. *Journal of Plant Physiology* 168(8): 807–815.

Nemhauser, J. L., Hong, F., and Chory, J. 2006. Different plant hormones regulate similar processes through largely nonoverlapping transcriptional responses. *Cell* 126(3): 467–475.

Nishiyama, R., Watanabe, Y., Fujita, Y., Le, D. T., Kojima, M., Werner, T., Vankova, R., Yamaguchi-Shinozaki, K., Shinozaki, K., Kakimoto, T., and Sakakibara, H. 2011. Analysis of cytokinin mutants and regulation of cytokinin metabolic genes reveals important regulatory roles of cytokinins in drought, salt and abscisic acid responses, and abscisic acid biosynthesis. *The Plant Cell* 23(6): 2169–2183.

Nuruzzaman, M., Sharoni, A. M., and Kikuchi, S. 2013. Roles of NAC transcription factors in the regulation of biotic and abiotic stress responses in plants. *Frontiers in Microbiology* 4: 248.

Pavlů, J., Novák, J., Koukalová, V., Luklová, M., Brzobohatý, B., and Černý, M. 2018. Cytokinin at the crossroads of abiotic stress signalling pathways. *International Journal of Molecular Sciences* 19(8): 2450.

Peña-Castro, J. M. et al. 2011. Expression of rice SUB1A and SUB1C transcription factors in Arabidopsis uncovers flowering inhibition as a submergence tolerance mechanism. The *Plant Journal* 67: 434–446.

Pérez-Llorca, M., Muñoz, P., Müller, M., and Munné-Bosch, S. 2019. Biosynthesis, metabolism and function of auxin, salicylic acid and melatonin in climacteric and non-climacteric fruits. *Frontiers in Plant Science* 10: 136.

Qin, F., Kodaira, K. S., Maruyama, K., Mizoi, J., Tran, L. S. P., Fujita, Y., Morimoto, K., Shinozaki, K., and Yamaguchi-Shinozaki, K. 2011. Spindly, a negative regulator of gibberellic acid signaling, is involved in the plant abiotic stress response. *Plant Physiology* 157(4): 1900–1913.

Reguera, M., Peleg, Z., Abdel-Tawab, Y. M., Tumimbang, E. B., Delatorre, C. A., and Blumwald, E. 2013. Stress-induced cytokinin synthesis increases drought tolerance through the coordinated regulation of carbon and nitrogen assimilation in rice. *Plant Physiology* 163(4): 1609–1622.

Ren, C. G., Kong, C. C., and Xie, Z. H. 2018. Role of abscisic acid in strigolactone-induced salt stress tolerance in arbuscular mycorrhizal *Sesbania cannabina* seedlings. *BMC Plant Biology* 18(1): 1–10.

Ren, D., Rao, Y., Wu, L., Xu, Q., Li, Z., Yu, H., Zhang, Y., Leng, Y., Hu, J., Zhu, L. and Gao, Z., 2016. The pleiotropic abnormal flower and dwarf1 affects plant height, floral development and grain yield in rice. *Journal of Integrative Plant Biology* 58(6): 529–539.

Rodríguez-Serrano, M. A. R. Í. A., Romero-Puertas, M. C., Zabalza, A. N. A., Corpas, F. J., Gomez, M., Del Rio, L. A., and Sandalio, L. M. 2006. Cadmium effect on oxidative metabolism of pea (*Pisum sativum* L.) roots. Imaging of reactive oxygen species and nitric oxide accumulation in vivo. *Plant, Cell & Environment* 29(8): 1532–1544.

Seo, P. J., Xiang, F., Qiao, M. et al. 2009.The MYB96 transcription factor mediates abscisic acid signaling during drought stress response in *Arabidopsis*. *Plant Physiology* 151: 275–289.

Shan, W., Kuang, J. F., Wj, L., and Chen, J. Y. 2014. Banana fruit NAC transcription factor MaNAC1 is a direct target of MaICE1 and involved in cold stress through interacting with MaCBF1. *Plant Cell & Environment* 37: 2116–2127.

Shang, Y., Dai, C., Lee, M. M., Kwak, J. M., and Nam, K. H. 2016. BRI1-associated receptor kinase 1 regulates guard cell ABA signaling mediated by open stomata 1 in Arabidopsis. *Molecular Plant* 9(3): 447–460.

Sharma, L., Priya, M., Kaushal, N., Bhandhari, K., Chaudhary, S., Dhankher, O. P., Prasad, P. V., Siddique, K. H., and Nayyar, H. 2020. Plant growth-regulating molecules as thermoprotectants: Functional relevance and prospects for improving heat tolerance in food crops. *Journal of Experimental Botany* 71(2): 569–594.

Shi, C., Qi, C., Ren, H., Huang, A., Hei, S., and She, X. 2015. Ethylene mediates brassinosteroid-induced stomatal closure via Gα protein-activated hydrogen peroxide and nitric oxide production in Arabidopsis. *The Plant Journal* 82(2): 280–301.

Singh, A. P., Dixit, G., Mishra, S., Dwivedi, S., Tiwari, M., Mallick, S., Pandey, V., Trivedi, P. K., Chakrabarty, D., and Tripathi, R. D. 2015. Salicylic acid modulates arsenic toxicity by reducing its root to shoot translocation in rice (*Oryza sativa* L.). *Frontiers in Plant Science* 6: 340.

Strader, L. C., and Bartel, B. 2011. Transport and metabolism of the endogenous auxin precursor indole-3-butyric acid. *Molecular Plant* 4(3): 477–486.

Takahashi, F., Hanada, K., Kondo, T., and Shinozaki, K., 2019. Hormone-like peptides and small coding genes in plant stress signaling and development. *Current Opinion in Plant Biology* 51: 88–95.

Takahashi, T., and Kakehi, J. I. 2010. Polyamines: Ubiquitous polycations with unique roles in growth and stress responses. *Annals of Botany* 105(1): 1–6.

Tanaka, Y., Sano, T., Tamaoki, M., Nakajima, N., Kondo, N., and Hasezawa, S., 2005. Ethylene inhibits abscisic acid-induced stomatal closure in Arabidopsis. *Plant Physiology* 138(4): 2337–2343.

Umehara, M., Hanada, A., Magome, H., Takeda-Kamiya, N., and Yamaguchi, S. 2010. Contribution of strigolactones to the inhibition of tiller bud outgrowth under phosphate deficiency in rice. *Plant and Cell Physiology* 51(7): 1118–1126.

Urquhart, S., Foo, E., and Reid, J. B. 2015. The role of strigolactones in photomorphogenesis of pea is limited to adventitious rooting. *Physiologia Plantarum* 153(3): 392–402.

Vercruyssen, L., Gonzalez, N., Werner, T., Schmülling, T., and Inzé, D. 2011. Combining enhanced root and shoot growth reveals cross talk between pathways that control plant organ size in Arabidopsis. *Plant Physiology* 155(3): 1339–1352.

Wang, H., Tang, J., Liu, J., Hu, J., Liu, J., Chen, Y., Cai, Z., and Wang, X. 2018. Abscisic acid signaling inhibits brassinosteroid signaling through dampening the dephosphorylation of BIN2 by ABI1 and ABI2. *Molecular Plant* 11(2): 315–325.

Wang, L., Hua, D., He, J., Duan, Y., Chen, Z., Hong, X., and Gong, Z. 2011. Auxin Response Factor2 (ARF2) and its regulated homeodomain gene HB33 mediate abscisic acid response in Arabidopsis. *PLoS Genetics* 7(7): e1002172.

Wang, X., Ruan, M., Wan, Q., He, W., Yang, L., Liu, X., He, L., Yan, L., and Bi, Y. 2020. Nitric oxide and hydrogen peroxide increase glucose-6-phosphate dehydrogenase activities and expression upon drought stress in soybean roots. *Plant Cell Reports* 39(1): 63–73.

Waters, M. T., Gutjahr, C., Bennett, T., and Nelson, D. C. 2017. Strigolactone signaling and evolution. *Annual Review of Plant Biology* 68: 291–322.

Watkins, J. M., Hechler, P. J., and Muday, G. K. 2014. Ethylene-induced flavonol accumulation in guard cells suppresses reactive oxygen species and moderates stomatal aperture. *Plant Physiology* 164(4): 1707–1717.

Wei, L., Deng, X. G., Zhu, T., Zheng, T., Li, P. X., Wu, J. Q., Zhang, D. W., and Lin, H. H. 2015. Ethylene is involved in brassinosteroids induced alternative respiratory pathway in cucumber (*Cucumis sativus* L.) seedlings response to abiotic stress. *Frontiers in Plant Science* 6: 982.

Werner, T., Holst, K., Pörs, Y., Guivarc'h, A., Mustroph, A., Chriqui, D., Grimm, B., and Schmülling, T. 2008. Cytokinin deficiency causes distinct changes of sink and source parameters in tobacco shoots and roots. *Journal of Experimental Botany* 59(10): 2659–2672.

Wilkinson, S., Davies, W. J. 2010. Drought, ozone, ABA and ethylene: New insights from cell to plant to community. *Plant Cell & Environ*ment 33: 510–525.

Xia, X. J., Gao, C. J., Song, L. X., Zhou, Y. H., Shi, K. A. I., and Yu, J. Q. 2014. Role of H_2O_2 dynamics in brassinosteroid-induced stomatal closure and opening in *Solanum lycopersicum*. *Plant, Cell & Environment* 37(9): 2036–2050.

Xie, X., Yoneyama, K., and Yoneyama, K. 2010. The strigolactone story. *Annual Review of Phytopathology* 48: 93–117.

Xu, Y., Gianfagna, T., and Huang, B. 2010. Proteomic changes associated with expression of a gene (ipt) controlling cytokinin synthesis for improving heat tolerance in a perennial grass species. *Journal of Experimental Botany* 61(12): 3273–3289.

Xu, Z. Y., Kim, D. H., and Hwang, I., 2013. ABA homeostasis and signaling involving multiple subcellular compartments and multiple receptors. *Plant cell Reports* 32(6): 807–813.

Xuan, L., Li, J., Wang, X., and Wang, C. 2020. Crosstalk between hydrogen sulfide and other signal molecules regulates plant growth and development. *International Journal of Molecular Sciences* 21(13): 4593.

Yamaguchi, S. 2008. Gibberellin metabolism and its regulation. *Annual Review of Plant Biology* 59: 225–251.

Yamamoto, Y., Kamiya, N., Morinaka, Y., Matsuoka, M., and Sazuka, T. 2007. Auxin biosynthesis by the YUCCA genes in rice. *Plant Physiology* 143(3): 1362–1371.

Yang, X., Yang, Y. N., Xue, L. J., Zou, M. J., Liu, J. Y., Chen, F., and Xue, H. W. 2011. Rice ABI5-Like1 regulates abscisic acid and auxin responses by affecting the expression of ABRE-containing genes. *Plant Physiology* 156(3):1397–1409.

Yao, L., Zhong, Y., Wang, B., Yan, J., and Wu, T. 2020. BABA application improves soybean resistance to aphid through activation of phenylpropanoid metabolism and callose deposition. *Pest Management Science* 76(1): 384–394.

Yin, C. C., Ma, B., Collinge, D. P., Pogson, B. J., He, S. J., Xiong, Q., Duan, K. X., Chen, H., Yang, C., Lu, X., and Wang, Y. Q. 2015. Ethylene responses in rice roots and coleoptiles are differentially regulated by a carotenoid isomerase-mediated abscisic acid pathway. *The Plant Cell* 27(4): 1061–1081.

Zandalinas, S. I., Balfagón, D., Arbona, V., Gómez-Cadenas, A., Inupakutika, M. A., and Mittler, R. 2016. ABA is required for the accumulation of APX1 and MBF1c during a combination of water deficit and heat stress. *Journal of Experimental Botany* 67(18): 5381–5390.

Zhang, Q., Li, J., Zhang, W., Yan, S., Wang, R., Zhao, J., Li, Y., Qi, Z., Sun, Z., and Zhu, Z. 2012. The putative auxin efflux carrier OsPIN3t is involved in the drought stress response and drought tolerance. *The Plant Journal* 72(5): 805–816.

Zhou, J., Wang, J., Li, X., Xia, X. J., Zhou, Y. H., Shi, K., Chen, Z., and Yu, J. Q. 2014. H_2O_2 mediates the crosstalk of brassinosteroid and abscisic acid in tomato responses to heat and oxidative stresses. *Journal of Experimental Botany* 65(15): 4371–4383.

Zhou, Y., Xiong, Q., Yin, C. C., Ma, B., Chen, S. Y., and Zhang, J. S. 2020. Ethylene biosynthesis, signaling, and crosstalk with other hormones in rice. *Small Methods* 4(8): 1900278.

Zhu, Y., Zuo, M., Liang, Y., Jiang, M., Zhang, J., Scheller, H. V., Tan, M., and Zhang, A. 2013. MAP65-1a positively regulates H2O2 amplification and enhances brassinosteroid-induced antioxidant defence in maize. *Journal of Experimental Botany* 64(12): 3787–3802.

10 Multifaceted Roles of Versatile LEA-II Proteins in Plants

Mughair Abdul Aziz and Khaled Masmoudi
Integrative Agriculture Department, College of Agriculture and Veterinary Medicine,
United Arab Emirates University

CONTENTS

10.1 INTRODUCTION

In plants, embryogenesis is a process that involves sequential phases of development that leads to a well-defined tiny plant within the seed. The process is instigated by fertilization forming a zygote that develops into a viable embryo. The embryo primarily remains dormant within the evolving seed to intercept early sprouting. As the seed passes through the maturation stage, the dormant embryo remains metabolically active and viable for at least two growing seasons. This is due to an increase in the embryo's dry mass with a progressive dehydration stage and accumulation of storage compounds.

The consecutive stages of embryonic development during the seed dehydration are underpinned by certain group of genes or gene products and their expression patterns. During the past two decades, the discovery of these genes, investigation of their by-products, and manner of regulation became a crucial scope of study in seed biology. In the recent past few years, much of the researches have been focused

DOI: 10.1201/9781003300564-10

on the roles of such genes in the acquisition of abiotic stress tolerance in plants. The manipulation of these genes can have direct agricultural relevance in the current scenario of climatic changes.

A specific class of embryonic proteins known as the late embryogenesis abundant (LEA) proteins are thought to be intimately related to the embryo viability under the conditions of dehydration (Zheng et al., 2019). They are hydrophilic proteins, which are also induced in the vegetative tissues of plants under severe environmental stresses and in response to the release of plant hormone, abscisic acid (ABA). These protein macromolecules are categorized into eight different classes of proteins based on their phylogenic relationships, sequences of amino acid, and conserved motifs. They are divided into eight LEA groups, from LEA-I to LEA-VI, dehydrin (DHN), and seed maturation protein (SMP) (Battaglia and Covarrubias, 2013). Among the different classes of LEA proteins, LEA-II proteins, or DHNs, are the most abundant groups and are presumed to play a larger role against plant injuries due to temperature, drought, and salinity stresses (Hara, 2010).

LEA-II proteins are key plant metabolites that assemble largely during the late phases of seed maturation and development, and play a counter-product for severe abiotic stress condition. One of the most distinct biochemical sets of LEA-II proteins are DHNs, which are also termed as LEA-D11 (Battaglia and Covarrubias, 2013). The functioning of LEA-II proteins is determined by a group of genes known as LEA-II genes. LEA-II proteins were primarily identified in the developing cotton embryos and are omnipresent in both gymnosperms and angiosperms (Jin et al., 2019). There are numerous investigations that indicate a constructive link between plant abiotic stress tolerance and the induction of LEA-II proteins (Zheng et al., 2019). However, in terms of the present climatic changes and with the advent of contemporary genomics, LEA-II proteins need to be examined for the enumeration of additional important literatures on their functional properties for the coping-up of the plant stresses.

The present chapter is essential for generating an advanced understanding about the biological roles of LEA-II proteins in plant tolerance reactions to climatic changes. This kind of study concedes the supposition of utilizing the LEA-II gene transcripts and proteins for abiotic stress tolerance development in plants through genetic engineering and biotechnology. Therefore, the current chapter overviews the classification, structure, occurrence, and functional characterization of LEA-II proteins under various abiotic stresses. The chapter also provides some insight into the omics approaches used for analyzing the LEA-II proteins and reinforcing their functional relevance in plants.

10.2 DOMAIN SEQUENCE AND CLASSIFICATION OF LEA-II PROTEINS

LEA-II proteins are differentiated from other groups of LEA proteins based on three major conserved motifs (Malik et al., 2017). They can be composed of a highly preserved motif consisting of lysine (Lys)-rich (EKKGIMDKIKEKLPG) 15 amino acid sequence, which is determined as the K-segment, it also contains a Y-segment ([T/V]D[E/Q]YGNP) located in the N-terminal end, and another motif of serine (Ser)-track repeats termed as S-segment. The fundamental element of LEA-II proteins is the substantial K-segment that is placed interspersed in one or more repeats between a non-preserved motif consisting of polar amino acids called the Φ-segments (Figure 10.1) (Zhang et al., 2018a). Depending on the replication and arrangement of these three preserved motif sequences, LEA-II proteins are present in five different subsets: K_n, K_nS, SK_n, Y_nK_n, and Y_nSK_n.

The K-segment comprising LEA-II proteins are categorized under the K-subgroup of LEA-II proteins, in which the K-segment are found to occur in 1–11 copies in the domain sequence (Malik et al., 2017). Y-subgroup of LEA-II proteins consists of the Y-segment in their amino acid chains that occurs in 1–35 jointly repeats, similar in order to the nucleotide binding sites of bacterial and plant protein chaperones (Zaman Khan et al., 2020). Those with an S-segment followed by a K-segment are grouped into the LEA-II proteins SK-subgroup. The Ser residues of LEA-II proteins are phosphorylated by protein kinase, promoting their movement from the nucleus to the cytoplasm. Additionally, a recent conserved motif was identified in LEA-II proteins as F-segment

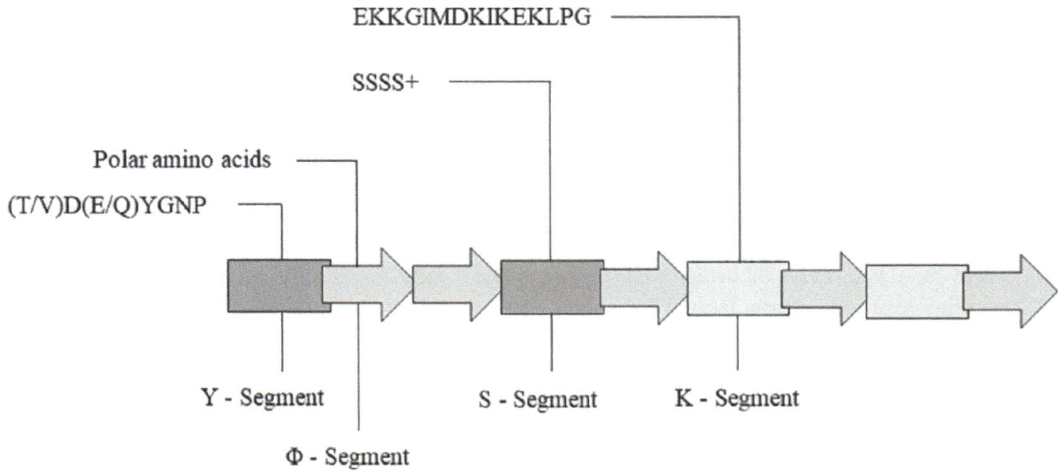

FIGURE 10.1 Linear chain showing a YSK$_n$-type LEA-II protein. The different sequence of the segments occurring in the chain includes Y – tyrosine segment with a sequence of [T/V]D[E/Q]YGNP, Φ – segment includes the polar amino acids, S – segment with Ser track, and the K – Lys segment with a sequence of EKKGIMDKIKEKLPG.

(DRGLFDFLGKK), present at N-terminus, which has possible functional ability to bind to membranes and other protein molecules (Strimbeck, 2017).

The diverse LEA-II genes determined in one plant species varied in terms of their domain sequences and had different modes of gene expression. LEA-II proteins in gymnosperms differed from the angiosperms in the amino acid sequences of the K-segment. The initial study about the identification of LEA-II proteins in Gymnospermae was through Close et al. (1993), which revealed a positive response on polyclonal antibodies that were elevated toward the K-segment consensus of angiosperms *Pinus taeda* and *Ginkgo biloba*. A recent study was performed using the phylogenetic and structural architecture analysis of 426 LEA-II genes from three gymnosperm and 53 angiosperm genomes (Riley et al., 2019). All the five domain sequences were identified in the angiosperms, while gymnosperms possessed only the K- and SK-segments in their domain sequences of amino acids. This indicates that LEA-II gene ancestral group in gymnosperms contained either a K- or SK-segment, whereas Y-segment of LEA-II gene sequences was found initially in angiosperms. The possible occurrence of Y-segment in the angiosperms could have been from the splitting of the YSK-segment out of a K- or SK-segment of LEA-II proteins. However, the occurrence of diverse LEA-II genes in many plant species due to duplication events has resulted in different LEA-II protein structures (Hundertmark and Hincha, 2008).

10.3 STRUCTURAL PROPERTIES OF LEA-II PROTEINS

LEA-II proteins are intrinsically disorder proteins (IDPs) that are immensely hydrophilic with a molecular weight between 9 and 200 KDa. IDPs are those proteins that do not possess a well-defined three-dimensional fold, which may play different biological roles by binding to their target molecules through folding mechanism (coupled folding and binding) (Zhang et al., 2018b). IDPs are part of a number of cellular functions, such as in the modulation of cell division, transcription and translation regulation, transduction of plant stress-signaling pathways, phosphorylating proteins, micro molecules storage, acting as chaperone, and in the regulation of large complexes of multiprotein molecules construction. IDPs lack hydrophobic amino acids such as Val, Leu, and Tyr, and are augmented with charged polar amino acids such as glutamine (Gln), Ser, and glycine (Gly) (Zhang et al., 2018b). Due to lower hydrophobic residues, IDPs lack a tertiary structure due to their inability to form stable hydrophobic core. LEA-II proteins tend to be unstructured and possess many structural properties of IDPs due to the lower occurrence of intramolecular hydrogen bonds between their various residues of amino acids. Similar to IDPs, LEA-II proteins change their structural conformation depending on their ambient microenvironment that results in functional variations (Jin et al., 2019).

In aqueous solutions, LEA-II proteins are present in random coil conformation, in which they form maximum number of hydrogen bonds with the surrounding water molecules and minimum number of hydrogen bonds with the amino acid residues within the protein molecules. Based on a number of studies, it has been confirmed that the decrease in the LEA-II protein hydration status in their encompassing microenvironment and the inclusion of elevated amount of compatible solutes such as NaCl and SDS changes their conformation. These changes can be controlled with the techniques of far-UV circular dichroism (CD). During the conditions of moisture loss, LEA-II protein's K-segments result in intramolecular hydrogen bonding among the chains of their amino acid residues rather than among the amino acid residues and water molecules in the surrounding environment. This predominates the conformation of random coil and turns it into an α-helical structure. In the random coil conformation, there is relative symmetry, which has no deviation from the linearity plane of polarization. However, the α-helix structure is hugely asymmetrical, which deviates from the horizon of linearity. The K-segment of LEA-II proteins determines the amphipathicity of the structure of α-helixes, an essential feature for LEA-II proteins function under the temperature stress (Yang et al., 2015). This structural conformation allows LEA-II proteins to perform as chaperones, which enables them to uphold membranes and also avert the protein aggregation through protein–protein and protein–lipid interactions under the stress conditions.

With the formation of α-helices structure, the negatively charged amino acids are placed on one side of helical structure, hydrophobic amino acids lie on the opposite side, and the positively charged amino acids of LEA-II proteins occur on both the polar and nonpolar side of the helical structure (Saavedra et al., 2006). It has been indicated that some LEA-II proteins found in soybean, GmDHN1, form a left-handed expanded poly (l-proline)-type II (PII) helical organization, rather than forming a classical structure of α-helix with the addition of detergents (Soulages et al., 2003). The study further pointed that such broadened helix structure of PII are due to the larger proportions of amino acids that have the capability to form intermolecular hydrogen bonds as secondary conformation motifs of LEA-II proteins. These protein molecules with extended PII helical conformations tend to contain a high content of Gly and proline (Pro) amino acid residues.

10.4 OCCURRENCE OF LEA-II PROTEINS IN PLANT KINGDOM

LEA-II proteins have been described in different plant species expanding from late 1980s until present time. These proteins were first identified in monocotyledonous and dicotyledonous angiosperms such as the rice and cotton. But later they were also found in other plant taxons such as mosses, lycopods, and gymnosperms. LEA-II proteins were sequenced in the moss plant *Tortula ruralis* and in *Physcomitrella patens* (Saavedra et al., 2006). In comparison to angiosperms and gymnosperms, only a single copy of LEA-II genes was indicated in mosses and lycopods, and multicopies in angiosperm and gymnosperm plant genomes. The LEA-II proteins were also discovered in different lycopods and ferns, such as, *Selaginella lepidophylla* and *Polypodium virginianum* (Iturriaga et al., 2006). Among dicotyledonous plants, LEA-II proteins were identified and examined in model plant species such as *Arabidopsis thaliana*.

A broad genomic study was conducted for the analysis of LEA genes across 60 genomes of different plant species (Hundertmark and Hincha, 2008). The results indicated eight polygene groups for the eight discrete of LEA proteins. The analysis identified the distribution of about 4,836 contrasting genes in the plant LEA genome, where an eminent distribution of LEA-II genes was identified, around 3,126 LEA-II genes found in angiosperms, bryophyte clades, and in lower plant species (Hundertmark and Hincha, 2008). However, the dissimilar copies of LEA-II genes among plant species suggest the duplication and disappearance of these genes among various plants (Strimbeck, 2017). Moreover, the replication of LEA-II genes is interconnected to stress tolerance of different plants, initial evolution of LEA-II genes was associated to the plant water stress adaptation (Riley et al., 2019).

Using GWAS, in both *Vitis vinifera* and *Vitis yeshanensis*, four LEA-II genes were detected. For the *Picea glauca*, 41 LEA-II genes were identified, and it was shown by a phylogenetic reconstruction that the duplication of these genes resulted in their movement to the clades specifically to Pinaceae (Sena et al., 2018). A GWAS on *Setaria italica* and *Zea mays* indicated the presence of seven LEA-II genes, and in *Sorghum bicolor*, five genes that were categorized into the different subclasses of LEA-II proteins (Nagaraju et al., 2019). There was one orthologue connection between LEA-II genes of *Sorghum* with rice and maize, and three with *Setaria italiac*. It was also indicated that *Sorghum bicolor* genes of LEA-II proteins had multiple phosphorylation sites (Nagaraju et al., 2019). In *Actinidia chinensis*, seven LEA-II genes of YSK and SK subgroups were identified using GWAS (Zhang et al., 2021).

LEA genes were studied in arid land woody plants such as in date palm. In a recent study for the identification of LEA genes in *Phoenix dactylifera*, their vegetative parts such as leaves were exposed to ABA treatments to build-up drought stress for the plants (Müller et al., 2017). It was reported by the study that an extensive overlap existed between the differentially expressed genes (DEGs) in date palms under ABA treatments and the drought stress-responsive genes of *Arabidopsis*. An immense amassing of date palm LEA proteins was observed that played a role in *Arabidopsis* abiotic stress tolerance. Moreover, the accumulation of LEA proteins in *P. dactylifera* revealed their feasible role in date palm's tolerance to various abiotic stresses in the arid environment.

Al-Mssallem et al. (2013) performed a whole genome sequencing of khalas date palms. The results from the whole sequencing of date palm genome enumerated identified LEA genes into 8 categories of LEA proteins and into 84 gene members (Müller et al., 2017). The study revealed an abundance of LEA-II genes in the genome assembly of date palm, which contained 62 forms of LEA-II genes. The data from the transcriptomic study displayed composite ABA-instigated LEA-II genes in various *Phoenix dactylifera* developmental stages and vegetative parts.

10.5 PLANT ABIOTIC STRESSES AND LEA-II GENES

LEA-II gene expressions are triggered by numerous abiotic stresses such as temperature, drought, salinity, heavy metals, and by phytohormones such as ABA. Cellular dehydration is a major component of all these stresses that is closely related to the functioning of the LEA-II genes. The overexpression of LEA-II genes enhancing the plant tolerance against abiotic stresses has been reported in several investigations (Hara, 2010). Stress-induced LEA-II genes can consist of myeloblastosis (MYB), NAC gene family transcription factor, ABA-responsive elements (ABREs), drought-responsive element (DRE), and heat shock factor (HSF), as transcriptional activators in the specific regions of promoter. The expression of LEA-II genes is governed by both the ABA-dependent and ABA-independent signaling transduction routes. In the ABA-dependent signaling mechanism, activators of transcription such as ABRE-binding factors (ABF) interact with LEA-II ABREs, and MYB promoter elements. Whereas, in the ABA-independent signaling pathways transcription activators such as drought-responsive element-binding factors (DREB2A and DREB2B) bind to DRE. Different abiotic stresses increase Ca^{2+} and inositol 1, 4, 5 triphosphate (IP3) concentrations within cytosol. Both the Ca^{2+} and IP3 perform as secondary signaling molecules, for triggering the mitogen-activated protein kinase (MAPK) cascades for the control of the phosphorylation of several transcription elements involving MYB, DREB/CFA, and NAC factors (Park et al., 2016) (Figure 10.2). This leads to overexpression of LEA-II genes that result in the stabilization of enzymes, protection of proteins, membranes, and cell nucleotides under the plant stress.

10.5.1 Drought Stress and LEA-II Genes

A consequential environmental stress that limits the plant productivity is the condition of drought. It is a state in which plants lack sufficient water availability for their physiological processes. Plants respond to drought through the induction of LEA-II proteins (Zhang et al., 2018a). Various

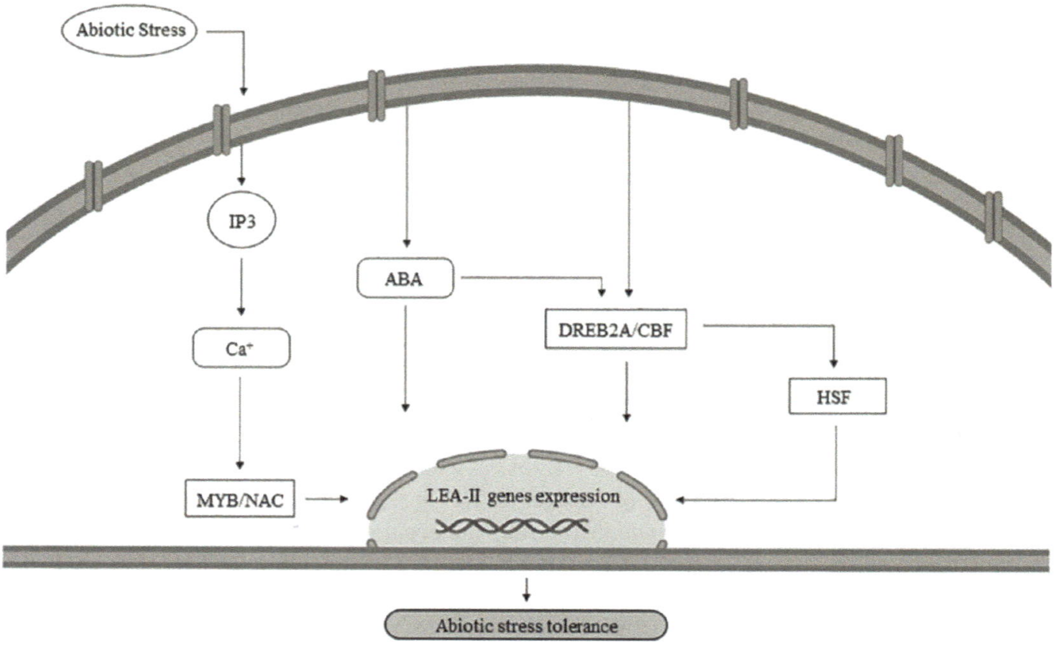

FIGURE 10.2 The abiotic stress factors involved in the signaling and regulation of LEA-II genes expression. The different signaling components IP3, Ca^{2+}, and ABA triggers transcription factors such as MYB/NAC, DREB2A/CBF, and HSFs that leads to overexpression of LEA-II genes.

physiological studies have indicated a positive correlation between the accumulation of LEA-II genes or proteins and the adaptation of plants with the stress of drought (Kosová et al., 2019). Most of the drought-tolerant plant cultivars have an elevated content of LEA-II or DHN transcripts and proteins compared to the drought-sensitive plant genotypes. However, there is an ambiguous interrelation between the assemblage of LEA-II proteins and stress tolerance.

In several studies, different transcription factors and activators play crucial roles for governing the drought-tolerant proteins in response to the cell water loss. In *Medicago truncatula*, autophagic degradation under the drought stress was prevented by the DHN, MtCAS31 (Li et al., 2020). The expression and role of *MtCAS31* in the autophagic degradation was determined by using an autophagy-specific inhibitor treatment in the green fluorescent protein (GFP) cleavage assay. Under drought conditions, DHN gene, *Wdhn13*, from *Triticum boeoticum*, exhibited a higher gene expression level in comparison to the lower expression in a tolerant cultivated variety (Sirvan) and other wild species (Mehrabad Pour-Benab et al., 2019). In wheat plants, a remarkable link was indicated among the activities of antioxidant enzymes such as superoxide dismutase and ascorbate peroxidase, and the LEA-II gene transcripts for the tolerance of drought. In rice plants, LEA-II genes were identified in the DEGs associated with the tolerance of drought stress.

It was further found that in *Solanum habrochaites*, a DHN gene was expressed to its peak level of 12-fold within 6 hours under the dehydration stress (Liu et al., 2015). Additionally, in *Saussurea involucrata*, a DHN gene, *SiDhn2*, reached the expression level of 12-fold within 3 hours of drought. However, it took 24 hours for DHN gene, *WZY2*, from winter wheat cv. Zhengyin 1 to achieve a maximum expression level under the drought condition (Zhu et al., 2014). Thus, it can be stated that a difference exists in the response duration of DHN genes against drought stress.

10.5.2 Temperature Stress and LEA-II Genes

The variation of the temperature within the plants' environment results in temperature stress. It impacts major developmental stages of plants. For instance, plant seed germination and flowering are sensitive to changes in the temperature. Certain fluctuations in the threshold temperature can limit the geographical distribution and reduce the productivity of crops. Temperature stress can cause a build-up of fluid imbalances within membranes and results in metabolic changes in temperature that damages plant cell membrane, causing metabolic disruption (Liliane and Charles, 2020). Heat and cold are the two major temperature stresses faced by the plants.

10.5.2.1 Heat Stress

Heat stress is a composite phenomenon that generates the aggregation and degradation of protein molecules. In wheat, DHN genes, *TaDHN1* and *TaDHN3*, have heat stress elements (HSEs) present on their promoters, playing a role in the reaction toward heat stress. These HSEs are identified precisely by the HSFs, due to their preserved N-terminal DNA-binding domain (DBD). However, low evidences exist about the occurrence of such HSEs in the promoters of some LEA-II genes.

In addition to the HSEs, promoters of LEA-II genes often consist of DREs with complex control mechanisms during the stress of heat. In *Arabidopsis thaliana*, the expression of *AtHSFA3* is governed due to the binding of DRE-binding protein 2A (DREB2A) directly to the DREs promoter region of *AtHSFA3*. Moreover, the DREB protein family are also involved in the control of DHNs under the stresses of drought and cold, inferring the involvement of similar strategies in the plant's tolerance to heat, cold, and drought stresses.

Evidences from several studies have reported a strong interrelation between LEA-II genes and plants' tolerance to heat stress. For example, *Cucumis sativus* and wheat LEA-II proteins, CsLEA11 and WZY2, protected the recombinant *Escherichia coli* under heat stress by shielding the activity of the lactate dehydrogenase (LDH) enzyme (Yang et al., 2015). Another wheat DHN protein, DHN-5, played a significant role of preserving the enzyme β-glucosidase (bglG) during the heat stress. Moreover, the truncation protein assay of DHN-5 showed the thermal protection of LEA-II proteins was due to the presence of the K-segments in their amino acid chains (Drira et al., 2013).

10.5.2.2 Cold Stress

Cold stress causes the formation of ice crystal particles in the plant extracellular spaces that decreases the moisture content in the cells. Cold stress-induced LEA-II genes were found in a number of plants such as soybean, *Arabidopsis,* and rice. In a study performed on *Arabidopsis*, overexpression of two DREs, *AtDREB1A* or *AtDREB2A*, resulted in the induction of cold stress interlinked genes of LEA-II proteins, involving *rd29A* and *COR47* (Narusaka et al., 2003). In another study on *Prunus mume*, overexpression of LEA-II gene, *PmLEAS*, improved the cold tolerance in tobacco and *E. coli* (Bao et al., 2017). Moreover, the overexpression of *Lycopersicon esculentum* LEA-II genes in yeast reduced the detrimental impacts of freezing and ionic stresses. It was also found that ABA treatment stimulated the cold stress and resulted in the overexpression of *Oryza sativa* LEA-II genes due to the presence of ABRE in the LEA-II genes.

Under the cold stress of cowpea, co-segregation of LEA-II gene indicated its role toward the cold tolerance during seedling emergence. It was initiated that in different plant species, the same abiotic stress did not cause an elevation of DHNs, but it was identified that different DHN genes are expressed in different plants under different abiotic stresses. Moreover, it was also reported that DREs composed of DHN genes did not respond against the stresses of cold (Yu et al., 2018). This indicates a composite pathway for the regulation of DHN genes against the cold stress that requires extensive examination for efficient regulation of LEA-II genes under cold stress.

10.5.3 Salinity Stress and LEA-II Genes

Salinity stress occurs in plants due to high salt ionic concentration in either the soil or water, which are taken up by the plants. It disrupts the growth of plants through nutritional imbalances caused by the cytotoxicity. Plants use the ATP-dependent ion channels in vacuole and endoplasmic reticulum membranes to regulate the ionic concentration of the cytoplasm particularly of Na^+, K^+, and Cl^- ions. In halophytes, there is a high build-up of compatible solutes such as polyamines and prolines, and hydrophilic proteins such as the LEA-II proteins. The salinity stress tolerance and its pathway has been largely studied in several plants, which is triggered by both ABA-dependent and ABA-independent signaling pathways.

The overexpression of durum wheat LEA-II genes in *Arabidopsis* improved the salt tolerance of the transgenic plants through the control of different protein interactions (Brini et al., 2007). In another study, the expression profile of an SK_3-type DHN gene from banana, *MusaDHN-1*, in leaves and roots, was correlated with enhanced salt tolerance in transgenic banana plants (Shekhawat et al., 2011). In addition, *Arabidopsis* gained higher salt tolerance than the control and wild-type plants, because of the differential expression of two LEA-II genes from *Physcomitrella patens*, *PpDHNA* and *PpDHNC* (Li et al., 2017). Furthermore, it was revealed that the overexpression of a DHN gene from *Capsicum annuum*, *CaDHN4*, improved the *Arabidopsis* salinity tolerance by displaying higher rate of seed germination and better root growth in the presence of NaCl (Zhang et al., 2020).

Moreover, the application of methyl jasmonate (MeJA) had resulted into a double-fold rise in the DHN levels within plant cells and also improved the lignin concentration of the cell wall, providing high salinity stress protection to the plants (Avalbaev et al., 2020). *Arabidopsis* salinity stress tolerance was also improved through the overexpression of *Hevea brasiliensis* DHN (Cao et al., 2017). In another study, it was identified that *Avicennia officinalis* DHN 1 gene, *AoDHN1*, belonged to LEA-II genes and was upregulated against salt treatment (Jyothi-Prakash et al., 2014). In several studies, the ability of DHN genes to reduce salinity stress was associated to their protease activity, which is essential for plants salinity stress responses (Li et al., 2017).

10.5.4 Osmotic Stress and LEA-II Genes

Osmotic stress is caused within the plants due to their exposure to different abiotic stresses of drought, salinity, and low temperature. It disrupts the productivity of plants (Liliane and Charles, 2020). Osmotic stress in plants declines exterior cells' chemical water potential and results in the transport of water molecules out of the cell membrane, which leads to a state of dehydration. Plants produced LEA-II proteins in response to osmotic stress due to high accumulation of ABA within the plant cytoplasm during the stress.

A reduced growth was observed under osmotic stress in a knockout DHN *Physcomitrella patens* mutant that was produced using homologous recombination (Saavedra et al., 2006). However, the overexpression of *Physcomitrella patens* DHN genes, such as *PpDHNA* and *PpDHNB* in transgenic *Arabidopsis* plants displayed enhanced root growth under osmotic stress. In another study, a *Pennisetum glaucum* DHN gene, *PgDHN*, enhanced the osmotic stress tolerance of recombinant *E. coli* cells and resulted in higher growth rate in contrast to control *E. coli* cells (Singh et al., 2015). Additionally, the non-homologous expression of *PgDHN* in transformed yeast improved osmotic stress tolerance. Furthermore, in a study, transformed *Arabidopsis thaliana* lines were generated with or without the K-segments of *DHN-5* gene from *Triticum durum* (Drira et al., 2016). The study showed that the *Arabidopsis* constructs consisting of single or double K-segments enhanced tolerance of the *Arabidopsis thaliana* seedlings to diverse stresses including osmotic stress. In comparison to the wild type and YS form *DHN-5* gene, elevated enzymatic activity of catalase and peroxide dismutase and reduced contents of malondialdehyde (MDA) and hydrogen peroxide (H_2O_2) were found for the transgenic plants with K-segment constructs (Drira et al., 2016). Also, in another investigation, the enhanced expression of a *Caragana korshinskii* LEA-II gene, *CkLEA2–3*, in *Arabidopsis*, provided enhanced protection at their seed germination stage during the osmotic stress (Yu et al., 2019).

Some essential protective properties under osmotic stress were identified from *Agapanthus praecox* DHN genes, *ApY2SK2* and *ApSK3* (Yang et al., 2019). For instance, in *Arabidopsis thaliana*, the overexpression of *ApY2SK2* and *ApSK3* reduced the reactive oxygen species (ROS) levels and the damage caused to the plasma membrane. These genes resulted in greater antioxidant activity and improved photosynthetic potential of transformed *Arabidopsis* plants than the wild types under the osmotic stress (Yang et al., 2019). In another study, there was a high level of LEA-II gene transcript accumulation under osmotic stress by an YSK_2-type DHN from *Sorghum bicolor*. The transformation of tobacco plants with *SbDhn1* gene resulted in an improved tolerance toward the osmotic stress, which also decreased the membrane damage and lowered the MDA levels. In addition,

Ammopiptanthus mongolicus DHN genes, *AmDHN132*, *AmDHN154*, and *AmDHN200*, improved the osmotic stress tolerance of transformed *Arabidopsis* plants, with *AmDHN132* gene resulting in higher tolerance (Cui et al., 2020). In another study, DHN gene, *CaDHN5* from *Capsicum annuum*, improved the *Arabidopsis thaliana* tolerance against the osmotic stresses in contrast to the wild species (Luo et al., 2019). Moreover, the overexpression of an intrinsically disordered DHN gene, *ERD14* from *Arabidopsis thaliana*, with increasing levels of H_2O_2 during osmotic stress showed enzymatic activity protection due to its chaperone-like properties (Nguyen et al., 2020).

10.5.5 Heavy Metal Stress and LEA-II Genes

Heavy metal stress severely alters the plant physiological processes. It takes place when there is high concentration of relatively dense metallic elements such as Cd and Cr in plant environment. These metallic elements belong to the non-biodegradable, persistent groups of inorganic chemical constituents that can cause mutagenic, cytotoxic, or genotoxic effects to plants. The toxic concentration of heavy metals hampers the normal functioning of plants by degrading the building blocks of different proteins. Both underground and aboveground plant parts can uptake heavy metals, which can cause an increased production of ROS, such as free radicals of superoxide.

In order to understand the link between LEA-II proteins and heavy metal tolerance, an examination of their regulatory networks is essential. The transcription factor DREB plays an important role in heavy metal signaling transduction pathway. In a study, the elevated expression of DREB in transformed plants displayed a greater expression of LEA-II genes of *Arabidopsis thaliana* such as *COR15a*, *RD29A*, and *RD29B* that resulted in enhanced tolerance against heavy metal toxicity (Li et al., 2014). An isolated SK_2-type LEA-II gene from *Phaseolus vulgaris*, *PvSR3*, was largely accumulated under heavy metal toxicity (Zhang et al., 2006). Furthermore, *BjDHN2* and *BjDHN3*, two DHN genes of *Brassica juncea*, when overexpressed in transformed tobacco plants, resulted in an improved tolerance to heavy metals (Xu et al., 2008). In vitro studies indicated metal binding ability of LEA-II proteins that displays their potential role in heavy metal stresses. An immobilized metal ion affinity chromatography (IMAC) showed that *Citrus unshiu* LEA-II protein, CuCOR15, was able to bind to Co^{2+} and Ni^{2+} ions, which implies the presence of specific sequence containing histidine (His) in these protein molecules for reducing the metal toxicity in plants (Hara et al., 2005). Some of the recent LEA-II genes identified for the prevention of several abiotic stresses are highlighted in the Table 10.1.

TABLE 10.1
Different Plant LEA-II Genes and Their Tolerance to Various Stresses

Plant Species	LEA-II Genes	Classification	Abiotic Stresses
Zea mays	*ZnDHN13*	KS	Osmotic
Olea europaea	*OesDHN*	SK_2	Drought
Solanum sogarandinum	*DHN24*	SK_3	Cold
Stipa purpurea	*SpDHN1*	SK_3	Desiccation
Ipomoea pes-caprae	*IpDHN*	SK_3	Salinity
Medicago truncatula	*CAS31*	Y_2K_2	Salinity
Oryza sativa	*OsDhn-Rab16D*	YSK_2	Drought
Capsicum annum	*CaDHN3*	YSK_2	Drought
Vitis vinifera	*VvDhn*	YSK_2	Drought & salinity
Triticum durum	*DHN-5*	YSK_2	Salinity
Agapanthus praecox	*APY2SK3*	Y_2SK_3	Salt, osmotic & Cold
Malus domestica	*MdoDHN11*	Y_3SK_2	Dehydration
Vitis vinifera	*DHN1*	Y_2SK_2	Heat
Cucumis melo	*CmLEA-S*	Y_3SK_2	Salinity & drought
Arabidopsis thaliana	*Lti30*	K_6	Cold
Opuntia streptacantha	*OpsDHN1*	SK_3	Freezing

10.6 MULTIFUNCTIONAL PROPERTIES OF LEA-II PROTEINS

Plants respond to stress in a variety of ways that involves multipath. One way of stress responses in plants is the accumulation of LEA-II proteins that perform various protective functions (Abdul Aziz et al., 2021). In relation to this, a number of studies have been undertaken to depict their physiological, biochemical, and biological functional characteristics under the abiotic stresses that are discussed in this chapter.

10.6.1 Membrane Stability

The structural characteristics of plasma membrane's lipid bilayer are closely related to temperature and moisture content. The effects of LEA-II proteins on the plasma membrane have been widely studied. The LEA-II proteins' K-segment shows a high membrane affinity (Figure 10.3A). In a study, *Arabidopsis thaliana* LEA-II protein, Lti30, under the cold stress, displayed a high membrane vesicle affinity due to its K-segment domain sequence. The membrane linking was further governed by pH-dependent histones and phosphorylation. Lti30 stabilized membranes by reducing the lipid phase transition temperature (Gupta et al., 2019). It was identified through a molecular-level investigation that K-segments of LEA-II proteins partially fold in a helix form on the membrane surface layers. There is also an interaction between LEA-II proteins and plasma membrane intrinsic proteins (PIPs). PIPs are crucial groups of plant aquaporin (AQP) family found on membranes that play a role in controlling the abiotic stress-related water transport (Feng et al., 2018). It was reported that *Arabidopsis* LEA-II proteins, AtCOR47 and AtERD10, formed a homodimer interaction, whereas *Arabidopsis* LEA-II proteins AtRAB18, resulted in the formation of a heterotrimeric complex with *Opuntia streptacantha* LEA-II protein, OpsDHN1, and their mutual interaction reduced damages to proteins caused by abiotic stresses.

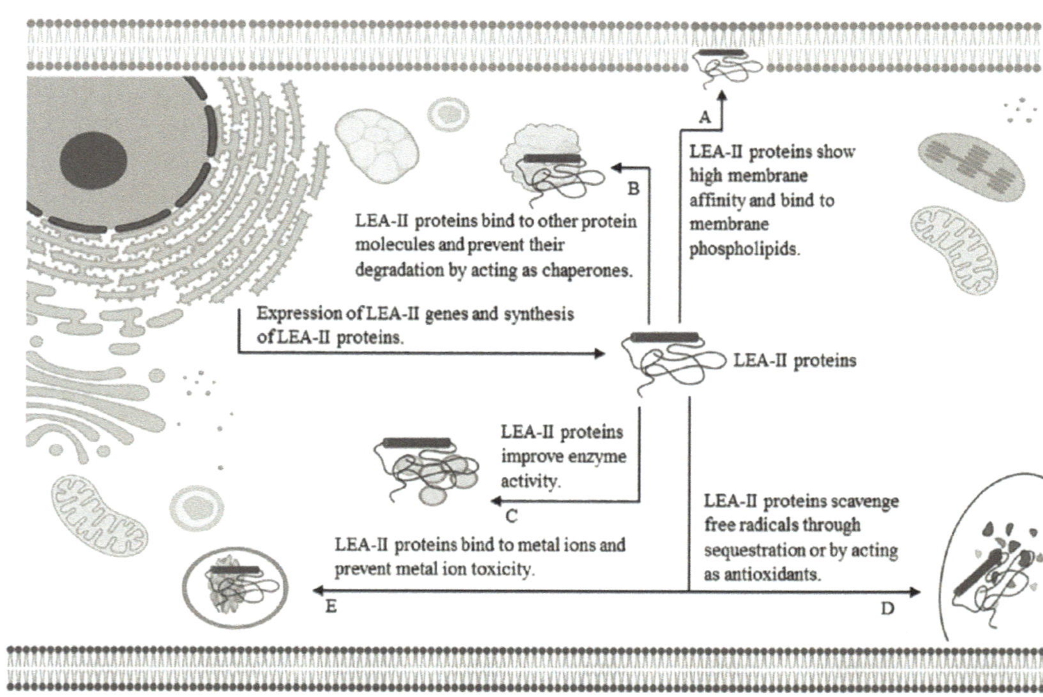

FIGURE 10.3 Protective functions of plant LEA-II proteins. (A) LEA-II proteins provide membrane stability by binding to and safeguarding membranes. (B) LEA-II proteins act as chaperones and protect macromolecules such as proteins and carbohydrates. (C) LEA-II proteins bind to different enzymes and prevent their loss of activity. (D) LEA-II proteins reduce the high ROS accumulation through their radical scavenging activity. (E) LEA-II proteins are involved in the reduction of metal ion toxicity by sequestration of large amount of metal ions.

Through bimolecular fluorescence complementation (BiFC) technology, it was indicated that *Arabidopsis* LEA-II proteins, AtCOR47, AtERD10, and AtRAB18, commonly target *Arabidopsis* AQPs such as AtPIP2Bs. These interactions prevent AtPIP2B inactivation and denaturation, and results in the stabilization of plasma membrane under the abiotic stress. It was demonstrated recently that *Medicago truncatula* LEA-II protein, MtCAS31, promotes the autophagous degradation of negative regulator of drought stress, MtPIP2;7, which lowers the root moisture loss and enhances drought tolerance (Li et al., 2020). Due to the presence of His motifs and rich K-segments, LEA-II proteins bind to phospholipids (Figure 10.3A). Such a binding triggers the induction of essential stress-signaling phospholipid and phosphatidic acid (PA) (Yu et al., 2018). In an inflated plasma membrane, the PA content are low, which increases during the loss of water or the release of ABA under the drought stress.

The LEA-II proteins bind to anionic phospholipids because of the certain amino acid residues such as arginine (Arg) and Lys in their polychain. This interaction prevents loss of water molecules and maintains a homeostasis for normal temperature within the cells. Since LEA-II proteins bind to acidic phospholipid, it is postulated that such a binding occurs at specific sites on the cell membrane. With the aid of electrostatic interaction, LEA-II proteins also bind to charged lipids. It is found by Clarke et al. (2015) that some LEA-II proteins transform into helical structure through their binding to acidic phospholipids, due to which they are able to form interlinkage with other biomolecules of cytosol and provide protection against stress damages.

10.6.2 Macromolecules Protection

LEA-II proteins are able to bind certain macromolecules under the stress condition (Figure 10.3B). They protect denatured proteins and prevent their aggregation during low water availability and cold stresses. The nuclear localization of LEA-II proteins suggests two functional characteristics: either binding to DNA molecule through Zn ions or bind to other protein molecules due to their chaperone property, thereby preserving plant growth under stress.

It was reported by Boddington and Graether (2019) that LEA-II genes in the nucleus of *Vitis riparia*, *VrDHN1*, interacts and protects DNA from high levels of ROS such as H_2O_2. A study indicated that two DHN genes, *CaDHN3* and *CaHIRD11*, of *Capsicum annuum*, provided protection under the stresses of salt and drought by their protein–protein-binding ability in the plasma membrane. In addition, LEA-II proteins protect the activities of specific enzymes such as malate dehydrogenase (MDH) and LDH during cold stress (Yang et al., 2015).

LEA-II proteins stabilize different cell structures and organelles through its amphipathic α-helical binding with macromolecules (Clarke et al., 2015). The overexpression of LEA-II genes in recombinant tomato plants resulted in the cell organelles stabilization, which prevented the loss of water from the cells under the stress (Liu et al., 2015). Similarly, *Rhododendron catawbiense* and *Vitis vinifera* LEA-II genes, *DHN5* and *DHN1a*, stabilized the cell structures under freezing and dehydration. Such an attribute of LEA-II genes was due to the presence of Φ- and K-segments. In recent studies, it has been indicated that LEA-II proteins form both homodimeric and heterodimeric complexes with macromolecules and shield them against the effects of abiotic stresses, and allow for regular cell processes (Yu et al., 2018).

Cryoprotective function is the ability of LEA-II proteins to protect the loss of activity of enzymes. LEA-II proteins preserve the activity of enzymes under the osmotic stress (Figure 10.3C). It has been identified that CuCOR19 from *Citrus unshiu* and DHN5 from barley act as cryoprotectants against various abiotic stresses. The LEA-II proteins' cryoprotective property was attributed to the presence of K-segment domain. In a study on *Arabidopsis* LEA-II protein, dehydrin ERD14 was found to activate the enzymes catalase (CAT) and Phi class glutathione transferase (Nguyen et al., 2020). The K-segments of ERD14 protected the activity of LDH. Furthermore, an *Arabidopsis thaliana* LEA-II protein, AtHIRD11, protected inactivation of LDH enzymes under heavy metals, due to the His residues. Moreover, another *Arabidopsis thaliana* LEA-II protein (FSKn type),

COR47, protected LDH activity due to their hydrophobic residues of the F-segments, under the freezing stress (Ohkubo et al., 2020). Thus, LEA-II proteins are multifunctional protectors of macromolecules that enhance the plant stress tolerance.

10.6.3 Radical Scavenging Activity

Abiotic stresses cause the accumulation of ROS in plant cells, thereby resulting in lipid, protein, and nucleic acid damages. There are numerous defense mechanisms developed by the plants for the maintenance of cellular redox homeostasis. In terms of ROS, H_2O_2 is largely produced within the plants that can signal the presence of stress. The increase in H_2O_2 concentration overexpresses LEA-II genes for the production of LEA-II proteins, which plays a role in lowering the levels of high H_2O_2 and prevents the damages caused to the cells (Hara, 2010). In stressed plants, LEA-II proteins are broadly dispersed in cells and induced within the cellular regions sensitive to ROS. For instance, nucleus is reactive to the high levels of ROS, due to which several LEA-II proteins are accumulated in the nucleus and interact with the DNA. This results in the chromosomal stability and preservation of transcriptional machinery. Moreover, LEA-II proteins are widely distributed in the chloroplast and mitochondria that are highly susceptible to the attacks of free radicals.

LEA-II proteins scavenge ROS through their ability to bind to free radicals and remove them through cell compartments under stress (Figure 10.3D). High H_2O_2 upregulated the expression of a bean gene, *PvSR3*, that plays a role in catalyzing the production of toxic ROS (Zhang et al., 2006). It was demonstrated that a LEA-II protein from *Citrus unshiu*, CuCOR19, which is a K_3S-type LEA-II protein, restricted the in vitro peroxidation of liposomes in transformed tobacco plants under the cold stress (Hara et al., 2003). It was also found that KS-type LEA-II proteins from the *Arabidopsis thaliana*, AtHIRD11, reduced the levels of ROS. However, in KS-type LEA-II proteins, the proportion of peptide bonds and the contents of His determined the reduction of ROS. It has also been postulated that LEA-II proteins can directly scavenge free radicals by acting as antioxidants.

10.6.4 Antioxidant Activity

The antioxidant activity of LEA-II proteins is associated to the presence of high content of amino acid residues, such as Lys, Gly, and His that are target amino acids for the radical-mediated oxidation in proteins. Some LEA-II proteins are found to accumulate near the plasma membrane that inhibits lipid peroxidation under environmental stress by acting as an antioxidant (Figure 10.3D).

In a study, it was indicated that citrus LEA-II protein, CuCOR19, had well-built inhibitory activity against free radicals that causes peroxidation, in contrast to some known antioxidants such as proline, albumin, and glutathione (Hara et al., 2003). Furthermore, the overexpression of *CuCOR19* in transgenic tobacco plants can inhibit lipid peroxidation under cold stress (Hara et al., 2003). Moreover, recombinant tobacco plants overexpressing rice LEA-II protein, RAB16A, and *Brassica juncea* DHNs gained an enhanced stress tolerance with remarkable antioxidative activity of peroxidation and membrane injury reduction (Xu et al., 2008).

There was an enhanced LEA-II gene expression in plants due to increased concentration of salicylic acid that acts as an H_2O_2 inducer (Zhang et al., 2006). Salicylic acid inhibits enzymatic activities of antioxidants. Due to this, there was an overexpression of LEA-II genes in plants under stress, which acts as an alternative antioxidant for plants. Thus, antioxidative property of LEA-II proteins is a novel putative function for abiotic stress tolerance in plants.

10.6.5 Metal Ion-Binding Activity

LEA-II proteins protect plants through their ability to bind to metal ions under abiotic stresses (Figure 10.3E). The catalytic metal ions, such as zinc and copper, largely occur as complexes of protein molecules and other metals under the desirable plant environment (Liliane and Charles, 2020). However, under the abiotic stress condition, these metal elements are released as free ions within

TABLE 10.2
Functional Properties of LEA-II Proteins in Different Plants

Plant Species	LEA-II Proteins	Functions	Abiotic Stresses
Arabidopsis thaliana	COR47	LDH protection	Cold
Capsicum annuum	CaDHN3	Antioxidant activity	Drought & salinity
Agapanthus praecox	DHN	Metal ion binding	Cold, drought, & salinity
Arabidopsis thaliana	ERD14	Membrane protection	Cold, drought, & salinity
Ammopiptanthus mongolicus	DHN154	Membrane protection	Cold & salinity
Arabidopsis thaliana	Lti30	Membrane protection	Cold
Cucumis sativus	DHN4	LDH protection	Heat
Zea mays	DHN11	Antioxidant enzymes protection	Osmotic
Medicago truncatula	CAS31	LDH protection	Drought
Saussurea involucrata	DHN	ROS scavenging activity	Cold & drought
Oryza sativa	DHN1	ROS scavenging activity	Salinity
Triticum durum	DHN5	LDH and bglG protection	Cold & salinity
Cucumis melo	LEA-S	CAT activity enhancement	Drought & salinity
Ipomoea pes-caprae	DHN	Antioxidant enzyme activity enhancement	Cold, drought, & salinity

the plant cells. Through the Haber–Weiss reaction, these free ions cause injury to the plant cells by increased ROS production. LEA-II proteins commonly target the free metal ions that are released from cell membranes and organelles. It has been postulated that LEA-II proteins get coupled to free metal ions and reduce the damage caused to the plants (Yu et al., 2018).

There were reports of LEA-II proteins binding to metal ions in AtHIRD11, LEA-II protein of *Arabidopsis thaliana* and CuCOR15, LEA-II protein of *Citrus unshiu*, both of which bind to cobalt and iron metal elements than to calcium and magnesium, and that prevents the dispersal of free ions (Liu et al., 2017). *Citrus unshiu* LEA-II protein, CuCOR15, also acts as a free radical scavenger through its binding to metal ions during the stress of moisture loss, which reduces the metal toxicity in plants. Moreover, in *Ricinus communis*, a KS-DHN, which is an ion transport protein (ITP), is able to bind and actively transport metal ions within the plants under stress. The functions of few LEA-II proteins against various environmental stresses are shown in Table 10.2.

10.7 OMICS APPROACHES FOR THE FUNCTIONAL CHARACTERIZATION OF LEA-II GENES

The three major functional omics technologies that have been used largely for identifying a wide range of stress-related genes in plants are the transcriptomics, proteomics, and epigenetics. Also, the next-generation sequencing techniques such as RNA-seq approaches have been broadly undertaken to extract putative environmental stress-reactive genes. These approaches also provided in-depth insights into the molecular events taking place in the plant genome during the normal and stress conditions.

10.7.1 Transcriptomics and Proteomics

The transcriptomics and proteomics analysis have displayed potential genetic variations in the stress-sensitive and -tolerant plant varieties. Through the use of these omics' approaches, it has been indicated that LEA-II proteins are ubiquitously present and expressed in plants under abiotic stresses. A transcriptomic analysis was conducted by Zeng et al. (2016) on Tibetan hulless barley under water-deficit conditions. The results of the study indicated alteration in the gene induction patterns under the signaling pathways of stress, both ABA-dependent and -independent pathways. However, the accumulation of MYB and LEA-II genes strikingly improved under severe state of drought. Du et al. (2011) performed a contrasting investigation of transcriptional and translational variations of the LEA-II genes between wild barley and Tibetan hulless barley. It was found that LEA-II genes had differential gene expression among the wild and Tibetan hulless barley. Moreover, through the use of polyclonal antibodies against LEA-II proteins, in comparison to the wild species, a higher content of LEA-II proteins was found in the Tibetan hulless barley. Another study on Tibetan hulless barley had particularly pointed Dhn6, a LEA-II protein, which was firmly associated with the stress of moisture loss (Liang et al., 2012).

In a proteomics study, it was detected that *Arabidopsis* LEA-II protein, ERD14, interlinks with GSTF9 and CAT, and accordingly actuates the activity of antioxidant enzymes, which plays an essential part in plants' tolerance to the conditions of stress (Nguyen et al., 2020). It was found that there was an upregulation of LEA-II genes through transcriptomic and proteomic analyses of *Agropyron mongolicum* under drought stress (Zhao et al., 2018). Incidentally, proteomics profiling of yard-dried grains of *Coffea Arabica* had an abundance of LEA-II genes expressed for preserving the quality of grains. However, a varied proteomic profile was observed when grains were dried at 60°C (do Livramento et al., 2017). Moreover, the LEA-II proteins were identified on mass spectroscopy gels of peach bark tissue in relation to low temperature and short photoperiod stress. This highlighted the essential role of LEA-II proteins in plant cold acclimation.

In another proteomic analysis study in peach fruit under cold stress, there was also an abundance of LEA-II proteins that were induced due to dehydration (Nilo et al., 2010). Similarly, in poplar plants, an abundance of LEA-II proteins occurred under cold stress, which further indicated the possible role of LEA-II proteins in plant acclimatization. A LEA proteome analysis was performed on two contrasting wheat genotypes under the drought stress. It indicated a unique expression of genes and greater upregulation of LEA-II genes in tolerant genotypes.

10.7.2 Epigenetics

Besides transcriptomics and proteomics, epigenetic regulations affect the gene expressions through DNA methylation, histone mutations, and posttranslational histone modifications. These alterations in DNA and histones play a crucial role for the regulations of gene expression and plant growth under different abiotic stresses. The epigenetic modifications can be either fixed in one generation or can be transferred to another generation through stress identification. Although the heritability of such traits is unknown, stress identification enables the next generation to have a higher stress tolerance.

Posttranslational histone modification is the most commonly observed epigenetic phenomenon that includes chemical variations in the histone structural pattern through methylation, acetylation, or phosphorylation of various amino acids, for instance presence of Arg and Lys in the N-terminus of histone protein tails (Mahrez et al., 2016). Two enzymes that are used for the maintenance of histone acetylation are histone acetyltransferases (HATSs) and histone deacetylases (HDACs) that enable the compaction of DNA and results in transcription activation or suppression (Figure 10.4). It was found that an *Arabidopsis thaliana* HD2 protein, AtHD2C, was involved in ABA and abiotic stress reactions. The overexpression of *AtHD2C* in transformed plants displayed higher tolerance to salt and drought stress in contrast to wild species (Figure 10.4A).

DNA methylation process also plays a role in the epigenetic process related to abiotic stress tolerance. A genome-wide analysis was performed by Zong et al. (2013) to identify the stress-linked differences in *Oryza sativa* plants under the stress of moisture. The levels of pattern related to trimethylated histone H3 lysine 4 (H3K4me3) under dehydration condition was analyzed through ChIP-Seq and RNA-seq. It was found that the modification of H3K4me3 was firmly correlated

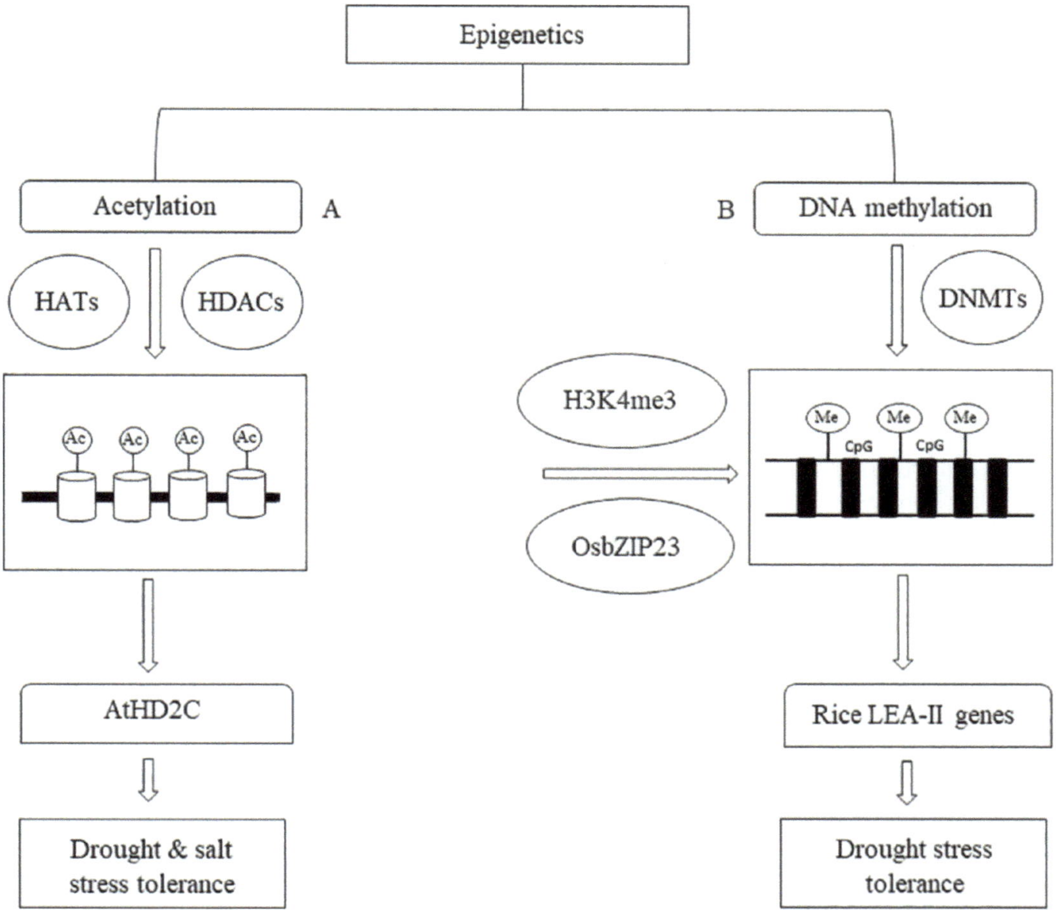

FIGURE 10.4 Epigenetic regulation of LEA-II gene expression for enhanced stress tolerance. (A) Overexpression of *AtHD2C* gene under drought stress through the acetylation with HATs and HDACs. (B) DNA methylation with H3K4me3 improved the rice LEA-II genes expression under drought stress.

with the drought-responsive gene induction. It was reported that there was an interaction between OsbZIP23 transcription factor and H3K4me3 modification, which were directly correlated to the regulation of rice LEA-II gene cluster under drought stress (Zong et al., 2020). The LEA-II gene expression patterns were enhanced due to the H3K4me3 alterations and larger interlinking of OsbZIP23 to the LEA-II genes' promoter region (Figure 10.4B). However, in OsbZIP23 mutant rice, the expression of LEA-II genes and H3K4me3 alteration was lower during the drought condition.

10.8 CONCLUDING REMARKS

The climatic constraints in the form of drought, heat, and salinity restricts the global crop productivity. A high focus is placed on the detection of genes and proteins that enable the plants to adapt to climatic changes without yield loss. LEA-II genes have been identified as efficient stress-responsive-genes with crucial functions in plants' adaptive responses to abiotic stresses. LEA-II genes encode the corresponding proteins determined by their disordered structure. LEA-II proteins are accumulated during the seed desiccation stage and environmental stresses for protecting the plants from injury.

The present chapter highlighted the classification, structure, occurrence, and functional properties of LEA-II proteins under various abiotic stresses. The LEA-II proteins are generally classified according to their conserved domain sequences that are determined as Y, S, and K segments. Structurally, LEA-II proteins are IDPs that are able to turn into an amphiphilic α-helix, an essential feature for their functioning under abiotic stresses. They protect the plants through various ways, such as protecting membranes and macromolecules, scavenging free radicals, acting as antioxidants, and binding to metal ions. The chapter also elaborates some of the omics approaches used for the analysis and regulation of LEA-II proteins. Both transcriptomics and proteomics studies indicate an overexpression of LEA-II genes under various abiotic stress conditions. In addition, the epigenetics modifications play a vital role in the upregulation of LEA-II genes for plant protection under stress.

The functional characteristics of LEA-II proteins were mainly derived through in vitro studies. To effectively breed plants for stress tolerance using LEA-II genes, in vivo studies related to LEA-II genes are required. The supplemental exploration of plants' morphological changes due to LEA-II genes can prescribe their other functions under stress. Conscientious research is also needed toward epigenetic processes, post-transcriptional and posttranslational modification of LEA-II genes, and their mRNA for a higher expression in plants. LEA-II proteins are essential components combating abiotic stresses and a comprehensive understanding of the LEA-II genes will lead to the cultivation and expansion of higher yielding crops that can withstand unpropitious climatic threats.

REFERENCES

Abdul Aziz, M., Sabeem, M., Mullath, S. K., Brini, F., and Masmoudi, K. 2021. Plant group II LEA proteins: Intrinsically disordered structure for multiple functions in response to environmental stresses. *Biomolecules* 11: 1662. doi: 10.3390/biom11111662.

Al-Mssallem, I. S., Hu, S., Zhang, X., Lin, Q., Liu, W., Tan, J., et al. 2013. Genome sequence of the date palm *Phoenix dactylifera* L. *Nature Communications* 4: 2274. doi: 10.1038/ncomms3274.

Avalbaev, A., Allagulova, C., Maslennikova, D., Fedorova, K., and Shakirova, F. 2020. Methyl jasmonate and cytokinin mitigate the salinity-induced oxidative injury in wheat seedlings. *Journal of Plant Growth Regulators* 40: 1741–1752 doi: 10.1007/s00344-020-10221-1.

Bao, F., Du, D., An, Y., Yang, W., Wang, J., Cheng, T., et al. 2017. Overexpression of *Prunus mume* dehydrin genes in tobacco enhances tolerance to cold and drought. *Frontiers in Plant Science* 8: 151. doi: 10.3389/fpls.2017.00151.

Battaglia, M. and Covarrubias, A. 2013. Late embryogenesis abundant (LEA) proteins in legumes. *Frontiers in Plant Science* 4: 190. doi: 10.3389/fpls.2013.00190.

Boddington, K. and Graether, S. 2019. Binding of a *Vitis riparia* dehydrin to DNA. *Plant Science* 287: 110172. doi: 10.1016/j.plantsci.2019.110172.

Brini, F., Hanin, M., Lumbreras, V., Amara, I., Khoudi, H., Hassairi, A., et al. 2007. Overexpression of wheat dehydrin *DHN-5* enhances tolerance to salt and osmotic stress in *Arabidopsis thaliana. Plant Cell Reports* 26: 2017–2026. doi: 10.1007/s00299-007-0412-x.

Cao, Y., Xiang, X., Geng, M., You, Q., and Huang, X. 2017. Effect of *HbDHN1* and *HbDHN2* genes on abiotic stress responses in Arabidopsis. *Frontiers in Plant Science* 8: 470. doi: 10.3389/fpls.2017.00470.

Clarke, M., Boddington, K., Warnica, J., Atkinson, J., McKenna, S., Madge, J., et al. 2015. Structural and functional insights into the cryoprotection of membranes by the intrinsically disordered dehydrins. *Journal of Biological Chemistry* 290: 26900–26913. doi: 10.1074/jbc.M115.678219.

Close, T. J., Fenton, R. D., Yang, A., Asghar, R., DeMason, D. A., Crione, D., et al. 1993. Dehydrin: The protein. In Close T. and Bray E. (Eds.) *Plant Responses to Cellular Dehydration during Environmental Stress*, 104–118. Rockville, MD: American Society of Plant Physiologists.

Cui, H., Wang, Y., Yu, T., Chen, S., Chen, Y., and Lu, C. 2020. Heterologous expression of three *Ammopiptanthus mongolicus* dehydrin genes confers abiotic stress tolerance in Arabidopsis thaliana. *Plants* 9: 193. doi: 10.3390/plants9020193.

do Livramento, K. G., Borém, F. M., José, A. C., Santos, A. V., Livramento, D. E., Alves, J. D., et al. 2017. Proteomic analysis of coffee grains exposed to different drying process. *Food Chemistry* 221: 1874–1882 doi: 10.1016/j.foodchem.2016.10.069.

Drira, M., Hanin, M., Masmoudi, K., and Brini, F. 2016. Comparison of full-length and conserved segments of wheat dehydrin *DHN-5* overexpressed in *Arabidopsis thaliana* showed different responses to abiotic and biotic stress. *Functional Plant Biology* 43: 1048–1060. doi: 10.1071/FP16134.

Drira, M., Saibi, W., Brini, F., Gargouri, A., Masmoudi, K., and Hanin, M. 2013. The K-segments of the wheat dehydrin DHN-5 are essential for the protection of lactate dehydrogenase and beta-glucosidase activities *in vitro*. *Molecular Biotechnology* 54: 643–650. doi: 10.1007/s12033-012-9606-8.

Du, J. B., Yuan, S., Chen, Y. E., Sun, X., Zhang, Z. W., Xu, F., et al. 2011. Comparative expression analysis of dehydrins between two barley varieties, wild barley and Tibetan hulless barley associated with different stress resistance. *Acta Physiologiae Plantarum* 33: 567–574 doi: 10.1007/s11738-010-0580-0.

Feng, Z. J., Xu S. C., Liu, N., Zhang, G. W., Hu, Q. Z., Xu, Z. S., et al. 2018. Identification of the AQP members involved in abiotic stress responses from Arabidopsis. *Gene* 646: 64–73. doi: 10.1016/j.gene.2017.12.048.

Gupta, A., Marzinek, J. K., Jefferies, D., Bond, P. J., Harryson, P., and Wohland, T. 2019. The disordered plant dehydrin Lti30 protects the membrane during water-related stress by cross-linking lipids. *Journal of Biological Chemistry* 294: 6468–6482. doi: 10.1074/jbc.RA118.007163.

Hara, M. 2010. The multifunctionality of dehydrins: An overview. *Plant Signaling and Behavior* 5: 503–508. doi: 10.4161/psb.11085.

Hara, M., Fujinaga, M., and Kuboi, T. 2005. Metal binding by citrus dehydrin with histidine-rich domains. *Journal of Experimental Botany* 56: 2695–2703. doi: 10.1093/jxb/eri262.

Hara, M., Terashima, S., Fukaya, T., and Kuboi, T. 2003. Enhancement of cold tolerance and inhibition of lipid peroxidation by citrus dehydrin in transgenic tobacco. *Planta* 217: 290–298. doi: 10.1007/s00425-003-0986-7.

Hundertmark, M. and Hincha, D. 2008. LEA (Late Embryogenesis Abundant) proteins and their encoding genes in *Arabidopsis thaliana*. *BMC Genomics* 9: 118. doi: 10.1186/1471-2164-9-118.

Iturriaga, G., Cushman, M., and Cushman, J. 2006. An EST catalogue from the resurrection plant *Selaginella lepidophylla* reveals abiotic stress-adaptive genes. *Plant Biology* 170: 1173–1184. doi: 10.1016/j.plantsci.2006.02.004.

Jin, X., Cao, D., Wang, Z., Ma, L., Tian, K., Liu, Y., et al. 2019. Genome-wide identification and expression analyses of the LEA protein gene family in tea plant reveal their involvement in seed development and abiotic stress responses. *Scientific Reports* 9: 14123. doi: 10.1038/s41598-019-50645-8.

Jyothi-Prakash, P., Mohanty, B., Wijaya, E., Lim, T., Lin, Q., Loh, C., et al. 2014. Identification of salt gland-associated genes and characterization of a dehydrin from the salt secretor mangrove *Avicennia officinalis*. *BMC Plant Biology* 14: 6. doi: 10.1186/s12870-014-0291-6.

Kosová, K., Prášil, I., and Vítámvás, P. 2019. Role of dehydrins in plant stress response. In Pessarakali, M. (Eds.) *Handbook of Plant and Crop Stress*, 239–285. Boca Raton, FL: CRC Press.

Li, Q., Zhang, X., Lv, Q., Zhu, D., Qiu, T., Xu, Y., et al. 2017. *Physcomitrella patens* dehydrins (*PpDHNA* and *PpDHNC*) confer salinity and drought tolerance to transgenic Arabidopsis plants. *Frontiers in Plant Science* 8: 1316. doi: 10.3389/fpls.2017.01316.

Li, X., Liu, Q., Feng, H., Deng, J., Zhang, R., Wen, J., et al. 2020. Dehydrin MtCAS31 promotes autophagic degradation under drought stress. *Autophagy* 16: 862–877 doi: 10.1080/15548627.2019.1643656.

Li, X., Zhang, D., Li, H., Wang, Y., Zhang, Y., and Wood, A. J. 2014. EsDREB2B, a novel truncated DREB2-type transcription factor in the desert legume *Eremosparton songoricum*, enhances tolerance to multiple abiotic stresses in yeast and transgenic tobacco. *BMC Plant Biology* 14: 44. doi: 10.1186/1471-2229-14-44.

Liang, J., Deng, G., Long, H., Pan, Z., Wang, C., Cai, P., et al. 2012. Virus-induced silencing of genes encoding LEA protein in Tibetan hulless barley (*Hordeum vulgare* ssp. *vulgare*) and their relationship to drought tolerance. *Molecular Breeding* 30: 441–451 doi: 10.1007/s11032-011-9633-3.

Liliane, T. and Charles, M. 2020. *Factors Affecting Yield of Crops*. London: Intech Open. doi: 10.5772/intechopen.90672.

Liu, H., Yu, C., Li, H., Ouyang, B., Wang, T., Zhang, J., et al. 2015. Overexpression of *ShDHN*, a dehydrin gene from *Solanum habrochaites* enhances tolerance to multiple abiotic stresses in tomato. *Plant Science* 231: 198–211 doi: 10.1016/j.plantsci.2014.

Liu, Y., Song, Q., Li, D., Yang, X., and Li, D. 2017. Multifunctional roles of plant dehydrins in response to environmental stresses. *Frontiers in Plant Science* 8: 1018. doi: 10.3389/fpls.2017.01018.

Luo, D., Hou, X., Zhang, Y., Meng, Y., Zhang, H., Liu, S., et al. 2019. *CaDHN5*, a dehydrin gene from pepper, plays an important role in salt and osmotic stress responses. *International Journal of Molecular Sciences* 20: 1989. doi: 10.3390/ijms20081989.

Mahrez, W., Arellano, M., Moreno-Romero, J., Nakamura, M., Shu, H., Nanni, P., et al. 2016. H3K36ac is an evolutionary conserved plant histone modification that marks active genes. *Plant Physiology* 170: 1566–1577. doi: 10.1104/pp.15.01744.

Malik, A., Veltri, M., Boddington, K., Singh, K., and Graether, S. 2017. Genome analysis of conserved dehydrin motifs in vascular plants. *Frontiers in Plant Science* 8: 709. doi: 10.3389/fpls.2017.00709.

Mehrabad Pour-Benab, S., Fabriki-Ourang, S., and Mehrabi, A. 2019. Expression of dehydrin and antioxidant genes and enzymatic antioxidant defense under drought stress in wild relatives of wheat. *Biotechnology and Biotechnological Equipment* 33: 1063–1073. doi: 10.1080/13102818.2019.1638300.

Müller, H., Schäfer, N., Bauer, H., Geiger, D., Lautner, S., Fromm, J., et al. 2017. The desert plant *Phoenix dactylifera* closes stomata via nitrate-regulated SLAC1 anion channel. *New Phytologist* 216: 150–162. doi: 10.1111/nph.14672.

Nagaraju, M., Kumar, S., Reddy, P., Kumar, A., Rao, D., and Kishor, P. 2019. Genome-scale identification, classification, and tissue specific expression analysis of late embryogenesis abundant (LEA) genes under abiotic stress conditions in *Sorghum bicolor* L. *PLoS ONE* 14: e0209980. doi: 10.1371/journal.pone.0209980.

Narusaka, Y., Nakashima, K., Shinwari, Z. K., Sakuma, Y., Furihata, T., Abe, H., et al. 2003. Interaction between two *cis*-acting elements, ABRE and DRE, in ABA-dependent expression of *Arabidopsis rd29A* gene in response to dehydration and high-salinity stresses. *Plant Journal* 34: 137–148 doi: 10.1046/j.1365-313x.2003.01708.x.

Nguyen, P., Tossounian, M., Kovacs, D., Thu T., Stijlemans, B., Vertommen, D., et al. 2020. Dehydrin ERD14 activates glutathione transferase Phi9 in *Arabidopsis thaliana* under osmotic stress. *Biochimica et Biophysica Acta - General Subjects* 1864: 129506. doi: 10.1016/j.bbagen.2019.129506.

Nilo, R., Saffie, C., Lilley, K., Baeza-Yates, R., Cambiazo, V., Campos-Vargas, R., et al. 2010. Proteomic analysis of peach fruit mesocarp softening and chilling injury using difference gel electrophoresis (DIGE). *BMC Genomics* 11: 43. doi: 10.1186/1471-2164-11-43.

Ohkubo, T., Kameyama, A., Kamiya, K., Kondo, M., and Hara, M. 2020. F-segments of Arabidopsis dehydrins show cryoprotective activities for lactate dehydrogenase depending on the hydrophobic residues. *Phytochemistry* 173: 112300. doi: 10.1016/j.phytochem.2020.112300.

Park, H. J., Kim, W. Y., and Yun, D. J. 2016. A new insight of salt stress signaling in plant. *Molecules and Cells* 39: 447. doi: 10.14348/molcells.2016.0083.

Riley, A., Ashlock, D., and Graether, S. 2019. Evolution of the modular, disordered stress proteins known as dehydrins. *PLoS ONE* 14: e0211813. doi: 10.1371/journal.pone.0211813.

Saavedra, L., Svensson, J., Carballo, V., Izmendi, D., Welin, B., and Vidal, S. 2006. A dehydrin gene in *Physcomitrella patens* is required for salt and osmotic stress tolerance. *Plant Journal* 45: 237–249. doi: 10.1111/j.1365-313X.2005.02603.x.

Sena, J., Giguère, I., Rigault, P., Bousquet, J., and Mackay, J. 2018. Expansion of the dehydrin gene family in the Pinaceae is associated with considerable structural diversity and drought-responsive expression. *Tree Physiology* 38: 442–456. doi: 10.1093/treephys/tpx125.

Shekhawat, U., Srinivas, L., and Ganapathi, T. 2011. *MusaDHN-1*, a novel multiple stress-inducible SK_3-type dehydrin gene, contributes affirmatively to drought- and salt-stress tolerance in banana. *Planta* 234: 915–932 doi: 10.1007/s00425-011-1455-3.

Singh, J., Reddy, P., Reddy, C., and Reddy, M. 2015. Molecular cloning and characterization of salt inducible dehydrin gene from the C_4 plant *Pennisetum glaucum*. *Plant Gene* 4: 55–63. doi: 10.1016/j.plgene.2015.08.002.

Soulages, J. L., Kim, K., Arrese, E. L., Walters, C., and Cushman, J. C. 2003. Conformation of a group 2 late embryogenesis abundant protein from soybean. Evidence of poly (L-proline) type II structure. *Plant Physiology* 131: 963–975 doi: 10.1104/pp.015891.

Strimbeck, G. 2017. Hiding in plain sight: The F segment and other conserved features of seed plant SK_n dehydrins. *Planta* 245: 1061–1066. doi: 10.1007/s00425-017-2679-7.

Xu, J., Zhang, Y. X., Wei, W., Han, L., Guan, Z. Q., Wang, Z., et al. 2008. BjDHNs confer heavy-metal tolerance in plants. *Molecular Biotechnology* 38: 91–98. doi: 10.1007/s12033-007-9005-8.

Yang, W., Zhang, L., Lv, H., Li, H., Zhang, Y., Xu, Y., et al. 2015. The K-segments of wheat dehydrin WZY2 are essential for its protective functions under temperature stress. *Frontiers in Plant Science* 6: 406. doi: 10.3389/fpls.2015.00406.

Yang, Z., Sheng, J., Lv, K., Ren, L., and Zhang, D. 2019. Y_2SK_2 and SK_3 type dehydrins from *Agapanthus praecox* can improve plant stress tolerance and act as multifunctional protectants. *Plant Science* 284: 143–160. doi: 10.1016/j.plantsci.2019.03.012.

Yu, X., Yue, W., Yang, Q., Zhang, Y., Han, X., Yang, F., et al. 2019. Identification of the LEA family members from *Caragana korshinskii* (Fabaceae) and functional characterization of CkLEA2-3 in response to abiotic stress in Arabidopsis. *Revista brasileira de botanica* 42: 227–238. doi: 10.1007/s40415-019-00529-y.

Yu, Z., Wang, X., and Zhang, L. 2018. Structural and functional dynamics of dehydrins: A plant protector protein under abiotic stress. *International Journal of Molecular Sciences* 19: 3420. doi: 10.3390/ijms19113420.

Zaman Khan, N., Lal, S., Ali, W., Aasim, M., Mumtaz, S., Kamil, A., et al. 2020. Distribution and classification of dehydrins in selected plant species using bioinformatics approach. *Iranian Journal of Biotechnology* 18: e2680. doi: 10.30498/ijb.2020.2680.

Zeng, X., Bai, L., Wei, Z., Yuan, H., Wang, Y., Xu, Q., et al. 2016. Transcriptome analysis revealed the drought-responsive genes in Tibetan hulless barley. *BMC Genomics* 17: 386. doi: 10.1186/s12864-016-2685-3.

Zhang, H., Liu, S., Ma, J., Wang, K., Meng, Y., Zhang, Y., et al. 2020. *CaDHN4*, a salt and cold stress-responsive dehydrin gene from pepper decreases abscisic acid sensitivity in Arabidopsis. *International Journal of Molecular Sciences* 21: 26. doi: 10.3390/ijms21010026.

Zhang, H., Zheng, J., Su, H., Xia, K., Jian, S., and Zhang, M. 2018a. Molecular cloning and functional characterization of the dehydrin (*IpDHN*) gene from *Ipomoea pescaprae. Frontiers in Plant Science* 9: 1454. doi: 10.3389/fpls.2018.01454.

Zhang, J., Xia, H., Liang, D., Lin, L., Deng, H., Lv, X., et al. 2021. Genome-wide identification and expression profiling of the dehydrin gene family in *Actinidia chinensis. Scientia Horticulturae* 280: 109930. doi: 10.1016/j.scienta.2021.109930.

Zhang, Y., Launay, H., Schramm, A., Lebrun, R., and Gontero, B. 2018b. Exploring intrinsically disordered proteins in *Chlamydomonas reinhardtii. Scientific Reports* 8: 6805. doi: 10.1038/s41598-018-24772-7.

Zhang, Y., Li, J., Yu, F., Cong, L., Wang, L., Burkard, G., et al. 2006. Cloning and expression analysis of SK_n-type dehydrin gene from bean in response to heavy metals. *Molecular Biotechnology* 32: 205–218. doi: 10.1385/MB:32:3:205.

Zhao, Y., Gao, C., Shi, F., Yun, L., Jia, Y., and Wen, J. 2018. Transcriptomic and proteomic analyses of drought responsive genes and proteins in *Agropyron mongolicum* Keng. *Current Plant Biology* 14: 19–29. doi: 10.1016/j.cpb.2018.09.005.

Zheng, J., Su, H., Lin, R., Zhang, H., Xia, K., Jian, S., et al. 2019. Isolation and characterization of an atypical LEA gene (*IpLEA*) from *Ipomoea pescaprae* conferring salt/drought and oxidative stress tolerance. *Scientific Reports* 9: 14838. doi: 10.1038/s41598-019-50813-w.

Zhu, W., Zhang, L., Lv, H., Zhang, H., Zhang, D., Wang, X., et al. 2014. The dehydrin wzy2 promoter from wheat defines its contribution to stress tolerance. *Functional and Integrative Genomics* 14: 111–125. doi: 10.1007/s10142-013-0354-z.

Zong, W., Yang, J., Fu, J., and Xiong, L. 2020. Synergistic regulation of drought-responsive genes by transcription factor OsbZIP23 and histone modification in rice. *Journal of Integrative Plant Biology* 62: 723–729 doi: 10.1111/jipb.12850.

Zong, W., Zhong, X., You, J., and Xiong, L. 2013. Genome-wide profiling of histone H3K4-tri-methylation and gene expression in rice under drought stress. *Plant Molecular Biology* 81: 175–188. doi: 10.1007/s11103-012-9990-2.

11 Molecular Understanding of Signaling Compounds for Optimizing Cell Signal Transduction Mechanism under Abiotic Stresses in Crop Plants

Pratikshya Mishra
Odisha University of Agriculture and Technology

Smruti Ranjan Padhan
ICAR – Indian Agricultural Research Institute

Jyoti Prakash Sahoo
Odisha University of Agriculture and Technology
CV Raman Global University

Ananya Anyatama Mohanty
Odisha University of Agriculture and Technology

Satyaprakash Barik
CV Raman Global University

Pranaya Pradhan and Kailash Chandra Samal
Odisha University of Agriculture and Technology

CONTENTS

DOI: 10.1201/9781003300564-11

11.1 INTRODUCTION

Climate change and pollution may exacerbate abiotic stress interactions in crop plants (Ramegowda and Senthil-Kumar, 2015). For biotic and abiotic interactions in plants, stress factors have been used to explore the underlying biochemical mechanisms, but this hasn't been done for interactions between living and nonliving species. Reactive oxygen species (ROS) are examples of interactions between nonliving and living organisms (Zheng et al., 2018). Plants in the wild have to deal with a wide range of abiotic and biotic issues that might occur simultaneously or in succession. Plants can modify their molecular, physiological, and metabolic processes when subjected to both biotic and abiotic stress. A complex network of signaling pathways was activated during plant development when drought and illness were present simultaneously (Ramegowda and Senthil-Kumar, 2015). Several elements govern how a species or cultivar of plant reacts when under stress, including its susceptibility to stress, its exposure period, and the kind of stress agent it is exposed to. This chapter summarizes the molecular knowledge of signaling molecules for improving cell signal transduction pathways under abiotic stresses in agricultural crop plants.

11.2 SIGNALING MOLECULES UNDER STRESSFUL CONDITIONS

More ROS are produced by stressed agricultural plants, resulting in protein, fatty acid, and carbohydrate oxidation and adversity to DNA molecules (Liang et al., 2018). ROS disrupt cellular metabolism. Small signaling molecules activate the oxidants, flavonoid production, low-molecular-weight compound accumulation, vitamin C, anthocyanin, and other defense-related activities (Ahmad et al., 2010). Signal transduction pathways activate many antitoxin proteins, osmolytes, and enzymes (Prasch and Sonnewald, 2015). As a result of the decreased efficacy of antioxidant mechanisms, stress may lead to cell death and membrane damage (Bandurska et al., 2013). Oxidative stress, vitamin C, mRNA synthesis, lipid, protein, and sugar homeostasis are only some of the signals involved in plant growth and development. They play a vital role in coordinating developmental processes and signaling networks in response to a wide range of abiotic stresses (Prasch and Sonnewald, 2015).

By activating NADPH oxidase and promoting H_2O_2 production, ABA helps to increase NO production and activate the defense mechanism via ethylene creation in the plant system. This causes the stomata to close (He et al., 2011). Signaling molecules must be present at small concentrations for signaling molecules to be effective. The antioxidant defense gene is activated when exposed to low amounts of UV-B light. The *UVR8*, *COP1*, and *HY/HYP5* transcription factors are part of this signaling pathway (He et al., 2011) (Figure 11.1).

Stress alters the apoplastic hexose accumulation. In plant tissues, carbohydrates serve an important role in enabling most plant stress links. Sugars are an important source of energy for plants and a structural component of their defense mechanisms. Both as hormone signaling molecules and signal molecules, these compounds are found throughout the plant immune system (Morkunas and Ratajczak, 2014). It was also found that ethylene and abscisic acid (ABA) affect the abiotic stress tolerance in crop plants. Some case studies are indexed in Figures 11.2 and 11.3. Figure 11.2 describes the response of ethylene, while Figure 11.3 explains the response of ABA in crop plants during various abiotic stresses.

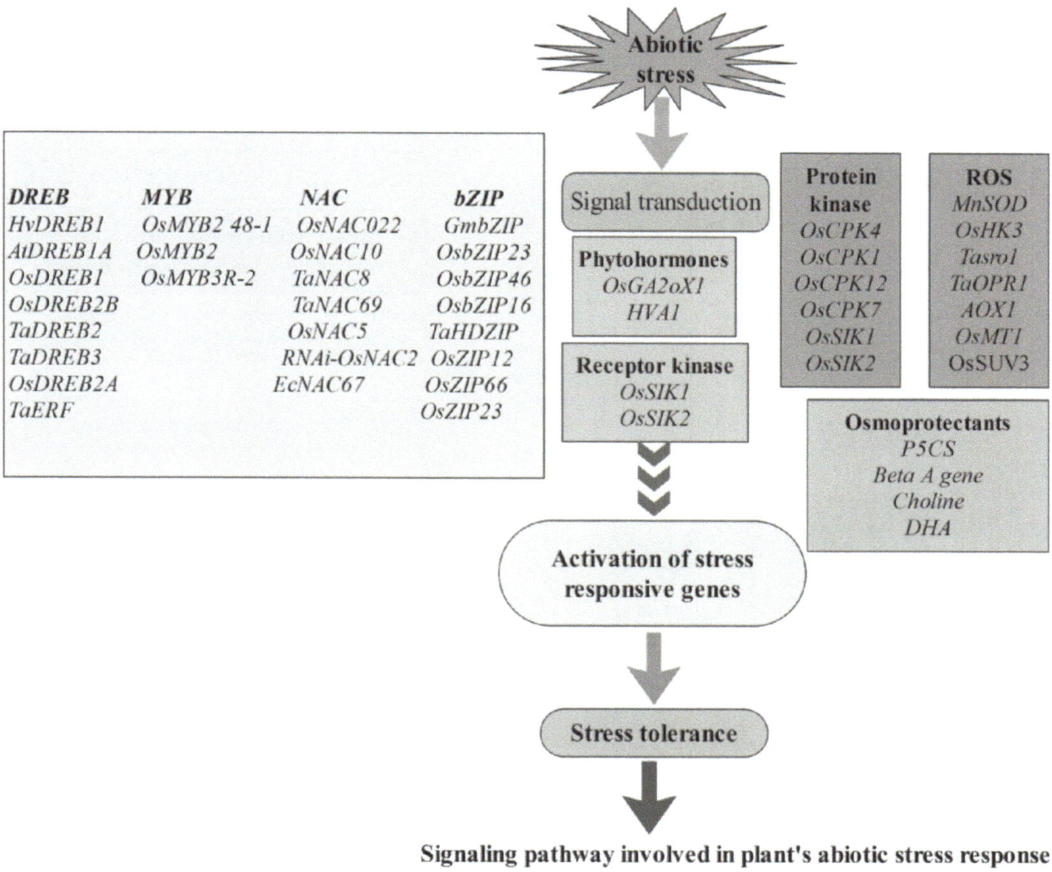

FIGURE 11.1 Signaling pathway during abiotic stress.

11.3 STRESS SIGNAL SENSORS

Gene expression, metabolic activity, and phenotypic response appear to be modulated by environmental challenges in plant cells (Zhu, 2016). Only a few stress sensors have been uncovered despite rigorous study. The genes that produce stress sensors display functional redundancy. Environmental stresses induce transient Ca21 (Ca21)cyt to enter the cell's cytoplasm (Xiong et al., 2002). Ca21 channels are the principal mechanism of transport for (Ca21) cyt. The osmotic Ca21 channels present in bacteria, yeast, and plant cells may be employed as osmotic sensors (Árnadóttir et al., 2010). An unbiased forward genetic search utilizing calcium imaging has found the Ca21 sensor gene *OSCA1* in *Arabidopsis thaliana* roots (Yuan et al., 2014). Sulfonylurea (Sul) and mannitol (Mn)-supplemented media led to greater levels of (Ca)cyt, as well as improved drought and salt stress responses in plants cultivated in these medium (Yuan et al., 2014). OSCA1 may serve as an osmotic sensor to induce (Ca21)cyt in response to environmental stress.

11.3.1 Sensors Monitoring the Impact of Salt Stress

Soil salt concentration (Na1) directly influences plant growth, promoting hyper reactive oxygen speciesmotic and oxidative stress (Zhu, 2007). Calcium-binding protein, SOS proteins, are employed as a substitute for calcineurin as plants lack it (Zhu, 2002). In response to the signal, the

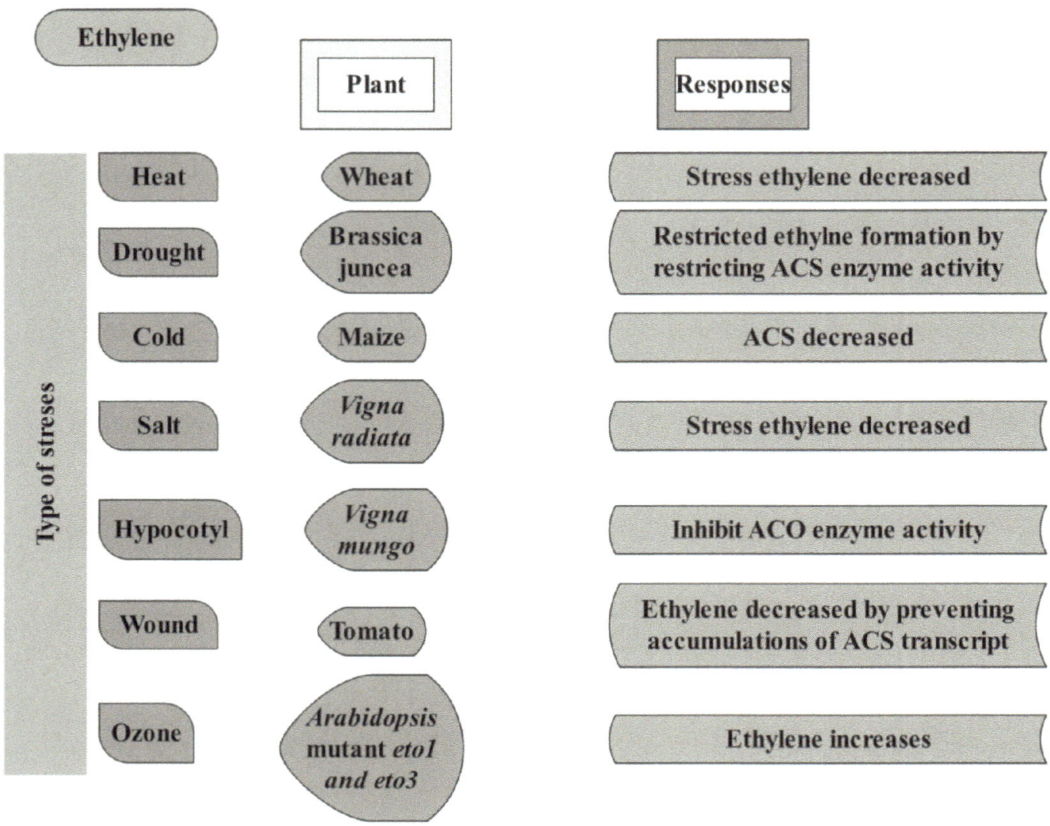

FIGURE 11.2 Ethylene response toward the abiotic stress tolerance in crop plants.

calcium sensor protein SOS3 activates a serine/threonine-protein kinase enzyme (SOS2), which in turn phosphorylates and positively controls the plasma membrane-bound sodium/proton antiporter SOS1, which is decreased in root cells when salt concentrations in the soil are excessive. Protecting cells from sodium toxicity, the root epidermis excretes Na1 into the soil, while Na1 retained in xylem cells excludes Na1 from leaf mesophyll cells for long-distance transport (Zhu, 2016). Salt damage prevention is provided by SOS3 and SOS3 homolog protein *SCaBBP8/calcineurin B-like 10 (CBL10)* (Du et al., 2008).

Research in *Arabidopsis thaliana* has found protein kinases related to the SOS2 protein kinases and connected to ROS nonfermenting 1 protein kinases. SOS3-like calcium-binding proteins (*ScaBPs*) interact with 37 *SnRK* members from three subfamilies (Guo et al., 2001). By interacting with PKS/CIPK, ScaBP lowers plasma membrane H1-ATPase activity (Fuglsang et al., 2007). D6-like cellulose synthases in *Arabidopsis thaliana* have exhibited osmotic tolerance by adjusting ROS levels in the plant (Zhu et al., 2010). The ability of *Arabidopsis thaliana* sos1 mutant cells to withstand high salt concentrations is increased by overexpression of SOS1-cDNA (Gronthos et al., 2000). Scientists showed for the first time that salinity stress tolerance and the distribution of Na1 across plant organs are both reliant on SOS1 (Olias et al., 2009). Constitutive overexpression of SOS2 demonstrated higher salinity stress tolerance and stronger antioxidative activity in transgenic *Arabidopsis thaliana*, tomato, and apple (Yang et al., 2015). Similar to SOS pathway overexpression alone, *AtNHX1* transgenic overexpression of SOS1, SOS2, and SOS3 enhanced salinity stress tolerance (Yang et al., 2009). Some studies examined the relationship between salt-stressed plants' stomatal activity and primary or secondary signaling molecules (Table 11.1).

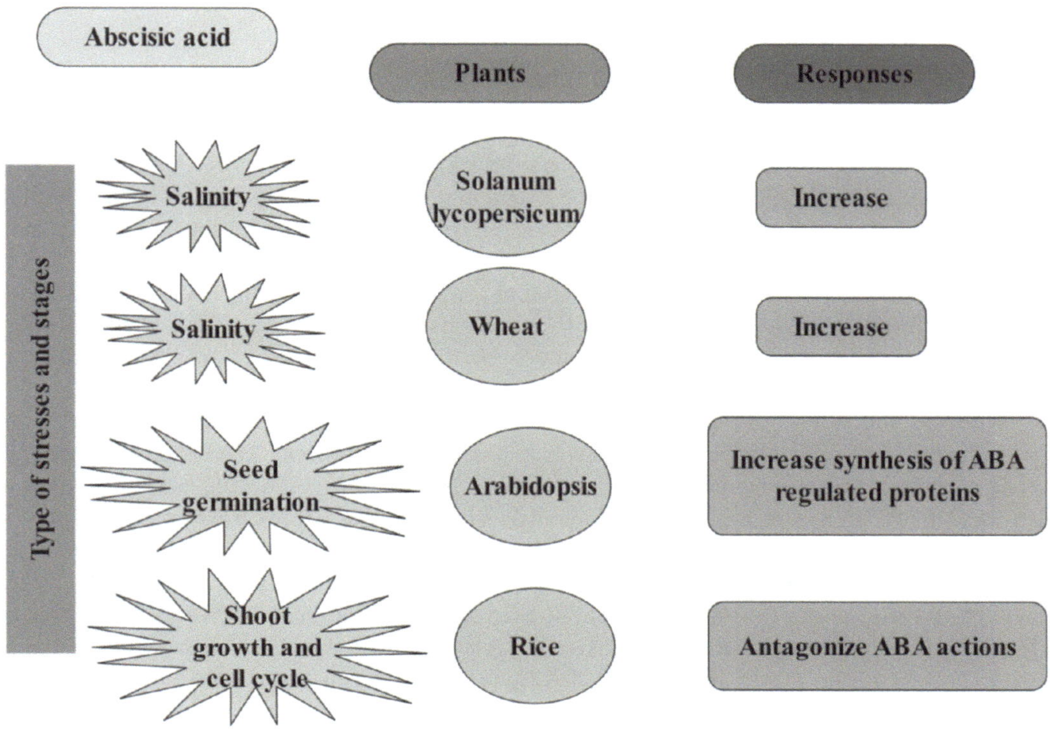

FIGURE 11.3 ABA response toward the abiotic stress tolerance in crop plants.

TABLE 11.1
Relationship between Salt-Stressed Plants' Stomatal Activity and Primary or Secondary Signaling Molecules

Signaling Molecules	Crop Species	Stomatal Activity	Measurement of other Parameters	Experiment Condition	References
Hydrogen peroxide (H_2O_2)	Dianthus superbus	Development of stoma and chloroplasts	Growth parameters; photosynthesis activity; antioxidant activity	Pot experiment	Ma et al. (2017)
	Common wheat	Development of stoma and chloroplasts	Growth parameters; water loss from leaf surface; accumulation of proline and soluble sugar	Liquid solution culture	Yang et al. (2016)
	Barrel medic	Conductance of stomata	Growth parameters; antioxidant activity; accumulation of proline	Pot experiment (perlite/sand)	Filippou et al. (2016)
	Arabidopsis thaliana	Osmotic potential of leaf	Biometric parameters; leaf water loss; enzyme activity	Hydroponic culture	Ozgur et al. (2014)
	Pea	Conductance of stomata	Growth; photosynthesis; antioxidant enzyme activity; mineral content	Hydroponic system	Martí et al. (2011)
Abscisic acid (ABA)	Arabidopsis thaliana	Movements of stomata	Cotyledons characteristics	MS agar medium	Singh et al. (2015)
	Sweet basil	Conductance of stomata	Growth; volatile compounds	Greenhouse/ growth chamber	Barbieri et al. (2012)
	Tomato	Conductance of stomata	Growth; yield; proline accumulation	Greenhouse experiment	Orsini et al. (2010)

11.3.2 Osmotic Stress Sensor

Osmotic stress is recognized by *OSCA1*, and cytosolic calcium levels are raised (Yuan et al., 2014). CPKs and ScaBP/CBL-PKS/CIPKs are activated (Zhu, 2016). Osmotic stress in *Arabidopsis thaliana* activated several protein kinases from the *SnRK2* family (Boudsocq et al., 2004). Yet, no one knows exactly how osmotic stress triggers the *SnRK2s*. *Arabidopsis thaliana* snrk2 mutants show reduced growth, gene expression, and accumulation of suitable osmolytes such as proline, secondary messenger inositol 1,4,5-triphosphate (IP3), and ABA concentration, but when osmotic stress is avoided, they grow like wild-type plants (Fujii et al., 2011). The involvement of SnRK2 in the accumulation of ABA and the regulation of osmotic pressure during osmotic stress signaling was shown in this study (Fujii et al., 2011). SnRK2s are activated by ABA and abiotic stress-responsive Raf-like kinase (ARK) in the moss *Physcomitrella patens* (Saruhashi et al., 2015; Wang et al., 2016a).

Osmotic and salt stress, for example, regulate MAP kinases, which are important signal transducers, both transcriptionally and pyrotechnically. At least one MAPK downstream cascade is activated by the osmotic signal received by an *Arabidopsis thaliana* osmosensor known as *AtHK1* (Wohlbach et al., 2008). When AtHK1 transgenic overexpression was applied to *Lycium barbarum*, the plant's ability to withstand salt and water stress was enhanced (Chen et al., 2010). In addition to MEKK1, the MEKK1 signaling pathway also activates MKK4 and MKK5 (Asai et al., 2002). Salt and cold stress trigger activation of MPK4 and MP6 via the MKK2 activator. Multiple stress-responsive genes were elevated in *Arabidopsis thaliana* plants with transgenic overexpression of the MKK2 gene, resulting in better cold and salinity tolerance (Teige et al., 2004).

11.3.3 Kinases

In plant cells, the protein kinase family of proteins regulates the ionic and osmotic balance, affecting plant metabolic and physiological functions. As with other forms of protein kinase in plants, MAPK and SnRK, which are ROS, non-fermenting-related protein kinases, are well-understood in plant research. Salt overly sensitive (SOS) controls ion homeostasis through SnRK1, a protein kinase, in response to salinity stress. The ionic and oxygen stress signaling pathways of another rice SnRK2 protein, SAPK4, were revealed to be involved in homeostasis via the involvement of SAPK4 (Diédhiou et al., 2008). Wheat SnRK2 serine/threonine kinase family members were involved in ABA signaling for the first time (Fujii and Zhu, 2012). An ABA-independent mechanism in finger millet relies on the action of these proteins (Lata and Prasad, 2011). The connection between physiological and cellular pathways illustrates how MAPK-driven signal transduction cascades play a significant role in stress response.

All eukaryotes have a MAPK stress-response system, which is highly conserved across various ROS. Abiotic stress tolerance in rice and the pathogen *Cochliobolus miyabeanus* were influenced by MPK5 overexpression (Beckers et al., 2009). Drought and salinity stress tolerance are improved in rice lines when OsCPK4 is overexpressed (Campo et al., 2014). During drought stress, the laccase gene and other oxidative stress-related genes were elevated in these transgenic lines (Campo et al., 2014). OsCDPK7 overexpression in rice increased drought and salt stresses, but not low-temperature difficulties. This shows that rice has two distinct signaling pathways, as has previously been established (Saijo et al., 2000). There has been a link between the development of drought stress tolerance in transgenic plants and the accumulation of osmoprotectants in OsCIPK12 transgenic rice lines that express OsPK12. A case study on the expression of kinases involved in signaling pathways through transgenic approach in rice is illustrated in Figure 11.4, which elaborates on the genes, promotors, and phenotypic expression employed for drought and salinity tolerance in rice.

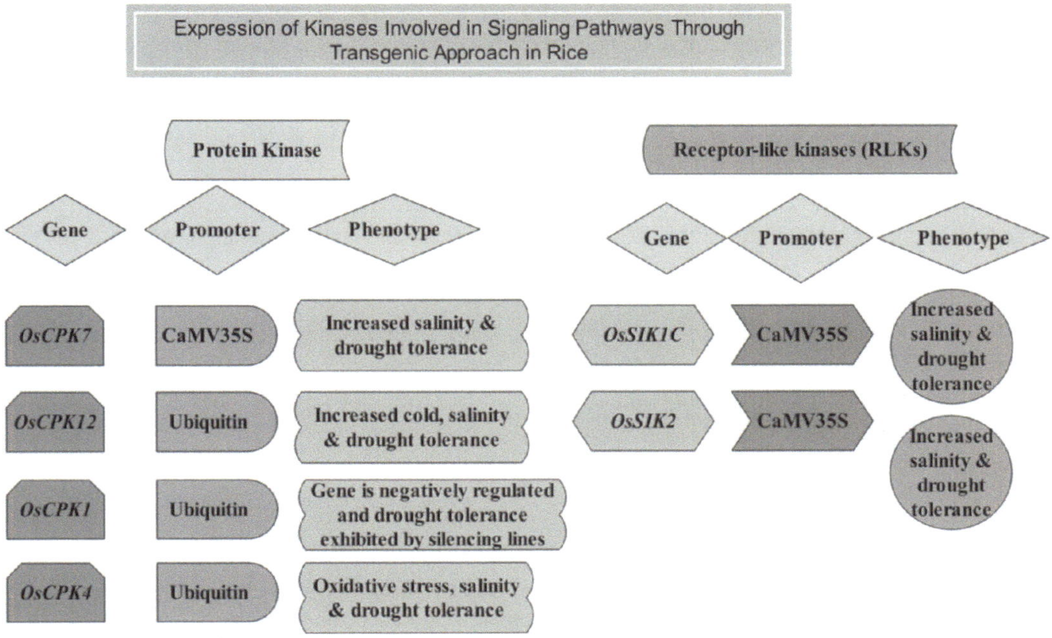

FIGURE 11.4 Expression of kinases involved in signaling pathways through transgenic approach in rice.

11.3.4 Calcium Sensors and Signaling

Various environmental variables, including biotic and abiotic stress and intracellular activities, may activate Ca21 (DeFalco et al., 2010). As a result of this research, two proteins have been identified as Ca21 sensors in plants. Ca21 has been detected in plants by *CMLs*, *CBLs*, and *CDPKs/CIPKs*. Since Ca21 may be found in both plants and animals, EF-hand domains can be found on both the N and C termini of the CaMs (Perochon et al., 2011). Osmotic stress activates a CaM-mediated signaling pathway in several plant species. One of them is AtCML8, a GmCaM4 ortholog (Magnan et al., 2008). Overexpression of the soybean CaM gene in *Arabidopsis thaliana* plants increased the transcriptional activity of the MYB2 transcription factor, promoting the expression of salt- and drought-sensitive genes (Yoo et al., 2005). When transgenic *Arabidopsis* plants exhibiting drought and salinity tolerance were overexpressed with the rice calmodulin-like OsMSR2 (multi-stress-responsive gene 2), it regulated the known stress-responsive genes in a way that was linked to the ABA system (Xu et al., 2011).

11.4 GENETICAL ALTERATION OF KINASES TO INCREASE DROUGHT AND SALINITY STRESS TOLERANCE

During periods of drought, auxin is important to the plant's capacity to adapt. There is complete control over all of these processes when there is abiotic stress (Shibasaki et al., 2009). Difficulty in adapting to drought and other environmental stresses has been linked to the auxin-responsive gene family, such as Auxin/Indole-3-Acetic Acid genes (*AUX/IAA*s act as repressors), Auxin Response Factor genes (*ARF*s act as transcription activators), Small Auxin Up RNA genes (*SAUR*s regulate the auxin-signaling pathway), and Gretchen Hagen 3 genes (*GH3*s), which encodes auxin conjugating enzymes IAA-amido synthetases (Kong et al., 2019). OsGH3.13 triggered the LEA (late embryogenesis abundant) genes, which improved rice seedlings' ability to withstand drought (Zhang et al., 2009). When rice's *OsGH3-2* gene was overexpressed, free IAA and carotenes and

ABA were reduced, which led to increased water loss and drought response in rice (Du et al., 2012). Transcriptome analyses suggest that dehydration affects several auxin-related genes. Drought resistance is improved by auxin, which alters the architecture of the root system (Shi et al., 2014). To improve drought tolerance, *Arabidopsis thaliana* activated the auxin biosynthesis gene *YUCCA7*, which produced more endogenous auxin (Lee and Luan, 2012). Auxin also contributed to the accumulation of sugars and sugar alcohols, both compatible solutes. Auxin's antioxidant enzymes (superoxide dismutase, catalase, and peroxide) that effectively detoxified ROS were positively regulated under drought conditions (Shi et al., 2014). IAA and other hormones may help the plant's response to drought. The ABA-mediated inhibition of LR development is regulated by lowering PAT levels. When ABI3 interacts with ARF or IAA proteins, LTP is boosted (Shkolnik-Inbar and Bar-Zvi, 2010). Under conditions of mild water shortage, ABA inhibits auxin transport, affecting proton secretion at the root tips and regulating root development (Sharma et al., 2015). Some case studies of transgenic plants overexpressing signaling molecules for abiotic stress tolerance are indexed in Table 11.2.

TABLE 11.2
Transgenic Plants Overexpressing Signaling Molecules for Abiotic Stress Tolerance

Functional Category	Genes	Origin	Transformation Receptor	Abiotic Stress Regulation	References
Protein kinase	SOS1	AT	AT	SST	Qiu et al. (2004)
	SOS2	AT	AT	SST	Ohta et al. (2003)
	SOS2	AT	AT	SST	Liu et al. (2011)
	SOS3	AT	AT	SST	Guo et al. 2001)
	SOS3	AT	AT	SST	Ishitani et al. (2000)
	SOS2-1 and SOS3-1	AT	AT	SST	Kamei et al., 2005)
	SOS6/AtCSLD5	AT	AT	SST and DST	Zhu et al. (2010)
	SCABP8/CBL10	AT	AT	SST	Quan et al. (2007)
	LKS1	AT	AT	IH	Wang et al. (2022)
	CBL-CIPK 3/9/23/26	AT	AT	MT	Tang et al. (2015)
	SnRK2	AT	AT	HOST	Boudsocq et al. (2004)
	SnRK2.2/2.3/2.6	AT	AT	ABAS	Fujii et al. (2009)
	SnRK2.2/2.3/2.6	AT	AT	ABAS	Fujii et al. (2009)
	SnRK2.6	AT	AT	ABAS	Wang et al. (2015)
	SnRK2	AT	AT	DHT	Yoshida et al. (2006)
	ARK	PP	PP	HOST	Saruhashi et al. (2015)
	ROP11	AT	AT	ABAS	Li et al. (2010)
	SLAC1/OST1	AT	AT	DST	Geiger et al. (2009)
	OST1	AT	AT	ROSS	Sirichandra et al. (2009)
	OST1	AT	AT	ABAS	Ding et al. (2015)
	OST1	AT	AT	ABAS	Mustilli et al. (2002)
	SLAC1	AT	AT	ABAS	Brandt et al. (2015)
	SLAC1	AT	AT	ABAS	Geiger et al. (2010)
	SLAC1	AT	AT	ABAS	Brandt et al. (2015)
	SLAC1	AT	AT	ABAS	Geiger et al. (2010)
	CIPK26	AT	NB	ROSS	Drerup et al. (2013)
	Tyr nitration	AT	AT	ABAS	Castillo et al. (2015)
	CBK3	AT	AT	HST	Liu et al. (2011)
	ABI1 and PP2CA	AT	AT	ABAS	Rodrigues et al. (2013)
	SrlK	MT	MT	SST	de Lorenzo et al. (2009)

(Continued)

TABLE 11.2 (*Continued*)
Transgenic Plants Overexpressing Signaling Molecules for Abiotic Stress Tolerance

His kinase	AtHKT1	AT	AT	SST	Maser et al. (2002)
	AtHKT1	AT	AT	SST	Rus et al. (2004)
	AtHKT1	AT	AT	SST	Rus et al. (2001)
	AtHKT1	AT	AT	VOSS	Wohlbach et al. (2008)
ABA receptor	PYL8	AT	AT	PLRG	Zhao et al. (2009)
	PYL6/RCAR9	AT	NB	ABAS	Aleman et al. (2016)
	PYR/PYL	AT	NB	ABAS	Park et al. (2009)
	PYR1 PYL1/2/4/5/8	AT	AT	ABAS	Gonzalez-Guzman et al. (2012)
	PYR/PYL/RCAR	AT	NB	ABAS	Rodriguez et al. (2014)
TFs	AREB1	AT	AT	ABAS	Furihata et al. (2006)
	COR	AT	AT	CST	Norén et al. (2016)
	AtMYC2 and AtMYB2	AT	AT	ABAS, DST	Abe et al. (2003)
	MBF1C	AT	AT	HST	Suzuki et al. (2011)
	HDSTART	AT	AT	HST	Yu et al. (2008)
MAPK	MPK9&12	AT	AT	ROSS	Jammes et al. (2009)
	MKK2	AT	AT	CST; SST	Yang et al. (2010)
	TPC1	AT	AT	ROSS; SST	Choi et al. (2014)
	MKK2	AT	AT	CST; SST	Teige et al. (2004)
HSFs	HSFA1S	AT	AT	HST	Liu et al. (2011)
	AtCaM3	AT	AT	ROSS	Zhang et al. (2009)

PP, Physcomitrella patens; AT, Arabidopsis thaliana; NB, Nicotiana benthamiana; MT, Medicago truncatula; SST, salinity stress tolerance; DST, drought stress tolerance; IH, ion homeostasis; MT, magnesium tolerance; HOST, hyperosmotic stress tolerance; ABAS, abscisic acid signaling; DHT, dehydration tolerance; ROSS, reactive oxygen species signaling; HST, heat stress tolerance; VOSS, vegetative osmotic stress signaling; PLRG, promote lateral root growth; CST, cold stress tolerance.

11.5 GENETIC ENGINEERING OF MOLECULAR CHAPERONES, HSPS, AND PLANT TRANSCRIPTION FACTORS FOR ABIOTIC STRESS TOLERANCE

Many plants produce proline as a consequence of environmental stress. 1-pyrroline-5-carboxylate synthetase was active when *Ailanthus altissima* was exposed to salt and drought (Filippou et al., 2014). Stress from salt, dryness, and heat led maize plants to produce more proline (Huang et al., 2018). It is possible to genetically alter plant TFs and molecular chaperones. The folding, assembly, translocation, and destruction of proteins are all dependent on molecular chaperones. A wide variety of proteins include chaperones. Molecular chaperones are stress proteins, and many of them were initially found as heat shock proteins (HSPs) (Wang et al., 2004). Cytoplasmic regulatory proteins (HSFs) may have a role in regulating gene expression in response to heat stress (HSPs). Sequential oligomerization and nucleus re-compartmentation occur in living organisms (Baniwal et al., 2004). Among the five most prominent HSP families in terms of molecular weight are the DnaK (Hsp70), the chaperonin (GroEL and Hsp60), and the Hsp90 and Hsp100 (Clp) families (Wang et al., 2004). *A. thaliana* overexpressed five Hsp90 genes in *Glycine max* when stressed by heat.

The amount of biomass produced reduced ROS, pod setting, and lipid peroxidation and chlorophyll loss were all the by-products of these changes (Xu et al., 2013). It was shown that when the OsHSP18.6 gene was overexpressed in transgenic rice, it was better able to withstand heat, drought, salt, and cold weather conditions. Exogenous expression of Hsp70 dramatically improved the plant's ability to tolerate salt (Hoang et al., 2015). Overexpression of PgHsc70 and PgHsp90 improves *Escherichia coli* cells' tolerance to salt and heat stress (Reddy et al., 2010). Overexpression of EaHSP70 in sugarcane transgenic plants resulted in higher water and chlorophyll content, salt stress tolerance, and photosynthetic efficacy (Augustine et al., 2015). *ZmHsf04*, a heat-activated maize *HsfA2* gene, was recently found and characterized. This gene has been demonstrated to function in the body's response to heat and salt. Some transcription factors associated with signaling mechanism are indexed in Table 11.3 as well as their promotors for expression and phenotype of the transgene.

TABLE 11.3
Transcription Factors Associated with Signaling Mechanism

Items	Gene	Promoter	Plant	Phenotype of Transgenics	References
DREB	HVDREB1	CaMV35S	Triticum aestivum	SST	Xu et al. (2009)
	OsDREB2B	CaMV35S rd29A	Oryza sativa	Non-functional transcript in *Oryza sativa*	Matsukura et al. (2010)
	TaDREB2 and TaDREB3	Double 35S, Maize Rab17	*Triticum aestivum* & *Hordeum vulgare*	Plant growth delay, flowering, grain yield; DST	Morran et al. (2011)
	TaERF	Ubiquitin	Triticum aestivum	Accumulation of proline during drought	Rong et al. (2014)
	OsDREB2A	Rd29A	Oryza sativa	DST, SST	Mallikarjuna et al. (2011)
	AtDREB1A/CBF3	Rd29A	Triticum aestivum	WST	Pellegrineschi et al. (2004)
	AtDREB1A/CBF3	35sCAMV	Oryza sativa	Stunted growth, DST and SST	Oh et al. (2005)
	OsDREB1/CBF	CaMV35S	Oryza sativa	CRGE	Ito et al. (2006)
NAC	OsNAC10	CaMV35S	Oryza sativa	DST and SST	Jeong et al. (2010)
	TaNAC8	CaMV35S	Triticum aestivum	Regulation of transcriptional gene	Xia et al. (2010)
	OsNAC5	CaMV35S	Oryza sativa	DST and SST	Song et al. (2011)
	EcNAC67	Rd29A	Oryza sativa	DST and SST	Rahman et al. (2016)
	ONAC022	CaMV35S	Oryza sativa	PA	Hong et al. (2016)
	OsNAC2	CaMV35S	Oryza sativa	SST	Shen et al. (2017)
bZIP	OsbZIP23	Ubiquitin	Oryza sativa	DST and SST	Xiang et al. (2008)
	GmbZIP	CaMV35S, rd29A	Triticum aestivum	DST and SST	Gao et al. (2011)
	OsbZIP46	Ubiquitin	Oryza sativa	DST and SST	Tang et al. (2012)
	OsbZIP16	CaMV35S, rd29A	Oryza sativa	DST and SST	Todaka et al. (2015)
	TaHDZIP	CaMV35S	Triticum aestivum	DST and SST	Yang et al. (2017)
	OsbZIP12	Cytochrome c	Oryza sativa	DST	Joo et al. (2014)
	OsbZIP66	GOS2 and RCc3	Oryza sativa	DST	Yoon et al. (2017)

SST, salinity stress tolerance; DST, drought stress tolerance; CST, cold stress tolerance; WST, water stress tolerance; CRGE, cold responsive genes expression; PA, proline accumulation.

11.6 EPIGENETIC REGULATION OF ABIOTIC STRESS TOLERANCE

Plants may adjust gene expression, which is epigenetically regulated but does not change the genetic sequence information in response to environmental stresses. There are several examples of how these non-coding non-proteins (miRNA and siRNA) impact gene transcription and posttranscriptional expression by altering the structure of DNA via DNA methylation (Kumar, 2016). Due to changes in the binding effectiveness of transcription factors, which resulted in the fluctuating expression of stress-sensitive genes, adaptation was made feasible by unfavorable stresses (Zhu et al., 2014). Abiotic stress-related genes may be activated or silenced by acetylation of histone lysine and arginine residues in the N-terminal region, known as the histone tail. Activation or muting may result from such a change (Kim et al., 2015a). Changes in histone modification have been linked to environmental stresses such as drought, heat, and salt.

Changes in the chromatin structure, such as enzyme modification or the insertion of histone variations, may affect transcriptional factors and influence gene expression (Asensi-Fabado et al., 2017). Under high-temperature conditions, DNA methylation has been demonstrated to alter miRNA gene expression in several ways. These histone changes influence the transcription of cell cycle genes under abiotic stress.

11.6.1 Abiotic Stress Tolerance and Protein-Coding Ubiquitination

Ubiquitination, an intrinsic posttranslational protein modification, is critical to plant growth and adaptation. Cellular defense proteins are regulated by modifying the number and activity of these proteins in the plant's metabolic system in response to external stresses such as drought or salt (Lyzenga and Stone, 2012). Ubiquitination may regulate protein abundance and stability and affect gene expression (Ding et al., 2015). In controlling the protein response to abiotic stress, protein stability, cell membrane integrity, and ROS all play a role (Dametto et al., 2015). The 26S proteasome destroys ubiquitin, a protein that serves as a mediator by covalently adhering to a target protein's lysine residue. The E3 family includes ubiquitin-activating enzymes El and E2 and ubiquitin-conjugating enzymes E3. These enzymes work together to destroy the targeted protein (Sharma et al., 2016). A key component of the Ubiquitin proteasome system, the E3, has been extensively investigated in terms of how it responds to abiotic stress (Stone, 2014). A list of ubiquitin-encoding genes in response to different abiotic stresses through genetic engineering approaches is indexed in Table 11.4.

11.6.2 Genome Editing Approaches to Modulate Abiotic Stress Signaling in Crop Plants

Gene editing, a new method, has made it possible to alter the structure of the genome with remarkable precision and efficiency. When a gene of interest is altered, researchers may do a functional study on it. An organism's genome may be activated or deactivated using site-specific nuclease (SSN) and site-specific recombinase (SSR). It has been employed extensively in abiotic stress tolerance, crop nutrition, and disease management. There are two types of DNA repair mechanisms in the cell: non-homologous end joining (NHEJ) and homology-directed repair (HDR). New gene functions are created utilizing HDR to repair DSBs. ZFN and transcription activator-like effector nucleases and the CRISPR-Cas9 system (Kim et al., 2015b). Gene knockout may be caused by the NHEJ repair mechanism, which causes frameshift mutations and destroys *cis*-regulation areas in the gene's coding region, rather than inserting the appropriate sequence via homologous repair (Kamburova et al., 2017). FokI (dimer) restriction nucleases, for example, have zinc finger DNA-binding protein domains to initiate double-strand break cleavage at certain loci (Petolino, 2015). ZFNs may be used to change gene expression by causing chromosomal double-strand breaks and nucleotide substitutions (Li et al., 2013).

TABLE 11.4
List of Ubiquitin-Encoding Genes in Response to Different Abiotic Stresses through Genetic Engineering Approaches

Plant	Abiotic Stress	Genes	Ubiquitin	Functional Expression Studies	References
Arabidopsis thaliana	Drought stress	PUB46 and PUB48	U box E3 ligase	Negatively control plant response to drought stress by degrading target proteins	Adler et al. (2017)
Arabidopsis thaliana	Salinity stress	SDIR1	finger E3 ligase	SDIRIP1 is a substrate for SDIR1, which adversely affects ABA and salt stress response	Zhang et al. (2015)
Malus pumila	Salinity and oxidative stress	MdMIEL1	RING E3 ligase	Antioxidative stress tolerance is negatively regulated by an increase in the concentration of reactive oxygen species	An et al. (2017)
Grimmia pilifera	Drought stress	GpDSR7	RING E3 ligase	Reduced water loss, which results from a water shortage, increases drought stress tolerance before and after recovery	Li et al. (2016b)
Citrullus lanatus	Drought stress	GdRZF1	RING E3 ligase	As a result of the antisense lines' high proline content and lower MDA and ion leakage in wild-type plants, the antisense lines' decreased expression increases their resistance to drought stress	Chung et al. (2017)
Triticum aestivum	Drought stress	TaSAP5	E3 ligase	DREB2A protein accumulates in drought-stressed wheat seedlings due to the degradation of DRIP proteins	Zhang et al. (2017)
Oryza sativa	Heat stress	OsHTAS	RING E3 ligase	Has the ability to increase the rice plant's resistance to extreme heat stress by activating an ABA-dependent and DST (drought and salinity stress tolerance)-mediated pathway in the leaf blade	Liu et al. (2016a)
Brassica napus	Heat stress	BnTR1	RINGv E3 ligase	Tolerance to heat may be conferred through modifying the expression of heat shock factors and heat shock proteins, which are thought to be controlled by calcium channels	Liu et al. (2014)
Zea mays	Drought stress	ZmAIRP4	RING finger E3 ligase	Using transgenic *Arabidopsis thaliana* to increase gene expression, reducing the pace at which water is lost, and modifying other stress-response genes may improve drought stress tolerance	Yang et al. (2018)
Piper nigrum	Drought stress	CaREL1	RING-type E3 ligase	Reduced sensitivity to abscisic acid in transgenic *Arabidopsis thaliana*, increased water loss, and reduced stomatal closure corroborate the role of overexpression in drought stress regulation	Lim et al. (2017)
Gossypium spp.	Salinity stress	GhSARP1	RING finger E3 ligase	A lower germination rate is associated with overexpressing transgenic *Arabidopsis thaliana*	Liu et al. (2016b)

11.7 CONCLUSION

Climate and environmental changes have made the issues more complicated and linked. Plants for agriculture have to be bred or developed to handle various stresses. When dealing with dual or multiple stresses, it is vital to consider the interplay between biotic and abiotic stresses. Genetic and hormonal stress-response signaling molecules, enzymes, antioxidants, and oxidant chemicals should be investigated. It is necessary to discuss the cellular and molecular mechanisms that underlie combined stress to understand it. Identification and segregation of stress pathways are necessary under single and multiple stressful circumstances. As demonstrated by research in this area, multiple stresses have unique molecular and metabolic responses that cannot be predicted from the responses to individual stress components. Even if physiological or biochemical parameters measure stress interactions, they must be recognized at the molecular level. When many stresses occur concurrently or sequentially in nature, agricultural plants that can withstand multiple stresses must be evaluated independently. There is a considerable gap between the processes of signaling molecules in the context of linked stress concerns. Each signaling molecule's mechanism and any putative linkages should be studied separately. Studies of gene expression and signaling molecules may provide new techniques for understanding the defense mechanisms that are now opaque. In general, stresses with synergistic or additive effects are more harmful, while those with antagonistic effects are less harmful. Research on *Arabidopsis thaliana* has been limited in its focus on the genome's response to various stress conditions. The ability of plants to recognize and send stress signals to cellular components, which in turn induce metabolic changes, is a critical factor in plant survival under various stressful conditions.

CONFLICT OF INTEREST

The authors declare no conflict of interest.

REFERENCES

Abe, H., Urao, T., Ito, T., Seki, M., Shinozaki, K., and Yamaguchi-Shinozaki, K. 2003. Arabidopsis AtMYC2 (bHLH) and AtMYB2 (MYB) function as transcriptional activators in abscisic acid signaling. *The Plant Cell* 15(1): 63–78.

Adler, G., Konrad, Z., Zamir, L., Mishra, A. K., Raveh, D., and Bar-Zvi, D. 2017. The Arabidopsis paralogs, PUB46 and PUB48, encoding U-box E3 ubiquitin ligases, are essential for plant response to drought stress. *BMC Plant Biology* 17(1): 1–12.

Ahmad, P., Jaleel, C. A., Salem, M. A., Nabi, G., and Sharma, S. 2010. Roles of enzymatic and nonenzymatic antioxidants in plants during abiotic stress. *Critical Reviews in Biotechnology* 30(3): 161–175. doi: 10.3109/07388550903524243.

Aleman, F., et al. 2016. An ABA-increased interaction of the PYL6 ABA receptor with MYC2 transcription factor: A putative link of ABA and JA signaling. *Scientific Reports* 6: 1–10. doi: 10.1038/srep28941.

An, J. P., Liu, X., Song, L. Q., You, C. X., Wang, X. F., and Hao, Y. J. 2017. Apple RING finger E3 ubiquitin ligase MdMIEL1 negatively regulates salt and oxidative stresses tolerance. *Journal of Plant Biology* 60(2): 137–145.

Árnadóttir, J., and Chalfie, M. 2010. Eukaryotic mechanosensitive channels. *Annual Review of Biophysics* 39: 111–137.

Asai, T., Tena, G., Plotnikova, J., Willmann, M.R., Chiu, W.L., and Gomez-Gomez, L., et al. 2002. MAP kinase signalling cascade in Arabidopsis innate immunity. *Nature* 415: 977–983. doi: 10.1038/415977a.

Asensi-Fabado, M. A., Amtmann, A., and Perrella, G. 2017. Plant responses to abiotic stress: the chromatin context of transcriptional regulation. *Biochimica et Biophysica Acta (BBA)-Gene Regulatory Mechanisms* 1860(1): 106–122.

Augustine, S. M., Narayan, J. A., Syamaladevi, D. P., Appunu, C., Chakravarthi, M., Ravichandran, V., and Subramonian, N. 2015. *Erianthus arundinaceus* HSP70 (EaHSP70) overexpression increases drought and salinity tolerance in sugarcane (*Saccharum* spp. hybrid). *Plant Science* 232: 23–34.

Bandurska, H., Niedziela, J., and Chadzinikolau, T. 2013. Separate and combined responses to water deficit and UV-B radiation. *Plant Science* 213: 98–105. doi: 10.1016/j.plantsci.2013.09.003.

Baniwal, S. K., Bharti, K., Chan, K. Y., Fauth, M., Ganguli, A., Kotak, S., ... and von Koskull-DÖring, P. 2004. Heat stress response in plants: a complex game with chaperones and more than twenty heat stress transcription factors. *Journal of Biosciences* 29(4): 471–487.

Barbieri, G., Vallone, S., Orsini, F., Paradiso, R., De Pascale, S., Negre-Zakharov, F., and Maggio, A. 2012. Stomatal density and metabolic determinants mediate salt stress adaptation and water use efficiency in basil (*Ocimum basilicum* L.). *Journal of Plant Physiology* 169(17): 1737–1746. doi: 10.1016/j.jplph.2012.07.001.

Beckers, G.J., Jaskiewicz, M., Liu, Y., Underwood, W.R., He, S.Y., Zhang, S., et al. 2009. Mitogen-activated protein kinases 3 and 6 are required for full priming of stress responses in *Arabidopsis thaliana*. *Plant Cell* 21(3): 944–953.

Boudsocq, M., Barbier-Brygoo, H., and Lauriere, C. 2004. Identification of nine sucrose nonfermenting 1-related protein kinases 2 activated by hyperosmotic and saline stresses in *Arabidopsis thaliana*. *Journal of Biological Chemistry* 279(40): 41758–41766.

Brandt, B., Munemasa, S., Wang, C., Nguyen, D., Yong, T., Yang, P. G., Poretsky, E., Belknap, T.F., Waadt, R., Aleman, F., et al. 2015. Calcium specificity signaling mechanisms in abscisic acid signal transduction in Arabidopsis guard cells. *eLife*. 4: e03599. doi: 10.7554/eLife.03599.

Campo, S., Baldrich, P., Messeguer, J., Lalanne, E., Coca, M., and San Segundo, B. 2014. Overexpression of a calcium-dependent protein kinase confers salt and drought tolerance in rice by preventing membrane lipid peroxidation. *Plant Physiology* 32: 739–747.

Castillo, M. C., Lozano-Juste, J., González-Guzmán, M., Rodriguez, L., Rodriguez, P. L., and León, J. 2015. Inactivation of PYR/PYL/RCAR ABA receptors by tyrosine nitration may enable rapid inhibition of ABA signaling by nitric oxide in plants. *Science Signaling* 8(392): ra89–ra89.

Chen, L., Zhang, L., and Yu, D. 2010. Wounding-induced WRKY8 is involved in basal defense in Arabidopsis. *Molecular Plant-Microbe Interactions* 23(5): 558–565.

Choi, J., Ravipati, A., Nimmagadda, V., Schubert, M., Castellani, R. J., and Russell, J. W. 2014. Potential roles of PINK1 for increased PGC-1α-mediated mitochondrial fatty acid oxidation and their associations with Alzheimer disease and diabetes. *Mitochondrion*, 18, 41–48.

Chung, J. S., Park, S. H., Min, J. H., Min, K. H., Lee, S., Lee, K. H., & Kim, C. S. 2017. Reduced expression of Gongdae Ring Zinc Finger 1 (GdRZF1) enhances drought stress tolerance in watermelon (*Citrullus lanatus*). *Horticultural Science and Technology*, 35(5), 637–646.

Dametto, A., Buffon, G., dos Reis Blasi, É. A., and Sperotto, R. A. 2015. Ubiquitination pathway as a target to develop abiotic stress tolerance in rice. *Plant Signaling & Behavior* 10(9): e1057369.

De Lorenzo, L., Merchan, F., Laporte, P., Thompson, R., Clarke, J., Sousa, C., and Crespi, M. 2009. A novel plant leucine-rich repeat receptor kinase regulates the response of *Medicago truncatula* roots to salt stress. *The Plant Cell* 21(2): 668–680. doi: 10.1105/tpc.108.059576.

DeFalco, T. A., Bender, K. W., and Snedden, W. A. 2010. Breaking the code: Ca^{2+} sensors in plant signalling. *Biochemical Journal* 425(1): 27–40.

Diédhiou, C. J., Popova, O. V., Dietz, K. J., and Golldack, D. 2008. The SNF1-type serine-threonine protein kinase SAPK4regulates stress-responsive gene expression in rice. *BMC Plant Biology* 8(1): 1–13.

Ding, Y., Li, H., Zhang, X., Xie, Q., Gong, Z., and Yang, S. 2015. OST1 kinase modulates freezing tolerance by enhancing ICE1 stability in Arabidopsis. *Developmental Cell* 32(3): 278–289.

Drerup, M. M., Schlücking, K., Hashimoto, K., Manishankar, P., Steinhorst, L., Kuchitsu, K., and Kudla, J. 2013. The calcineurin B-like calcium sensors CBL1 and CBL9 together with their interacting protein kinase CIPK26 regulate the Arabidopsis NADPH oxidase RBOHF. *Molecular Plant* 6(2): 559–569. doi: 10.1093/mp/sst009.

Du, H., Wu, N., Fu, J., Wang, S., Li, X., Xiao, J., and Xiong, L. 2012. A GH3 family member, OsGH3-2, modulates auxin and abscisic acid levels and differentially affects drought and cold tolerance in rice. *Journal of Experimental Botany* 63(18): 6467–6480.

Du, Y. Y., Wang, P. C., Chen, J., and Song, C. P. 2008. Comprehensive functional analysis of the catalase gene family in Arabidopsis thaliana. *Journal of Integrative Plant Biology* 50(10): 1318–1326.

Filippou, P., Antoniou, C., Obata, T., Van Der Kelen, K., Harokopos, V., et al. 2016. Kresoxim-methyl primes Medicago truncatula plants against abiotic stress factors via altered reactive oxygen and nitrogen species signalling leading to downstream transcriptional and metabolic readjustment. *Journal of Experimental Botany* 67(5): 1259–1274. doi: 10.1093/jxb/erv516.

Filippou, P., Bouchagier, P., Skotti, E., and Fotopoulos, V. 2014. Proline and reactive oxygen/nitrogen species metabolism is involved in the tolerant response of the invasive plant species *Ailanthus altissima* to drought and salinity. *Environmental and Experimental Botany* 97: 1–10.

Fuglsang, A. T., Guo, Y., Cuin, T. A., Qiu, Q., Song, C., et al. 2007. Arabidopsis protein kinase PKS5 inhibits the plasma membrane H+-ATPase by preventing interaction with 14-3-3 protein. *The Plant Cell* 19(5): 1617–1634.

Fujii, H., and Zhu, J. K. 2012. Osmotic stress signaling via protein kinases. *Cellular and Molecular Life Sciences* 69(19): 3165–3173.

Fujii, H., Chinnusamy, V., Rodrigues, A., Rubio, S., Antoni, R., et al. 2009. In vitro reconstitution of an abscisic acid signalling pathway. *Nature* 462(7273): 660–664.

Fujii, H., Verslues, P. E., and Zhu, J. K. 2011. Arabidopsis decuple mutant reveals the importance of SnRK2 kinases in osmotic stress responses in vivo. *Proceedings of the National Academy of Sciences* 108(4): 1717–1722.

Furihata, T., Maruyama, K., Fujita, Y., Umezawa, T., Yoshida, R., Shinozaki, K., and Yamaguchi-Shinozaki, K. 2006. Abscisic acid-dependent multisite phosphorylation regulates the activity of a transcription activator AREB1. *Proceedings of the National Academy of Sciences* 103(6): 1988–1993.

Gao, S. Q., Chen, M., Xu, Z. S., Zhao, C. P., Li, L., et al. 2011. The soybean GmbZIP1 transcription factor enhances multiple abiotic stress tolerances in transgenic plants. *Plant Molecular Biology* 75(6): 537–553.

Geiger, D., Scherzer, S., Mumm, P., Marten, I. A. P. P. L., et al. 2010. Guard cell anion channel SLAC1 is regulated by CDPK protein kinases with distinct Ca2+ affinities. *Proceedings of the National Academy of Sciences* 107(17): 8023–8028. doi: 10.1073/pnas.0912030107.

Geiger, D., Scherzer, S., Mumm, P., Stange, A., et al. 2009. Activity of guard cell anion channel SLAC1 is controlled by drought-stress signaling kinase-phosphatase pair. *Proceedings of the National Academy of Sciences* 106(50): 21425–21430. doi: 10.1073/pnas.0912021106.

Gonzalez-Guzman, M., Pizzio, G. A., Antoni, R., Vera-Sirera, F., Merilo, E., Bassel, G.W., et al. 2012. Arabidopsis PYR/PYL/RCAR receptors play a major role in quantitative regulation of stomatal aperture and transcriptional response to abscisic acid. *The Plant Cell* 24(6): 2483–2496. doi: 10.1105/tpc.112.098574.

Gronthos, S., Mankani, M., Brahim, J., Robey, P. G., and Shi, S. 2000. Postnatal human dental pulp stem cells (DPSCs) in vitro and in vivo. *Proceedings of the National Academy of Sciences*, 97(25), 13625-13630.

Guo, Y., Halfer, U., Ishitani, M., and Zhu, J. K., 2001. Molecular characterization of functional domains in the protein kinase SOS2 that is required for plant salt tolerance. *Plant Cell* 13: 1383–1399.

He, Y., Cui, J., Lee, J. C., Ding, S., Chalimoniuk, M., Simonyi, A., and Sun, G. Y. 2011. Prolonged exposure of cortical neurons to oligomeric amyloid-β impairs NMDA receptor function via NADPH oxidase-mediated ROS production: protective effect of green tea (-)-epigallocatechin-3-gallate. *ASN neuro*, 3(1), AN20100025.

Hoang, T. M. L., Moghaddam, L., Williams, B., Khanna, H., Dale, J., and Mundree, S. G. 2015. Development of salinity tolerance in rice by constitutive-overexpression of genes involved in the regulation of programmed cell death. *Frontiers in Plant Science* 6: 175.

Hong, Y., Zhang, H., Huang, L., Li, D., and Song, F. 2016. Overexpression of a stress-responsive NAC transcription factor gene ONAC022 improves drought and salt tolerance in rice. *Frontiers in Plant Science* 7: 4.

Huang, Q., Wang, M., and Xia, Z. 2018. The SULTR gene family in maize (*Zea mays* L.): gene cloning and expression analyses under sulfate starvation and abiotic stress. *Journal of Plant Physiology* 220: 24–33.

Ishitani, M., Liu, J., Halfter, U., Kim, C.-S., Shi, W., and Zhu, J. K. 2000. SOS3 Function in plant salt tolerance requires N-myristoylation and calcium binding. *The Plant Cell* 12(9): 1667–1678.

Ito, Y., Katsura, K., Maruyama, K., Taji, T., Kobayashi, M., Seki, M., et al. 2006. Functional analysis of rice DREB1/CBF-type transcription factors involved in cold-responsive gene expression in transgenic rice. *Plant Cell Physiology* 47(1): 141–153.

Jammes, F., et al. 2009. MAP kinases MPK9 and MPK12 are preferentially expressed in guard cells and positively regulate ROS mediated ABA signaling. *Proceedings of the National Academy of Sciences USA* 106: 20520–20525. doi: 10.1073/pnas.0907205106.

Jeong, J. S., Kim, Y. S., Baek, K. H., Jung, H., Ha, S. H., Do Choi, Y., et al. 2010. Root-specific expression of OsNAC10 improves drought tolerance and grain yield in rice under field drought conditions. *Plant Physiology* 153(1): 185–197.

Joo, J., Lee, Y. H., and Song, S. I. 2014. Overexpression of the rice basic leucine zipper transcription factor OsbZIP12 confers drought tolerance to rice and makes seedlings hypersensitive to ABA. *Plant Biotechnology Reports* 8(6): 431–441.

Kamburova, V. S., Nikitina, E. V., Shermatov, S. E., Buriev, Z. T., Kumpatla, S. P., Emani, C., et al. 2017. Genome editing in plants: an overview of tools and applications. *International Journal of Agronomy*, 2017: Article ID 7315351, doi: https://doi.org/10.1155/2017/7315351. 15.

Kamei, A., Seki, M., Umezawa, T., Ishida, J., Satou, M., Akiyama, K., et al., 2005. Analysis of gene expression profiles in Arabidopsis salt overly sensitive mutants sos2-1 and sos3-1. *Plant, Cell & Environment* 28(10): 1267–1275. doi: 10.1111/j.1365-3040.2005.01363.x.

Kim, H., Kim, S. T., Kim, S. G., and Kim, J. S. 2015a. Targeted genome editing for crop improvement. *Plant Breeding and Biotechnology* 3(4): 283–290.

Kim, J.-M., Sasaki, T., Ueda, M., Sako, K., and Seki, M. 2015b. Chromatin changes in response to drought, salinity, heat, and cold stresses in plants. *Frontiers in Plant Science* 6. doi: https://doi.org/10.3389/fpls.2015.00114.

Kong, W., Zhong, H., Deng, X., Gautam, M., Gong, Z., Zhang, Y., and Li, Y. 2019. Evolutionary analysis of GH3 genes in six Oryza species/subspecies and their expression under salinity stress in Oryza sativa ssp. japonica. *Plants*, 8(2), 30. doi: https://doi.org/10.3390/plants8020030.

Kumar, S. 2016. Epigenetic regulation of abiotic stress tolerance in plants. *Advances in Plants & Agriculture Research* 5(5): 00179.

Lata, C., and Prasad, M. 2011. Role of DREBs in regulation of abiotic stress responses in plants. *Journal of Experimental Botany* 62: 4731–4748.

Lee, S. C., and Luan, S. 2012. ABA signal transduction at the crossroad of biotic and abiotic stress responses. *Plant, Cell & Environment* 35(1): 53–60.

Li, M., Li, Y., Zhao, J., Liu, H., Jia, S., Li, J., et al. 2016b. GpDSR7, a novel E3 ubiquitin ligase gene in *grimmia pilifera* is involved in tolerance to drought stress in Arabidopsis. *PLoS One* 11(5): e0155455.

Li, W. T., He, M., Wang, J., and Wang, Y. P. 2013. Zinc finger protein (ZFP) in plants-A review. *Plant Omics* 6(6), 474–480, [i]–[viii].

Li, Y., Han, D., Hu, G., Sommerfeld, M., and Hu, Q. 2010. Inhibition of starch synthesis results in overproduction of lipids in *Chlamydomonas reinhardtii. Biotechnology and Bioengineering*, *107*(2): 258–268.

Liang, D., Shen, Y., Ni, Z., Wang, Q., Lei, Z., Xu, N., et al. 2018. Exogenous melatonin application delays senescence of kiwifruit leaves by regulating the antioxidant capacity and biosynthesis of flavonoids. *Frontiers in Plant Science* 9: 426. doi: 10.3389/fpls.2018.00426.

Lim, C. W., Park, C., Kim, J. H., Joo, H., Hong, E., and Lee, S. C. 2017. Pepper CaREL1, a ubiquitin E3 ligase, regulates drought tolerance via the ABA-signalling pathway. *Scientific Reports* 7(1): 1–12.

Liu, J., Zhang, C., Wei, C., Liu, X., Wang, M., et al. 2016a. The RING finger ubiquitin E3 ligase OsHTAS enhances heat tolerance by promoting H2O2-induced stomatal closure in rice. *Plant Physiology* 170(1): 429–443.

Liu, Q., Sun, Y., Yang, T., Feng, W., Li, C., & Li, F. 2011. Sub-10nm hexagonal lanthanide-doped NaLuF4 upconversion nanocrystals for sensitive bioimaging in vivo. *Journal of the American Chemical Society*, 133(43): 17122–17125.

Liu, Y., Zhang, X., Zhu, S., Zhang, H., Li, Y., Zhang, T., and Sun, J. 2016b. Overexpression of GhSARP1 encoding a E3 ligase from cotton reduce the tolerance to salt in transgenic Arabidopsis. *Biochemical and Biophysical Research Communications* 478(4): 1491–1496.

Liu, Z. B., Wang, J. M., Yang, F. X., Yang, L., Yue, Y. F., et al. 2014. A novel membrane-bound E 3 ubiquitin ligase enhances the thermal resistance in plants. *Plant Biotechnology Journal* 12(1): 93–104.

Lyzenga, W. J., and Stone, S. L. 2012. Abiotic stress tolerance mediated by protein ubiquitination. *Journal of Experimental Botany* 63(2): 599–616.

Ma, X., Zheng, J., Zhang, X., Hu, Q., and Qian, R. 2017. Salicylic acid alleviates the adverse effects of salt stress on *Dianthus superbus* (Caryophyllaceae) by activating photosynthesis, protecting morphological structure, and enhancing the antioxidant system. *Frontiers in Plant Science* 8: 600. doi: 10.3389/fpls.2017.00600.

Magnan, F., Ranty, B., Charpenteau, M., Sotta, B., Galaud, J. P., and Aldon, D. 2008. Mutations in AtCML9, a calmodulin-like protein from Arabidopsis thaliana, alter plant responses to abiotic stress and abscisic acid. *The Plant Journal* 56: 575–589. doi: 10.1111/j.1365-313X.2008.03622.x.

Mallikarjuna, G., Mallikarjuna, K., Reddy, M. K., and Kaul, T. 2011. Expression of OsDREB2A transcription factor confers enhanced dehydration and salt stress tolerance in rice (*Oryza sativa* L.). *Biotechnology Letters* 33(8): 1689–1697.

Martí, M. C., Florez-Sarasa, I., Camejo, D., Ribas-Carbó, M., Lázaro, J. J., Sevilla, F., and Jiménez, A. 2011. Response of mitochondrial thioredoxin PsTrx o 1, antioxidant enzymes, and respiration to salinity in pea (*Pisum sativum L.*) leaves. *Journal of Experimental Botany* 62(11): 3863–3874. doi: 10.1093/jxb/err076.

Mäser, P., Eckelman, B., Vaidyanathan, R., Horie, T., Fairbairn, D. J., Kubo, M., and Schroeder, J. I. 2002. Altered shoot/root Na+ distribution and bifurcating salt sensitivity in Arabidopsis by genetic disruption of the Na+ transporter AtHKT1. *FEBS letters*, 531(2): 157–161.

Matsukura, S., Mizoi, J., Yoshida, T., Todaka, D., Ito, Y., et al. 2010. Comprehensive analysis of rice DREB2-type genes that encode transcription factors involved in the expression of abiotic stress-responsive genes. *Molecular Genetics and Genomics* 283(2): 185–196.

Morkunas, I., and Ratajczak, L. 2014. The role of sugar signaling in plant defense responses against fungal pathogens. *Acta Physiologiae Plantarum* 36(7): 1607–1619. doi: 10.1007/s11738-014-1559-z.

Morran, S., Eini, O., Pyvovarenko, T., Parent, B., Singh, R., et al. 2011. Improvement of stress tolerance of wheat and barley by modulation of expression of DREB/CBF factors. *Plant Biotechnology Journal* 9(2): 230–249.

Mustilli, A. C., Merlot, S., Vavasseur, A., Fenzi, F., and Giraudat, J. 2002. Arabidopsis OST1 protein kinase mediates the regulation of stomatal aperture by abscisic acid and acts upstream of reactive oxygen species production. *The Plant Cell* 14(12): 3089–3099. doi: 10.1105/tpc.007906.

Norén, L., Kindgren, P., Stachula, P., Rühl, M., Eriksson, M. E., et al. 2016. Circadian and plastid signaling pathways are integrated to ensure correct expression of the CBF and COR genes during photoperiodic growth. *Plant Physiology* 171(2): 13921406. doi: 10.1104/pp.16.00374.

Oh, S. J., Song, S. I., Kim, Y. S., Jang, H. J., Kim, S. Y., Kim, M., et al. 2005. Arabidopsis CBF3/DREB1A and ABF3 in transgenic rice increased tolerance to abiotic stress without stunting growth. *Plant Physiology* 138(1): 341–351.

Ohta, M., Guo, Y., Halfter, U., and Zhu, J. K. 2003. A novel domain in the protein kinase SOS2 mediates interaction with the protein phosphatase 2C ABI2. *Proceedings of the National Academy of Sciences* 100(20): 11771–11776. doi: 10.1073/pnas.2034853100.

Olias, R., Eljakaoui, Z., Li, J. U. N., De Morales, P. A., Marin-Manzano, M. C., Pardo, J. M., and Belver, A. 2009. The plasma membrane Na+/H+ antiporter SOS1 is essential for salt tolerance in tomato and affects the partitioning of Na+ between plant organs. *Plant, Cell & Environment* 32(7): 904–916.

Orsini, F., Cascone, P., De Pascale, S., Barbieri, G., Corrado, G., Rao, R., et al. 2010. Systemin-dependent salinity tolerance in tomato: evidence of specific convergence of abiotic and biotic stress responses. *Physiologia Plantarum* 138(1): 10–21. doi: 10.1111/j.1399-3054.2009.01292.x.

Ozgur, R., Turkan, I., Uzilday, B., and Sekmen, A. H. 2014. Endoplasmic reticulum stress triggers ROS signalling, changes the redox state, and regulates the antioxidant defence of Arabidopsis thaliana. *Journal of Experimental Botany* 65(5): 1377–1390. doi: 10.1093/jxb/eru034.

Park, S. Y., Fung, P., Nishimura, N., Jensen, D. R., Fujii, H., et al. 2009. Abscisic acid inhibits type 2C protein phosphatases via the PYR/PYL family of START proteins. *Science* 324(5930): 1068–1071.

Pellegrineschi, A., Reynolds, M., Pacheco, M., Brito, R. M., Almeraya, R., Yamaguchi-Shinozaki, K., and Hoisington, D. 2004. Stress-induced expression in wheat of the Arabidopsis thaliana DREB1A gene delays water stress symptoms under greenhouse conditions. *Genome* 47(3): 493–500.

Perochon, A., Aldon, D., Galaud, J. P., and Ranty, B. 2011. Calmodulin and calmodulin-like proteins in plant calcium signaling. *Biochimie* 93: 2048–2053. doi: 10.1016/j.biochi.2011.07.012.

Petolino, J. F. 2015. Genome editing in plants via designed zinc finger nucleases. *In Vitro Cellular & Developmental Biology-Plant* 51(1): 1–8.

Prasch, C. M., and Sonnewald, U. 2015. Signaling events in plants: stress factors in combination change the picture. *Environmental and Experimental Botany* 114: 4–14. doi: 10.1016/j.envexpbot.2014.06.020.

Qiu, Q. S., Guo, Y., Quintero, F. J., Pardo, J. M., Schumaker, K. S., and Zhu, J. K. 2004. Regulation of vacuolar Na1/H1 exchange in *Arabidopsis thaliana* by the salt-overly-sensitive (SOS) pathway. *Journal of Biological Chemistry* 279(1): 207–215. doi: 10.1074/jbc.M307982200.

Quan, R., Lin, H., Mendoza, I., Zhang, Y., Cao, W., Yang, Y., and Guo, Y. 2007. SCABP8/CBL10, a putative calcium sensor, interacts with the protein kinase SOS2 to protect Arabidopsis shoots from salt stress. *The Plant Cell, 19*(4): 1415–1431.

Rahman, H., Ramanathan, V., Nallathambi, J., Duraialagaraja, S., and Muthurajan, R. 2016. Over-expression of a NAC 67 transcription factor from finger millet (*Eleusine coracana L.*) confers tolerance against salinity and drought stress in rice. *BMC Biotechnology* 16(1): 7–20.

Ramegowda, V., and Senthil-Kumar, M. 2015. The interactive effects of simultaneous biotic and abiotic stresses on plants: mechanistic understanding from drought and pathogen combination. *Journal of Plant Physiology* 176: 47–54. doi: 10.1016/j.jplph.2014.11.008.

Reddy, P. S., Mallikarjuna, G., Kaul, T., Chakradhar, T., Mishra, R. N., Sopory, S. K., et al. 2010. Molecular cloning and characterization of gene encoding for cytoplasmic Hsc70 from *Pennisetum glaucum* may play a protective role against abiotic stresses. *Molecular Genetics and Genomics* 283(3): 243–254.

Rodrigues, A., Adamo, M., Crozet, P., Margalha, L., Confraria, A., Martinho, C., et al. 2013. ABI1 and PP2CA phosphatases are negative regulators of snf1-related protein kinase1 signaling in arabidopsis. *The Plant Cell* 25(10): 3871–3884. doi: 10.1105/tpc.113.114066.

Rodriguez, L., Gonzalez-Guzman, M., Diaz, M., Rodrigues, A., Izquierdo-Garcia, A. C., Peirats-Llobet, M., Fernandez, M. A., Antoni, R., Fernandez, D., Marquez, J. A., et al. 2014. C2-domain abscisic acid-related proteins mediate the interaction of pyr/ pyl/rcar abscisic acid receptors with the plasma membrane and regulate abscisic acid sensitivity in arabidopsis. *The Plant Cell* 26(12): 48024820. doi: 10.1105/tpc.114.129973.

Rong, W., Qi, L., Wang, A., Ye, X., Du, L., Liang, H., et al. 2014. The ERF transcription factor TaERF3 promotes tolerance to salt and drought stresses in wheat. *Plant Biotechnology Journal* 12(4): 468–479.

Rus, A., Lee, B., Muñoz-Mayor, A., Sharkhuu, A., Miura, K., Zhu, J.-K., Bressan, R. A., and Hasegawa, P. M. 2004. AtHKT1 facilitates Na1 homeostasis and K1 nutrition in planta. *Plant Physiology* 136(1): 2500–2511. doi: 10.1104/pp.104.042234.

Rus, A., Yokoi, S., Sharkhuu, A., Reddy, M., Lee, B. H., Matsumoto, T. K., Koiwa, H., Zhu, J. K., Bressan, R.A., and Hasegawa, P.M. 2001. AtHKT1 is a salt tolerance determinant that controls Na1 entry into plant roots. *Proceedings of the National Academy of Sciences USA* 98: 14150–14155. doi: 10.1073/pnas.241501798.

Saijo, Y., Hata, S., Kyozuka, J., Shimamoto, K., and Izui, K. 2000. Overexpression of a single Ca2 1-dependent protein kinase confers both cold and salt/drought tolerance on rice plants. *The Plant Journal* 23(3): 319–327.

Saruhashi, M., Kumar Ghosh, T., Arai, K., Ishizaki, Y., Hagiwara, K., Komatsu, K., et al., 2015. Plant Raf-like kinase integrates abscisic acid and hyperosmotic stress signaling upstream of SNF1-related protein kinase2. *Proceedings of the National Academy of Sciences* 112(46): E6388–E6396.

Sharma, B., Joshi, D., Yadav, P. K., Gupta, A. K., and Bhatt, T. K. 2016. Role of ubiquitin-mediated degradation system in plant biology. *Frontiers in Plant Science* 7. Doi: https://doi.org/10.3389/fpls.2016.00806.

Sharma, E., Sharma, R., Borah, P., Jain, M., and Khurana, J. P. 2015. Emerging roles of auxin in abiotic stress responses. In: Pandey, G. (ed) *Elucidation of Abiotic Stress Signaling in Plants*. New York, NY: Springer, 299–328.

Shen, J., Lv, B., Luo, L., He, J., Mao, C., Xi, D., and Ming, F. 2017. The NAC-type transcription factor OsNAC2 regulates ABA-dependent genes and abiotic stress tolerance in rice. *Scientific reports*, *7*(1): 40641.

Shi, H., Chen, L., Ye, T., Liu, X., Ding, K., and Chan, Z. 2014. Modulation of auxin content in Arabidopsis confers improved drought stress resistance. *Plant Physiology and Biochemistry* 82: 209–217.

Shibasaki, K., Uemura, M., Tsurumi, S., and Rahman, A. 2009. Auxin response in Arabidopsis under cold stress: underlying molecular mechanisms. *The Plant Cell* 21: 3823–3838.

Shkolnik-Inbar, D., and Bar-Zvi, D. 2010. ABI4 mediates abscisic acid and cytokinin inhibition of lateral root formation by reducing polar auxin transport in Arabidopsis. *The Plant Cell* 22: 3560–3573.

Singh, A., Jha, S. K., Bagri, J., and Pandey, G. K. 2015. ABA inducible rice protein phosphatase 2C confers ABA insensitivity and abiotic stress tolerance in Arabidopsis. *PLoS One* 10. doi: 10.1371/journal.pone.0125168.

Sirichandra, C., Gu, D., Hu, H. C., Davanture, M., Lee, S., Djaoui, M., et al. 2009. Phosphorylation of the Arabidopsis AtrbohF NADPH oxidase by OST1 protein kinase. *FEBS Letters* 583: 2982–2986.

Song, S. Y., Chen, Y., Chen, J., Dai, X. Y., and Zhang, W. H. 2011. Physiological mechanisms underlying OsNAC5-dependent tolerance of rice plants to abiotic stress. *Planta* 234(2): 331–345.

Stone, S. L. 2014. The role of ubiquitin and the 26S proteasome in plant abiotic stress signaling. *Frontiers in Plant Science* 5: 135.

Suzuki, N., Sejima, H., Tam, R., Schlauch, K., and Mittler, R. 2011. Identification of the MBF1 heat-response regulon of *Arabidopsis thaliana. The Plant Journal for Cell and Molecular Biology* 66(5): 844–851. doi: 10.1111/j.1365-313X.2011.04550.x.

Tang, N., Zhang, H., Li, X., Xiao, J., and Xiong, L. 2012. Constitutive activation of transcription factor OsbZIP46 improves drought tolerance in rice. *Plant Physiology* 158: 1755–1768.

Tang, R.-J., Zhao, F.-G., Garcia, V. J., Kleist, T. J., Yang, L., Zhang, H.-X., and Luan, S. 2015. Tonoplast CBLCIPK calcium signaling network regulates magnesium homeostasis in Arabidopsis. *Proceedings of the National Academy of Sciences* 112(10): 3134–3139. doi: 10.1073/pnas.1420944112.

Teige, M., Scheikl, E., Eulgem, T., Doczi, R., Ichimura, K., Shinozaki, K., et al. 2004. The MKK2 pathway mediates cold and salt stress signaling in Arabidopsis. *Molecular Cell* 15: 141–152. doi: 10.1016/j.molcel.2004.06.023.

Todaka, D., Shinozaki, K., and Yamaguchi-Shinozaki, K. 2015. Recent advances in the dissection of drought-stress regulatory networks and strategies for development of drought-tolerant transgenic rice plants. *Frontiers in Plant Science* 6: 84.

Wang, C., Lu, W., He, X., Wang, F., Zhou, Y., and Guo, X. 2016a. The cotton mitogen-activated protein kinase 3 functions in drought tolerance by regulating stomatal responses and root growth. *Plant and Cell Physiology* 57(8): 1629–1642.

Wang, M., Ji, Q., Liu, P., & Liu, Y. 2022. Guarding and hijacking: stomata on the move. *Trends in Plant Science*. *27*(8): 736–738.

Wang, P., Du, Y., Hou, Y.-J., Zhao, Y., Hsu, C. C., Yuan, F., Zhu, X., Tao, W. A., Song, C. P., and Zhu, J. K. 2015. Nitric oxide negatively regulates abscisic acid signaling in guard cells by S-nitrosylation of OST1. *Proceedings of the National Academy of Sciences* 112(2): 613–618. doi: 10.1073/pnas.1423481112.

Wang, W., Vinocur, B., Shoseyov, O., and Altman, A. 2004. Role of plant heat-shock proteins and molecular chaperones in the abiotic stress response. *Trends in Plant Science* 9: 244–252.

Wohlbach, D. J., Quirino, B. F., and Sussman, M. R. 2008. Analysis of the Arabidopsis histidine kinase ATHK1 reveals a connection between vegetative osmotic stress sensing and seed maturation. *Plant Cell* 20(4): 1101–1117.

Xia, N., Zhang, G., Sun, Y. F., Zhu, L., Xu, L. S., Chen, X. M., et al. 2010. TaNAC8, a novel NAC transcription factor gene in wheat, responds to stripe rust pathogen infection and abiotic stresses. *Physiological and Molecular Plant Pathology* 74(5–6): 394–402.

Xiang, Y., Tang, N., Du, H., Ye, H., and Xiong, L. 2008. Characterization of OsbZIP23 as a key player of the basic leucine zipper transcription factor family for conferring abscisic acid sensitivity and salinity and drought tolerance in rice. *Plant Physiology* 148(4): 1938–1952.

Xiong, L., Schumaker, K. S., and Zhu, J. K. 2002. Cell signaling during cold, drought, and salt stress. *The Plant Cell* 14(suppl_1): S165–S183.

Xu, J., Tian, Y. S., Peng, R. H., Xiong, A. S., Zhu, B., Jin, X. F., et al. 2010. AtCPK6, a functionally redundant and positive regulator involved in salt/drought stress tolerance in Arabidopsis. *Planta* 231: 1251–1260.

Xu, Z. S., Chen, M., Li, L. C., Ma, Y. Z. 2011. Functions and application of the AP2/ERF transcription factor family in crop improvement. *Journal of Integrative Plant Biology* 53(7): 570–585.

Xu, Z. S., Ni, Z. Y., Li, Z. Y., Li, L. C., Chen, M., Gao, D. Y., et al. 2009. Isolation and functional characterization of HvDREB1—a gene encoding a dehydration-responsive element binding protein in *Hordeum vulgare*. *Journal of Plant Research* 122(1): 121–130.

Yang, C., Wang, R., Gou, L., Si, Y., and Guan, Q. 2017. Overexpression of *Populus trichocarpa* mitogen-activated protein kinase kinase 4 enhances salt tolerance in tobacco. *International Journal of Molecular Sciences* 18(10): 2090. doi: 10.3390/ijms18102090.

Yang, L., Wu, L., Chang, W., Li, Z., Miao, M., Li, Y., et al. 2018. Overexpression of the maize E3 ubiquitin ligase gene ZmAIRP4 enhances drought stress tolerance in Arabidopsis. *Plant Physiology and Biochemistry* 123: 34–42.

Yang, Q., Chen, Z. Z., Zhou, X. F., Yin, H. B., Li, X., Xin, X. F., et al. 2009. Overexpression of SOS (salt overly sensitive) genes increases salt tolerance in transgenic Arabidopsis. *Molecular Plant* 2: 22–31. doi: 10.1093/mp/ssn058.

Yang, T., Shad Ali, G., Yang, L., Du, L., Reddy, A. S. N., and Poovaiah, B. W. 2010. Calcium/calmodulin-regulated receptor-like kinase CRLK1 interacts with MEKK1 in plants. *Plant Signaling & Behavior* 5(8): 991–994. doi: 10.4161/psb.5.8.12225.

Yang, T., Yao, S., Hao, L., Zhao, Y., Lu, W., and Xiao, K. 2016. Wheat bHLH-type transcription factor gene TabHLH1 is crucial in mediating osmotic stresses tolerance through modulating largely the ABA-associated pathway. *Plant Cell Reports* 35: 2309–2323. doi: 10.1007/s00299-016-2036-5.

Yang, Y., Tang, R. J., Jiang, C. M., Li, B., Kang, T., Liu, H., et al. 2015. Overexpression of the PtSOS2 gene improves tolerance to salt stress in transgenic poplar plants. *Plant Biotechnology Journal* 13(7): 962–973. doi: 10.1111/pbi.12335.

Yoo, J. H., Park, C. Y., Kim, J. C., Do Heo, W., Cheong, M. S., Park, H. C., et al. 2005. Direct interaction of a divergent CaM isoform and the transcription factor, MYB2, enhances salt tolerance in Arabidopsis. *Journal of Biological Chemistry* 280(5): 3697–3706. doi: 10.1074/jbc.M408237200.

Yoon, S., Lee, D. K., Yu, I. J., Kim, Y. S., Do Choi, Y., and Kim, J. K. 2017. Overexpression of the OsbZIP66 transcription factor enhances drought tolerance of rice plants. *Plant Biotechnology Reports* 11(1): 53–62.

Yoshida, R., Umezawa, T., Mizoguchi, T., Takahashi, S., Takahashi, F., Shinozaki, K. 2006. The regulatory domain of SRK2E/OST1/ SnRK2.6 interacts with ABI1 and integrates abscisic acid (ABA) and osmotic stress signals controlling stomatal closure in Arabidopsis *Journal of Biological Chemistry* 281(8): 5310–5318. doi: 10.1074/jbc.M509820200.

Yu, H., Chen, X., Hong, Y. Y., Wang, Y., Xu, P., Ke, S. D., and Xiang, C. B. 2008. Activated expression of an Arabidopsis HD-START protein confers drought tolerance with improved root system and reduced stomatal density. *The Plant Cell*, 20(4): 1134–1151.

Yuan, F., Yang, H., Xue, Y., Kong, D., Ye, R., Li, C., et al. 2014. OSCA1 mediates osmotic-stress evoked Ca2 1 increases vital for osmosensing in Arabidopsis. *Nature* 514: 367–371.

Zhang, C., Li, C., Liu, J., Lv, Y., Yu, C., Li, H., et al. 2017. The OsABF1 transcription factor improves drought tolerance by activating the transcription of COR413-TM1 in rice. *Journal of Experimental Botany* 68(16): 4695–4707.

Zhang, F., Yao, J., Ke, J., Zhang, L., Lam, V. Q., Xin, X. F., et al. 2015. Structural basis of JAZ repression of MYC transcription factors in jasmonate signalling. *Nature* 525: 269–273.

Zhang, S. W., Li, C. H., Cao, J., Zhang, Y. C., Zhang, S. Q., Xia, Y. F., and Sun, Y. 2009. Altered architecture and enhanced drought tolerance in rice via the down-regulation of indole-3-acetic acid by TLD1/OsGH3. 13 activation. *Plant Physiology*, *151*(4): 1889–1901.

Zhao, X., Cheng, K. and Liu, D. (2009). Organosolv pretreatment of lignocellulosic biomass for enzymatic hydrolysis. *Applied Microbiology and Biotechnology*, 82: 815–827.

Zheng, J., Ma, X., Zhang, X., Hu, Q., and Qian, R. 2018. Salicylic acid promotes plant growth and salt-related gene expression in *Dianthus superbus* L. (Caryophyllaceae) grown under different salt stress conditions. *Physiology and Molecular Biology of Plants* 24(2): 231–238. doi: 10.1007/s12298-017-0496-x.

Zhu, J. K. 2007. Plant salt stress. *Encyclopedia of Life Sciences*, doi: 10.1002/9780470015902.a0001300.pub2.

Zhu, J. K. 2016. Abiotic stress signaling and responses in plants. *Cell* 167(2): 313–324.

Zhu, M., Chen, G., Zhang, J., Zhang, Y., Xie, Q., Zhao, Z., and Hu, Z. 2014. The abiotic stress-responsive NAC-type transcription factor SlNAC4 regulates salt and drought tolerance and stress-related genes in tomato (*Solanum lycopersicum*). *Plant Cell Reports*, 33: 1851–1863.

Zhu, Q., Zhang, J., Gao, X., Tong, J., Xiao, L., Li, W., et al. 2010. The Arabidopsis AP2/ERF transcription factor RAP2. 6 participates in ABA, salt and osmotic stress responses. *Gene* 457(1–2): 1–12.

12 Co-Overexpression of Two or More Genes to Achieve Higher Tolerance to Single as well as Multiple Stresses in Plants

From Arabidopsis to Cotton

Nardana Esmaeili
University of Pennsylvania

Inosha Wijewardene and Hong Zhang
Texas Tech University

CONTENTS

12.1 INTRODUCTION

The world population is rapidly increasing, and it is projected to pass 9 billion by 2050 (FAO, 2009). According to a recent projection, around 660 million people worldwide will face starvation in 2030 (FAO, IFAD, UNICEF, WFP and WHO, 2021). In contrast, global agricultural production is negatively affected by environmental stresses such as water deprivation, heat, and salinity, which cause 51%–82% annual loss in crop production (Abdelraheem et al., 2019; Mantri et al., 2012). Climate change is making these environmental stresses more severe. According to climate models, the surface temperature will rise by 3°C–5°C in 50 years, resulting in unforeseeable climate patterns around the world (Solomon et al., 2007). The concurrence of climate change and world population growth will jeopardize global food and water security. Therefore, there is an urgency to increase crop production by 70%–100% to meet the future need of the growing population (Edgerton, 2009). Abiotic stresses affect crops at different growth and development stages, leading to detrimental impacts on agricultural production. Individual environmental stresses such as drought, heat, and

DOI: 10.1201/9781003300564-12

salinity have been the main focus of numerous research studies over the past few decades. However, the co-occurrence of different abiotic stresses in the field is the primary reason for huge crop losses worldwide (Mittler, 2006). The lack of adequate studies on the underlying mechanisms in plant response to multiple stresses could be one of the main reasons for the poor performance of transgenic plants in the field. Therefore, it is necessary to focus on different aspects of combined stresses to successfully develop crops that could withstand multiple stresses and produce high yields under harsh environmental stress conditions (Esmaeili et al., 2019; Mittler, 2006). Plant genomes were altered using the traditional breeding method for many years, yet this approach has been slow and inefficient in selecting abiotic stress-tolerant cultivars. With the application of recombinant DNA technology in agriculture in recent years, genes from one species could be easily incorporated into another species, which greatly facilitated the transfer of agronomically important genes into crops. The manipulation of functional genes among different plant species was named genetic engineering, which has benefited the agricultural industry through increased yields, improved nutritional values, reduced need for pesticides, improved resistance to pests and disease, and medical benefits to people. Manipulating crops through genetic engineering holds great prospects in improving multi-stress tolerance in crops. The advantages of 'stacking' or 'pyramiding' of stress-related genes in crops offer a great potential to prepare futural crops for a fast-changing environment. Therefore, this chapter focuses on improving the performance of plants under salt as well as multiple stress conditions by introducing multiple genes into plants to overcome ongoing and future challenges that threaten agricultural production.

12.2 AN OVERVIEW OF THE LITERATURE ON CO-OVEREXPRESSION OF MULTIPLE GENES IN ENHANCING MULTIPLE STRESS TOLERANCE IN PLANTS

Drought, heat, and salinity are the three primary abiotic stresses that usually occur in combination in nature. Plant response to combined stresses cannot be extrapolated from the response to individual stress (Esmaeili et al., 2019). Therefore, adopting the multi-gene expression strategy in plants seems to be an excellent solution for plants grown in a complex natural environment. Manipulation of genes encoding chaperones, ion transporters, enzymes, and osmoregulatory factors has been employed to improve plant abiotic stress tolerance (Hu and Xiong, 2014). For example, in a recent study by Chang et al. (2017), improved drought and heat stress tolerance was achieved in transgenic rice by co-overexpressing the genes *OsbZIP46* and *SAPK6*. The performance of transgenic rice plants was improved under heat and drought stress conditions at both the seedling and the reproductive stages (Chang et al., 2017). Plants utilize different strategies to cope with changing environments and among these, transcription factors play an essential role. Several transcription factors have been discovered as potential candidates that can be manipulated to enhance plant response to environmental stresses (Joshi et al., 2016). For instance, improved drought tolerance was achieved in transgenic tomato plants when *AtDREB1A* from *Arabidopsis thaliana* and *BcZAT12* from *Brassica carinata* were co-overexpressed (Krishna et al., 2021). The transgenic tomato plants co-overexpressing *AtDREB1A* and *BcZAT12* showed higher relative water content and proline content than wild-type plants. A higher yield was also observed in transgenic plants, which was attributed to the elevated chlorophyll color index and photosynthetic efficiency (Krishna et al., 2021). Co-overexpression of *AtMYC* and *AtMYB2* in *Arabidopsis* resulted in improved osmotic stress tolerance via upregulation of several downstream genes such as *RD22* and *ADH1* (Abe et al., 2003). In addition, co-overexpression of abiotic stress-specific transcription factors *AtDREB2A*, *AtHB7*, and *AtABF3* led to enhanced salinity, drought, and oxidative stress tolerance in peanut plants (Pruthvi et al. 2014). Co-overexpression of transcription factor genes *TsHD1* and *TsNAC1* from *Thellungiella halophila* showed improved heat and drought tolerance in transgenic plants (Liu et al., 2019).

Heat-shock factors (HSFs) are a group of transcriptional factors that control the expression of heat-shock protein (HSP) genes, which play important roles in plant response to abiotic stresses, particularly heat stress. In a study by Person et al. (2014), improved tolerance against dehydration and oxidative stress was observed in transgenic sunflower plants when *HaHSFA9* and *HaHSFA4a* were co-overexpressed. Dehydration-responsive element-binding protein genes (*DREB*) and ethylene-responsive factor genes (*ERF*) also play important roles in regulating plant response to abiotic stresses. It was demonstrated that enhanced drought and salt tolerance was achieved in *Lotus corniculatus* transgenic plants when *PeDREB2a* and *KcERF* were co-overexpressed (Wang et al., 2018). When *DREB1A* and PHYTOCHROME-INTERACTING FACTOR-LIKE 1 gene (*OsPIL1*) were co-overexpressed in *Arabidopsis* and rice, improved drought tolerance and cell elongation were observed (Kudo et al., 2017). Co-overexpression of a diacylglycerol acyltransferase gene (*DGAT*) and a MYB transcriptional factor gene (*MYB96A*) in *Camelina sativa* resulted in enhanced tolerance to drought stress and increased fatty acid content when compared with non-transgenic plants (Kim et al., 2019).

Aquaporins such as plasma membrane intrinsic proteins (PIPs) are involved in the translocation of water across the membrane. Guo et al. (2006) showed that overexpression of the rice *PIP1* and *PIP2* in *Arabidopsis* improves salt and drought tolerance. Glutamine biosynthesis from inorganic ammonium is an ATP-dependent process involving glutamine synthase (GS) whose activity and expression are negatively affected by abiotic stresses. In transgenic rice, co-overexpression of *OsGS1* and *OsGS2* encoding the cytosolic GS and chloroplastic GS, respectively, improved tolerance to osmotic and salinity stresses. In addition, transgenic rice plants showed higher fresh weight, chlorophyll content, and relative water content than non-transgenic rice (James et al., 2018). Nitrogen is an essential nutrient in plants, and NO_3^- and NH_4^+ are two primary sources of inorganic nitrogen. In rice, the nitrate transporter 2 gene *NRT2* and the nitrate reductase 2 gene *NAR2* are highly expressed in the presence of nitrate (Liu et al., 2015). Recently, Chen et al. (2020) co-overexpressed *OsNAR2.1* and *OsNRT2.3a* in transgenic rice, and they achieved a 24.6% increase in grain yield in the field associated with higher nitrogen use efficiency.

Salinity is a growing problem worldwide, and it negatively impacts crop production. Around 71% of the water on earth is saline, and roughly 25% of the earth's outermost layer is affected by salt (Rao et al., 2006). Accumulation of salt in soil inhibits water uptake by roots, disrupts ion homeostasis, and inversely affects plant growth and development (Abdelraheem et al., 2019). Co-overexpression of the vitamin E biosynthetic pathway genes such as the homogentisate-phytyltransferase gene *AtHPT* and the γ-tocopherol-methyltransferase gene *AtγTMT* in potato increased salt tolerance in transgenic plants, which also produced 173%–258% more vitamin E than non-transgenic plants (Upadhyaya et al., 2020). Reactive oxygen species (ROS) are the by-products of most environmental stresses, which cause oxidative damage in plant cells (Miller et al., 2010). Thus, overexpression of antioxidant enzyme genes is essential for improving plant stress tolerance. Overexpression of the superoxide dismutase gene *GhSOD1* and the catalase gene *GhCAT1* in cotton enhanced salt tolerance (Luo et al., 2013). Transgenic tobacco plants overexpressing the glutathione-*S*-transferase gene *GST* and the glutathione reductase gene *GR* also showed improved cold and salt stress tolerance (Le Martret et al., 2011). Improved drought stress tolerance was achieved when the *Arabidopsis* Na^+/H^+ antiporter gene *AtNHX1* and the DEAD-box RNA helicase gene *eIF4A1* were co-overexpressed in sweet potato, and transgenic potato produced a larger root system and showed delayed senescence upon exposure to 18 days of drought stress treatment (Zhang et al., 2019).

12.3 FURTHER INCREASING SALT STRESS TOLERANCE

12.3.1 Overexpression of Single Genes to Confer Increased Salinity Stress Tolerance

Among the many genes that contribute toward making up the complex regulatory network to elicit salt tolerance in plants, the salt overly sensitive (SOS) signaling pathway genes play a significant role in sodium exclusion at the cellular level and maintaining ion homeostasis in plant cells, thereby promoting salt tolerance (Shi et al., 2000). In the SOS signaling pathway, SOS1 is a plasma membrane-located Na^+/H^+ antiporter, while SOS2 and SOS3 are a serine/threonine-type protein kinase belonging to the SnRK3 kinase family and a calcineurin B-like calcium-binding protein, respectively (Qiu et al., 2002). When plant roots encounter salt stress, the Ca^{2+} level increases in the root cell cytosol, triggering the SOS signaling pathway, where this high level of calcium ions in the cytosol is perceived by SOS3 (Ji et al., 2013; Zhu et al., 1998). Next, SOS3 interacts with SOS2, and the activated SOS2 then activates SOS1 as a result of phosphorylation by SOS2, leading to the extrusion of Na^+ from the cytosol (Ji et al., 2013). With *sos1* mutant showing hypersensitivity to both Na^+ and K^+ under salt stress conditions, functioning SOS1 exports sodium ions to the extracellular space from the cytosol to reduce Na^+ accumulation to toxic levels inside root cells (Shi et al., 2003). Additionally, a study related to *sos* mutants has shown how *SOS* genes are critical for cellular potassium homeostasis and SOS2, particularly, functions in potassium nutrition (Zhu et al., 1998). Containing 12 transmembrane domains and showing considerable sequence similarities to the Na^+/H^+ antiporters present in animals and microorganisms, *SOS1* expression is induced upon salt stress (Shi et al., 2000). Further, since SOS1 upregulation is largely observed near and around the xylem, this shows how SOS1 functions in the xylem-loading of Na^+, thereby removing the sodium ions from the xylem to reduce its accumulation in the shoots (Shi et al., 2003). Successful results with improved salt tolerance have been reported with the overexpression of *SOS* genes in poplar (Yang et al. 2015), *Arabidopsis* (Nie et al., 2015; Yang et al., 2009), and tomato (Baghour et al., 2019; Huertas et al., 2012). Recent work related to salinity stress has shown that other regulatory proteins assist in the SOS pathway-mediated salt tolerance due to the different salinity-sensitive phenotypes displayed by *sos1* and *sos2* mutants (Yin et al., 2019). Quan et al. (2017) suggested that ethylene insensitive 3 (EIN3), a transcription factor in the ethylene signaling pathway, interacts with SOS2, thus demonstrating a possible link between the salt tolerance regulation through ethylene signaling and SOS regulatory pathways.

Similar to the SOS1 present on the plasma membrane, the vacuolar Na^+/H^+ antiporter (NHX) also plays a key role in conferring salt tolerance through sequestering sodium ions into the vacuole (Apse et al. 1999). Extensive research has been conducted on plant NHX, where more than 60 NHX proteins have been identified across various plant species (Zhang et al., 2015). Located on the tonoplast, NHX actively sequesters Na^+ into the vacuole with the help of H^+-ATPase and H^+-PPase, thus reducing the salt toxicity in the cytosol (Bassil et al., 2018; Yarra, 2019). When *NHX1* overexpressed in different plant species, results showed that transgenic plants consistently displayed increased ion compartmentalization in vacuoles, higher amounts of sodium ions accumulated in root and leaf tissues, unhampered photosynthesis, better ROS scavenging capacity, better water use efficiency, and overall improvement of salt tolerance when compared to wild-type plants (Apse et al., 1999; Fan et al., 2015; Lu et al., 2014; Sahoo et al., 2016; Yarra and Kirti, 2019; Zhang et al., 2015).

In addition to playing a crucial role in plant drought tolerance, the *Arabidopsis* vacuolar pyrophosphatase 1 (AVP1) is also an important player in salt stress tolerance in plants (Pasapula et al., 2011). AVP1 was initially identified and isolated from the plant vacuolar membrane (Blumwald, 2000; Gaxiola et al., 1999), where its function in pumping protons into the vacuole, thereby acidifying the vacuole lumen, contributes to maintaining a proton electrochemical gradient across the tonoplast membrane. Due to this function, the efficiency of Na^+/H^+ antiporters located on the vacuolar membrane could be improved since the H^+ pumped by AVP1 could be utilized as cellular currency in exchange for the cytosolic Na^+ that can now be sequestered into the vacuolar lumen through the activity of tonoplast Na^+/H^+ antiporters (Gaxiola et al., 2001). In addition to being present on the tonoplast, AVP1 was subsequently identified in the plasma membrane of phloem companion cells, where unlike its role in the tonoplast, this plasma membrane AVP1 utilizes the proton electrochemical gradient across the membrane to synthesize inorganic pyrophosphate (PPi) (Khadilkar et al., 2016; Pizzio et al., 2015). Consequently, this aids in better phloem loading, transport, and biomass partitioning, which contributes to an overall better and healthier phenotype in plants overexpressing *AVP1* even under non-stressed conditions, leading to better tolerance to salinity as well as drought stresses (Gaxiola et al., 2016b; Schilling et al., 2017).Therefore, depending on the cell type and where AVP1 is located, it plays opposite roles either as a pyrophosphatase or a PPi synthase, providing an edge to transgenic plants overexpressing *AVP1* over wild-type plants. Successful results have been consistently obtained over the years through overexpressing *AVP1* in a variety of plant species with improved tolerance to both salinity and drought stresses and higher productivity (Arif et al., 2013; Gaxiola et al., 2001, 2012, 2016a; Kim et al., 2014; Kumar et al., 2014; Pasapula et al., 2011; Qin et al., 2013; Yang et al., 2014).

12.3.2 Stacking Genes to Further Improve Salt Stress Tolerance in Plants

Despite the success obtained over the past decades with overexpressing single genes to confer tolerance to salt stress, to further improve plant tolerance and yield when grown in saline soil, it is more practical to engineer plants with stacked transgenes to increase productivity under more severe saline conditions. Having this notion in mind, co-overexpressing or stacking genes to further increase salt tolerance was adopted in recent years. For instance, when either *AtNHX1* or *AVP1* alone was overexpressed in cotton, the salt tolerance in plants was increased to around 150 mM NaCl (He et al., 2005; Pasapula et al., 2011), suggesting that it could be possible to further improve the salt tolerance in transgenic plants by co-overexpressing *AVP1* and *AtNHX1* due to the coupled actions of enhanced proton electrochemical gradient and Na^+/H^+ antiporter activity across the tonoplast membrane. Indeed, transgenic cotton co-overexpressing *AVP1* and *AtNHX1* performed significantly better than wild-type and single-gene overexpressing plants, that is, *AVP1*-overexpressing and *AtNHX1*-overexpressing cotton plants, in terms of photosynthetic rates, fresh and dry shoot biomass, boll number, and fiber yield under 200 mM NaCl treatment (Shen et al., 2015). It appears that the AtNHX1's capacity of sequestering sodium ions into vacuole is further enhanced due to the increased activity of AVP1 through its overexpression. Moreover, under low-irrigation and dryland field conditions, the *AVP1/AtNHX1* co-overexpressing cotton plants produced significantly higher amounts of fiber than wild-type, *AVP1*-overexpressing and *AtNHX1*-overexpressing cotton plants. This shows how pyramiding transgenes could confer higher salt stress tolerance, thereby withstanding the natural environmental conditions better than their wild-type and single-gene overexpressing counterparts.

Similarly, higher salt tolerance was achieved by co-overexpressing *AtNHX1* and *SOS1* in *Arabidopsis* (Pehlivan et al., 2016), thus providing an excellent proof-of-concept that tolerance to salinity could be further increased by stacking multiple genes. Previous success was achieved in conferring improved salt tolerance through overexpressing single salt stress tolerance genes where transgenic plants were shown to be able to withstand around 150 mM NaCl (Banjara et al., 2012; Pasapula et al., 2011; Shi et al., 2003). However, the *AtNHX1/SOS1* co-overexpressing *Arabidopsis* plants were able to outperform wild-type, *AtNHX1*-overexpressing, and *SOS1*-overexpressing plants in terms of seed yield even at 250 mM NaCl concentrations (Pehlivan et al., 2016), hence suggesting how this strategy could be adopted to improve the salt tolerance of agricultural crops to a greater extent and reclaim the saline affected soil across the globe. AtNHX1 and SOS1 work together to mitigate the toxic Na^+ accumulation in the cytosol by sequestering sodium ions into vacuole or exporting sodium ions to extracellular space, respectively, thus minimizing sodium toxicity in plant cells. Therefore, this approach provides an additive tolerance to salinity compared to single-gene overexpressing plants.

In addition to these well-known candidate genes that are primarily involved in plant salinity stress response pathways, recent research has focused on the regulatory functions of the plant protein phosphatase 2A (PP2A), which belongs to the Ser/Thr phosphatase family of phosphoprotein phosphatases (Bheri and Pandey, 2019). The PP2A holoenzyme consists of a scaffolding subunit A, a regulatory subunit B, and a catalytic subunit C. Of special interest was the PP2A-C5 subunit, which is one of the C subunits classified under the subfamily I of the catalytic subunit C of PP2A (Farkas et al., 2007). Research has shown that PP2A-C5 is a positive regulator in conferring salt tolerance in *Arabidopsis*. Compared to wild-type *Arabidopsis* plants, the *pp2a-c5-1* mutant displayed a salt-sensitive phenotype, while the *PP2A-C5* overexpressing plants had a salt-tolerant phenotype (Hu et al. 2017a). Moreover, the yeast two-hybrid analysis showed that several chloride ion channel (CLC) proteins located on the tonoplast membrane such as AtCLCa, AtCLCb, AtCLCc, and

AtCLCg interacted with PP2A-C5, implying a possibility of these CLC proteins acting as substrates of PP2A-C5 in vivo (Hu et al. 2017a, b). Hence, based on the hypothesis that PP2A-C5 could increase or trigger the CLC enzyme activities and previous results on *AVP1* conferring salt tolerance when overexpressed, the capacity of enhanced salt tolerance was tested in *Arabidopsis* plants co-overexpressing *AVP1* and *PP2A*-C5 (Sun et al., 2018). In MS media as well as in soil, *AVP1/PP2A-C5* co-overexpressing *Arabidopsis* plants showed superior growth and tolerance to NaCl when compared to wild-type, *AVP1*-overexpressing, and *PP2A-C5* overexpressing plants, where *AVP1/PP2A-C5* co-overexpressing plants were able to withstand 200mM NaCl in soil with relatively the least chlorotic phenotype than the other genotypes tested. Furthermore, the *AVP1/PP2A-C5* co-overexpressing plants and the *PP2A-C5* overexpression plants were more tolerant to varying concentrations of KNO_3 solutions, with longer roots, a higher number of lateral roots, and greener cotyledons since the *PP2A-C5* overexpressing plants previously showed enhanced tolerance to higher concentrations of KNO_3 (Hu et al., 2017a). Lithium contamination in soils, particularly near lithium mining and lithium battery factories, is a serious environmental problem as lithium is extremely toxic to plants even at low concentrations. It was observed that in the presence of 15mM LiCl the *AVP1/PP2A-C5* co-overexpressing plants performed significantly better than wild-type and *AVP1*-overexpressing *Arabidopsis* plants. This study indicates that in addition to improved tolerance to NaCl and KNO_3, *AVP1/PP2A-C5* co-overexpression has a great potential in bioremediation of contaminated soils and fertilizer overuse through removing lithium, Na^+, Cl^-, K^+, and NO_3^- ions (Sun et al., 2018).

A recent study shows improved salt stress tolerance in transgenic cotton co-overexpressing *AtNHX1* and *TsVP* (Cheng et al., 2018). The *AtNHX1-TsVP* transgenic cotton plants produced higher seed yield when grown in saline soil, attributed to the accumulation of Na^+, K^+, and Ca^{2+} in leaves and maintaining better ion homeostasis in transgenic plants compared to wild-type plants (Cheng et al., 2018). Co-overexpression of *LeNHX2* and *SlSOS2* in tomato resulted in improved salt tolerance, and transgenic tomato plants also produced higher yields with enhanced fruit quality (Baghour et al., 2019). Genes *LEA* and *bZIP,* encoding stress-related late embryogenesis abundant protein and basic leucine zipper transcriptional factor, respectively, were overexpressed in tobacco plants, leading to enhanced tolerance to salt and osmotic stresses with increased seed production (Qu et al., 2012). Co-overexpression of the *Arabidopsis* genes *SOS1*, *SOS2*, and *SOS3* improved salt tolerance in tall fescue plants when transgenic plants were exposed to 350mM of NaCl (Ma et al., 2014). These studies together show how co-overexpression of genes related to salinity tolerance across a diverse range of plant species contributes to improved tolerance to salt stress, leading to higher yields under high saline conditions.

12.4 CO-OVEREXPRESSION OF TWO GENES TO CONFER MULTIPLE STRESS TOLERANCE

Apart from increasing the tolerance to a single stress through gene stacking, this approach could be applied to increase tolerance to multiple stresses, since in nature, a single stress rarely comes alone; rather, plants are more often exposed to multiple abiotic stresses at a given time such as heat and drought or drought and salinity. Thus, several research studies have focused on enhancing tolerance to multiple abiotic stresses over recent years.

A recent study achieved increased tolerance to three abiotic stresses, drought, salinity, and heat, by co-overexpressing the *Arabidopsis* gene *AVP1* and the creosote Rubisco activase gene *RCA* in *Arabidopsis* (Wijewardene et al., 2020). Compared to the well-known *AVP1*, the application of *RCA* to produce transgenic plants with improved heat tolerance is a strategy less explored. The key function of RCA is to remove inhibitory sugar phosphates from Rubisco active site at the expense of ATP, thus maintaining Rubisco in its active state, resulting in continuous RuBP regeneration (Portis Jr et al., 2008; Salvucci and Ogren, 1996). Due to the high heat sensitivity of most RCA enzymes in the temperate region (Salvucci et al., 2001; Salvucci and Crafts-Brandner, 2004), it was hypothesized that introducing an RCA with high thermo-tolerance to plants would make transgenic plants more thermo-stable, which would lead to an overall better plant performance and higher productivity under heat stress conditions. Therefore, overexpressing *RCA* derived from creosote (*Larrea tridentata*), a shrub commonly grown in Arizona and Mexico deserts where the summer temperatures often exceed 40°C–45°C, along with *AVP1*-overexpression appeared to be an effective approach in simultaneously increasing plant tolerance to heat, drought, and salinity (Wijewardene et al., 2020). This proof-of-concept was successful, where not only the transgenic *Arabidopsis* plants co-overexpressing *AVP1* and *RCA* outperformed wild-type plants when exposed to drought, heat, or salt as single stress, but these plants also showed superior performance in terms of biomass and seed yield under combined drought and heat stresses as well as all three stresses together. In fact, the exceptional results obtained under combined drought and heat stresses imply that this approach holds a great potential in improving the yield of agriculturally important crops in the face of the incoming global climate change.

12.5 FURTHER INCREASING DROUGHT, HEAT, AND SALT TOLERANCE

As mentioned earlier in this chapter, heat, drought, and salinity cause significant yield loss in agriculture. The heterogenic condition of the field environment and concurrence of different abiotic stresses make the stress effects far more severe. Combined stresses such as heat and drought, drought and salinity, salinity and heat, and a distinct combination of heat, drought, and salinity are common in many agricultural fields and cause a more significant crop loss worldwide, compared to the reduction of productivity in plants when exposed to a single stress. It is well-known that plant response to combined stresses is unique, and it is neither additive nor subtractive of a single stress (Esmaeili et al., 2021). In the case of combined stresses, plant responses are often antagonistic. For instance, as a response to heat stress, plants open stomata, while under drought stress, plants need to close stomata to avoid water loss from the leaves. However, plants subjected to combined drought and heat stresses respond differently than those exposed to these stresses separately (Zurbriggen et al., 2010). To date, different approaches have been taken to combat the negative impacts of changing environment, and gene pyramiding is one of the most effective strategies.

We previously showed that overexpression of *AVP1* in cotton could increase drought and salt tolerance in transgenic cotton (Pasapula et al., 2011). Then we showed that overexpression of *OsSIZ1* in *Arabidopsis* and cotton could lead to increased tolerance to drought, heat, and salt stresses in transgenic *Arabidopsis* and cotton plants (Mishra et al., 2017, 2018). To further increase tolerance to drought, heat, and salt stresses simultaneously, we co-overexpressed these two genes in *Arabidopsis* to provide the proof-of-concept that stacking *AVP1* and *OsSIZ1* might have the potential to not only enhance plant tolerance to multiple stresses but also result in higher yield under multi-stress conditions. Our experiments confirmed our hypothesis that *AVP1/OsSIZ1* co-overexpressing plants outperformed *AVP1*-overexpressing, *OsSIZ1*-overexpressing, and wild-type plants under single stress (drought, heat, or salt) as well as under combined stress conditions (all three stresses combined or in any two combinations) (Esmaeili et al., 2019). An example of these plants grown under combined drought, heat, and salt stresses is shown in Figure 12.1.

Based on this experiment performed in *Arabidopsis*, we created transgenic cotton plants co-overexpressing *AVP1* and *OsSIZ1*, and we analyzed their performance under single and multiple stresses in the growth chamber, greenhouse, and field conditions (Esmaeili et al., 2021). In every condition that we tested, *OsSIZ1/AVP1* co-overexpressing cotton displayed significantly improved stress tolerance compared to *AVP1*-overexpressing, *OsSIZ1*-overexpressing, and non-transgenic plants. In two consecutive field trials under dryland conditions, we found that *OsSIZ1/AVP1* co-overexpressing cotton outperformed wild-type cotton with an increased fiber production of 133% and 81% in the first and second field trials, respectively (Esmaeili et al., 2021) (Figure 12.2). Although overexpression of some stress-responsive genes may result in growth repression or yield loss, *OsSIZ1/AVP1* co-overexpressing cotton plants did not show any adverse effects in either growth or yield production under normal growth conditions.

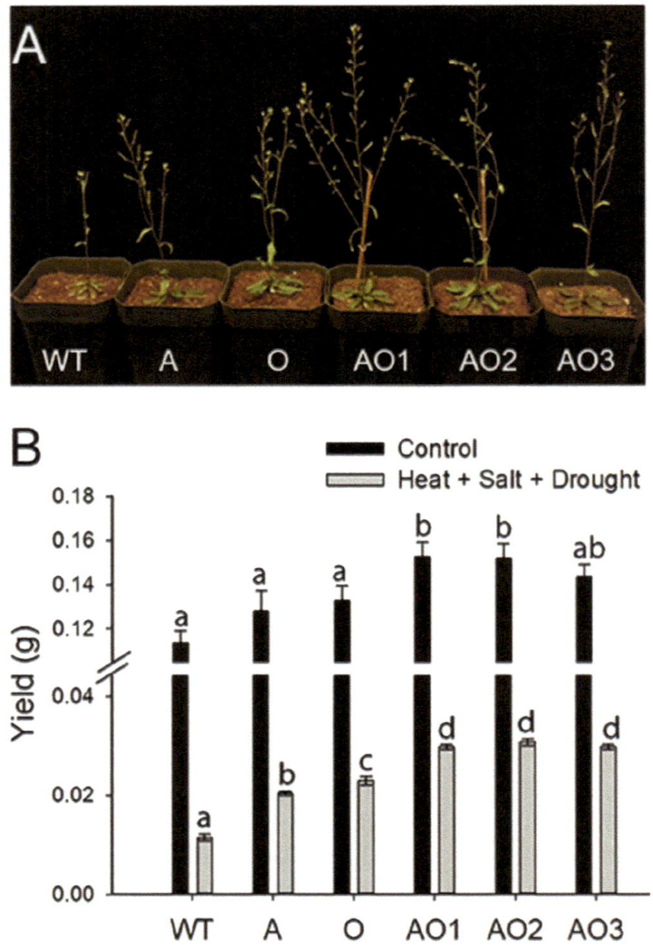

FIGURE 12.1 Analysis of wild-type, *AVP1*-overexpressing, *OsSIZ1*-overexpressing, and *AVP1/OsSIZ1* co-overexpressing plants under combined drought, heat, and salt stresses. (a) Phenotypes of wild-type and transgenic plants grown under combined heat, salt, and drought stresses in soil. (b) Total seed yields of plants grown under normal growth conditions (black bar) and combined heat, salt, and drought stresses (gray bar). Data are means ± SE (n = 9). WT, wild-type plant; A, *AVP1*-overexpressing plant; O, *OsSIZ1*-overexpressing plant; AO1 to AO3, three independent *AVP1/OsSIZ1* co-overexpressing plants. The figure is obtained from Esmaeili et al. (2019) with permission.

12.6 CONCLUSIONS AND PERSPECTIVES

Plants respond to the changing environment through an orchestrated cellular, molecular, and physiological manifestation. Different strategies are employed to improve plant tolerance to environmental stresses, and among them, genetic engineering plays an important role. The extensive damages caused by combined stresses underscore the urgent need to generate plants that can withstand the detrimental effects of combined stresses. On the other hand, understanding plant response mechanisms to abiotic stresses and identifying genes and regulatory pathways involved in plant acclimation to the complex environment are needed. Although numerous studies have been conducted on single stress tolerance, our relatively limited knowledge on underlying mechanisms of plant response to combined stresses remains a restricting factor in improving stress tolerance in plants (Zurbriggen et al., 2010). This lack of understanding might explain why laboratory-developed plants do not perform well in their full capacity when grown under the heterogeneous conditions of the field (Mittler, 2006). Therefore, gaining a better insight and unraveling the regulatory networks that govern the responses to multiple stress conditions as well as developing transgenic plants under conditions that mimic the field conditions is of great interest.

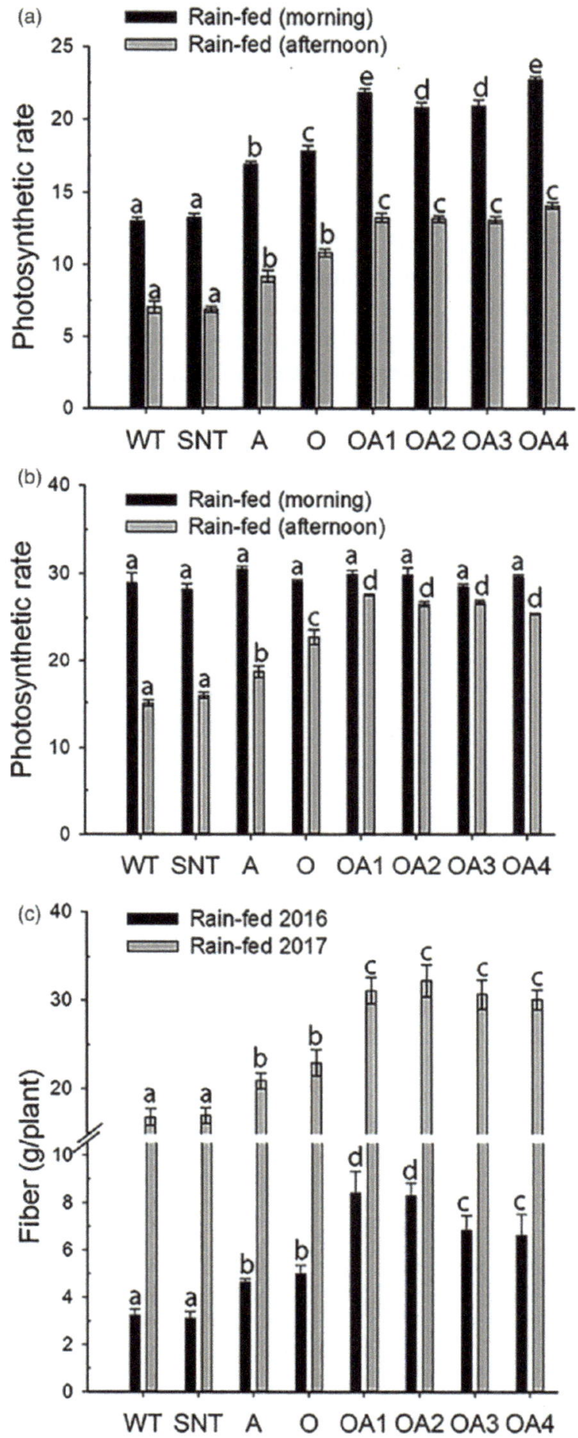

FIGURE 12.2 The performance of control and *OsSIZ1/AVP1* co-overexpressing cotton plants under rain-fed conditions in 2016 and 2017. (a) Photosynthetic rates of cotton plants under the rain-fed condition in 2016. (b) Photosynthetic rates of cotton plants under the rain-fed condition in 2017. Results are the means ± SE ($n=8$). (c) Fiber yield produced per plant under the rain-fed conditions in 2016 (black bar) and 2017 (gray bar). Results are the means ± SE. WT, wild-type plant; SNT, segregated non-transgenic plant; A, *AVP1*-overexpressing plant; O, *OsSIZ1*-overexpressing plant; OA1 to OA4, four independent *OsSIZ1/AVP1* co-overexpressing plants. Samples denoted by different letters are significantly different ($P<0.05$, Tukey correction). The figure is obtained from Esmaeili et al. (2021) with permission.

Using 'Omics' technologies such as proteomics, genomics, and transcriptomics would assist us in discovering mechanisms involved in plant response to abiotic stresses (Gokce et al., 2020). Functional genomics provides valuable data on functions of stress-related genes identified (Rhee and Mutwil, 2014) while transcriptomic profiling of plants subjected to environmental stresses provides detailed features of gene expression in plants (Unamba et al., 2015). Although most transcriptomic studies mainly focus on plant responses to single stresses, more studies concentrating on response to combined stress treatments are required (Gokce et al., 2020). Therefore, high-throughput technologies offer promising approaches to better understand the underlying mechanisms of plant response to unfavorable environments. In addition, examination of plant species that are highly adapted to harsh environments, such as halophytes or desert plants, could provide more profound and invaluable information on plant stress tolerance (Gokce et al., 2020). Since the emergence of CRISPR/Cas9 technology, it has been widely used to improve abiotic stress tolerance in different crops, including rice (Liu et al., 2020) and maize (Shi et al., 2017). The powerful CRISPR/Cas9 editing technology along with genetic engineering will assist us in tailoring crop tolerance to combined abiotic stresses and ensure global food security. Other non-transgenic approaches such as genome editing can also be employed to alter those genes that are involved in combined stress tolerance (Luo et al., 2015).

The strategy that we followed, that is, testing the proof-of-concept in *Arabidopsis* first and then applying it to crops such as cotton and peanut, appears successful in experiments conducted in growth chamber, greenhouse, and small-scale field testing (Banjara et al., 2012; Esmaeili et al., 2021; He et al., 2005; Kuppu et al., 2013; Mishra et al., 2017; Pasapula et al., 2011; Qin et al., 2011 and 2013; Shen et al., 2015; Zhu et al., 2018), but it is not proving that these transgenic cotton or peanut could produce a higher yield in large-scale field testing. It is still a big challenge to demonstrate that we could substantially increase crop yield in large-scale field testing under severe environmental stress conditions. More gene combinations should be tested, and more large-scale field trials are needed in the future.

REFERENCES

Abdelraheem, A., Esmaeili, N., O'Connell, M. and Zhang, J. 2019. Progress and perspective on drought and salt stress tolerance in cotton. *Industrial Crops and Products* 130: 118–129. doi: 10.1016/j.indcrop.2018.12.070.

Abe, H., Urao, T., Ito, T., Seki, M., Shinozaki, K. and Yamaguchi-Shinozaki, K. 2003. Arabidopsis *AtMYC2* (bHLH) and *AtMYB2* (MYB) function as transcriptional activators in abscisic acid signaling. *Plant Cell* 15: 63–78. doi: 10.1105/tpc.006130.

Apse, M. P., Aharon, G. S., Snedden, W. A. and Blumwald, E. 1999. Salt tolerance conferred by overexpression of a vacuolar Na^+/H^+ antiport in Arabidopsis. *Science* 285(5431): 1256–1258. doi: 10.1126/science.285.5431.1256.

Arif, A., Zafar, Y., Arif, M. and Blumwald, E. 2013. Improved growth, drought tolerance, and ultrastructural evidence of increased turgidity in tobacco plants overexpressing Arabidopsis vacuolar pyrophosphatase (*AVP1*). *Molecular Biotechnology* 54(2): 379–392. doi: 10.1007/s12033-012-9577-9.

Baghour, M., Gálvez, F. J., Sánchez, M. E., Aranda, M. N., Venema, K. and Rodríguez-Rosales, M. P. 2019. Overexpression of *LeNHX2* and *SlSOS2* increases salt tolerance and fruit production in double transgenic tomato plants. *Plant Physiology and Biochemistry* 135: 77–86. doi: 10.1016/j.plaphy.2018.11.028.

Banjara, M., Zhu, L., Shen, G., Payton, P. and Zhang, H. 2012. Expression of an Arabidopsis sodium/proton antiporter gene (*AtNHX1*) in peanut to improve salt tolerance. *Plant Biotechnology Reports* 6(1): 59–67. doi: 10.1007/s11816-011-0200-5.

Bassil, E., Zhang, S., Gong, H., Tajima, H. and Blumwald, E. 2018. Cation specificity of vacuolar NHX-type cation/H^+ antiporters. *Plant Physiology* 179(2): 616–629. doi: 10.1104/pp.18.01103.

Bheri, M. and Pandey, G. K. 2019. PP2A phosphatases take a giant leap in the post-genomics era. *Current Genomics* 20(3): 154–171. doi: 10.2174/1389202920666190517110605.

Blumwald, E. 2000. Sodium transport and salt tolerance in plants. *Current Opinion in Cell Biology* 12(4): 431–434. doi: 10.1016/s0955-0674(00)00112-5.

Chang, Y., Nguyen, B. H., Xie, Y., Xiao, B., Tang, N., Zhu, W., et al. 2017. Co-overexpression of the constitutively active form of *OsbZIP46* and ABA-activated protein kinase *SAPK6* improves drought and temperature stress resistance in rice. *Frontiers in Plant Science* 8: 1112. doi: 10.3389/fpls.2017.01102.

Chen, J., Liu, X., Liu, S., Fan, X., Zhao, L., Song, M., et al. 2020. Co-overexpression of OsNAR2.1 and OsNRT2.3a increased agronomic nitrogen use efficiency in transgenic rice plants. *Frontiers in Plant Science* 11: 1245. doi: 10.3389/fpls.2020.01245.

Cheng, C., Zhang, Y., Chen, X., Song, J., Guo, Z., Li, K. and Zhang, K. 2018. Co-expression of *AtNHX1* and *TsVP* improves the salt tolerance of transgenic cotton and increases seed cotton yield in a saline field. *Molecular Breeding* 38: 19. doi: 10.1007/s11032-018-0774-5.

Edgerton, M. D. 2009. Increasing crop productivity to meet global needs for feed, food, and fuel. *Plant Physiology* 149: 7–13. doi: 10.1104/pp.108.130195.

Esmaeili, N., Cai, Y., Tang, F., Zhu, X., Smith, J., Mishra, N., et al. 2021. Towards doubling fibre yield for cotton in the semiarid agricultural area by increasing tolerance to drought, heat and salinity simultaneously. *Plant Biotechnology Journal* 19: 462–476. doi: 10.1111/pbi.13476.

Esmaeili, N., Yang, X., Cai, Y., Sun, L., Zhu, X., Shen, G., et al. 2019. Co-overexpression of *AVP1* and *OsSIZ1* in Arabidopsis substantially enhances plant tolerance to drought, salt, and heat stresses. *Scientific Reports* 9: 7642. doi: 10.1038/s41598-019-44062-0.

Fan, W., Deng, G., Wang, H., Zhang, H. and Zhang, P. 2015. Elevated compartmentalization of Na^+ into vacuoles improves salt and cold stress tolerance in sweet potato (*Ipomoea batatas*). *Physiologia Plantarum* 154(4): 560–571. doi: 10.1111/ppl.12301.

FAO. 2009. How to feed the world in 2050. Available at: https://www.fao.org/fileadmin/templates/wsfs/docs/expert_paper/How_to_Feed_the_World_in_2050.pdf.

FAO, IFAD, UNICEF, WFP and WHO. 2021. *The State of Food Security and Nutrition in the World 2021. Transforming Food Systems for Food Security, Improved Nutrition and Affordable Healthy Diets for all.* Rome: FAO. Available at: https://www.fao.org/publications/sofi/2021.

Farkas, I., Dombradi, V., Miskei, M., Szabados, L. and Koncz, C. 2007. Arabidopsis PPP family of serine/threonine phosphatases. *Trends in Plant Science* 12(4): 169–176. doi: 10.1016/j.tplants.2007.03.003.

Gaxiola, R. A., Li, J., Undurraga, S., Dang, L. M., Allen, G. J., Alper, S. L. et al. 2001. Drought- and salt-tolerant plants result from overexpression of the AVP1 H^+-pump. *Proceedings of National Academy of Science of the USA* 98: 11444–11449. doi: 10.1073/pnas.191389398.

Gaxiola, R. A., Rao, R., Sherman, A., Grisafi, P., Alper, S. L. and Fink, G. R. 1999. The *Arabidopsis thaliana* proton transporters, *AtNHX1* and *AVP1*, can function in cation detoxification in yeast. *Proceedings of the National Academy of Sciences of the USA* 96(4): 1480–1485. doi: 10.1073/pnas.96.4.1480.

Gaxiola, R. A., Regmi, K. and Hirschi, K. D. 2016a. Moving on up: H^+-PPase mediated crop improvement. *Trends in Biotechnology* 34(5): 347–349. doi: 10.1016/j.tibtech.2015.12.016.

Gaxiola, R. A., Regmi, K., Paez-Valencia, J., Pizzio, G. and Zhang, S. 2016b. Plant H^+-PPases: reversible enzymes with contrasting functions dependent on membrane environment. *Molecular Plant* 9(3): 317–319. doi: 10.1016/j.molp.2015.09.008.

Gaxiola, R. A., Sanchez, C. A., Paez-Valencia, J., Ayre, B. G. and Elser, J. J. 2012. Genetic manipulation of a "vacuolar" H^+-PPase: from salt tolerance to yield enhancement under phosphorus-deficient soils. *Plant Physiology* 159(1): 3–11. doi: 10.1104/pp.112.195701.

Gokce, Z. N., Akbas, S., Ayten, S., Azimi, M. H., Das, R., Guven, S. B., et al. 2020. Abiotic stress tolerance in field crops: integration of omics approaches. In *Agronomic Crops*, 503–526. Singapore: Springer. doi: 10.1007/978-981-15-0025-1_24.

Guo, L., Wang, Z. Y., Lin, H., Cui, W. E., Chen, J., Liu, M., et al. 2006. Expression and functional analysis of the rice plasma-membrane intrinsic protein gene family. *Cell Research* 16: 277–286. doi: 10.1038/sj.cr.7310035.

He, C., Yan, J., Shen, G., Fu, L., Holaday, A. S., Auld, D., et al. 2005. Expression of an Arabidopsis vacuolar sodium/proton antiporter gene in cotton improves photosynthetic performance under salt conditions and increases fiber yield in the field. *Plant and Cell Physiology* 46(11): 1848–1854. doi: 10.1093/pcp/pci201.

Hu, H., and Xiong, L. 2014. Genetic engineering and breeding of drought-resistant crops. *Annual Review of Plant Biology* 65: 715–741. doi: 10.1146/annurev-arplant-050213-040000.

Hu, R., Zhu, Y., Shen, G., and Zhang, H. 2017b. Overexpression of the *PP2A-C5* gene confers increased salt tolerance in Arabidopsis thaliana. *Plant Signaling and Behavior* 12(2): e1276687. doi: 10.1080/15592324.2016.1276687.

Hu, R., Zhu, Y., Wei, J., Chen, J., Shi, H., Shen, G., et al. 2017a. Overexpression of *PP2A-C5* that encodes the catalytic subunit 5 of protein phosphatase 2A in Arabidopsis confers better root and shoot development under salt conditions. *Plant, Cell and Environment* 40(1): 150–164. doi: 10.1111/pce.12837.

Huertas, R., Olías, R., Eljakaoui, Z., Gálvez, F. J., Li, J., De Morales, P. A., et al. 2012. Overexpression of *SlSOS2* (*SlCIPK24*) confers salt tolerance to transgenic tomato. *Plant, Cell and Environment* 35(8): 1467–1482. doi: 10.1111/j.1365-3040.2012.02504.x.

James, D., Borphukan, B., Fartyal, D., Ram, B., Singh, J., Manna, M., et al. 2018. Concurrent overexpression of *OsGS1*;1 and *OsGS2* genes in transgenic rice (*Oryza sativa* L.): impact on tolerance to abiotic stresses. *Frontiers in Plant Science* 9: 786. doi: 10.3389/fpls.2018.00786.

Ji, H., Pardo, J. M., Batelli, G., Van Oosten, M. J., Bressan, R. A. and Li, X. 2013. The salt overly sensitive (SOS) pathway: established and emerging roles. *Molecular Plant* 6(2): 275–286. doi: 10.1093/mp/sst017.

Joshi, R., Wani, S. H., Singh, B., Bohra, A., Dar, Z. A., Lone, A. A., et al. 2016. Transcription factors and plants response to drought stress: current understanding and future directions. *Frontiers in Plant Science* 7: 1029. doi: 10.3389/fpls.2016.01029.

Khadilkar, A. S., Yadav, U. P., Salazar, C., Shulaev, V., Paez-Valencia, J., Pizzio, G. A., et al. 2016. Constitutive and companion cell-specific overexpression of *AVP1*, encoding a proton-pumping pyrophosphatase, enhances biomass accumulation, phloem loading, and long-distance transport. *Plant Physiology* 170(1): 401–414. doi: 10.1104/pp.15.01409.

Kim, R. J., Kim, H. U. and Suh, M. C. 2019. Development of camelina enhanced with drought stress resistance and seed oil production by co-overexpression of *MYB96A* and *DGAT1C*. *Industrial Crops and Products* 138: 111475. doi: 10.1016/j.indcrop.2019.111475.

Kim, Y. S., Kim, I. S., Choe, Y. H., Bae, M. J., Shin, S. Y., Park, S. K., et al. 2014. Overexpression of the Arabidopsis vacuolar H^+-pyrophosphatase *AVP1* gene in rice plants improves grain yield under paddy field conditions. *Journal of Agricultural Science* 152(6): 941–953. doi: 10.1017/S0021859613000671.

Krishna, R., Ansari, W. A., Jaiswal, D. K., Singh, A. K., Verma, J. K. and Singh, M. 2021. Co-overexpression of AtDREB1A and BcZAT12 increases drought tolerance and fruit production in double transgenic tomato (*Solanum lycopersicum*) plants. *Environmental and Experimental Botany* 184: 104396. doi: 10.1016/j.envexpbot.2021.104396.

Kudo, M., Kidokoro, S., Yoshida, T., Mizoi, J., Todaka, D., Fernie, A. R., et al. 2017. Double overexpression of *DREB* and *PIF* transcription factors improves drought stress tolerance and cell elongation in transgenic plants. *Plant Biotechnology Journal* 15: 458–471. doi: 10.1111/pbi.12644.

Kumar, T., Uzma, Khan, M. R., Abbas, Z. and Ali, G. M. 2014. Genetic improvement of sugarcane for drought and salinity stress tolerance using Arabidopsis vacuolar pyrophosphatase (*AVP1*) gene. *Molecular Biotechnology* 56(3): 199–209. doi: 10.1007/s12033-013-9695-z.

Kuppu, S., Mishra, N., Hu, R., Sun, L., Zhu, X., Shen, G., et al. 2013. Water-deficit inducible expression of a cytokinin biosynthetic gene *IPT* improves drought tolerance in cotton. *PLOS One* 8(5): e64190. doi: 10.1371/journal.pone.0064190.

Le Martret, B., Poage, M., Shiel, K., Nugent, G. D. and Dix, P. J. 2011. Tobacco chloroplast transformants expressing genes encoding dehydroascorbate reductase, glutathione reductase, and glutathione-S-transferase, exhibit altered antioxidant metabolism and improved abiotic stress tolerance. *Plant Biotechnology Journal* 9: 661–673. doi: 10.1111/j.1467-7652.2011.00611.x.

Liu, C., Ma, H., Zhou, J., Li, Z., Peng, Z., Guo, F. et al. 2019. *TsHD1* and *TsNAC1* cooperatively play roles in plant growth and abiotic stress resistance of *Thellungiella halophile*. *Plant Journal* 99: 81–97. doi: 10.1111/tpj.14310.

Liu, X. Q., Feng, H. M., Huang, D. M., Fan, X. R. and Xu, G. H. 2015. Two short sequences in the promoter of *OsNAR2.1* are necessary for fully activating the nitrate induced gene expression in rice. *Scientific Reports* 5: 11950. doi: 10.1038/srep11950.

Liu, X. H., Lyu, Y. S., Yang, W., Yang, Z. T., Lu, S. J. and Liu, J. X. 2020. A membrane-associated NAC transcription factor OsNTL3 is involved in thermotolerance in rice. *Plant Biotechnology Journal* 18: 1317–1329. doi: 10.1111/pbi.13297.

Lu, W., Guo, C., Li, X., Duan, W., Ma, C., Zhao, M., et al. 2014. Overexpression of *TaNHX3*, a vacuolar Na^+/H^+ antiporter gene in wheat, enhances salt stress tolerance in tobacco by improving related physiological processes. *Plant Physiology and Biochemistry* 76: 17–28. doi: 10.1016/j.plaphy.2013.12.013.

Luo, S., Li, J., Stoddard, T. J., Baltes, N. J., Demorest, Z. L., Clasen, B. M., et al. 2015. Non-transgenic plant genome editing using purified sequence specific nucleases. *Molecular Plant* 8(9): 1425–1427. doi: 10.1016/j.molp.2015.05.012.

Luo, X., Wu, J., Li, Y., Nan, Z., Guo, X., Wang, Y., et al. 2013. Synergistic effects of *GhSOD1* and *GhCAT1* overexpression in cotton chloroplasts on enhancing tolerance to methyl viologen and salt stresses. *PLoS One* 8(1): e54002. doi: 10.1371/journal.pone.0054002.

Ma, D. M., Xu, W. R., Li, H. W., Jin, F. X., Guo, L. N., Wang, J., et al. 2014. Co-expression of the *Arabidopsis SOS* genes enhances salt tolerance in transgenic tall fescue (*Festuca arundinacea Schreb.*). *Protoplasma* 251: 219–231. doi: 10.1007/s00709-013-0540-9.

Mantri, N., Patade, V., Penna, S., Ford, R. and Pang, E. 2012. Abiotic stress in plants: present and future. In: Ahmad, P. and Prasad, M. N. V (eds). *Abiotic Stress Responses in Plants: Metabolism, Productivity and Sustainability*, 1–19. New York, NY: Springer. doi: 10.1007/978-1-4614-0634-1_1.

Miller, G., Suzuki, N., Ciftci-Yilmaz, S. and Mittler, R. 2010. Reactive oxygen species homeostasis and signaling during drought and salinity stresses. *Environment* 33: 453–467. doi: 10.1111/j.1365-3040.2009.02041.x.

Mishra, N., Srivastava, A. P., Esmaeili, N., Hu, W. and Shen, G. 2018. Overexpression of the rice gene *OsSIZ1* in Arabidopsis improves drought-, heat-, and salt-tolerance simultaneously. *PLoS One* 13: e0201716. doi: 10.1111/j.1365-3040.2009.02041.x.

Mishra, N., Sun, L., Zhu, X., Smith, J., Srivastava, P. A., Yang, X., et al. 2017. Overexpression of the rice SUMO E3 ligase gene OsSIZ1 in cotton enhances drought and heat tolerance, and substantially improves fiber yields in the field under reduced irrigation and rainfed conditions. *Plant and Cell Physiology* 58: 735–746. doi: 10.1093/pcp/pcx032.

Mittler, R. 2006. Abiotic stress, the field environment and stress combination. *Trends in Plant Science* 11(1): 15–19. doi: 10.1016/j.tplants.2005.11.002.

Nie, W. X., Xu, L. and Yu, B. J. 2015. A putative soybean *GmsSOS1* confers enhanced salt tolerance to transgenic Arabidopsis sos1-1 mutant. *Protoplasma* 252(1): 127–134. doi: 10.1007/s00709-014-0663-7.

Pasapula, V., Shen, G., Kuppu, S., Paez-Valencia, J., Mendoza, M., Hou, P., et al. 2011. Expression of an Arabidopsis vacuolar H^+-pyrophosphatase gene (*AVP1*) in cotton improves drought- and salt tolerance and increases fibre yield in the field conditions. *Plant Biotechnology Journal* 9: 88–99. doi: 10.1111/j.1467-7652.2010.00535.x.

Pehlivan, N., Sun, L., Jarrett, P., Yang, X., Mishra, N., Chen, L., et al. 2016. Co-overexpressing a plasma membrane and a vacuolar membrane sodium/proton antiporter significantly improves salt tolerance in transgenic Arabidopsis plants. *Plant and Cell Physiology* 57(5): 1069–1084. doi: 10.1093/pcp/pcw055.

Person, J-M., Tejedor-Cano, J., Prieto-Dapena, P., Almoguera, C. and Jordano, J. 2014. Co-overexpression of two heat shock factors results in enhanced seed longevity and in synergistic effects on seedling tolerance to severe dehydration and oxidative stress. *BMC Plant Biology* 14: 56. doi: 10.1186/1471-2229-14-56.

Pizzio, G. A., Paez-Valencia, J., Khadilkar, A. S., Regmi, K., Patron-Soberano, A., Zhang, S. J., et al. 2015. Arabidopsis type I proton-pumping pyrophosphatase expresses strongly in phloem, where it is required for pyrophosphate metabolism and photosynthate partitioning. *Plant Physiology* 167(4): 1541–1553. doi: 10.1104/pp.114.254342.

Portis Jr, A. R., Li, C., Wang, D. and Salvucci, M. E. 2008. Regulation of Rubisco activase and its interaction with Rubisco. *Journal of Experimental Botany* 59(7): 1597–1604. doi: 10.1093/jxb/erm240.

Pruthvi, V., Narasimhan, R. and Nataraja, K. N. 2014. Simultaneous expression of abiotic stress responsive transcription factors, *AtDREB2A, AtHB7* and *AtABF3* improves salinity and drought tolerance in peanut (*Arachis hypogaea* L.). *PLoS One* 9(12): e111152. doi: 10.1371/journal.pone.0111152.

Qin, H., Gu, Q., Kuppu, S., Sun, L., Zhu, X., Mishra, N., et al. 2013. Expression of the Arabidopsis vacuolar H^+-pyrophosphatase gene *AVP1* in peanut to improve drought and salt tolerance. *Plant Biotechnology Reports* 7(3): 345–355. doi: 10.1007/s11816-012-0269-5.

Qin, H., Gu, Q., Zhang, J., Sun, L., Kuppu, S., Zhang, Y., et al. 2011. Regulated expression of an isopentenyltransferase gene (*IPT*) in peanut significantly improves drought tolerance and increases yield under field conditions. *Plant and Cell Physiology* 52: 1904–1914. doi: 10.1093/pcp/pcr125.

Qiu, Q. S., Guo, Y., Dietrich, M. A., Schumaker, K. S. and Zhu, J. K. 2002. Regulation of SOS1, a plasma membrane Na^+/H^+ exchanger in *Arabidopsis thaliana*, by SOS2 and SOS3. *Proceedings of the National Academy of Sciences of the USA* 99(12): 8436–8441. doi: 10.1073/pnas.122224699.

Qu, G. Z., Zang, L., Xilin, H. Gao, C., Zheng, T. and Li, K. L. 2012. Co-transfer of *LEA* and *bZip* Genes from *Tamarix* confers additive salt and osmotic stress tolerance in transgenic Tobacco. *Plant Molecular Biology Reporter* 30: 512–518. doi: 10.1007/s11105-011-0371-9.

Quan, R., Wang, J., Yang, D., Zhang, H., Zhang, Z. and Huang, R. 2017. EIN3 and SOS2 synergistically modulate plant salt tolerance. *Scientific Reports* 7(1): 44637. doi: 10.1038/srep44637.

Rao, K. M., Raghavendra, A. and Reddy, K. J. 2006. *Physiology and Molecular Biology of Stress Tolerance in Plants*. Dordrecht: Springer. doi: 10.1007/1-4020-4225-6.

Rhee, S. Y. and Mutwil, M. 2014. Towards revealing the functions of all genes in plants. *Trends in Plant Science* 19: 213–221. doi: 10.1016/j.tplants.2013.10.006.

Sahoo, D. P., Kumar, S., Mishra, S., Kobayashi, Y., Panda, S. K. and Sahoo, L. 2016. Enhanced salinity tolerance in transgenic mungbean overexpressing Arabidopsis antiporter (*NHX1*) gene. *Molecular Breeding* 36(10): 144. doi: 10.1007/s11032-016-0564-x.

Salvucci, M. E. and Crafts-Brandner, S. J. 2004. Inhibition of photosynthesis by heat stress: the activation state of Rubisco as a limiting factor in photosynthesis. *Physiologia Plantarum* 120(2): 179–186. doi: 10.1111/j.0031-9317.2004.0173.x.

Salvucci, M. E. and Ogren, W. L. 1996. The mechanism of rubisco activase: insights from studies of the properties and structure of the enzyme. *Photosynthesis Research* 47(1): 1–11. doi: 10.1007/bf00017748.

Salvucci, M. E., Osteryoung, K. W., Crafts-Brandner, S. J. and Vierling, E. 2001. Exceptional sensitivity of Rubisco activase to thermal denaturation *in vitro* and *in vivo*. *Plant Physiology* 127(3): 1053–1064. doi: 10.1104/pp.010357.

Schilling, R. K., Tester, M., Marschner, P., Plett, D. C. and Roy, S. J. 2017. AVP1: one protein, many roles. *Trends in Plant Science* 22(2): 154–162. doi: 10.1016/j.tplants.2016.11.012.

Shen, G., Kuppu, S., Sun, L., Payton, P., and Zhang, H. 2015. Double overexpression of *AVP1* and *AtNHX1* in cotton further improves drought- and salt-tolerance in transgenic cotton plants. *Plant Molecular Biology Reporter* 33: 167–177. doi: 10.1007/s11105-014-0739-8.

Shi, H., Ishitani, M., Kim, C. and Zhu, J. K. 2000. The *Arabidopsis thaliana* salt tolerance gene SOS1 encodes a putative Na^+/H^+ antiporter. *Proceedings of the National Academy of Sciences of the USA* 97(12): 6896–6901. doi: 10.1073/pnas.120170197.

Shi, H., Lee, B. H., Wu, S. J. and Zhu, J. K. 2003. Overexpression of a plasma membrane Na^+/H^+ antiporter gene improves salt tolerance in *Arabidopsis thaliana*. *Nature Biotechnology* 21(1): 81–85. doi: 10.1038/nbt766.

Shi, J., Gao, H., Wang, H., Laftte, H. R., Archibald, R. L., Yang, M., et al. 2017. ARGOS 8 variants generated by CRISPR-Cas9 improve maize grain yield under field drought stress conditions. *Plant Biotechnology Journal* 15: 207–216. doi: 10.1111/pbi.12603.

Solomon, S., Qin, D., Manning, M., Averyt, K. and Marquis, M. 2007. *IPCC, 2007: Climate Change 2007: The Physical Science Basis. Contribution of Working Group I to the Fourth Assessment. Report of the Intergovernmental Panel on Climate Change*. Cambridge: Cambridge University Press.

Sun, L., Pehlivan, N., Esmaeili, N., Jiang, W., Yang, X., Jarrett, P., et al. 2018. Co-overexpression of *AVP1* and *PP2A-C5* in Arabidopsis makes plants tolerant to multiple abiotic stresses. *Plant Science* 274: 271–283. doi: 10.1016/j.plantsci.2018.05.026.

Unamba, C. I. N., Nag, A. and Sharma, R. K. 2015. Next generation sequencing technologies: the doorway to the unexplored genomics of non-model plants. *Frontiers in Plant Science* 6: 1074. doi: 10.3389/fpls.2015.01074.

Upadhyaya, D. C., Bagri, D. S., Upadhyaya, C. P., Kumar, A., Thiruvengadam, M. and Jain, S. K. 2020. Genetic engineering of potato (*Solanum tuberosum* L.) for enhancedα-tocopherols and abiotic stress tolerance. *Physiologia Plantarum* 173: 116–128. doi: 10.1111/ppl.13252.

Wang, D., Luo, W., Khurshid, M., Gao, L., Sun, Z., Zhou, M. et al. 2018. Co-expression of *PeDREB2a* and *KcERF* improves drought and salt tolerance in transgenic *Lotus corniculatus*. *Journal of Plant Growth Regulation* 37: 550–559. doi: 10.1007/s00344-017-9753-z.

Wijewardene, I., Mishra, N., Sun, L., Smith, J., Zhu, X., Payton, P., et al. 2020. Improving drought-, salinity-, and heat-tolerance in transgenic plants by co-overexpressing Arabidopsis vacuolar pyrophosphatase gene *AVP1* and Larrea Rubisco activase gene *RCA*. *Plant Science* 110499. doi: 10.1016/j.plantsci.2020.110499.

Yang, H. B., Zhang, X., Gaxiola, R. A., Xu, G. H., Peer, W. A. and Murphy, A. S. 2014. Over-expression of the Arabidopsis proton-pyrophosphatase *AVP1* enhances transplant survival, root mass, and fruit development under limiting phosphorus conditions. *Journal of Experimental Botany* 65(12): 3045–3053. doi: 10.1093/jxb/eru149.

Yang, Q., Chen, Z. Z., Zhou, X. F., Yin, H. B., Li, X., Xin, X. F., et al. 2009. Overexpression of *SOS* (Salt Overly Sensitive) genes increases salt tolerance in transgenic Arabidopsis. *Molecular Plant* 2(1): 22–31. doi: 10.1093/mp/ssn058.

Yang, Y., Tang, R. T., Jiang, C. M., Li, B., Kang, T., Liu, H., et al. 2015. Overexpression of the $PtSOS_2$ gene improves tolerance to salt stress in transgenic poplar plants. *Plant Biotechnology Journal* 13(7): 962–973. doi: 10.1111/pbi.12335.

Yarra, R. 2019. The wheat NHX gene family: potential role in improving salinity stress tolerance of plants. *Plant Gene* 18: 100178. doi: 10.1016/j.plgene.2019.100178.

Yarra, R. and Kirti, P. B. 2019. Expressing class I wheat NHX (*TaNHX2*) gene in eggplant (*Solanum melongena* L.) improves plant performance under saline condition. *Functional and Integrative Genomics* 19(4): 541–554. doi: 10.1007/s10142-019-00656-5.

Yin, X., Xia, Y., Xie, Q., Cao, Y., Wang, Z., Hao, G., et al. 2019. The protein kinase complex CBL10–CIPK8–SOS1 functions in Arabidopsis to regulate salt tolerance. *Journal of Experimental Botany* 71(6): 1801–1814. doi: 10.1093/jxb/erz549.

Zhang, Y., Deng, G., Fan, W., Yuan, L., Wang, H. and Zhang, P. 2019. *NHX1* and *eIF4A1*-stacked transgenic sweetpotato shows enhanced tolerance to drought stress. *Plant Cell Reports* 38: 1427–1438. doi: 10.1007/s00299-019-02454-6.

Zhang, Y. M., Zhang, H. M., Liu, Z. H., Li, H. C., Guo, X. L. and Li, G. L. 2015. The wheat NHX antiporter gene *TaNHX2* confers salt tolerance in transgenic alfalfa by increasing the retention capacity of intracellular potassium. *Plant Molecular Biology* 87(3): 317–327. doi: 10.1007/s11103-014-0278-6.

Zhu, J. K., Liu, J. and Xiong, L. 1998. Genetic analysis of salt tolerance in Arabidopsis: evidence for a critical role of potassium nutrition. *Plant Cell* 10(7): 1181–1191. doi: 10.1105/tpc.10.7.1181.

Zhu, X., Sun, L., Kuppu, S., Hu, R., Mishra, N., Smith, J., et al. 2018. The yield difference between wild-type cotton and transgenic cotton that expresses *IPT* depends on when water-deficit stress is applied. *Scientific Reports* 8: 2538. doi: 10.1038/s41598-018-20944-7.

Zurbriggen, M. D., Hajirezaei, M. and Carrillo, N. 2010. Engineering the future. Development of transgenic plants with enhanced tolerance to adverse environments. *Biotechnology and Genetic Engineering Reviews* 27: 33–56. doi: 10.1080/02648725.2010.10648144.

13 Molecular Understanding of Transcription Factors Involved in Multiple Stress Tolerance in Plants

Sougata Bhatacharjee, Asha Kumari, and Krishnayan Paul
ICAR-Indian Agricultural Research Institute

Suresh Gali
Chaudhary Charan Singh Haryana Agricultural University

CONTENTS

DOI: 10.1201/9781003300564-13

13.1 INTRODUCTION

Stress in agricultural terms is defined as nonideal conditions, which shows negative impact on plant growth, development, or productivity (Gull et al., 2019). Plant stress factors are classified into two categories: abiotic and biotic (Umar et al., 2021). Abiotic factors are environmental factors that affect plant growth such as heat stress, drought stress, salinity stress, and heavy metal stress, whereas biotic factors are other organisms that share the environment and interact with the plants such as pathogens and pests (Mosa et al., 2017). Abiotic stresses are interconnected with the oxidative stress, which causes adverse effects on the ion distribution and imbalance in cell homeostasis (Wang et al., 2003). These changes result from the expression of group of genes leading to creation of negative impact on plant growth and production so, therefore must need knowledge about stress-responsive mechanism in plants to find these genes that are responsible for plant abiotic stress (Wang et al., 2003). Stress-tolerant plants, in response to stress, establish a new metabolic homeostasis and can thus continue to grow without suffering stress-induced injury. By adjusting growth, development, and cellular and molecular activities, tolerance mechanisms are coordinated and fine-tuned (Levitt, 1980; Shavalli Khan et al., 2014). The formation of reactive oxygen species (ROS), strong oxidants that can cause significant damage to membrane systems and DNA, is a common theme of stress at the cellular and molecular level (Redondo-Gómez, 2013). Superoxide (O^{2-}), hydrogen peroxide (H_2O_2), and super hydroxide (OOH) are examples of ROS (Redondo-Gómez, 2013). Antioxidant systems, including enzymatic superoxide dismutases, catalase, peroxidases, phenol oxidase, and ascorbic acid oxidase, as well as nonenzymatic systems (compounds that are strong reductants such as glutathione, phenols, flavonoids, and polyamines), are important in balancing and preventing oxidative damage (Redondo-Gómez, 2013) (Figure 13.1).

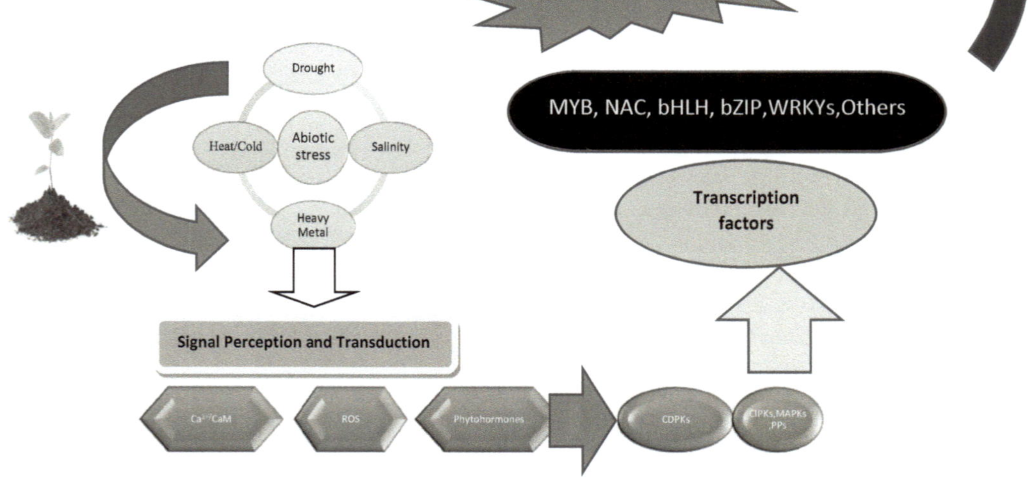

FIGURE 13.1 Schematic representation showing mechanism of abiotic stress tolerance in plants.

13.2 DROUGHT STRESS TOLERANCE

Plants are under water stress (deficiency) when transpiration from leaf surfaces exceeds water uptake by roots. This imbalance occurs primarily when the water potential of the soil is lower than the water potential of the roots of plants. As multidimensional stress, it causes changes in plants' physiology, morphology, biochemistry, and molecular characteristics. During a period of water deficiency, plants adopt measures to reduce transpiration loss, such as changing stomatal conductance and distribution, leaf rolling, changes in root-to-shoot ratios, root length increment, accumulating compatible solutes, enhancing transpiration efficiency, and delaying senescence. Drought induces oxidative stress by overproducing ROS, causing the membranes to rupture and stimulating stress pathways (ROS, mitogen-activated protein kinases, calcium, and hormone signaling). Stress signaling and responses in plants are coordinated by drought-induced transcription factors activation and abscisic acid concentration. In addition to functional genomic approaches, genome modification techniques such as the CRISPR/Cas9 system can help characterize genes separated from stress-related candidate genes and quantitative trait loci.

Various transcription factors (TFs) are known to be involved in plant drought response. Both conventional and transcriptome-based analyses are used to identify these TFs. These TFs are organized into large gene families, including AP2/ERF, bZIP, NAC, MYB, MYC, Cys2His2 zinc finger, and WRKY. The TFs regulate downstream genes that act as *cis*-elements and directly modulate drought response. The stress-responsive regulatory pathway may be categorized in two ways: ABA-dependent (e.g., ABF/AREB TFs, acting on genes carrying the ABRE) and ABA-independent (e.g., CBF/DREB TFs acting on genes carrying the CRT/DRE–C-repeat/dehydration-responsive elements). Several essential regulatory genes are associated with detoxification, osmolyte biosynthesis, proteolysis of cellular substrates, water channels, ion transporters, heat shock proteins (HSPs), and late embryogenesis abundant proteins (LEAs). It has been demonstrated that WRKY TFs, such as WRKY60, WRKY40, and WRKY18, co-regulate during drought stress (Yan et al., 2013). In addition, some MYB TF members are regulated by autoregulation or cross-regulation. Accordingly, the MYB10 expression in apple is autoregulated, enabling it to regulate further anthocyanin accumulation, associated with stress endurance (Espley et al., 2009; Brendolise et al., 2017). Some of the DREB TFs are regulated by cross-regulation as well. When exposed to abiotic stress, epigenetic and posttranscriptional regulatory mechanisms regulate TF expression. TFs such as ZmNAC111 are negatively regulated by methylation of its promoter by H3K9 methylase, which alleviates drought stress tolerance (Mao et al., 2015). Antioxidative defense systems comprise both enzymatic and nonenzymatic components to maintain ROS detoxification. The ascorbate–glutathione cycle is coordinated by several enzymes, including SOD (superoxide dismutase) and CAT (catalase), as well as MDHAR (monodehydroascorbate reductase), DHAR (dehydroascorbate reductase), and GR (glutathione reductase). In addition to enzymes, low molecular mass compounds such as ascorbic acid, glutathione, carotenoid, tocopherol, and polyphenols are present. During drought stress, the intricate network of protein turnover is significantly altered. Among these are HSP, late embryogenesis abundant protein (LEA), and galactinol synthase (GolS) (Campalans et al., 1999; Kaur and Asthir, 2015; Salvi et al., 2020). Other proteins repairing enzymes include protein l-ISOASPARTYL METHYLTRANSFERASE (PIMT) and METHIONINE SULFOXIDE REDUCTASE (MSR). Furthermore, ABA plays a central role in regulating drought stress adaptation by influencing the induction of genes that respond to drought stress. It assists cellular transcriptional reprogramming, primarily governed by TFs, to adapt to stress by acting as a link between perception of stress and cellular transcriptional response. Signal transduction via secondary messengers (ROS, cyclic nucleotides (cAMP, cGMP), sugars, calcium ions, inositol phosphate and nitric oxide, etc.) activates protein kinases such as calcium-dependent protein kinases (CDPKs) and mitogen-activated protein kinases (MAPKs), and phosphatases, which either activate or repress the activity of the targeted TFs through phosphorylation or dephosphorylation (Danquah et al., 2014; Huang et al., 2012).

13.2.1 Basic Leucine Zipper

Basic leucine zipper (bZIP) TFs (bZIP TFs) are a very large class of TFs that are widely distributed in eukaryotes. A distinctive feature of this transcription factor is the presence of a highly conserved bZIP domain, which is 60–80 amino acids in length and consists of two functional domains: a conserved basic region near the N-terminus and a variable leucine zipper region near the C-terminus (Glover and Harrison, 1995). bZIP TFs have been shown to respond to abiotic stresses such as drought (Xiang et al., 2008), high salinity (Hartmann et al., 2015), and extreme temperature (Liao et al. 2008) in plants. Yoon et al. (2017) reported that transgenic rice plants overexpressing OsbZIP66 were drought-tolerant. bZIP TFs, including AREB2, ABF4, ABF3, and ABF4 in *Arabidopsis thaliana*, positively regulate ABA-dependent gene expression (Choi et al. 2000).

13.2.2 Dehydration-Responsive Element Binding

It has been shown that several dehydration-responsive element bindings (DREBs) play a part in abiotic stress tolerance, both ABA-independently and ABA-dependently. The DREB genes contain a highly conserved AP2/ERF DNA-binding domain (50–60 amino acids long) (Lata and Prasad, 2011). While most members of the DREB recognize the DRE, some have been shown to recognize the *cis*-acting element AGCCGCC, which is often referred to as the GCC box (Sakuma et al., 2002). During ABA treatment and drought stress, AtCBF4, which encodes the DREB TF, rapidly increased (Haake et al., 2002). The report says that overexpression of rice OsDREB1A in *Arabidopsis* led to increased expression of six stress-related genes (Kasuga et al., 1999), indicative of the principal stress-responsive TF family with multiple stress tolerance ability.

13.2.3 DNA Binding with One Finger

This DNA binding with one finger (DOF) domain consists of 50–53 amino acid residues of the Cys2Cys2 (C2C2)-type zinc finger pattern that has remained conserved throughout evolution (Yanagisawa, 2002), having the transcriptional regulatory C-terminal of this protein composed of a highly variable amino acid sequence. In addition to its intrinsic ability to homodimerize and heterodimerize, DOF-TF interacts with other TFs (myb-TF and TCP-TF) in a complex regulatory network (Shigyo et al., 2007; Gupta et al., 2015). DOF-TFs (including CDF3) is highly upregulated during drought stress in *Arabidopsis* (Corrales et al., 2017), and their overexpression can modulate other TFs such as CBFs, DREB2A, and ZAT12. DOF-TF has been shown to interact with MYB, bZIP, and TCP proteins during drought stress. Under drought stress, identifying the critical player can be achieved by comprehensively analyzing the molecular mechanisms underlying the interaction between DOF and drought-responsive TF.

13.2.4 Heat Shock Factor

The heat shock factor (HSF) consists of modules containing DNA-binding domains (DBDs) with a central helix–turn–helix motif. Another important domain is the oligomerization domain (OD) connected to the DBD with a 15–80 amino acid residues flexible linker. HSF includes a nuclear localization signal (NLS), nuclear export signal (NES), activator motifs, repressor domain, and functional and signature sequences. The length of the flexible linker and type of amino acid residues between DBD and OD have been used to classify HSFs into three types: HSFA, HSFB, and HSFC. Several dehydration-responsive genes, including GALACTINOL SYNTHASE (AtGolS) and RAFFINOSE SYNTHASE (RafS), were found more abundant in transgenic lines overexpressing HsfA2. Increased GolS and Rafs mRNAs led to an overabundance of galactinol and raffinose, which prevents the generation of hydroxyl radicals in response to oxidative damage (Panikulangara et al., 2004; Nishizawa et al., 2008). Li et al. (2015) demonstrated that overexpression of the heat- and drought-responsive ZmHSF06 gene in *Arabidopsis* improved drought and heat stress tolerance. During drought stress treatment, *Chenopodium quinoa* CqHSF5 and CqHSF12 expression were upregulated.

13.2.5 MYB Superfamily

The MYB TFs participate in ABA-dependent drought response (Lata et al., 2011). In *Arabidopsis*, MYB2 TF was found to interact with a dehydration-responsive gene (RD22) and induce several ABA-responsive genes. In *Arabidopsis*, by downregulating PP2C genes, AtMYB44 inhibited ABA signaling (Jung et al., 2008). Among the ABA and auxin signaling pathways, AtMYB96 is required for imparting drought tolerance to plants (Cominelli et al., 2005), and AtMYB60 controls stomatal movement. OsMYB48-1 overexpression in rice-induced ABA-related gene upregulation, suggesting that MYB can confer drought and salt tolerance in an ABA-dependent fashion (Xiong et al. 2014). Therefore, MYB TFs regulate drought stress tolerance primarily through an ABA-dependent mechanism.

13.2.6 NAC Transcription Factors

NAC-TF are named after three functional genes that contain NAC domains: NAM (no apical meristem), ATAF (*Arabidopsis* transcription activation factor), and CUC (cup-shaped cotyledon). The NAC proteins contain a 150–160 amino acid-long DNA-binding domain (BD) that assists DNA–protein interaction and nuclear localization. The transcriptional regulatory region of the C-terminus has posttranslational regulation under specific conditions. Generally, NAC TFs bind to *cis*-regulatory elements, such as CGT(G/A) or CACG. In rice plants, overexpression of root-specific NAC-TF (OsNAC5, OsNAC6, OsNAC9, and OsNAC10) altered root architecture and led to drought tolerance (Tran et al., 2004; Hu et al., 2006; Nakashima et al., 2007; Redillas et al., 2012). Under drought stress, OsNAC1 overexpression resulted in a more fertile spikelet and protection against drought stress. In drought stress, the OsNAC1-OE plants retained large amounts of water by mediating the stomatal closure without affecting the photosynthetic rate by facilitating osmolyte accumulation, signal transduction, and redox homeostasis. In *Arabidopsis*, the NAC-TF AtJUB1 (JUNGBRUNNEN 1; ANAC042) regulates ROS and mediates drought stress response (Wu et al., 2012). AtJUB1 directly induces the expression of DREB2A, an AP2-type transcription factor associated with drought stress. The tomato homolog of AtJUB1 (SlJUB1, Solyc05G021090) binds directly to the promoter region of genes encoding SlDREB and DELLA to regulate ROS homeostasis and stress-associated stress gene modulation, and SlJUB1 is evident in drought stress tolerance (Thirumalaikumar et al., 2018). Even though they are abundant in the nucleus, some NAC TFs localize to the membrane region of the endoplasmic reticulum or the plasma membrane because they contain a transmembrane domain. Thus, they are NTLs (NAC-TF associated with transmembrane motif-1-like). During stress, these NTLs are allocated into the nucleus through different posttranslational modifications, such as phosphorylation, membrane proteolysis, alternative splicing, etc. Duan et al. (2017) reported that MfNACsa was found in the membrane of *Medicago* under normal conditions, but it was translocated to the nucleus during drought stress. Thus, these NAC TFs could improve drought stress tolerance in various crops by integrating with various pathways.

13.2.7 TCP Transcription Factors

The TEOSINTE BRANCHED1, CYCLOIDEA, and PCF (TCP) families are plant-specific and regulate many metabolic processes of plants. The TCP TFs generally bind to *cis*-regulatory elements: GGNCCCAC, G(T/C)GGNCCC (Kosugi and Ohashi, 2002). Few studies indicated the role of specific TCP-TF in drought stress response (Mukhopadhyay and Tyagi, 2015; Lei et al., 2017; Ding et al., 2019). For instance, it has been reported that the overexpression of OsTCP19 imparted drought stress tolerance in *Arabidopsis*. In the ABA-induced drought stress response, OsTCP19 and OsABI4 interacted significantly (Mukhopadhyay and Tyagi, 2015). Similarly, TCP14, a positive seed germination regulator, regulates ABA-related genes in a specific set of seed germination genes by acting antagonist to DOF6 TF. Downregulation of TCP by OsamiR319 in *Agrostis stolonifera* enhanced drought stress tolerance. Higher leaf wax content and improved water retention capacity support OsamiR319-mediated stress tolerance (Zhou et al., 2013). A more comprehensive analysis is necessary to understand better the TCP-TF family's role and regulation in drought stress.

13.2.8 WRKY Transcription Factors

The first WRKY gene was identified in sweet potatoes (Ishiguro and Nakamura, 1994). WRKY proteins contain 60 amino acids of highly conserved sequences: WRKYGQK and the zinc finger motif: Cys(2)-His(2) or Cys(2)-HisCys. WRKY TFs primarily bind to W-box [or TTGAC(C/T)] *cis*-element present in the promoter of the desired genes (Bakshi and Oelmüller, 2014). WRKY TFs can either activate or repress the expression of a gene in their various homo- and heterodimer combinations (Bakshi and Oelmüller, 2014). WRKY TF genes activate ABA signaling, and the phytohormone integrates various stress signals to control downstream stress responses further. During drought stress, WRKY TF positively controls ABA-mediated closure of stomatal apertures. In *Arabidopsis*, the expression of the sunflower HaWRKY76 induced drought stress adaptation (Raineri et al., 2015). In rice, overexpression of OsWRKY45 and OsWRKY72 was associated with improved drought and salt tolerance (Song et al., 2010; Jiang et al., 2017). In *Arabidopsis thaliana*, overexpression of GmWRKY13/54 increased the plant's tolerance to salt and drought (Zhou et al., 2008). According to Wang et al. (2015), TaWRKY44 positively regulates drought stress response by either reducing ROS or regulating the expression of stress-related genes. *Gossypium hirsutum* WRKY TFs GhWRKY25 positively influenced the salt stress-mediated dehydration tolerance (Liu et al., 2016).

13.3 SALT STRESS TOLERANCE

Globally, 7% of the land, 20% of arable land, and 50% of irrigated land are affected by salinity (Dajic, 2006). In another report, salinity-affected lands have been estimated at more than 9 billion ha (Botella et al., 2005). More than 20% of all cultivated land worldwide is affected by salinity due to several reasons such as high salinity of irrigation water, exhaustive use of groundwater, poor drainage, and the amount is increasing with time, understanding the control of ion homeostasis and osmotic regulation and using the knowledge to engineer crop plants with enhanced salt tolerance. The most common symptoms of salt stress are growth inhibition, accelerated development, and death. There may also be programmed cell death under severe salinity shock, although growth inhibition is the primary injury leading to other symptoms. The cytosolic enzymes of halophytes, which accumulate large quantities of sodium inside the cell, are just as sensitive to sodium as enzymes of glycophytes. As a result, halophytes must compartmentalize sodium in the vacuole, away from cytosolic enzymes. By comparison, glycophytes are severely inhibited or even killed by salinities of 100–200 mmol/L NaCl; halophytes can survive salinities of over 300 mmol/L. Salinity and drought exaggerate each other as the soil gets saturated by the salinity; its water potential is lowered. This situation restricts plants to get water and nutrients from the soil. Under such conditions, plants respond in two ways. First, the higher concentration of Na+ in roots causes osmotic stress, and second, disturbance in the balance of essential nutrients causes ionic stress (Munns et al., 2006). Beyond a limit, a higher concentration of Na^+ and Cl^- in the soil solution may lead to toxic ionic ratios of Na^+/Ca^{2+}, Na^+/K^+, Ca^{2+}/Mg^{2+}, and Cl^-/NO^{3-} that replace essential nutrients and generate competition for nutrient uptake in plants (Munns et al., 2006). The external Na^+ can negatively impact the intracellular K^+ influx. However, K^+ ions are one of the essential elements required for growth. The limited supply of K^+ ions may disturb the osmotic balance leading to the loss of stomatal functioning as well as other enzymes. Moreover, excess of Na^+ inside the cell has been reported to cause various physiological disorders such as reduced seed germination, seedling growth, flowering, fruiting, etc. (Parihar et al., 2015; Singh et al., 2015). In addition to this, induction of Na^+ and Cl^- toxicity in cells also causes plasmolysis and enhances the risk of oxygen toxicity induced by the acceleration of ROS formation (Parida and Das, 2005; Parihar et al., 2015). Aldesuquy et al. (2014) have reported a decline in leaf area, photosynthetic pigments, light reactions, as well as overall photosynthetic performance after irrigation of wheat plants with seawater. In another study performed by Kalaji et al. (2011), a significant fall in photosynthetic efficiency due to limited photosystem II (PS II) activity in two Syrian barley landraces was observed after a higher accumulation of salt in cells. Thus, salinity affects almost every aspect of the physiology and biochemistry of plants and significantly reduces crop yield and productivity.

13.3.1 Na Tolerance Mechanism

In response to salinity, plants have evolved several mechanisms to modify their morphological, physiological, and molecular behavior. Several mechanisms are involved in salt tolerance, including ion compartmentalization, osmolyte production, germination, adaptation to osmosis, succulence, selectivity of ion transport and uptake, enzyme responses, salt excretion, and genetic control (Koyro et al., 2011). Genetic analysis has shown that maintaining a low concentration of cytoplasmic sodium is the key to sodium tolerance (Zhu, 2002). In plant cells, the sensor protein for this calcium signal is termed SOS3. It is designated because a loss-of-function mutation in this protein renders the *Arabidopsis thaliana* plant overly sensitive to salt. SOS3 makes a complex with the serine/threonine-protein kinase SOS2. Upon receiving a calcium signal, the kinase complex phosphorylates target proteins such as SOS1. The SOS1 gene encodes a sodium/proton antiporter that removes sodium from the cells (Zhu, 2002). SOS1 is dependent on both SOS3 and SOS2 for sodium extrusion activity. The vacuolar compartment is an important mechanism for storing cytosolic sodium. Vacuolar compartmentation of sodium is achieved by the action of sodium/proton antiporters on the tonoplast, the vacuolar membrane. The tonoplast proton-ATPase and pyrophosphatase generate the proton gradient that drives the antiport. Compartmentation of sodium into the vacuole separates sodium from cytosolic enzymes and helps offset the low extracellular osmotic potential created by salt stress. The protein kinase SOS2 responsible for activating the plasma membrane sodium/proton antiporter is also important for activating vacuolar sodium/proton antiporters. Salt-tolerant plants may adapt to the prevailing salty environment by excluding, excreting, or accumulating salt. It is perhaps even more important for salt-tolerant plants such as halophytes to compartmentalize sodium in their vacuoles. Those halophytes cannot tolerate sodium inhibition of their cytosolic enzymes, so they must store most of the accumulated sodium in the vacuole. As a matter of fact, salt concentrations of approximately 0.5 mol/L or more are not uncommon in the vacuoles of halophytes. The counterions are typically chloride and malate. Na^+ and Cl^- enter inside cells via ion channels, Na^+ and Cl^- transporters, or pinocytosis. Na^+/H^+ antiporters reside in the tonoplast, and with the help of H+ATPases and PPIases, Na^+ enters into the vacuole. Leakage of Na^+ back to the cytoplasm is prevented by the modified lipid composition of tonoplasts of halophyte (Glenn et al., 1999). Further, the root system of plants function as ultrafiltration mechanism, causing these plants to become dominant components of mangrove vegetation. Some halophytes possess multicellular salt glands and salt hairs; those are common in many halophytes such as *Atriplex*, *Plumbago*, and some mangrove species, such as *Avicennia* and *Acanthus* (Hasanuzzaman et al., 2013). By connecting the mesophyll cells through numerous plasmodesmata, these glands secrete salt crystals at the leaf surface (Hasanuzzaman et al., 2013). Energy (ATP) is required for ion pumping during salt excretion. Halophytes obtain their energy from the active respiration of glandular cells (Marcum and Murdoch, 1992). TFs include many families, such as AP2/ERF, bZIP, NAC, MYB, and WRKY, which exhibit higher correlations with salinity. Different TFs have been shown to enhance the expression of different genes in numerous studies. There is evidence that changes in *GmDREB2*, *GmDREB3*, and *MtCBF4* genes contribute to resistance against drought and salinity in *Arabidopsis thaliana*, *Tamarix hispid* (Wang et al., 2010), *Caragana korshinskii* (Wang et al., 2011), and alfalfa. A study using transgenic alfalfa also demonstrated overexpression of *MsAlfn1*, which induced expression of MsPRP2, providing tolerance to salinity (Winicov and Bastola, 1999).

13.3.2 NAC

Moreover, it has been reported that the NAC domain of TF is not only involved in DNA binding but also helps in localization (related to the nucleus), synthesis of dimers (including homodimers and heterodimers) (along with many other NAC proteins), and regulation of transcription processes (Puranik et al., 2012). Additionally, it has also been found that NAC TFs interact with NACRS (NAC recognition sequences) and promoter genes of the CACG core DNA-binding motif, thus helping control downstream transcription. Ecotypic expression of NAC and overexpression of OsNAC5 were seen in the case of drought and salinity stress tolerance. However, RNAi could knock this gene out and make it susceptible to salt stress. Similarly, in *Oryza sativa*, drought and salinity stress tolerance are increased by the overexpression of the specific transcription factor gene OsNAC04. It has also been found that during increased abscisic acid (ABA) and a high level of salinity, the expression of an *Arabidopsis* plant-specific transcription NAC gene (*ATAF1*) occurs. Furthermore, AKIN11 and AKIN10 catalytic subunits of SnRK1 interact with ATAF1 and play a vital role in the signaling pathways of ABA. NAC overexpression leads to increased *Arabidopsis* plant sensitivity to salt and ABA. Transgenic rice having ATAF1 was generated to investigate salinity tolerance. This transgenic *Oryza sativa* showed improved salt tolerance. Liu et al. (2016) found that ATAF1 functions as a response to salinity stress and may help improve the salinity tolerance in *Oryza sativa*. Moreover, induction of many genes occurs under stress conditions such as phosphatases, dehydrins, kinases, detoxification proteins, and TFs (Yang et al., 2012; Zhang et al., 2012). Yang et al. (2011) described that VNI2 is a positive regulator acting against the salinity stress in *Arabidopsis* induced by ABA and high salinity stress. However, they found that VNI2 transgenic plants are comparatively less sensitive to high salinity. There are many members of NAC TFs that play a positive role in response to salinity stress, as evident from SNAC1 (Hu et al., 2006), ONAC045 (Zheng et al., 2009), OsNAC5/6 (Takasaki et al., 2010; Song et al., 2011), ONAC106, OsNAP, ONAC022, etc.

13.3.3 bZIP

Plant bZIP TFs have a great affinity to bind with ACGT core DNA sequences such as G-box (CACGTG), C-box (GAC GTC), A-box (TACGTA), PB-like (TGA AAA), ABRE (ABA-responsive element), that is, CCACGTGG) and GLM (GCN4-like motif, i.e., GTGAGTCAT) motif that occurs at the promoter regions of the stress-signaling genes (Jakoby et al., 2002; Liao et al., 2008). The sensor senses salt signals, and signal transduction starts, which causes alterations in electron transport systems of the cell (chloroplast, mitochondria, peroxisome) and results in an excess production of ROS that ultimately causes oxidative damage in cells. The activation of TFs in the nucleus also depends upon signal transduction. The abiotic stress-mediated TFs are ABA dependent on MYB2, MYC2, NAC, AREB/ABF, and others are ABA-independent NAC (HD-ZIP) DREB2. The ABA-responsive element (ABRE) is an expression regulation system whose function is highly dependent upon ABA production. Many transgenic plants with the overexpression of bZIP genes seem to be able to overcome cell dehydration caused by salinity stress. Huang et al. (2010) conducted a study on *Poncirus trifoliata* and showed the involvement of bZIP TF genes, including LEA, CDPK, and DREB, in stress tolerance. Moreover, another bZIP gene (ThbZIP1) isolated from *Tamarix hispida* prevented salt-induced cell death and showed remarkable salt resistance in transgenic tobacco (Wang et al., 2010). Another member of the bZIP family, GmbZIP1 from soybean, has been shown to promote tolerance to different abiotic stresses such as salinity, drought, and heat/low temperatures in transgenic *Arabidopsis* and tobacco plants (Gao et al., 2011).

13.3.4 MYB

ABA-dependent gene expression under drought and salt stress is regulated by *AtMYC2* and *AtMYB2* genes (Abe et al., 2003). Besides this, AtMYB41 regulates the short-term transcriptional responses against osmotic stress. In addition, overexpression of OsMYB2 promotes the formation of osmolytes such as soluble sugars and proline by upregulating genes involved in proline synthesis and transport and reducing the risk of oxidative injury caused by the accumulation of H_2O_2 and MDA in cells under salt stress (Yang et al., 2012). Furthermore, stress-mediated overexpression of MYB TF genes *ScMYBAS1–3* in sugarcane (*Saccharum officinarum*) also provides tolerance against water-deficit and salt stress conditions (Prabu and Prasad, 2012). Yang et al. (2012) have demonstrated the importance of the *R2R3-MYB* gene in rice. There is a positive correlation between OsMYB2 and salt, cold, and dehydration tolerance in rice.

13.3.5 AP2/ERFBP

Another group of plant-specific TFs is AP2/ERFBP family, which contains a highly conserved DNA-BD AP2/ethylene-responsive element-binding factor (ERF). In response to abiotic and biotic stresses such as salinity, UV, drought, pathogens, heat, and cold, ERF and AP2 (APETALA2) play crucial roles (Mizoi et al., 2012). Five types of drought-responsive element-binding (DREBs) named DREB1A, DREB B, DREB C, DREB D, and DREB 2A are present. Moreover, overexpression of OsDREB1A in *Arabidopsis thaliana* led to salinity tolerance (Dubouzet et al., 2003).

13.3.6 WRKY

The function of the *Arabidopsis* WRKY33 transcription factor is reported in targeting the salt-responsive downstream genes, which help in the detoxification of ROS and regulation of LOX1 (lipoxygenase), GSTU11 (glutathione-*S*-transferase), and peroxidases. Zheng et al. (2013) described that ThWRKY4 mediated cellular protective mechanism against the toxic level of ROS under salinity. Studies showed that the group II transcription factor of WRKY, that is, WRKY58 increased salt tolerance. Besides this, WRKY25 and WRKY33 are two WRKY factors that have been shown to enhance salinity stress tolerance in *Arabidopsis thaliana* (Jiang and Deyholos, 2009). In addition to this, AREB1/ABF2 transcription factor expressions and stomatal closure were affected, indicating WRKY factors' contribution to ABA-dependent salinity and drought adaptation pathways (Ren et al., 2010).

13.4 TEMPERATURE STRESS TOLERANCE IN PLANTS

High- or low-temperature stress affects plants in many forms such as ROS generation at mitochondria and chloroplast, calcium ion influx and cytoskeletal reorganizations, protein denaturation, etc., through sensing the environmental pressure by plasmalemma followed by changing membrane fluidity. Immediately after recognizing stress, the cytoplasmic MAPK (mitogen-activated protein kinase) cascade pathway and chaperon signaling help to activate different TFs, especially the HSPs, to retaliate to the stress effect. In *Arabidopsis* mutants, apart from heat shock proteins (HSP32, HSP101, etc.), the ABA, ROS, and salicylic acid (SA) pathways are also actively involved in developing and maintaining acquired heat tolerance (Larkindale and Huang, 2005).

13.4.1 Heat Shock Factors

Plant thermotolerance is mainly modulated by HSFs, and the protective properties of HSPs are attributable to the network of the chaperone machinery, in which many chaperones work in concert (Wang et al., 2004). Many heat-inducible genes are attributed to having conserved heat shock elements (HSEs) located in the upstream (TATA box proximal 5′ flanking regions) of heat shock genes. The expression of heat shock genes in plants is also affected by a number of other sequence motifs. For example, the CCAAT box and AT-rich sequences. In higher eukaryotes, the activation of the HSF requires multiple steps to become active. In response to heat stress, HSF is transformed into its trimeric form. There is a high affinity of trimeric HSF for binding to HSEs, and it is primarily localized in the nucleus. As a result of interaction between HSF and HSP70 or other HSPs, HSF is believed to be activated by conformational changes involving monomer to trimer transition and nuclear targeting (Schöffl et al., 1999). For example, it has also been demonstrated that tomato MT-sHSPact as a molecular chaperone (Lui and Shono, 1999) exhibits thermotolerance in transformed tobacco.

In posttranscriptional processes, HSFs can change their activities. The CaM-binding protein kinase (CBK) was found to interact with HSFA1a (Liu et al., 2008). For example, *Arabidopsis* knockout lines of the *AtCBK3* gene exhibited impaired basal thermotolerance and defects in the transcription of HSP genes. Study shows that AtCBK3 might control the binding activity of HSFs to HSEs by phosphorylating AtHSFA1a. Like CAMTA (discussed later in cold stress response) for cold induction, CBK3 is a potential connection between calcium signaling and heat-induced transcription of HSFs. During pollen tube thermotolerance, TMS1 (thermosensitive male sterile 1) is essential as an Hsp40 homolog.

13.5 COLD RESPONSE

Upon undergoing a prior nonfreezing treatment, plants acquire freezing tolerance through cold acclimation (Thomashow, 1999). During cold acclimation, transcriptional reprogramming occurs to induce cold-regulated (COR) genes, some of which are responsible for producing cryoprotective molecules. A major component of this transcriptional regulation is the CBF [C-repeat (CRT)/dehydration-responsive element (DRE)-binding factor] genes, which encode AP2/ERF family TFs (Thomashow, 1999). It was seen that three CBF genes (*CBF1–3*) are induced in *Arabidopsis* when exposed to cold stress and confer more freezing tolerance when got overexpressed. The CBF proteins generally bind to the conserved CCGAC motif of CRT/DRE, present in the promoters of several COR genes, named the CBF regulon. The significance of the CBF pathway in cold tolerance is pragmatic in both monocots and dicots. Several natural variations in the CBF family also exist. Notably, the CBF3/DREB1A gene is majorly activated by the MYC bHLH transcription factor ICE1 (inducer of CBF expression 1) and its interacting protein AtMYB15 (Lissarre et al., 2010). The report says that ICE1 binds to an MYC element CANNTG in the upstream promoter of the *CBF3* gene. Another player, ICE2, an ICE1 homolog, is also involved in cold stress tolerance, probably through modulating the CBF1 expression (Fursova et al., 2009).

Additionally, HOS1 (*High Expression of Osmotically Responsive Genes* 1), an E3 ubiquitin ligase and negative regulator of ICE1 is responsible for the ubiquitination and degradation of ICE1 through proteosomal degradation. In contrast, SIZ1, a SUMO E3 ligase, is a positive regulator of ICE1 (Miura et al., 2007). The siz1 mutant is reportedly defective in cold induction of the CBF and COR genes and has an increased chilling and freezing sensitivity. SIZ1 may post-translationally increase the CBF induction by sumoylating ICE1 to antagonize ubiquitination mediated by HOS1.

Indeed, SIZ1 expression led to both an increased level of sumoylation and a decreased level of polyubiquitination of ICE1. Further, a sumoylation-defective mutant of ICE1 represses cold induction of CBF3 and makes the plant freezing sensitive. Both ICE1 and SIZ1 have wider roles than merely regulating CBF expression. Interestingly, ICE1 has a role in stomata development also. SIZ1, being the only SUMO E3 ligase in *Arabidopsis*, also plays different roles including thermotolerance, SA accumulation, drought resistance, flowering time determination, etc. pertaining to plant growth and development (Miura et al., 2007).

The calmodulin-binding transcription activators, CAMTA, activate the *CBF2* gene. Additionally, the CM2 element, which matches the CG-1 element (CGCG) and is recognized and bound by the CG-1 domain present in the CAMTA proteins, is also responsible for the activation of *CBF2*. The increased concentration of calcium in the cytosol is an early event of cold response that makes a connection between calcium signaling and cold-regulated gene expression. Obviously, both ICE1 and CAMTA-binding sites are present in the promoter region of the CBF2 gene. Interesting to note that, the acclimatization to cold is not merely just dependent on the CBF gene activation. Freezing tolerance is controlled at numerous levels. Reportedly, the loss of function of the TFs HOS9 and HOS10 was found to be stronger or earlier activation of CBF and they may play a general inhibitory role in COR gene expression. Another independent regulator of the COR genes is SFR6 (SENSITIVE TO FREEZING 6) but not involved in CBF induction. SFR6 works downstream of the CBFs, as it is essential for CBF1 and CBF2 overexpression to trigger COR genes. The *SFR6* gene encodes a new protein that is found in the nucleus, implying that it may directly affect gene expression to influence a variety of responses. A study taking a set of 50 *Arabidopsis* accessions were performed considering the natural variations of plants revealed the complexity of freezing tolerance as the expression levels of CBF and COR genes do not correlate (McKhann et al., 2008). The CBF pathway is also get stimulated by signals other than cold (4°C). A sharp decrease of temperature from 20°C to 17°C,

14°C, 12°C, 10°C, or 8°C activates similar responses at the level of transcripts, enzyme activities, and metabolites (Usadel et al., 2008). Interestingly, these small decreases in temperature in the non-extreme range can trigger the expression of most of the genes induced majorly by cold. As a result, the CBF pathway is activated in response to minor temperature drops, which could be an adaptive strategy for plants to prepare for high temperatures. More fascinatingly, light also plays a crucial role in CBF expression (Franklin and Whitelam, 2007). CBF and downstream COR genes are upregulated at 16°C by a low red to far-red (R: FR) ratio, but not at 22°C. CBF expression at 16°C remarkably gets elevated in response to low R:FR levels. This link between light quality and cold response shows that plants may be able to integrate many environmental signals in order to predict seasonal changes and activate adaptive responses.

13.6 HEAVY METAL STRESS TOLERANCE IN PLANTS

Based on density (>5 g/cm^3), 53 d-block elements have been categorized as 'heavy metals' (HMs). Cadmium (Cd), lead (Pb), nickel (Ni), silver (Ag), zinc (Zn), cobalt (Co), chromium (Cr), iron (Fe), arsenic (As), and mercury (Hg) are some of the most frequent heavy metals present in the environment (Nagajyoti et al., 2010). Some HMs, for example, Fe, Zn, Mn, and Cu, are essential for vital metabolic processes of plants, such as co-factors of enzymes, redox homeostasis, biochemical reactions, etc. in lower concentrations (Rai et al., 2021). However, they are poisonous to plants in supra-optimal quantities. On the other hand, HMs such as As, Pb, and Cd have no apparent functions in plants and are toxic for cellular function.

The negative potential of the plasma membrane facilitates metal ion absorption in cells. Metal ions in the soil bind to the root cell wall during HM uptake, and subsequently, the HMs are absorbed with the help of transporters in the plasma membrane. Plants produce a variety of chelating agents and undergo various redox processes to facilitate metal ion intake. Different gene families are implicated in HMs transport from the extracellular environment to the cytoplasm and further sequestration in the vacuole (Table 13.1).

TABLE 13.1
List of Major Heavy Metal Transporters and Their Cellular Localization

Plasma Membrane	Major Transporter Family	HMs Transported	References
	ZIP family (Zn-regulated transporters)	Uptake of Zn^{2+} and Cu^{2+}	Guerinot (2000)
	HMA (heavy metal-associated domain) family	Gr. A – Pb/Co/Zn/Cd Gr. B – Ag/Cu	Senoura et al. (2017)
	YSL (yellow stripe 1-like) family	Fe, Cu-NA complex	González-Guerrero et al. (2016)
	NRAMP (natural resistance-associated macrophage protein)	Zn, Fe, Mn, Cu, Al, Ni, Cd, Co, Pb	Nevo and Nelson (2006)
	IRT (iron-regulated transporter)	Fe, Zn, Cd	Lee and An (2009)
	Phosphate transporter (PHT)	As	Wu et al. (2011); Zvobgo et al. (2018)
		Tonoplast	
	CDF family or MTP (metal-tolerance proteins)	Zn^{2+}, Mn^{2+}, Fe^{2+}, Cd^{2+}, Co^{2+}, and Ni^{2+}	Ricachenevsky et al. (2013)
	Vacuolar iron transporters (VIT) and VIT-likes (VTL	Iron, manganese, and zinc into the vacuole	Zhang et al. (2012)

13.6.1 MAPK Signaling in HMs Stress

In the MAPK cascade, when a stress signal is perceived, MAPKKKs (MAPK kinase kinases) stimulate MAPKKs, which activate MAPKs that mediate the phosphorylation and posttranslational modification of target proteins, mostly TFs. These TFs activate genes for metal transport, sequestration, and other defense-related compounds. The most well-known MAPKs, MPK3, and MPK6 are known to be expressed and activated by a broad spectrum of metals. Cadmium (Cd) and copper (Cu) are demonstrated to have a significant impact on these MAPKs in various animals. Upregulation of MPK3 and MPK6 in heavy metal stress provides insight into their role in metal homeostasis, as they either regulate downstream metal transporters or chelators in response to Cd and Cu.

13.6.2 Calcium Signaling in HMs Stress

Calcium signaling is an important phenomenon common in most stress-signaling pathways. Calcium ions (Ca^{2+}) are key signaling molecules that also play a role in signaling pathways in response to HM stressors. Cadmium (Cd) has been shown to reduce Ca^{2+} ion influx by competing with Ca^{2+} ions for uptake, increasing cytoplasmic Cd^{2+}, which inhibits root growth (Wilkins et al., 2016). The cytosolic free Ca^{2+} concentration changes in response to different stress stimuli and initiate a complex cascade of signaling (Rudd and Franklin-Tong, 2001). Ca^{2+} act as a secondary messenger in external stimuli. Calmodulin, along with other Ca-binding proteins, functions as a Ca^{2+} sensor, interacting with other proteins and regulating their activity in response to Ca binding (McAinsh and Pittman, 2009). Cd having similar physiochemical properties to Ca interferes in Ca^{2+} signaling (Choong et al., 2014). The Ca^{2+}/calmodulin system also plays a role in toxicity caused by heavy metals other than Cd, such as Pb^{2+} and nickel (Ni^{2+}) (Ahmad et al., 2015).

13.6.3 Plant Growth Regulators and HM Stress

HMs have a profound effect on plant growth and development. Plant hormones, such as gibberellins (GAs), cytokinins (CKs), auxins, ABA, and ethylene, as well as jasmonate (JA), brassinosteroids (BR), and SA, are known to regulate a variety of physiological processes and heavy metal stress resistance (Vriet et al., 2012; Peleg and Blumwald, 2011). Cd stress significantly reduced chlorophyll content and root elongation by altering the expression of auxin transporters, PINs and AUX1, and forming lateral roots in *Arabidopsis*. The ability of rice to withstand Cd poisoning is aided by ethylene signaling. Under normal conditions, blocking the ethylene signal-aided root elongation, but under Cd stress, it resulted in significant oxidative damage and root growth suppression. They further proved that EIN2 (ethylene insensitive 2) promotes flavonoid biosynthesis and peroxidase activity to induce ROS scavenging and form an apoplastic transport barrier to decrease Cd uptake. Exogenous application of selenium (Se) and auxin helps rice plants to mitigate arsenic (As) stress (Pandey and Gupta, 2015). It has been reported that SA provides protective action against Cu toxicity in cucumber and tobacco. According to Zhou et al. (2009), 0.2 mM of SA reduces mercury (Hg) toxicity in *Medicago sativa* by raising the activity of ascorbate peroxidase (APX), peroxidase (POD), and NADPH oxidase, as well as the levels of ascorbate, glutathione, and proline.

13.6.4 ROS and HMs

The term "reactive oxygen species" (ROS) refers to oxygen-containing molecules with a higher chemical reactivity than O_2. Singlet oxygen ($^1O_2$), superoxide anion (O_2^-), hydrogen peroxide (H_2O_2), and hydroxyl radical ($HO^·$) are the most common ROS in plants (Waszczak et al., 2018). ROS are produced during normal metabolic processes, electron transport chain in protoplast, mitochondria, peroxisome, and apoplastic spaces. Cu, Cd, Fe, and Zn are redox-active heavy metals that can cause ROS generation either directly by participating in Haber–Weiss and Fenton processes or indirectly by blocking enzymes in the cellular antioxidant defense network (Bhaduri and Fulekar, 2012). ROS can cause damage to biological molecules such as lipids, proteins, or nucleic acid. On the other hand, it can act as a signaling molecule involved in the physiological process, that is, cell growth, stomatal movement, and regulated by cellular antioxidant defense system enzymatic, for example, SOD, catalase (CAT), glutathione peroxidase (GPX), ascorbate peroxidase (APX), and nonenzymatic antioxidants, for example, glutathione, ascorbate, proline, and α-tocopherol. Studies have shown that heavy metal treatment increases ROS production, resulting in a significant rise in SOD, CAT, and APX activity. GPX activity was increased in rice roots under Cd stress emphasizing its role in ROS detoxification (Zhang and Ying, 2008). Oxidative stress caused by lead (Pb) toxicity also increases ROS activity (Verma and Dubey, 2003). Furthermore, chromium (Cr) (Zeng et al., 2012) also enhances GPX activity in rice. It has been reported that glutathione levels increased in rice seedlings positively with exposure to arsenic (As) stress.

13.6.5 Transcription Factors

When plants detect high levels of heavy metals, signaling networks are activated, resulting in the activation of several metal-sensitive TFs. These TFs assist the plant in coping with stress by regulating the expression of metal-responsive and other stress-related genes. TFs such as WRKY (containing a conserved WRKYGQK domain and a zinc finger-like motif), MYB (myeloblastosis), ZAT (C2H2-type zinc finger transcription factor), AP2 (activator protein 2), bZIP (basic region leucine zipper), ERF (ethylene-responsive factor), NAC, and DREB protein are reported to be involved in the HMs' stress responses. OsDREB1B, OsDREB1A, and WRKY09 TFs were found to be upregulated in rice roots under Cd stress (Ogawa et al., 2009). MYB TFs, such as MYB4, MYB10, MYB72, and bHLH protein, such as ORG2, ORG3, and bHLH100, are highly expressed during Cd and Zn metal stress (van de Mortel et al., 2008). They further showed that MYB72 loss-of-function mutants of *Arabidopsis* was more sensitive to excess Zn or iron (Fe) deficiency compared to wild type. At the same time, *Arabidopsis* lines overexpressing bHLH100 provided better tolerance to high Zn and nickel (Ni), confirming their role in metal homeostasis in *Arabidopsis.* Basic domain leucine zipper (bZIP) TFs are a major family of TFs that controls a key role in growth, development, and stress signaling in plants. The *Brassica juncea* BjCdR15, a bZIP transcription found to regulate cadmium (Cd) uptake by roots, long-distance transport, and accumulation in shoots to confer cadmium (Cd) tolerance in transgenic plants (Farinati et al., 2010). A bZIP transcription factor of Ramie BnbZIP2 was involved in drought, salt, and HM stress response; specifically, during germination, overexpression of BnbZIP2 makes the plant more sensitive toward Cd (Huang et al., 2016). WRKY TFs bind to the W-box element in the promoter of target genes and regulate plants' biotic and abiotic stress responses. We highly upregulated upon short-term Cu exposures to *Arabidopsis* root WRKY family TFs, WRKY22, WRKY25, and WRKY29. In contrast, only the expression of WRKY25 and WRKY29 was affected upon exposure to Cd for 24 hours (Opdenakker et al., 2012). Transgenic tobacco lines overexpressing ZmbHLH105 showed improved Mn tolerance by significantly increasing antioxidant enzymes and reduced ROS generation.

13.7 FUTURE PROSPECTS

Plants' response to abiotic stresses comprising drought, salinity, temperature, heavy metal, etc. is a complex process that involves many changes from morphological to molecular level. Under abiotic stress many TFs work both independently and collaboratively, generating a web of interactions. Due to intensive agricultural practices and global warming, crop production faces challenges in meeting the demands of the increasing global population. TFs are potential candidates for engineering stress-tolerant plants, as a single transcription factor can modulate a large set of genes. bZIPs, bHLHs, NACs, AP2/ERFs, MYBs, and WRKYs are among the stress-responsive TFs studied for their role in modulating plant response and tolerance to abiotic stress. Other families of plants, on the other hand, HD-ZIP and zinc finger TFs, for example, whose expressions are regulated under abiotic stress. Despite the fact that many questions about the molecular mechanisms underlying transcriptional and translational interactions between multiple TFs remain unanswered; these findings suggest that transcription modulation or manipulation is involved in plant stress response factors that would be a good strategy for increasing crop tolerance to abiotic stress. Numerous studies have identified and characterized the roles of stress-responsive TFs in plant response and tolerance to abiotic stress, but most of these studies were performed in laboratory conditions. Stress-responsive TFs could be used to improve crop abiotic stress tolerance if they are functionally validated in the field.

REFERENCES

Ahmad, A., Hadi, F., and Ali, N. 2015. Effective phytoextraction of cadmium (Cd) with increasing concentration of total phenolics and free proline in *Cannabis sativa* (L) plant under various treatments of fertilisers, plant growth regulators and sodium salt. *International Journal of Phytoremediation* 17(1): 56–65.

Aldesuquy, H., Baka, Z., and Mickky, B. 2014. Kinetin and spermine mediated induction of salt tolerance in wheat plants: Leaf area, photosynthesis and chloroplast ultrastructure of flag leaf at ear emergence. *Egyptian Journal of Basic and Applied Sciences* 1(2): 77–87.

Bakshi, M., and Oelmüller, R. 2014. WRKY transcription factors: Jack of many trades in plants. *Plant Signaling & Behavior* 9(2): e27700.

Bhaduri, A. M., and Fulekar, M. H. 2012. Antioxidant enzyme responses of plants to heavy metal stress. *Reviews in Environmental Science and Bio/Technology* 11(1): 55–69.

Botella, M. A., Rosado, A., Bressan, R. A., and Hasegawa, P. M. 2005. Plant adaptive responses to salinity stress. *Plant Abiotic Stress* 21: 38–70.

Brendolise, C., Espley, R. V., Lin-Wang, K., Laing, W., Peng, Y., McGhie, T., and Allan, A. C. 2017. Multiple copies of a simple MYB-binding site confers trans-regulation by specific flavonoid-related R2R3 MYBs in diverse species. *Frontiers in Plant Science* 8: 1864.

Campalans, A., Messeguer, R., Goday, A., and Pagès, M. 1999. Plant responses to drought, from ABA signal transduction events to the action of the induced proteins. *Plant Physiology and Biochemistry* 37(5): 327–340.

Choi, H. I., Hong, J. H., Ha, J. O., Kang, J. Y., and Kim, S. Y. 2000. ABFs, a family of ABA-responsive element binding factors. *Journal of Biological Chemistry* 275(3): 1723–1730.

Choong, G., Liu, Y., and Templeton, D. M. 2014. Interplay of calcium and cadmium in mediating cadmium toxicity. *Chemico-Biological Interactions* 211: 54–65.

Cominelli, E., Galbiati, M., Vavasseur, A., Conti, L., Sala, T., Vuylsteke, M., and Tonelli, C. 2005. A guard-cell-specific MYB transcription factor regulates stomatal movements and plant drought tolerance. *Current Biology* 15(13): 1196–1200.

Corrales, A. R., Carrillo, L., Lasierra, P., Nebauer, S. G., Dominguez-Figueroa, J., Renau-Morata, B., and Medina, J. 2017. Multifaceted role of cycling DOF factor 3 (CDF3) in the regulation of flowering time and abiotic stress responses in Arabidopsis. *Plant, Cell & Environment* 40(5): 748–764.

Dajic, Z. 2006. Salt stress. In: Madhava Rao, K., Raghavendra, A., Janardhan Reddy, K. (eds) *Physiology and Molecular Biology of Stress Tolerance in Plants*, 41–99. Dordrecht: Springer.

Danquah, A., de Zelicourt, A., Colcombet, J., and Hirt, H. 2014. The role of ABA and MAPK signaling pathways in plant abiotic stress responses. *Biotechnology Advances* 32(1): 40–52.

Ding, S., Cai, Z., Du, H., and Wang, H. 2019. Genome-wide analysis of TCP family genes in Zea mays L. identified a role for ZmTCP42 in drought tolerance. *International Journal of Molecular Sciences* 20(11): 2762.

Duan, M., Zhang, R., Zhu, F., Zhang, Z., Gou, L., Wen, J., and Wang, T. 2017. A lipid-anchored NAC transcription factor is translocated into the nucleus and activates glyoxalase I expression during drought stress. *The Plant Cell* 29(7): 1748–1772.

Espley, R. V., Brendolise, C., Chagne, D., Kutty-Amma, S., Green, S., Volz, R., and Allan, A. C. 2009. Multiple repeats of a promoter segment causes transcription factor autoregulation in red apples. *The Plant Cell* 21(1): 168–183.

Farinati, S., DalCorso, G., Varotto, S., and Furini, A. 2010. The Brassica juncea BjCdR15, an ortholog of Arabidopsis TGA3, is a regulator of cadmium uptake, transport and accumulation in shoots and confers cadmium tolerance in transgenic plants. *New Phytologist* 185(4): 964–978.

Franklin, K. A., and Whitelam, G. C. 2007. Light-quality regulation of freezing tolerance in Arabidopsis thaliana. *Nature Genetics* 39(11): 1410–1413.

Fursova, O. V., Pogorelko, G. V., and Tarasov, V. A. 2009. Identification of ICE2, a gene involved in cold acclimation which determines freezing tolerance in *Arabidopsis thaliana*. *Gene* 429(1–2): 98–103.
Gao, S. Q., Chen, M., Xu, Z. S., Zhao, C. P., Li, L., Xu, H. J., and Ma, Y. Z. 2011. The soybean GmbZIP1 transcription factor enhances multiple abiotic stress tolerances in transgenic plants. *Plant Molecular Biology* 75(6): 537–553.
Glenn, E. P., Brown, J. J., and Blumwald, E. 1999. Salt tolerance and crop potential of halophytes. *Critical Reviews in Plant Sciences* 18(2): 227–255.
Glover, J. N., and Harrison, S. C. 1995. Crystal structure of the heterodimeric bZIP transcription factor c-Fos–c-Jun bound to DNA. *Nature* 373(6511): 257–261.
González-Guerrero, M., Escudero, V., Saéz, Á., and Tejada-Jiménez, M. 2016. Transition metal transport in plants and associated endosymbionts: Arbuscular mycorrhizal fungi and rhizobia. *Frontiers in Plant Science* 7: 1088.
Guerinot, M. L. 2000. The ZIP family of metal transporters. *Biochimica et Biophysica Acta (BBA)-Biomembranes* 1465(1–2): 190–198.
Gull, A., Lone, A., and Wani N. 2019. Biotic and abiotic stresses in plants. doi: 10.5772/intechopen.85832.
Gupta, S., Malviya, N., Kushwaha, H., Nasim, J., Bisht, N. C., Singh, V. K., and Yadav, D. 2015. Insights into structural and functional diversity of Dof (DNA binding with one finger) transcription factor. *Planta* 241(3): 549–562.
Haake, V., Cook, D., Riechmann, J., Pineda, O., Thomashow, M. F., and Zhang, J. Z. 2002. Transcription factor CBF4 is a regulator of drought adaptation in Arabidopsis. *Plant Physiology* 130(2): 639–648.
Hartmann, L., Pedrotti, L., Weiste, C., Fekete, A., Schierstaedt, J., Göttler, J., and Dröge-Laser, W. 2015. Crosstalk between two bZIP signaling pathways orchestrates salt-induced metabolic reprogramming in Arabidopsis roots. *The Plant Cell* 27(8): 2244–2260.
Hasanuzzaman, M., Nahar, K., and Fujita, M. 2013. Plant response to salt stress and role of exogenous protectants to mitigate salt-induced damages. In: Ahmad, P., Azooz, M., Prasad, M. (eds).*Ecophysiology and Responses of Plants under Salt Stress*, 25–87. New York, NY: Springer.
Hu, H., Dai, M., Yao, J., Xiao, B., Li, X., Zhang, Q., and Xiong, L. 2006. Overexpressing a NAM, ATAF, and CUC (NAC) transcription factor enhances drought resistance and salt tolerance in rice. *Proceedings of the National Academy of Sciences* 103(35): 12987–12992.
Huang, C., Zhou, J., Jie, Y., Xing, H., Zhong, Y., Yu, W., She, W., Ma, Y., Liu, Z. and Zhang, Y. 2016. A ramie bZIP transcription factor BnbZIP2 is involved in drought, salt, and heavy metal stress response. *DNA and Cell Biology* 35(12): 776–786.
Huang, G. T., Ma, S. L., Bai, L. P., Zhang, L., Ma, H., Jia, P., and Guo, Z. F. 2012. Signal transduction during cold, salt, and drought stresses in plants. *Molecular Biology Reports* 39(2): 969–987.
Ishiguro, S., and Nakamura, K. 1994. Characterization of a cDNA encoding a novel DNA-binding protein, SPF1, that recognizes SP8 sequences in the 5′ upstream regions of genes coding for sporamin and β-amylase from sweet potato. *Molecular and General Genetics MGG* 244(6): 563–571
Jakoby, M., Weisshaar, B., Dröge-Laser, W., Vicente-Carbajosa, J., Tiedemann, J., Kroj, T., and Parcy, F. 2002. bZIP transcription factors in Arabidopsis. *Trends in Plant Science* 7(3): 106–111.
Jiang, J., Ma, S., Ye, N., Jiang, M., Cao, J., and Zhang, J. 2017. WRKY transcription factors in plant responses to stresses. *Journal of Integrative Plant Biology* 59(2): 86–101.
Jung, C., Seo, J. S., Han, S. W., Koo, Y. J., Kim, C. H., Song, S. I., and Cheong, J. J. 2008. Overexpression of AtMYB44 enhances stomatal closure to confer abiotic stress tolerance in transgenic Arabidopsis. *Plant Physiology* 146(2): 623.
Kalaji, H. M., Bosa, K., Kościelniak, J., and Żuk-Gołaszewska, K. 2011. Effects of salt stress on photosystem II efficiency and CO_2 assimilation of two Syrian barley landraces. *Environmental and Experimental Botany* 73: 64–72.
Kasuga, M., Liu, Q., Miura, S., Yamaguchi-Shinozaki, K., and Shinozaki, K. 1999. Improving plant drought, salt, and freezing tolerance by gene transfer of a single stress-inducible transcription factor. *Nature Biotechnology* 17(3): 287–291.
Kaur, G., and Asthir, B. J. B. P. 2015. Proline: A key player in plant abiotic stress tolerance. *Biologia Plantarum* 59(4): 609–619.
Kosugi, S., and Ohashi, Y. 2002. DNA binding and dimerization specificity and potential targets for the TCP protein family. *The Plant Journal* 30(3): 337–348.
Koyro, H. W., Khan, M. A., and Lieth, H. 2011. Halophytic crops: A resource for the futureto reduce the water crisis? *Emirates Journal of Food and Agriculture* 23(1): 1–16.
Larkindale, J., and Huang, B. 2005. Abscisic acid, salicylic acid, ethylene, and hydrogen peroxide affect thermotolerance and recovery for creeping bentgrass. *Plant Growth Regulation* 47(1): 17–28.

Lata, C., and Prasad, M. 2011. Role of DREBs in regulation of abiotic stress responses in plants. *Journal of Experimental Botany* 62(14): 4731–4748.

Lata, C., Yadav, A., and Prasad, M. 2011. Role of plant transcription factors in abiotic stress tolerance. *Abiotic Stress Response in Plants, INTECH Open Access Publishers* 10: 269–296.

Lee, S., and An, G. 2009. Over-expression of OsIRT1 leads to increased iron and zinc accumulations in rice. *Plant, Cell & Environment* 32(4): 408–416.

Lei, N., Yu, X., Li, S., Zeng, C., Zou, L., Liao, W., and Peng, M. 2017. Phylogeny and expression pattern analysis of TCP transcription factors in cassava seedlings exposed to cold and/or drought stress. *Scientific Reports* 7(1): 1–13.

Levitt, J. 1980. *Responses of Plant to Environmental Stress: Water, Radiation, Salt and Other Stresses.* New York, NY: Academic Press, 365.

Li, D., Fu, F., Zhang, H., and Song, F. 2015. Genome-wide systematic characterization of the bZIP transcriptional factor family in tomato (*Solanum lycopersicum* L.). *BMC Genomics* 16(1): 1–18.

Liao, Y., Zou, H. F., Wei, W., Hao, Y. J., Tian, A. G., Huang, J., and Chen, S. Y. 2008. Soybean GmbZIP44, GmbZIP62 and GmbZIP78 genes function as negative regulator of ABA signaling and confer salt and freezing tolerance in transgenic Arabidopsis. *Planta* 228(2): 225–240.

Lissarre, M., Ohta, M., Sato, A., and Miura, K. 2010. Cold-responsive gene regulation during cold acclimation in plants. *Plant Signaling & Behavior* 5(8): 948–952.

Liu, H. T., Gao, F., Li, G. L., Han, J. L., Liu, D. L., Sun, D. Y., and Zhou, R. G. 2008. The calmodulin-binding protein kinase 3 is part of heat-shock signal transduction in *Arabidopsis thaliana. The Plant Journal* 55(5): 760–773.

Liu, X., Song, Y., Xing, F., Wang, N., Wen, F., and Zhu, C. 2016. GhWRKY25, a group I WRKY gene from cotton, confers differential tolerance to abiotic and biotic stresses in transgenic *Nicotiana benthamiana. Protoplasma* 253(5): 1265–1281.

Liu, Y., Sun, J., and Wu, Y. 2016. Arabidopsis ATAF1 enhances the tolerance to salt stress and ABA in transgenic rice. *Journal of Plant Research* 129(5): 955–962.

Lui, J., and Shono, M. 1999. Characterization of mitochondria-located small heat shock protein from tomato (*Lycopersicon esculentum*). *Plant and Cell Physiology* 40(12): 1297–1304.

Mao, H., Wang, H., Liu, S., Li, Z., Yang, X., Yan, J., and Qin, F. 2015. A transposable element in a NAC gene is associated with drought tolerance in maize seedlings. *Nature Communications* 6(1): 1–13.

Marcum, K. B., and Murdoch, C. L. 1992. Salt tolerance of the coastal salt marsh grass, *Sporobolus virginicus* (L.) Kunth. *New Phytologist* 120(2): 281–288.

McAinsh, M. R., and Pittman, J. K. 2009. Shaping the calcium signature. *New Phytologist* 181(2): 275–294.

McKhann, H. I., Gery, C., Bérard, A., Lévêque, S., Zuther, E., Hincha, D. K., and Téoulé, E. 2008. Natural variation in CBF gene sequence, gene expression and freezing tolerance in the Versailles core collection of *Arabidopsis thaliana. BMC Plant Biology* 8(1): 1–18.

Miura, K., Jin, J. B., and Hasegawa, P. M. 2007. Sumoylation, a post-translational regulatory process in plants. *Current Opinion in Plant Biology* 10(5): 495–502.

Mizoi, J., Shinozaki, K., and Yamaguchi-Shinozaki, K. 2012. AP2/ERF family transcription factors in plant abiotic stress responses. *Biochimica et Biophysica Acta (BBA)-Gene Regulatory Mechanisms* 1819(2): 86–96.

Mosa, K., Ismail, A., and Helmy, M. 2017. Introduction to plant stresses. In *Plant Stress Tolerance.* Cham: Springer. doi: 10.1007/978-3-319-59379-1_1.

Mukhopadhyay, P., and Tyagi, A. K. 2015. OsTCP19 influences developmental and abiotic stress signaling by modulatingABI4-mediated pathways. *Scientific Reports* 5(1): 1–12.

Munns, R., James, R. A., and Läuchli, A. 2006. Approaches to increasing the salt tolerance of wheat and other cereals. *Journal of Experimental Botany* 57(5): 1025–1043.

Nagajyoti, P. C., Lee, K. D., and Sreekanth, T. V. M. 2010. Heavy metals, occurrence and toxicity for plants: A review. *Environmental Chemistry Letters* 8(3): 199–216.

Nakashima, K., Tran, L. S. P., Van Nguyen, D., Fujita, M., Maruyama, K., Todaka, D., and Yamaguchi-Shinozaki, K. 2007. Functional analysis of a NAC-type transcription factor OsNAC6 involved in abiotic and biotic stress-responsive gene expression in rice. *The Plant Journal* 51(4): 617–630.

Nevo, Y., and Nelson, N. 2006. The NRAMP family of metal-ion transporters. *Biochimica et Biophysica Acta (BBA)-Molecular Cell Research* 1763(7): 609–620.

Nishizawa, A., Yabuta, Y., and Shigeoka, S. 2008. Galactinol and raffinose constitute a novel function to protect plants from oxidative damage. *Plant Physiology* 147(3): 1251–1263.

Ogawa, I., Nakanishi, H., Mori, S., and Nishizawa, N. K. 2009. Time course analysis of gene regulation under cadmium stress in rice. *Plant and Soil* 325(1): 97–108.

Opdenakker, K., Remans, T., Keunen, E., Vangronsveld, J., and Cuypers, A. 2012. Exposure of *Arabidopsis thaliana* to Cd or Cu excess leads to oxidative stress mediated alterations in MAPKinase transcript levels. *Environmental and Experimental Botany* 83: 53–61.

Pandey, C., and Gupta, M. 2015. Selenium and auxin mitigates arsenic stress in rice (*Oryza sativa* L.) by combining the role of stress indicators, modulators and genotoxicity assay. *Journal of Hazardous Materials* 287: 384–391.

Panikulangara, T. J., Eggers-Schumacher, G., Wunderlich, M., Stransky, H., and Schöffl, F. 2004. Galactinol synthase1. A novel heat shock factor target gene responsible for heat-induced synthesis of raffinose family oligosaccharides in Arabidopsis. *Plant Physiology* 136(2): 3148–3158.

Parida, A. K., and Das, A. B. 2005. Salt tolerance and salinity effects on plants: A review. *Ecotoxicology and Environmental Safety* 60(3): 324–349.

Parihar, P., Singh, S., Singh, R., Singh, V. P., and Prasad, S. M. 2015. Effect of salinity stress on plants and its tolerance strategies: A review. *Environmental Science and Pollution Research* 22(6): 4056–4075.

Peleg, Z., and Blumwald, E. 2011. Hormone balance and abiotic stress tolerance in crop plants. *Current Opinion in Plant Biology* 14(3): 290–295.

Prabu, G., and Theertha Prasad, D. 2012. Functional characterization of sugarcane MYB transcription factor gene promoter (PScMYBAS1) in response to abiotic stresses and hormones. *Plant Cell Reports* 31(4): 661–669.

Puranik, S., Sahu, P. P., Srivastava, P. S., and Prasad, M. 2012. NAC proteins: Regulation and role in stress tolerance. *Trends in Plant Science* 17(6): 369–381.

Rai, S., Singh, P. K., Mankotia, S., Swain, J., and Satbhai, S. B. 2021. Iron homeostasis in plants and its crosstalk with copper, zinc, and manganese. *Plant Stress* 1: 100008.

Raineri, J., Ribichich, K. F., and Chan, R. L. 2015. The sunflower transcription factor HaWRKY76 confers drought and flood tolerance to *Arabidopsis thaliana* plants without yield penalty. *Plant Cell Reports* 34(12): 2065–2080.

Redillas, M. C., Jeong, J. S., Kim, Y. S., Jung, H., Bang, S. W., Choi, Y. D., and Kim, J. K. 2012. The overexpression of OsNAC9 alters the root architecture of rice plants enhancing drought resistance and grain yield under field conditions. *Plant Biotechnology Journal* 10(7): 792–805.

Redondo-Gómez, Susana. 2013. Abiotic and biotic stress tolerance in plants. In Rout, G. and Das, A. (eds), *Molecular Stress Physiology of Plants*. Springer. doi: 10.1007/978-81-322-0807-5_1.

Ren, X., Chen, Z., Liu, Y., Zhang, H., Zhang, M., Liu, Q., and Gong, Z. 2010. ABO3, a WRKY transcription factor, mediates plant responses to abscisic acid and drought tolerance in Arabidopsis. *The Plant Journal* 63(3): 417–429.

Ricachenevsky, F. K., Menguer, P. K., Sperotto, R. A., Williams, L. E., and Fett, J. P. 2013. Roles of plant metal tolerance proteins (MTP) in metal storage and potential use in biofortification strategies. *Frontiers in Plant Science* 4: 144.

Rudd, J. J., and Franklin-Tong, V. E. 2001. Unravelling response-specificity in Ca2+ signalling pathways in plant cells. *New Phytologist* 151(1): 7–33.

Sakuma, Y., Liu, Q., Dubouzet, J. G., Abe, H., Shinozaki, K., and Yamaguchi-Shinozaki, K. 2002. DNA-binding specificity of the ERF/AP2 domain of Arabidopsis DREBs, transcription factors involved in dehydration-and cold-inducible gene expression. *Biochemical and Biophysical Research Communications* 290(3): 998–1009.

Salvi, P., Kamble, N. U., and Majee, M. 2020. Ectopic over-expression of ABA-responsive Chickpea galactinol synthase (CaGolS) gene results in improved tolerance to dehydration stress by modulating ROS scavenging. *Environmental and Experimental Botany* 171: 103957.

Senoura, T., Sakashita, E., Kobayashi, T., Takahashi, M., Aung, M. S., Masuda, H., Nakanishi, H. and Nishizawa, N. K. 2017. The iron-chelate transporter OsYSL9 plays a role in iron distribution in developing rice grains. *Plant Molecular Biology* 95(4): 375–387.

ShaValli Khan, P.S., Nagamallaiah, G.V., Dhanunjay Rao, M., Sergeant, K., Hausman, J.F. 2014. Abiotic stress tolerance in plants: Insights from proteomics. *Emerging Technologies and Management of Crop Stress Tolerance* 2: 23–68.

Shigyo, M., Tabei, N., Yoneyama, T., and Yanagisawa, S. 2007. Evolutionary processes during the formation of the plant-specific Dof transcription factor family. *Plant and Cell Physiology* 48(1): 179–185.

Singh, M., Kumar, J., Singh, S., Singh, V. P., and Prasad, S. M. 2015. Roles of osmoprotectants in improving salinity and drought tolerance in plants: A review. *Reviews in Environmental Science and Bio/Technology* 14(3): 407–426.

Song, Y., Chen, L., Zhang, L., and Yu, D. 2010. Overexpression of OsWRKY72 gene interferes in the abscisic acid signal and auxin transport pathway of Arabidopsis. *Journal of Biosciences* 35(3): 459–471.

Takasaki, H., Maruyama, K., Kidokoro, S., Ito, Y., Fujita, Y., Shinozaki, K., and Nakashima, K. 2010. The abiotic stress-responsive NAC-type transcription factor OsNAC5 regulates stress-inducible genes and stress tolerance in rice. *Molecular Genetics and Genomics* 284(3): 173–183.

Thirumalaikumar, V. P., Devkar, V., Mehterov, N., Ali, S., Ozgur, R., Turkan, I., and Balazadeh, S. 2018. NAC transcription factor JUNGBRUNNEN 1 enhances drought tolerance in tomato. *Plant Biotechnology Journal* 16(2): 354–366.

Thomashow, M. F. 1999. Plant cold acclimation: Freezing tolerance genes and regulatory mechanisms. *Annual Review of Plant Biology* 50(1): 571–599.

Umar, O. B., Ranti, L. A., Abdulbaki, A. S., Bola, A. L., Abdulhamid, A. K., Biola, M. R., and Victor, K. O. 2021. Stresses in plants: Biotic and abiotic. In *Current Trends in Wheat Research., Intechopen*: 1–8.

Usadel, B., Blaesing, O. E., Gibon, Y., Poree, F., Hoehne, M., Guenter, M., and Stitt, M. 2008. Multilevel genomic analysis of the response of transcripts, enzyme activities and metabolites in Arabidopsis rosettes to a progressive decrease of temperature in the non-freezing range. *Plant, Cell & Environment* 31(4): 518–547.

van de Mortel, J. E., Schat, H., Moerland, P. D., Van Themaat, E. V. L., Van Der Ent, S. J. O. E. R. D., Blankestijn, H., Ghandilyan, A., Tsiatsiani, S. and Aarts, M. G. 2008. Expression differences for genes involved in lignin, glutathione and sulphate metabolism in response to cadmium in *Arabidopsis thaliana* and the related Zn/Cd- hyperaccumulator *Thlaspi caerulescens*. *Plant Cell & Environment* 31: 301–324.

Verma, S., and Dubey, R. S. 2003. Lead toxicity induces lipid peroxidation and alters the activities of antioxidant enzymes in growing rice plants. *Plant Science* 164(4): 645–655.

Vriet, C., Russinova, E., and Reuzeau, C. 2012. Boosting crop yields with plant steroids. *The Plant Cell* 24(3): 842–857.

Wang, W., Vinocur, B., Shoseyov, O., and Altman, A. 2004. Role of plant heat-shock proteins and molecular chaperones in the abiotic stress response. *Trends in Plant Science* 9(5): 244–252.

Wang, Wang-Xia, Vinocur, Basia, and Altman, Arie. 2003. Plant responses to drought, salinity and extreme temperatures: Towards genetic engineering for stress tolerance. *Planta*. 218: 1–14. doi: 10.1007/s00425-003-1105-5.

Wang, X., Chen, X., Liu, Y., Gao, H., Wang, Z., and Sun, G. 2011. CkDREB gene in *Caragana korshinskii* is involved in the regulation of stress response to multiple abiotic stresses as an AP2/EREBP transcription factor. *Molecular Biology Reports* 38(4): 2801–2811.

Wang, X., Zeng, J., Li, Y., Rong, X., Sun, J., Sun, T., and He, G. 2015. Expression of TaWRKY44, a wheat WRKY gene, in transgenic tobacco confers multiple abiotic stress tolerances. *Frontiers in Plant Science* 6: 615.

Wang, Y., Gao, C., Liang, Y., Wang, C., Yang, C., and Liu, G. 2010. A novel bZIP gene from *Tamarixhispida* mediates physiological responses to salt stress in tobacco plants. *Journal of Plant Physiology* 167(3): 222–230.

Waszczak, C., Carmody, M., and Kangasjärvi, J. 2018. Reactive oxygen species in plant signaling. *Annual Review of Plant Biology* 69: 209–236.

Wilkins, K. A., Matthus, E., Swarbreck, S. M., and Davies, J. M. 2016. Calcium-mediated abiotic stress signaling in roots. *Frontiers in Plant Science* 7: 1296

Winicov, I., and Bastola, D. R. 1999. Transgenic overexpression of the transcription factor Alfin1 enhances expression of the endogenous MsPRP2 gene in alfalfa and improves salinity tolerance of the plants. *Plant Physiology* 120(2): 473–480.

Wu, A., Allu, A. D., Garapati, P., Siddiqui, H., Dortay, H., Zanor, M. I., and Balazadeh, S. 2012. JUNGBRUNNEN1, a reactive oxygen species–responsive NAC transcription factor, regulates longevity in Arabidopsis. *The Plant Cell* 24(2): 482–506.

Wu, Z., Ren, H., McGrath, S. P., Wu, P., and Zhao, F. J. 2011. Investigating the contribution of the phosphate transport pathway to arsenic accumulation in rice. *Plant Physiology* 157(1): 498–508.

Xiang, Y., Tang, N., Du, H., Ye, H., and Xiong, L. 2008. Characterization of OsbZIP23 as a key player of the basic leucine zipper transcription factor family for conferring abscisic acid sensitivity and salinity and drought tolerance in rice. *Plant Physiology* 148(4): 1938–1952.

Xiong, H., Li, J., Liu, P., Duan, J., Zhao, Y., Guo, X., and Li, Z. 2014. Overexpression of OsMYB48-1, a novel MYB-related transcription factor, enhances drought and salinity tolerance in rice. *PLoS One* 9(3): e92913.

Yan, L., Liu, Z. Q., Xu, Y. H., Lu, K., Wang, X. F., and Zhang, D. P. 2013. Auto-and cross-repression of three Arabidopsis WRKY transcription factors WRKY18, WRKY40, and WRKY60 negatively involved in ABA signaling. *Journal of Plant Growth Regulation* 32(2): 399–416.

Yanagisawa, S. 2002. The Dof family of plant transcription factors. *Trends in Plant Science* 7(12): 555–560.

Yang, A., Dai, X., and Zhang, W. H. 2012. A R2R3-type MYB gene, OsMYB2, is involved in salt, cold, and dehydration tolerance in rice. *Journal of Experimental Botany* 63(7): 2541–2556.

Yang, S. D., Seo, P. J., Yoon, H. K., and Park, C. M. 2011. The Arabidopsis NAC transcription factor VNI2 integrates abscisic acid signals into leaf senescence via the COR/RD genes. *The Plant Cell* 23(6): 2155–2168.

Yoon, S., Lee, D. K., Yu, I. J., Kim, Y. S., Choi, Y. D., and Kim, J. K. 2017. Overexpression of the OsbZIP66 transcription factor enhances drought tolerance of rice plants. *Plant Biotechnology Reports* 11(1): 53–62.

Zeng, F., Qiu, B., Wu, X., Niu, S., Wu, F., and Zhang, G. 2012. Glutathione-mediated alleviation of chromium toxicity in rice plants. *Biological Trace Element Research* 148(2): 255–263.

Zhang, C. H., and Ying, G. E. 2008. Response of glutathione and glutathione S-transferase in rice seedlings exposed to cadmium stress. *Rice Science* 15(1): 73–76.

Zhang, Y., Xu, Y. H., Yi, H. Y., and Gong, J. M. 2012. Vacuolar membrane transporters OsVIT1 and OsVIT2 modulate iron translocation between flag leaves and seeds in rice. *The Plant Journal* 72(3): 400–410.

Zhang, Z., Wang, J., Zhang, R., and Huang, R. 2012. The ethylene response factor AtERF98 enhances tolerance to salt through the transcriptional activation of ascorbic acid synthesis in Arabidopsis. *The Plant Journal* 71(2): 273–287.

Zheng, L., Liu, G., Meng, X., Liu, Y., Ji, X., Li, Y., and Wang, Y. 2013. A WRKY gene from *Tamarixhispida*, ThWRKY4, mediates abiotic stress responses by modulating reactive oxygen species and expression of stress-responsive genes. *Plant Molecular Biology* 82(4): 303–320.

Zheng, X., Chen, B., Lu, G., and Han, B. 2009. Overexpression of a NAC transcription factor enhances rice drought and salt tolerance. *Biochemical and Biophysical Research Communications* 379(4): 985–989.

Zhou, M., Li, D., Li, Z., Hu, Q., Yang, C., Zhu, L., and Luo, H. 2013. Constitutive expression of a miR319 gene alters plant development and enhances salt and drought tolerance in transgenic creeping bentgrass. *Plant Physiology* 161(3): 1375–1391.

Zhou, Q. Y., Tian, A. G., Zou, H. F., Xie, Z. M., Lei, G., Huang, J., and Chen, S. Y. 2008. Soybean WRKY-type transcription factor genes, GmWRKY13, GmWRKY21, and GmWRKY54, confer differential tolerance to abiotic stresses in transgenic Arabidopsis plants. *Plant Biotechnology Journal* 6(5): 486–503.

Zhou, Z. S., Guo, K., Elbaz, A. A., and Yang, Z. M. 2009. Salicylic acid alleviates mercury toxicity by preventing oxidative stress in roots of *Medicago sativa. Environmental and Experimental Botany* 65(1): 27–34.

Zhu, J. K. 2002. Salt and drought stress signal transduction in plants. *Annual Review of Plant Biology* 53(1): 247–273.

Zvobgo, G., Lwalaba, J.L.W., Sagonda, T., Mapodzeke, J. M., Muhammad, N., Shamsi, I. H., and Zhang, G. 2018. Phosphate alleviates arsenate toxicity by altering expression of phosphate transporters in the tolerant barley genotypes. *Ecotoxicology and Environmental Safety* 147: 832–839.

14 Molecular Understanding of Seed Priming Approaches for Combined and Multiple Stresses Tolerance

Maryam Razzaq, Umer Hayat,
Muhammad Kamran, and Irfan Afzal
University of Agriculture

CONTENTS

DOI: 10.1201/9781003300564-14

14.1 INTRODUCTION

Crop plants face a varied range of abiotic stresses successively or simultaneously in the field. The response to stress by plants occurs as a result of the activation of a variety of cellular, molecular, and morphological mechanisms that promote stress resilience (Hossain et al., 2020). Plants can acquire resistance after an episode of mild stress shocks and develop tolerance against upcoming stresses. Such mild stress shocks permit plants to acclimatize to a plethora of recurring field stresses (Hossain et al., 2020). Scientists have termed such phenomena as multiple stress tolerance, cross-tolerance, combined stress tolerance, or cross-resistance (Llorens et al., 2020).

Stress memory in response to mild stress plays a vital role in the induction of acclimatizing resilience against each specific stress in plants. Such memory can help a plant to perform vigorously against later stress events. The most common technique to develop stress memory in the plants to attain vigor even under stressful environments is considered seed priming (Llorens et al., 2020). Seed priming includes the manipulation of seed water relations, to complete the metabolic and cellular events associated with germination. However, the definition of seed priming against multiple or combined abiotic stresses could be quite different from the sole imbibition perspective.

Priming usually combats stress in three distinct phases, that is, pre-challenge priming stage, stress induction priming stage, and *trans*-generational priming stage (Llorens et al., 2020). Some epigenetic studies also declared that priming induced changes that could be heritable (Hossain et al., 2020). The major attribute of priming is the lag phase, which can be distinguished as the priming activation phase to the secondary stress exposures. Priming stimulus is very crucial for defense-related mechanisms against secondary stresses and usually triggers changes in seeds at the molecular, biochemical, and epigenetic levels. However, the back of a fag packet calculations from complete defensive response shows that the lag phase induces changes at a very low fitness cost. In this manner, any kind of stress that occurs after another via looming behind the lag phase would be defined as 'priming-induced multiple stress tolerance or cross-tolerance'.

There are various priming approaches and methods that have been studied in seed species against multiple abiotic stress (MAS) tolerances. *Cis-* and *trans*-seed priming approach with conjugation of foliar spray with γ-aminobutyric acid (GABA)-enhanced antioxidant enzymatic activity, sucrose synthesis, chlorophyll concentrations, as well as photosynthetic efficiency in plants and saplings against heat stress (Bhardwaj et al., 2021). Hussain et al. (2020) reported a case of treated wheat seeds at high temperatures and identified 1,430 gene probes in the primed seed that were involved in different transcription, signal transduction, defense, and protein destination. *Lysine-specific histone demethylase 1 (LSD1)* was the major cause of histone demethylation, and responsible for epigenetic modification. Similarly, examples of many *trans*-priming approaches include the wide use of chemical elicitors such as probendazole and hexanoic acid against the multiple or cross-abiotic stresses (Hossain et al., 2020).

The effect, intensity, duration, and type of primary stress as a priming mediator is very crucial. Cross-tolerances or multiple stress tolerances were detected in eggplants after exposure to chilling shocks right after the radicle protrusion stage. This phenomenon resulted in enhanced acclimation in the next generation (Sękara et al., 2016). Priming-mediated stress tolerance was also reported in coffee beans that were subjected to drought and cold shock priming on seedling stages mitigated subsequent cold stresses via activating antioxidant defense mechanisms (Vallad and Goodman, 2004). Studies on single and multiple stresses have revealed various biomolecular mechanisms that were mediated by induced or inherent priming via reactive oxygen species (ROS) and nitrogen species, heat shock proteins, calcium, potassium, and phosphorus gradients, as well as plant hormonal signaling (Vallad and Goodman, 2004).

Although MAS tolerance induced via seed priming is a common phenomenon but very minutiae details are available regarding its molecular, cellular, and physiological bedrock events. The chapter covers all the biological, molecular, and biochemical pathways for induction of stress tolerances via seed priming in crop plants against MASs.

14.2 SEED PRIMING: A BRIEF HISTORY AND TYPES

Priming for germination enhancement is a very old practice and has been used eons ago by Egyptian and Mayan civilizations. Later, ancient Greeks also used these techniques to overcome germination-related maladies (Gour et al., 2019). Heydecker used this term for the very first time in 1973. A famous pupil of Aristotle, Theophrastus (372–287 BC) used cucumber seeds and primed them with milk and water to stimulate germination (Gour et al., 2019).

The conviction of primed seeds found in Egyptian scriptures is very strange though: as seeds primed with urine were used to test pregnancies at that time. Different pieces of papyrus purged from the pyramids revealed that barley and wheat seeds were applied with the urine of pregnant women: and in this way, sprouted seeds were used to reveal pregnancy (Ghalioungui et al., 1963). Possibly elevated levels of estrogen in urine could have accelerated the germination of wheat or barley seed. In *cis*-priming seeds, salinity can be ameliorated via the urine rich in salts as well as proteins. Recent research could also be attributed in favor of this con's rational theory, in which protein hydrolysates-based seed priming tutelage the adverse effects of salt stress (Sorrentino et al., 2021). Other origination of priming is documented by Heydecker and Coolbear (1977) against dormancy breach and exasperated germination events via imparting stress tolerance. Later, Taylor et al. (1998) introduced the term 'seed enhancement' for improving germination and stand establishment of agronomic and horticultural seeds. Halmer, and subsequently Bray, worked on the taxonomy of categorization as well as exploration of physiological aspects of the priming approaches (Bray, 2017; Halmer, 2000). Till 2016, more than 1,200 research papers have been documented on seed primeomics globally. Per reports by Afzal et al. (2016), each year almost 100 articles on seed priming are published since 2010. However, the number of reviews on the molecular aspect of stress seed priming is scarce.

14.2.1 Types of Seed Priming on the Bases of Multiple Abiotic Stress Tolerance

Seed priming types against MASs are discussed subsequently (Figure 14.1).

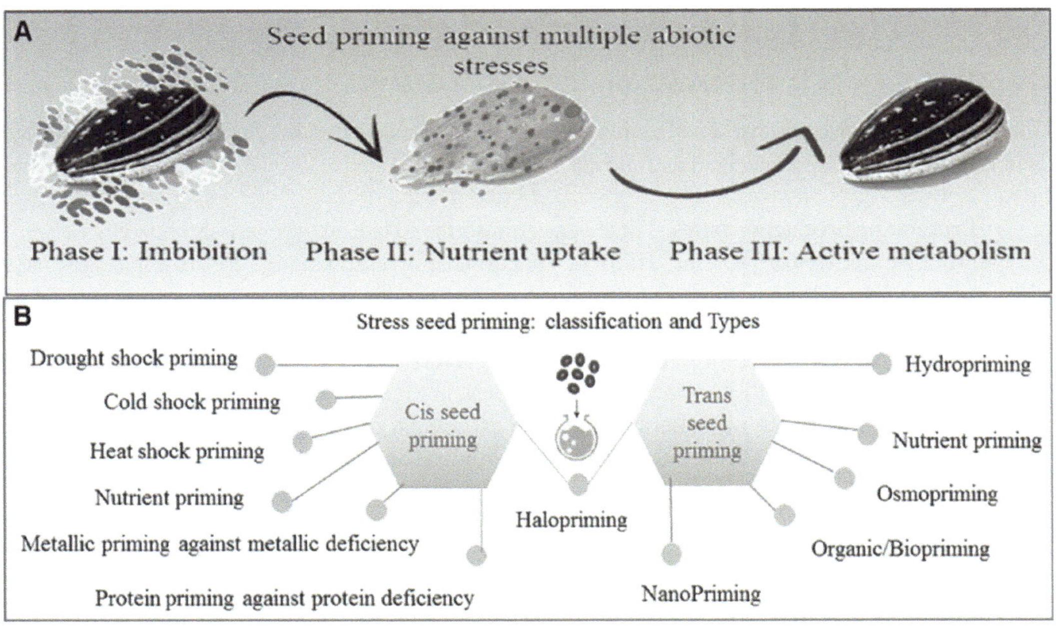

FIGURE 14.1 Phases of water uptake in seed (A) and types of seed priming on the basis of multiple abiotic stress tolerance (B).

14.2.1.1 *Cis*-Priming Approaches

A simple stress stimulus is necessary sometimes to enhance abiotic stress tolerance, and when the stimuli and stress are the same, the priming approach to enhance seed quality is termed as *cis*-priming of the seeds (Hossain et al., 2020). Pragmatic investigations have been performed under *cis*-priming seed to cope with abiotic stresses. Drought-stimulated, heat-induced, and cold-primed seeds can preserve the memory of former stress and provide antioxidant protection by activating the cellular defense mechanism immediately, as well as simultaneously reduce imbibition time, increase the germination stimulants, and upregulate the osmotic pressure (Marchin et al., 2020).

14.2.1.2 *Trans*-Priming Approaches

The stress priming approach with opposite priming stimuli applied to the seed is termed as *trans*-seed priming (Hossain et al., 2020). *Trans*-priming using chemical compounds with their evocative connotations has been documented in most of the crop and tree seeds. An example of *trans*-priming includes the use of β-sitosterol in tomato seeds that exhibited higher tolerance flux when subjected to low and high temperatures.

14.2.1.3 *Trans*-Generational Seed Priming

Seed priming within a parental generation is usually termed as *trans*-generational seed priming. Rampant efforts have been made to explore the potential of chemical and other biological stimuli for *trans*-generational seed enhancements against stresses. Hussain et al. (2018) evaluated thermo tolerance induced by drought priming stimuli for up to three generations in wheat crop. Higher seed yield, enhanced antioxidants, and membrane damage accompanied by better leaf photosynthesis were observed in progeny plants.

14.3 MOLECULAR MECHANISM OF SEED PRIMING

Molecular responses to stresses have been investigated in various plant species to identify the complex cadre of pathways regulated in protection and acclimation against different environmental stresses.

14.3.1 Genomics Underpinning Multiple Abiotic Stresses

The genetic response against abiotic stresses enables the plant to continue reasonable plant growth, and development are categorized as follows.

1. **Protectant-producing genes.** These genes transcribe to late embryogenesis abundant (LEA) proteins, antifreeze proteins, HSPs or chaperones, or antioxidants, for example, glycine betaine (gbsAB) upregulate for osmotic adjustment under salinity.
2. **Signaling genes.** Such genes are involved in transcriptional signaling of complex molecules; for instance, the expression of transcriptional factors, phospholipases, calcium-dependent protein kinases (CDPKs), mitogen-activated protein kinases (MAPKs), and SOS kinases.
3. **Essential substance up takers genes.** The expression of such genes involves the uptake and transport of ions and water, for instance ion transporters and aquaporins (Ciarmiello et al., 2011).

Similarly, some other researchers classified stress-sensitive genes into two types as follows.

1. **Early induced genes.** Early induced genes are activated shortly after stress signals are transmitted, and they usually express for a shorter duration.
2. **Late induced genes.** Late induced genes, on the other hand, are expressed gradually under stress, that is, after hours of absorbing stress signals, and their transcription is permanent (Rao et al., 2016).

14.3.2 Genomics Underlying Seed Priming

14.3.2.1 Gene Targeted Priming

The seed yields and crop establishment can be well enhanced via the genome-based priming of seeds, which can also mediate the negative effects of abiotic stresses usually a seed faces during its early growing days. For this purpose, various climate engineers have devised seed enhancement strategies. Among them, the most efficient fast neutron mutagenized practice is used for deleting central carbon metabolic genes in the genetic background of different cultivars (Kambhampati et al., 2019). This approach enhanced protein and oil contents in soybean cultivars up to 10%. When compared to the soybean cultivar, the deferred carbon allocation to RFO biosynthesis resulted in higher crude oil without compromising protein levels within mutant lines. Additionally, Kanai et al. (2019) identified four soybean genes that encode TAG lipases, *Sugar depend1*, including *GmSDP1-1*, *GmSDP1–2*, *GmSDP1–3*, and *GmSDP1–4*, with quite similar sequence homologies to *Arabidopsis*

TABLE 14.1
Key Genes Responsible for Seed Enhancements via Different Genomic Priming Approaches

Key Genes	Stress Type	Priming Approach	Traits	Seed Crop	References
Central carbon metabolic genes	Nil	Gene targeted primeomics – fast neutron mutagenesis	Enhanced crude protein and oil content – seed filling	Soybean	Kambhampati et al. (2019)
Sugar depend1 GmSDP1-1; GmSDP1-2; GmSDP1-3; GmSDP1-4	Nil	Gene knockdown technique	Enhanced tag lipase – seed filling	*Arabidopsis thaliana* L.	Miray et al. (2021); Han et al. (2020); Eshkiki et al. (2020); Kanai et al. (2019)
Raffinose synthase, *ABI5*, and *Galactinol synthase*	Oxidative stress	Genetic mapping	Photosynthesis-associated nuclear genes triggers photosynthates	Soybean, pea, and chickpea	Dwivedi et al. (2021)
HSP18.2	Heat shock and oxidative stress	Proteome-based primeomics	Seed vigor and longevity	*Arabidopsis* and soybean	Han et al. (2020)
Reduced ABA 1 (AtRABA1) and *speed of germination 1 (AtSOG1)*	Heat and oxidative stress	Gene mapping	Seed germination and vigor	Kale	Han et al. (2020)
miR-156-IPA1	Oxidative stress	Gene pyramiding	Seed yield	Arabidopsis	Dwivedi et al. (2021)
GNI-A1	Temperature stress	Genetic mapping	Seed weight and number	Wheat	Dwivedi et al. (2021)
OsALDH2B1	Multiple abiotic stresses	Gene pyramiding	Homogenous seed growth, defense, and reproduction	*Arabidopsis* and rice	Dwivedi et al. (2021)
bZIP transcription factor *ABI5*	Oxidative stress	QTL analysis of RIL2/genome imprinting	Enhanced seed longevity	*Medicago truncatula* Gaertn.	Dwivedi et al. (2021)
Ability after storage 1–5 (GAAS1–5)	High temperature, humidity, and oxygen stress during storage	Genetic mapping	Enhanced seed longevities	*Arabidopsis thaliana* L.	Han et al. (2020)

thaliana (L.) Heynh *(AtSDP1)*. When RNAi-depleted lines (overexpressing these genes) were compared to wild type, the seed oil yield was considerably higher while linoleic acid was significantly reduced. While oleic acid continued to increase in transgenic soybean seeds as compared to wild type, implying that GmSDP1 proteins govern both fatty acid composition and TAG content during seed filling (Miray et al., 2021). Such knockdown techniques provide a unique way for increasing both the quality and quantity of soybean seed oil (Han et al., 2020).

14.3.2.2 Genes Responsive to Seed Priming

Seed priming-mediated abiotic stresses could be linked with the strategies that enhance seed vigor on immediate bases (Table 14.1). Because of its close relationship, seed vigor is frequently researched in conjunction with seed longevity, which relates to the capacity of seeds to preserve viability under dry storage conditions. As a result, artificial or natural seed aging has frequently been employed to elucidate genetic components underpinning seed vigor and their use in marker-aided breeding. For example, in terms of genetic mapping, RILs (recombinant inbred lines) in the model plant *A. thaliana* have found five loci linked with seed lifespan known as *Ability after storage 1–5 (GAAS1–5)*, and these loci were found responsive to seed priming against reducing longevities (Han et al., 2020).

The *bZIP* transcription factor (TF) *ABI5* has been revealed as a modulator of seed longevity acquisition in mature seeds of the legume barrel medic, *Medicago truncatula* Gaertn., due to its co-localization with a seed lifespan QTL discovered through genetic analysis of RIL populations. ABI5 is involved in controlling seed lifetime and is linked to the regulation of RFO as well as LEA protein levels, with the expression of photosynthesis-associated nuclear genes. Manipulation of RFO levels consistently influences seed vigor as well as the expression of genes regulating its syntheses, such as raffinose synthase, *ABI5*, and galactinol synthase, which have also been linked to the control of seed longevity and vigor during seed germination and maturation in dicots. RFO's impact on increasing seed vigor and lifespan has been linked to the suppression of ROS generation and lipid peroxidation (Dwivedi et al., 2021).

14.3.3 Genomics of Priming-Induced Stress Tolerance

14.3.3.1 Role of Transcriptional Factors in Regulating Post-Priming Stress Tolerance

The transcriptional factors that respond to stresses and the tolerance induced via priming can aid to alleviate stress damage to the plants. Many priming elicitors have been documented that showed tolerances toward environmental shock exposures via overexpressing stress-regulated genes. In the cases of abiotic stresses, drought is mainly influenced by the action mechanism of ABA-regulated genes. These genes have their notorious growth inhibition mechanism and can be mediated by mannitol. The TFs responsible for ABA sensitivity have been identified in ABA-deficient mutants where tolerance-responsive genes have been expressed (Xiong et al., 2006). Thus, ABA is associated with heat shock protein regulations and the key genes responsible for the expression of heat stress tolerance. ABA signaling pathways have also been inducing tolerance against cold and the genes *GLR1.2* or *GLR1.3*, responsible for cold acclimation, were associated with ABA. Increased endogenous levels of jasmonate have been found to mediate the expression of these genes in grapevines (Ritonga and Chen, 2020). Similarly, salinity tolerance has been mediated by the exogenous application of brassinosteroids and salicylic acid. Both chemicals are responsible for the expression of genes responsible for the salicylic acid biosynthetic enzyme. Similarly, genes responsible for GORK (guard cell outward rectifying K^+ channels) as well as upregulation of TFs involved in H^+-ATPase activity are also identified in *Arabidopsis* (Gupta and Huang, 2014). Brassinosteroids are also known to enhance the genetic expressions of important genes responsible for the activity of antioxidant enzymes in the seeds (Gupta and Huang, 2014).

Polyamines (spermine and spermidine) are used as seed priming agents in salt-stressed seedlings of Indica rice varieties. Some most important priming responsive genes against salt stress were expressed as a result of the action of osmolyte (*PDH, P5CS, BADH1*), enzymatic and non-enzymatic antioxidants (*CAT, SOD, ANS, APX, GR*), ABA biosynthetic enzyme (*NCED3*), LEA (*OsEm*), PA-metabolic enzymes (*SAMDC, DAO, SPDS, SPMS, PAO*), transcription factors (*WRKY71, TRAB1*), and ion transporters (*NHX1*) (Paul and Roychoudhury, 2017). Brassinosteroids were also responsible for the expression of TFs involved in the aggregation of non-enzymatic antioxidant compounds. Some genes that regulate the auxin flux against salt stress in the seedling stage are documented in rice, namely *Os09g0491740* and *Os01g0976100* for auxin flux and transportation (Islam et al., 2022). Osmopriming elucidated SNP markers have been identified in wheat that controls stomatal closure and the exclusion of sodium ions during salt-based stresses. One marker, *wsnp_Ex_rep_c101323_86702546 (5A)*, was known to show a pleiotropic effect when wheat seeds were exposed to salt stress. A gene that responds to osmoticum during priming has also been identified as *TraesCS2B02G135900* (potassium transporter) against salinity (Hasseb et al., 2022). Similarly, in heavy metal-tolerant *Noccaea goesingense* overly expressed *NgSAT* resulted in improved levels of GSH that initiated tolerance against Ni, Co, Zn, and Cd toxicity (Singh et al., 2016). At seed or sapling stages, dormancy induced by abiotic stresses is mainly influenced by gene *CYP707A2*, which exhibits an expression of the ABA 8′-hydroxylase gene (Rao et al., 2016).

14.3.3.2 Seed Priming Provoked Epigenetic Changes

Epigenetics is described as a temporary to permanent molecular change of DNA that does not affect the genomic sequence. By changing chromatin architecture, epigenetic control allows expression at specified regions of the genome. This can also be employed as a biomarker for the establishment of stress-tolerant agricultural plants that have been genetically modified. During the early stages of crop growth, epigenetic alterations caused by histone modifications and DNA methylation occur (Li et al., 2015). The gene encoding the lysine-specific histone demethylasel (LSD1), which is involved in histone demethylation associated with epigenetic change, was increased in primed cells relative to unprimed cells (Li et al., 2015). It was also shown that the expression levels encoding RNA methyl transferase, putated nucleic acid methyl transferases, LSD1, and binding proteins were higher in the progeny of primed plants compared to non-primed plants. According to the findings of this study, histone demethylation and DNA methylation have a role in the control of *trans*-generational memory. Primed plants have *trans*-generational memory and may so respond aggressively and quickly to future stimuli. Because epigenetic changes in DNA may be inherited, there is a chance that the epigenetic modifications in DNA obtained by priming will be handed down to the next generation. For instance, BABA-treated *Arabidopsis thaliana* progeny were also resistant to biotic stress (Li et al., 2015).

14.4 PHYSIOLOGY OF PRIMING-INDUCED STRESS TOLERANCE

Physiological responses such as stomatal closure, photosynthetic inhibition, and enhanced leaf temperature were identified in tobacco leaves under drought and heat stress. Additionally, proteome analysis of the combined drought and heat stress in *A. thaliana* revealed 45 proteins that accumulate particularly under MASs. Seed stress priming investigations have encompassed enzymes intricated in the Calvin cycle, malate metabolism, and the detoxification of ROS. Cytosolic ascorbate peroxidase 1 (APX1), in specific, has been recognized in response to MASs as well as oxidative stress (Koussevitzky et al., 2008). Several heat shock proteins (HSPs) with increased abundance under MAS were identified in both poplar and *Carissa spinarum*, that co-regulates and safeguard the photosynthetic machinery of these plants (Li et al., 2014). Metabolic alterations caused by the MASs in stamens, as well as in pollinated and unpollinated pistils were examined in rice (Li et al., 2015). Under combined stress, the levels of amino acids, proteins, and sugars were dramatically altered in floral tissues. There were nine metabolites discovered in tolerant rice genotypes that may offer protection against simultaneous heat and drought stress. Significantly, MAS reduced sucrose levels in the anthers of susceptible genotype (Li et al., 2015). The major negative impacts of heat stress (HS) during grain filling are related to the reduced buildup of storage compounds, which can have a poor impact on both seed quality and eventual yield. HSPs can play an important role in gluten production in wheat by acting as a 'glue' between the storage proteins (glutenins and gliadins) as well as starch (Khan et al., 2018).

The role of proteins in metallic stresses cannot be neglected. The involvement of amino acids in hyper-accumulators has been discovered to be crucial owing to the development of stable compounds with bivalent cations, which aids greatly in metal cation sequestration. Histidine (His), for instance, is implicated in Ni hyperaccumulation and has been found in high concentrations in the roots of Ni hyper-accumulators. Additionally, there are diverse metallic transporter subfamilies that are usually composed of membranous proteins, which further translocate metallic ions to epidermal cells from mesophyll or roots to the aerial plant parts. At the cellular stage, transporter proteins usually control the uptake and translocation of metallic ions across cellular membranes, via vacuoles through tonoplast pathways, as well as other membranous organelles. A few significant nickel transporters include heavy metal ATPase (HMA), ATP-binding cassette (ABC), natural resistance associated macrophage proteins (NRAMPS), CPx-type ATPase, cation diffusion facilitator family (CDF), and ZRT-IRT-like protein families (Hasanuzzaman, 2020).

14.4.1 Seed Priming and Germination Events

Water imbibition by seeds prior to germination can be split into three phases: free water absorption (phase I), improved water uptake (phase II), and water uptake concurrent with growth start (phase III). Seed priming is based on phase II events, in which regulated hydration leads to dormancy break even in an adverse environment. In phase II, critical germination activities such as cellular integrity restoration, DNA repair, mitochondrial repair, breakdown of stored reserves, synthesis of germination-associated RNAs and proteins, and the commencement of respiration occur. In addition to proteins involved in metabolic activities, protein synthesis required for cell communication and stress tolerance is increased during the initial stage II (24 hours) of imbibition. Osmopriming of *Spinacia oleracea* resulted in higher aquaporin (AQP) expression and improved germination. During phase II imbibition, the genes *SoPIP1;1*, *SoPIP1;2*, *SoPIP2;1* (PIP-plasma membrane intrinsic proteins), and *SoTIP* (TIP-tonoplast intrinsic proteins) were up activated (Chen et al., 2012).

14.4.2 Seed Priming-Mediated Stress Tolerance via Memory Induction

The mechanism of stress tolerance is linked to several biochemical, morphophysiological, and molecular processes (Figure 14.2). Stress exposure during seed priming maintains stress memory in crops, allowing them to endure a variety of challenges. The effect of 'priming memory' is an improvement in stress tolerance during the emergence of primed seeds (Hossain et al., 2020).

14.5 SEED PRIMING AND PLANT ABIOTIC STRESSES

Plants raised from primed seeds performed better against abiotic stress, and various metabolic and biochemical pathways are involved in enhancing resistance to abiotic stresses (Table 14.2).

14.5.1 Temperature Stress Tolerance

Plant species sensitive to high temperatures exhibit basic physiological disorders such as altered membrane fluidity, reduced photosynthetic efficiency, respiration, and poor water relations, which ultimately leads to a reduction in growth (Banerjee and Roychoudhury, 2016). Additionally, pollen sterility was reported (Cao et al., 2008), together with enzyme inactivation degrades chromatin, promotes premature senescence, and decreases Rubisco-binding proteins (RBPs), sucrose phosphate synthase, ADP-glucose pyro-phosphorylase, and invertase (Hasanuzzaman et al., 2013). HS can be reduced by pre-treating seeds with chemical agents prior to planting (Ibrahim, 2016). In

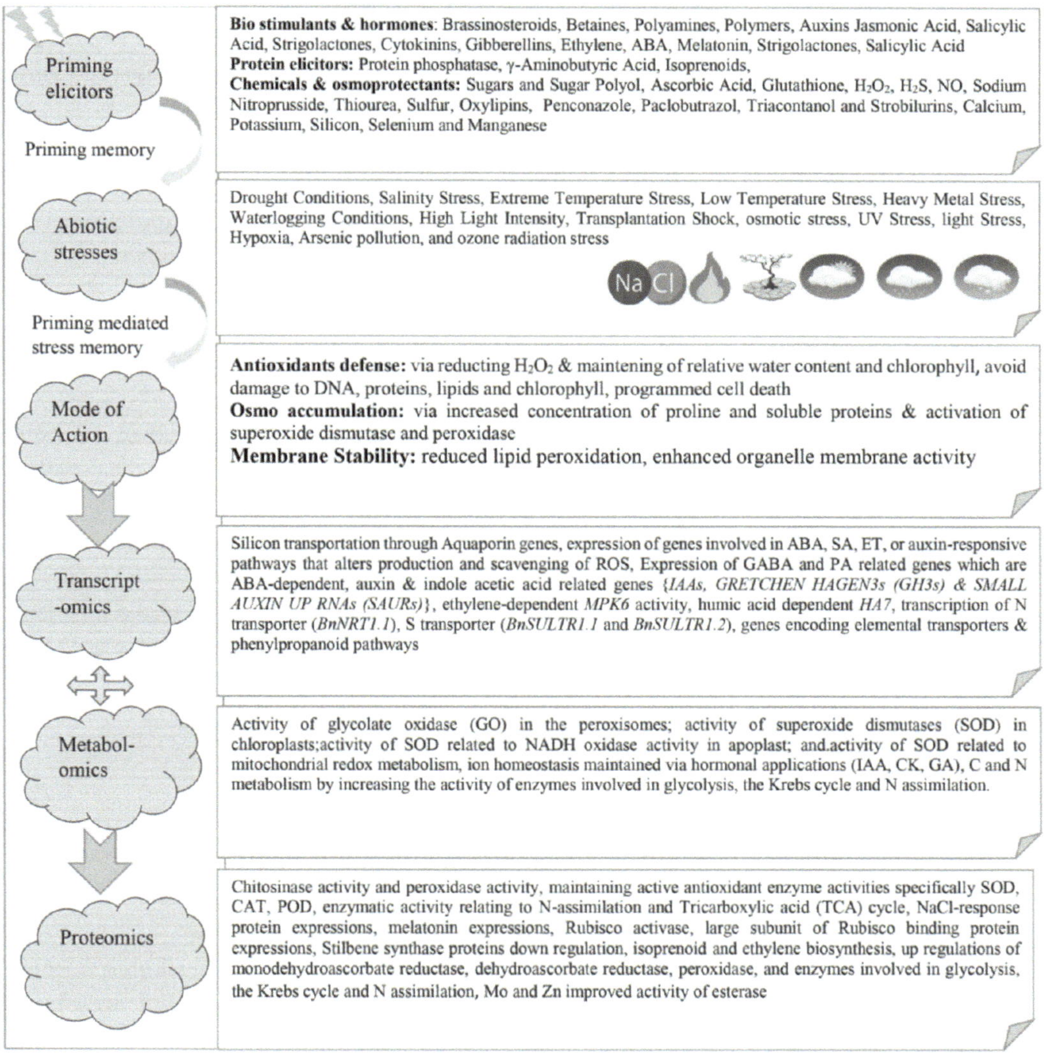

FIGURE 14.2 Mechanism of priming-mediated combined stress tolerance in seeds and plants (adapted from Hossain et al., 2020; Li et al., 2015; Sękara et al., 2016).

general, heat stress causes the upregulation of many heat-inducible genes known as 'heat shock genes' (HSGs) that encode HSPs (Chang et al., 2007), and act as molecular chaperones to protect intracellular proteins from denaturation via preserving functional stability (Baniwal et al., 2004). Alternatively, chilling stress reduces seed germination and affects plant growth during the early growth phase. It also induces pollen sterility, delays flowering, and ultimately results in severe output losses during the reproductive phase (Hussain et al., 2018).

Seed priming works as an adaptable strategy to mitigate the negative effects of chilling stress (Hussain et al., 2018) via amplified RNA and DNA synthesis, increased respiration rates, enhanced metabolite synthesis, and accumulation, membrane stability maintenance, and increased antioxidant activities. It also affects the expression of cold-responsive genes as well as proteins such as ZAT10, COR15A, DHNs, RAB18, and AQPs (Chen et al., 2012) to induce low-temperature tolerance (Hussain et al., 2018). Similarly, plants that originated from primed seeds perform better under cold stress due to the upregulation of LEAs (van Buer et al., 2016). Cold responsive-15A (COR15A; At2g42540), zinc finger of *Arabidopsis thaliana*-10 (ZAT10; At1g27730), phenyl ammonium lyase (PAL1; At2g37040), and chalcone synthase (CHS; At5g13930) are examples of LEA proteins (van Buer et al., 2016). The application of

TABLE 14.2
Various Priming Techniques and Associated Traits Improvement under Different Abiotic Stress Conditions

Abiotic Stress	Priming Technique	Priming Agent	Crop	Traits Improved	Reference
Heat stress	Hormonal priming; halopriming; hormonal priming	Abscisic acid; nitrate; thiourea	*Triticum aestivum*, *Lens culinaris*, and *Brassica napus*	An increase in chlorophyll, nitrogen, protein, and proline contents. NR and SOD activities, and improved harvest index, biological yield, grain yield, and 1,000-seed weight	Sehgal et al. (2019); Ahmad et al. (2021); Kosakivska et al. (2022)
Cold stress	Hormonal priming; matrix priming; nanoparticle priming	Gibberellic acid; chitosan; ZnO	*Oryza sativa* and *Vigna radiata*	Increase the α-amylase activity and total soluble sugar content by enhancing the starch degradation. Enhanced GA_3 and decreased ABA levels	Hussain et al. (2016); Song et al. (2021); Nie et al. (2022)
Drought stress	Osmopriming; hormonal priming; ultraviolet priming	$CaCl_2$; gibberellic acid; UV-B	*Lens culinaris*, *Brassica napus*, and *Oryza sativa*	Early, uniform germination, accumulate sugar, calcium, reduced lipid peroxidation. Improved the activity of PSI and PSII, upregulated expression of SOD, CAT, and APX genes	Khan et al. (2020); Farooq et al. (2020); Thomas et al. (2020)
Flooding stress	Nutripriming; hormonal priming; nanoparticle Priming	Selenium; melatonin; Al_2O_3 NPs	*Oryza sativa*, and *Glycine max*	Improvement of crop yield by modulating metabolism, regulating energy metabolism, cell death, and expression of genes	Mustafa et al. (2015); Wang et al. (2021); Hu et al. (2022)
Salinity stress	Halopriming; magneto priming; nanoparticle priming	NaCl; static magnetic field; ZnO	*Solanum lycopersicon*, *Glycine max*, and *Zea mays*	Improved osmotic adjustment, maximum quantum yield of photosystem II and, partially, the water use efficiency. Enhanced activities of the antioxidant enzyme	Kataria et al. (2019); González-Grande et al. (2020); Sun et al. (2021)
Alkalinity stress	Ultraviolet priming; osmopriming; hydro-priming	UV-B; $CaCl_2$; water	*Oryza sativa*, *Vigna unguiculata*, and *Capsicum annum* L.	Alteration in antioxidant capacity and improved photosynthetic electron transport. Reduces ROS accumulation	Thomas and Puthur (2019); Farooq et al. (2020)
Heavy metal stress	Hormonal priming; nanoparticle priming; osmopriming	Melatonin; ZnO; $CaCl_2$	*Triticum aestivum*, *Glycine max*, and *Vicia faba* L.	Increased Chl amount in plants affected by heavy metal toxicity, thus increased photosynthetic efficiency. Promoted growth under heavy metals with increased α-amylase activity	Nouairi et al. (2019); Lei et al. (2021); Li et al. (2021)

natural and synthetic growth-promoting substances as priming agents (Afzal et al., 2020) or foliar spray (Afzal et al., 2022) also proved to be the most successful strategy in conferring better thermotolerance at early sowing through improved antioxidant defense system and physiological mechanism of crops.

14.5.2 Water-Deficit and Waterlogging Stress Tolerance

Plant species sensitive to water deficiency exhibit basic physiological and biochemical changes including the closure of stomata, increased oxidative stress, changed cell wall integrity, reduction in leaf water potential, decreased stomatal conductance, reduced internal CO_2, ceased Rubisco activity, declined photochemical efficiency, decreased chlorophyll content, and high ROS production, which subsequently leads to a reduction in growth and yield of plants (Seleiman et al., 2021). Seed priming overcomes the deleterious effects of water deficiency in plants and promotes uniform germination via carbohydrate accumulation (Brocklehurst and Dearman, 1983), increased metabolite production, repaired deteriorated DNA, and increased antioxidant capability and osmolyte buildup. Drought-induced primed seeds activate the PRGs that translate proteins for the production of LEAs and formamidopyrimidine DNA glycosylase (FPGase). Similarly, the upregulation of drought-responsive genes *RD1* and *RD2* of the *AP2/ERF* family (Imperiali et al., 2017) and overexpression of the aquaporin gene *MaPIP2–7* improved stress tolerance via maintaining better plant aqua status (Xu et al., 2020) even in a severe water shortage. Moreover, Sirhindi et al. (2020) mentioned the upregulated expression of *PSbI* and *PSbM* upon jasmonic acid and methyl jasmonate priming in *Brassica*. These genes were involved in the formation of PSII dimer and maintenance of photosystem stability for rapid photosynthetic recovery on re-watering after drought stress (Sirhindi et al., 2020).

In contrast to water shortage, waterlogging (WL) is also a serious threat to plant survival, which interferes with normal growth and development by disrupting physiological and metabolic events. Seed priming promotes growth under WL stresses due to increased mineral ion uptake, especially N, P, and K (Ahanger and Ahmad, 2019). It also maintains plant homeostasis and upregulates genes associated with HSPs, amylase activity (Liu et al., 2021), Rubisco (Hussain et al., 2017), and antioxidants (Mei et al., 2017).

14.5.3 Salinity Stress Tolerance

Salinity affects plant morphological and metabolic functioning deleteriously (Zhang and Dai, 2019). Mainly photosynthetic machinery, transpiration, chloroplast ultrastructure, and stomatal conductance are affected (Pan et al., 2021). Moreover, salinity increases the concentration of ROS, which induces lipid peroxidation, membrane degradation, as well as DNA and protein damage. Seed priming allows early DNA replication and repair, enhances RNA and de novo protein synthesis (Sharma et al., 2014), activates germination-related genes (Sharma et al., 2014), minimizes metabolite leakage experienced by the plant from oxidative damage, and reduces lipid peroxidation (Wahid et al., 2008).

In addition to salt stress, saline–alkaline (SA) stress is one of the most important abiotic factors limiting plant growth and grain yield (Wei et al., 2020). Soil alkalinity induces a significant reduction in the accumulation of total soluble proteins and free amino acids. Alternatively, priming significantly increases antioxidant activities and accelerates the expression of PRGs responsible for enhanced proline content, and reduced ROS (Abdel Latef and Tran, 2016). For instance, priming with ABA significantly decreased Na^+ content via increased expression of stress tolerance genes *NHX1* and *AVP* (Lei et al., 2018).

14.5.4 Heavy Metal and Metallic Stressors

Toxic metal exposure causes growth disruptions and developmental anomalies in the majority of plant species, affecting biomass accretion, radical protrusion due to impaired oxidative status, photosynthetic performance, and changes in macromolecule synthesis and activity (Hasan et al., 2017).

Plants evolve complex molecular and physiological defense mechanisms by seed priming to withstand HM stresses, such as reduced HM uptake through root architecture modifications, metal sequestration into vacuoles, binding to phytochelatins/metallothioneins, and activation of various antioxidants and metabolites for improved adaptability, tolerance, and survival (Lal et al., 2021).

Priming agents primarily impact antioxidant activity and can result in increased expression of genes encoding antioxidant enzymes (Cuypers et al., 2016), such as cysteine desulfhydrolase (CAS), catalyzing H_2S formation, nitric oxide synthases (NOS), catalyzing NO synthesis, and isochorismate synthase (ICS) to combat HM stress tolerance (Lefevere et al., 2020). Similarly, seed priming-mediated upstream signaling enhances serine acetyltransferase (SAT) activity, GSH levels, and upregulates metal transporters, notably vacuolar transporters such as ABC proteins (Kabała et al., 2019), metal-binding peptides: metallothioneins (MTs) as well as activates enzymes responsible for metal chelator synthesis and transport, phytochelatins, and low-molecular-weight organic acids (LMWOA). Priming responsive compounds/proteins such as proline, HSPs, dehydrins, PRGs, and transcription factors, *MYB* and *WRKY*, were activated to induce stress memory in response to priming (Hameed et al., 2021).

14.6 CONCLUDING REMARKS

Seed priming is an effective strategy for mitigating the detrimental effects of a wide range of abiotic stimuli such as salt, drought, heavy metal, and temperature. Seed priming is a realistic, effective, and shotgun approach for improving germination rates, seedling vigor, growth, and productivity under various biotic and abiotic stresses. Biological seed enhancement is a very popular technique in disease control and abiotic stress tolerance and farmers are practicing this technology with the support of the seed industry in developing countries. However, the farmers in these countries are not fully utilizing these techniques as compared to developed countries due to a lack of resources and an increase in input costs. Nanotechnology is being used as a modern agricultural technique, and seed priming with nanoparticles (nanopriming) has received a lot of attention in the previous decade (Abbasi Khalaki et al., 2021). However, the long-term effects of such nanoparticle application on soil microbiota and human health must be unveiled.

The shelf life of the primed seed is still challenging for researchers. Investigations are needed to explore the biochemical and molecular pathways that affect the viability of primed during storage need to be elucidated. Moreover, there is still a room to investigate more priming agents and genomics to mitigate the stress episodes with more severity than in ancient times.

The knowledge of seed priming should be utilized for better stand establishment, growth, and productivity of crop plants under changing climate conditions. This will require the development of new cost-effective priming agents with complementary priming equipment and technologies. These goals can be accomplished by investigating the physiological, biochemical, and molecular mechanisms of new priming materials against abiotic stresses. Lastly, this technology can be adopted in developed and third-world countries through a partnership between academia with industry.

REFERENCES

Abbasi Khalaki, M., Moameri, M., Asgari, L. B., and Astatkie, T. 2021. Influence of nano-priming on seed germination and plant growth of forage and medicinal plants. *Plant Growth Regulation* 93: 13–28. doi: 10.1007/s10725-020-00670-9.

Abdel Latef, A. A., and Tran, L. S. P. 2016. Impacts of priming with silicon on the growth and tolerance of maize plants to alkaline stress. *Frontiers in Plant Science* 7: 243. doi: 10.3389/fpls.2016.00243.

Afzal, I., Imran, S., Javed, T., and Basra, S. M. A. 2020. Evaluating the integrative response of moringa leaf extract with synthetic growth promoting substances in maize under early spring conditions. *South African Journal of Botany* 132: 378–387. doi: 10.1016/j.sajb.2020.04.025.

Afzal, I., Imran, S., Javed, T., Tahir, A., Kamran, M., Shakeel, Q., et al. 2022. Alleviation of temperature stress in maize by integration of foliar applied growth promoting substances and sowing dates. *PLoS One* 17: e0260916. doi: 10.1371/journal.pone.0260916.

Afzal, I., Rehman, H. U., Naveed, M., and Basra, S. M. A. 2016. Recent advances in seed enhancements. In Araujo, S. and Balestrazzi, A. (eds), *New Challenges in Seed Biology-Basic and Translational Research Driving Seed Technology*, 47–74. London: InTech Open. doi: 10.5772/64791.

Ahanger, M. A., and Ahmad, P. 2019. Role of mineral nutrients in abiotic stress tolerance: Revisiting the associated signaling mechanisms. In Khan, M. I. R., et al. (eds), *Plant Signaling Molecules: Role and Regulation Under Stressful Environments*, 269–285. Elsevier. doi: 10.1016/B978-0-12-816451-8.00016-2.

Ahmad, M., Waraich, E. A., Hussain, S., Ayyub, C. M., Ahmad, Z., and Zulfiqar. U. 2021. Improving heat stress tolerance in Camelina sativa and Brassica napus through thiourea seed priming. *Journal of Plant Growth Regulation*. doi: 10.1007/s00344-021-10482-4.

Banerjee, A., and Roychoudhury, A. 2016. Group II late embryogenesis abundant (LEA) proteins: Structural and functional aspects in plant abiotic stress. *Plant Growth Regulation* 79: 1–17. doi: 10.1007/s10725-015-0113-3.

Baniwal, K. S., Bharti, K., Chan, K., Fauth, M., Ganguli, A., Kotak, S., et al. 2004. Heat stress response in plants: A complex game with chaperones and more than twenty heat stress transcription factors. *Journal of BioSciences* 29: 471–487. doi: 10.1007/BF02712120.

Bhardwaj, A., Sita, K., Sehgal, A., Bhandari, K., Shiv, K., Prasad, P. V. V., et al. 2021. Heat priming of lentil (Lens culinaris Medik.) seeds and foliar treatment with γ-aminobutyric acid (GABA), confers protection to reproductive function and yield traits under high-temperature stress environments. *International Journal of Molecular Sciences* 22: 5825. doi: 10.3390/ijms22115825.

Bray, C. M. 2017. Biochemical processes during the osmopriming of seeds. In Kigel, J. and Gad, G. (eds), *Seed Development and Germination*. New York, NY: Routledge. doi: 10.1201/9780203740071.

Brocklehurst, P. A., and Dearman, J. 1983. Interactions between seed priming treatments and nine seed lots of carrot, celery and onion. I. Laboratory germination. *Annals of Applied Biology* 102: 577–584. doi: 10.1111/j.1744-7348.1983.tb02729.x.

Cao, Y., Duan, H., Yang, L. N., Wang, Z. Q., Zhou, S. C., and Yang, J. C. 2008. Effect of heat stress during meiosis on grain yield of rice cultivars differing in heat tolerance and its physiological mechanism. *Acta Agronomica Sinica* 34: 2134–2142. doi: 10.1016/S1875-2780(09)60022-5.

Chang, H., Tang, Y., Hayar-hartl, M., and Hartl, F. U. 2007. SnapShot: Molecular chaperones, Part I. *Cell* 128: 212. doi: 10.1016/j.cell.2007.01.001.

Chen, K., Fessehaie, A., and Arora, R. 2012. Dehydrin metabolism is altered during seed osmopriming and subsequent germination under chilling and desiccation in Spinacia oleracea L. cv. Bloomsdale: Possible role in stress tolerance. *Plant Science* 183: 27–36. doi: 10.1016/j.plantsci.2011.11.002.

Ciarmiello, L. F., Woodrow, P., Fuggi, A., Pontecorvo, G., and Carillo, P. 2011. Plant genes for abiotic stress. In Shanker, A. and Venkateswarlu, B. (eds), *Abiotic Stress in Plants–Mechanisms and Adaptations*, 283–308. London: InTech Open. doi: 10.5772/22465.

Cuypers, A., Hendrix, S., Reis, A. D. R., de Smet, S., Deckers, J., Gielen, H., et al. 2016. Hydrogen peroxide, signaling in disguise during metal phytotoxicity. *Frontiers in Plant Science* 7: 470. doi: 10.3389/fpls.2016.00470.

Dwivedi, S. L., Reynolds, M. P., and Ortiz, R. 2021. Mitigating tradeoffs in plant breeding. *Science* 24: 102965. doi: 10.1016/j.isci.2021.102965.

Dwivedi, S. L., Spillane, C., Lopez, F., Ayele, B. T., and Ortiz, R. 2021. First the seed: Genomic advances in seed science for improved crop productivity and food security. *Crop Science* 61: 1501–1526. doi: 10.1002/csc2.20402.

Eshkiki, E. M., Hajiahmadi, Z., Abedi, A., Kordrostami, M., and Jacquard, C.2020. In silico analyses of autophagy-related genes in rapeseed (*Brassica napus* L.) under different abiotic stresses and in various tissues. *Plants* 9: 1393. doi: 10.3390/plants9101393.

Farooq, M., Rehman, A., Al-Alawi, A. K. M., Al-Busaidi, W. M., and Lee, D. 2020. Integrated use of seed priming and biochar improves salt tolerance in cowpea. *Scientia Horticulturae* 272: 109507. doi: 10.1016/j.scienta.2020.109507.

Farooq, M., Romdhane, L., Sulti, M. K. A., Rehman, A., Al-Busaidi, W. M., and Lee, D. 2020. Morphological, physiological and biochemical aspects of osmopriming-induced drought tolerance in lentil. *Journal of Agronomy and Crop Science* 206: 176–186. doi: 10.1111/jac.12384.

Ghalioungui, P., Khalil, S. H., and Ammar, A. R. 1963. On an ancient Egyptian method of diagnosing pregnancy and determining foetal sex. *Medical History* 7: 241–246. doi: 10.1017/s0025727300028386.

González-Grande, S., Surez, P. N., and Marin, O. 2020. Effect of salinity and seed salt priming on the physiology of adult plants of Solanum lycopersicum cv.'Río Grande'. *Brazilian Journal of Botany* 43: 775–787. doi: 10.1007/s40415-020-00636-1.

Gour, L., Lokesh, G., Ramakrishnan, R. S., Panwar, N. K., Sharma, R., Pathak, N., et al. 2019. Seed priming: An old empirical technique with new contemporary perspectives in respect to Pisum sativum L: A review. *Agricultural Reviews* 40: 136–142. doi: 10.18805/ag.R-1906.

Gupta, B., and Huang, B. 2014. Mechanism of salinity tolerance in plants: Physiological, biochemical, and molecular characterization. *International Journal of Genomics* 2014: 701596. doi: 10.1155/2014/701596.

Halmer, P. 2000. Commercial seed treatment technology. In Black, M. and Bewley, J. D. (eds), *Seed Technology and its Biological Basis*, 257–286. Sheffield: Sheffield Academic Press.

Hameed, A., Farooq, T., Hameed, A., and Sheikh, M. A. 2021. Sodium nitroprusside mediated priming memory invokes water-deficit stress acclimation in wheat plants through physio-biochemical alterations. *Plant Physiology and Biochemistry* 160: 329–340. doi: 10.1016/j.plaphy.2019.09.016.

Han, B., Xu, H., Feng, Y., Xu, W., Cui, Q., and Liu, A. 2020. Genomic characterization and expressional profiles of autophagy-related genes (ATGs) in oilseed crop castor bean (*Ricinus communis* L.). *International Journal of Molecular Sciences* 21: 562. doi: 10.3390/ijms21020562.

Hasan, M. K., Cheng, Y., Kanwar, M. K., Chu, X., Ahammed, G. J. and Zhen-Yu, Q. 2017. Responses of plant proteins to heavy metal stress: A review. *Frontiers in Plant Science* 8: 1492. doi: 3389/fpls.2017.01492.

Hasanuzzaman, M. 2020. *Plant Ecophysiology and Adaptation under Climate Change: Mechanisms and Perspectives I: General Consequences and Plant Responses*. Singapore: Springer Nature. doi: 10.1007/978-981-15-2172-0.

Hasanuzzaman, M., Nahar, K., Alam, M., Roychowdhury, R., and Fujita, M. 2013. Physiological, biochemical, and molecular mechanisms of heat stress tolerance in plants. *International Journal of Molecular Sciences* 14: 9643–9684. doi: 10.3390/ijms14059643.

Hasseb, N. M., Sallam, A., Karam, M. A., Gao, L., Wang, R. R. C., and Moursi, Y. S. 2022. High-LD SNP markers exhibiting pleiotropic effects on salt tolerance at germination and seedlings stages in spring wheat. *Plant Molecular Biology* 108: 585–603. doi: 10.1007/s11103-022-01248-x.

Heydecker, W., and Coolbear, P. 1977. Seed treatments for improved performance survey and attempted prognosis. *Seed Science and Technology* 5: 353–425. doi: 10.4236/abb.2011.21007.

Hossain, M. A., Liu, F., Burritt, D., Fujita, M., and Huang, B. (eds) 2020. *Priming-Mediated Stress and Cross-Stress Tolerance in Crop Plants*, 340. Academic Press. doi: 10.1016/C2018-0-02626-2.

Hu, F., Jiang, S., Wang, Z. et al. 2022. Seed priming with selenium: Effects on germination, seedling growth, biochemical attributes, and grain yield in rice growing under flooding conditions. *Plant Direct* 6: e378. doi: 10.1002/pld3.378.

Hussain, H. A., Hussain, S., Abdul-Khaliq, et al. 2018. Chilling and drought stresses in crop plants: Implications, cross talk, and potential management opportunities. *Frontiers in Plant Science* 9: 393. doi: 10.3389/fpls.2018.00393.

Ibrahim, E. A. 2016. Seed priming to alleviate salinity stress in germinating seeds. *Journal of Plant Physiology* 192: 38–46. doi: 10.1016/j.jplph.2015.12.011.

Imperiali, N., Chiriboga, X., Schlaeppi, K., et al. 2017. Combined field inoculations of Pseudomonas bacteria, arbuscular mycorrhizal fungi, and entomopathogenic nematodes and their effects on wheat performance. *Frontiers in Plant Science* 8: 1809. doi: 10.3389/fpls.2017.01809.

Islam, M. R., Naveed, S. A., Zhang, Y., et al. 2022. Identification of candidate genes for salinity and anaerobic tolerance at the germination stage in rice by genome-wide association analyses. *Frontiers in Genetics* 8: 822516. doi: 10.3389/fgene.2022.822516.

Kabała, K., Zboińska, M., Głowiak, D., Reda, M., Jakubowska, D., and Janicka, M. 2019. Interaction between the signaling molecules hydrogen sulfide and hydrogen peroxide and their role in vacuolar H+-ATPase regulation in cadmium-stressed cucumber roots. *Physiologia Plantarum* 166: 688–704. doi: 10.1111/ppl.12819.

Kambhampati, S., Aznar-Moreno, J. A., Hostetler, C., et al. 2019. On the inverse correlation of protein and oil: Examining the effects of altered central carbon metabolism on seed composition using soybean fast neutron mutants. *Metabolites* 10: 18. doi: 10.3390/metabo10010018.

Kanai, M., Yamada, T., Hayashi, M., Mano, S., and Nishimura, M. 2019. Soybean (*Glycine max* L.) triacylglycerol lipase GmSDP1 regulates the quality and quantity of seed oil. *Scientific Reports* 9: 1–10. doi: 10.1038/s41598-019-45331-8.

Kataria, S., Baghel, L., Jain, M., and Guruprasad, K. N. 2019. Magnetopriming regulates antioxidant defense system in soybean against salt stress. *Biocatalysis and Agricultural Biotechnology* 18: 101090. doi: 10.1016/j.bcab.2019.101090.

Khan, A., Pan, X., Najeeb, U., et al. 2018. Coping with drought: Stress and adaptive mechanisms, and management through cultural and molecular alternatives in cotton as vital constituents for plant stress resilience and fitness. *Biological Rrsearch* 51. doi: 10.1186/s40659-018-0198-z.

Khan, M. N., Nauman, M., Khan, Z., Luo, T., et al. 2020. Seed priming with gibberellic acid and melatonin in rapeseed: Consequences for improving yield and seed quality under drought and non-stress conditions. *Industrial Crops and Products* 156: 112850. doi: 10.1016/j.indcrop.2020.112850.

Kosakivska, I. V., Voytenko, L. V., Vasyuk, V. A., and Shcherbatiuk, M. M. 2022. Effect of pre-sowing priming of seeds with exogenous abscisic acid on endogenous hormonal balance of spelt wheat under heat stress. *Zemdirbyste-Agriculture* 109. doi: 10.13080/z-a.2022.109.003.

Koussevitzky, S., Suzuki, N., Huntington, S., et al. 2008. Ascorbate peroxidase 1 plays a key role in the response of Arabidopsis thaliana to stress combination. *Journal of Biological Chemistry* 283: 34197–34203. doi: 10.1074/jbc.M806337200.

Lal, D., Dadrwal, B. K., Saha, D., et al. 2021. Molecular advances in plant root system architecture response and redesigning for improved performance under unfavorable environments. In Aftab, T. and Hakeem, K. R. (eds), *Frontiers in Plant-Soil Interaction*, 49–82. Elsevier. doi: 10.1016/B978-0-323-90943-3.00013-4.

Lefevere, H., Bauters, L., and Gheysen, G. 2020. Salicylic acid biosynthesis in plants. *Frontiers in Plant Science* 11: 338. doi: 10.3389/fpls.2020.00338.

Lei, K., Sun, S., Zhong, K., et al. 2021. Seed soaking with melatonin promotes seed germination under chromium stress via enhancing reserve mobilization and antioxidant metabolism in wheat. *Ecotoxicology and Environmental Safety* 220: 112241. doi: 10.1016/j.ecoenv.2021.112241.

Lei, Y., Xu, Y., Hettenhausen, C., et al. 2018. Comparative analysis of alfalfa (Medicago sativa L.) leaf transcriptomes reveals genotype-specific salt tolerance mechanisms. *BMC Plant Biology* 18: 35. doi: 10.1186/s12870-018-1250-4.

Li, X., Lawas, L. M. F., Malo, R., et al. 2015. Metabolic and transcriptomic signatures of rice floral organs reveal sugar starvation as a factor in reproductive failure under heat and drought stress. *Plant, Cell and Environment* 38: 2171–2192. doi: 10.1111/pce.12545.

Li, X., Yang, Y., Sun, X., et al. 2014. Comparative physiological and proteomic analyses of poplar (Populus yunnanensis) plantlets exposed to high temperature and drought. *PLoS One* 9: e107605. doi: 10.1371/journal.pone.0107605.

Li, Y., Liang, L., Li, W., et al. 2021. ZnO nanoparticle-based seed priming modulates early growth and enhances physio-biochemical and metabolic profiles of fragrant rice against cadmium toxicity. *Journal of Nanobiotechnology* 19: 75. doi: 10.1186/s12951-021-00820-9.

Liu, Y., Ma, M., Ran, Y., Yi, X., Wu, S., and Huang, P. 2021. Disentangling the effects of edaphic and vegetational properties on soil aggregate stability in riparian zones along a gradient of flooding stress. *Geoderma* 385: 114883. doi: 10.1016/j.geoderma.2020.114883.

Llorens, E., González-Hernández, A. I., Scalschi, L., et al. 2020. Priming mediated stress and cross-stress tolerance in plants: Concepts and opportunities. In Hossain, M. A., et al. (eds), *Priming-Mediated Stress and Cross-Stress Tolerance in Crop Plants*, 1–20. Elsevier. doi: 10.1016/B978-0-12-817892-8.00001-5

Marchin, R. M., Ossola, A., Leishman, M. R., and Ellsworth, D. S. 2020. A simple method for simulating drought effects on plants. *Frontiers in Plant Science* 10: 1715. doi: 10.3389/fpls.2019.01715.

Mei, J., Wang, W., Peng, S., and Nie, L. 2017. Seed pelleting with calcium peroxide improves crop establishment of direct-seeded rice under waterlogging conditions. *Scientific Reports* 7: 1–12. doi: 10.1038/s41598-017-04966-1.

Miray, R., Kazaz, S., To, A., and Baud, S. 2021. Molecular control of oil metabolism in the endosperm of seeds. *International Journal of Molecular Sciences* 22: 1621. doi: 10.3390/ijms22041621.

Mustafa, G., Sakata, K., and Komatsu, S. 2015. Proteomic analysis of flooded soybean root exposed to aluminum oxide nanoparticles. *Journal of Proteomics* 128: 280–297. doi: 10.1016/j.jprot.2015.08.010.

Nie, L., Song, S., Yin, Q., et al. 2022. Enhancement in seed priming-induced starch degradation of rice seed under chilling stress via GA-mediated α-amylase expression. *Rice* 15: 1–13. doi: 10.1186/s12284-022-00567-3.

Nouairi, I., Jalali, K., Essid, S., Zribi, K., and Mhadhbi, H. 2019. Alleviation of cadmium-induced genotoxicity and cytotoxicity by calcium chloride in faba bean (*Vicia faba* L. var. Minor) roots. *Physiology and Molecular Biology of Plants* 25: 921–931. doi: 10.1007/s12298-019-00681-5.

Pan, T., Liu, M., Kreslavski, V. D., et al. 2021. Non-stomatal limitation of photosynthesis by soil salinity. *Critical Reviews in Environmental Science and Technology* 51: 791–825. doi: 10.1080/10643389.2020.1735231.

Paul, S., and Aryadeep Roychoudhury, A. 2017. Seed priming with spermine and spermidine regulates the expression of diverse groups of abiotic stress-responsive genes during salinity stress in the seedlings of indica rice varieties. *Plant Gene* 11: 124–132. doi: 10.3390/antiox10101544.

Rao, A., Salah-ud-Din, Akhtar, S., et al. 2016. Genomics of salinity tolerance in plants. In Abdurakhmonov, I. Y. (ed), *Plant Genomic*, 273–299. London: InTech Open. doi: 10.5772/63361.

Ritonga, F. N., and Chen, S. 2020. Physiological and molecular mechanism involved in cold stress tolerance in plants. *Plants* 9: 560. doi: 10.3390/plants9050560.

Sehgal, A., Sita, K., Bhandari, K., et al. 2019. Influence of drought and heat stress, applied independently or in combination during seed development, on qualitative and quantitative aspects of seeds of lentil (Lens culinaris Medikus) genotypes, differing in drought sensitivity. *Plant, Cell and Environment* 42: 198–211. doi: 10.1111/pce.13328.

Sękara, A., Bączek-Kwinta, R., Gawęda, M., Kalisz, A., Pokluda, R., and Jezdinský, A. 2016. Sequential abiotic stress applied to juvenile eggplant modifies the seedlings parameters, plant ontogeny and yield. *Horticultural Science* 43: 149–157. doi: 10.17221/162/2015-HORTSCI.

Seleiman, M. F., Al-Suhaibani, N., Ali, N., et al. 2021. Drought stress impacts on plants and different approaches to alleviate its adverse effects. *Plants* 10: 259. doi: 10.3390/plants10020259.

Sharma, A. D., Rathore, S. V. S., Srinivasan, K., and Tyagi, R. K. 2014. Comparison of various seed priming methods for seed germination, seedling vigour and fruit yield in okra (Abelmoschus esculentus L. Moench). *Scientia Horticulturae* 165: 75–81. doi: 10.1016/J.SCIENTA.2013.10.044.

Singh, S., Parihar, P., Singh, R., Singh, V. P., and Prasad, S. M. 2016. Heavy metal tolerance in plants: Role of transcriptomics, proteomics, metabolomics, and ionomics. *Frontiers in Plant Science* 6: 1143. doi: 10.3389/fpls.2015.01143.

Sirhindi, G., Mushtaq, R., Gill, S. S., Sharma, P., Abdu-Allah, E. F., and Ahmad, P. 2020. Jasmonic acid and methyl jasmonate modulate growth, photosynthetic activity and expression of photosystem II subunit genes in Brassica oleracea L. *Scientific Reports* 10: 9322. doi: 10.1038/s41598-020-65309-1.

Song, Y.; Jiang, M., Zhang, H., Li, R. 2021. 2021. Zinc oxide nanoparticles alleviate chilling stress in rice (*Oryza sativa* L.) by regulating antioxidative system and chilling response transcription factors. *Molecules* 26: 2196. doi: 10.3390/molecules26082196.

Sorrentino, M., Diego, N. D., Ugena, L., et al. 2021. Seed priming with protein hydrolysates improves Arabidopsis growth and stress tolerance to abiotic stresses. *Frontiers in Plant Science* 12: 626301. doi: 10.3389/fpls.2021.626301.

Sun, L., Song, F., Zhu, X., et al. 2021. Nano-ZnO alleviates drought stress via modulating the plant water use and carbohydrate metabolism in maize. *Archives of Agronomy and Soil Science* 67: 245–259. doi: 10.1080/03650340.2020.1723003.

Taylor, A. G., Allen, P. S., Bennett, M. A., Bradford, K. J., Burris, J. S., and Misra, M. K. 1998. Seed enhancements. *Seed Science Research* 8(2): 245–256. doi: https://doi.org/10.1017/S0960258500004141.

Thomas, D. T., and Puthur, J. T. 2019. Amplification of abiotic stress tolerance potential in rice seedlings with a low dose of UV-B seed priming. *Functional Plant Biology* 46: 455–466. doi: 10.1071/FP18258.

Vallad, G. E., and Goodman, R. M. 2004. Systemic acquired resistance and induced systemic resistance in conventional agriculture. *Crop Science* 44: 1920–1934. doi: 10.2135/cropsci2004.1920.

Van Buer, J., Cvetkovic, J., and Baier, M. 2016. Cold regulation of plastid ascorbate peroxidases serves as a priming hub controlling ROS signaling in Arabidopsis thaliana. *BMC Plant Biology* 16: 163. doi: 10.1186/s12870-016-0856-7.

Wahid, A., Noreen, A., Basra, S. M. A., Gelani, S., and Farooq, M. 2008. Priming-induced metabolic changes in sunflower (Helianthus annuus) achenes improve germination and seedling growth. *Botanical Studies* 49: 343–350.

Wang, X., Li, F., Chen, Z., Yang, B., Komatsu, S., and Zhou, S. 2021. Proteomic analysis reveals the effects of melatonin on soybean root tips under flooding stress. *Journal of Proteomics* 232: 104064. doi: 10.1016/j.jprot.2020.104064.

Wei, T., Jiang, C., Jin, Y., Zhang, G., Wang, M., and Liang, Z. 2020. Ca2+/Na+ ratio as a critical marker for field evaluation of saline-alkaline tolerance in alfalfa (Medicago sativa L.). *Agronomy* 10: 191. doi: 10.3390/agronomy10020191.

Xiong, L., Wang, R., Mao, G., and Koczan, M. J. 2006. Identification of drought tolerance determinants by genetic analysis of root response to drought stress and abscisic acid. *Plant Physiology* 142: 1065–1074. doi: 10.1104/pp.106.084632.

Xu, W., Tang, W., Wang, C., et al. 2020. SiMYB56 confers drought stress tolerance in transgenic rice by regulating lignin biosynthesis and ABA signaling pathway. *Frontiers in Plant Science* 11: 785. doi: 10.3389/fpls.2020.00785.

Zhang, Q., and Dai, W. 2019. Plant response to salinity stress. In Dai, W. (ed), *Stress Physiology of Woody Plants*. Boca Raton, FL: CRC Press. doi: 10.1201/9780429190476.

15 Genetic Resources of Potato with Tolerance to Elevated Temperature, Frost, and Drought

Elena V. Rogozina
N.I.Vavilov All-Russian Institute of Plant Genetic Resources (VIR)

CONTENTS

15.1 INTRODUCTION

The potato is the most important non-cereal food crop in the world cultivated in 149 countries, and a staple food for 1.3 billion people (with consumption of more than 50 kg/person per year) (Devaux et al., 2020). In 2020, potatoes occupied about 16.5 million hectares in the global agricultural production, the harvest amounted to more than 359 million tons, of which almost a half (178.5 million tons) was harvested in Asia, the largest producers being China and India (78.2 and 51.3 million tons, respectively) (https://www.fao.org/faostat). By 2050, the projected global population of 9.7 billion people will require 70% more food than is produced today (Devaux et al., 2020). When the climate is unstable and natural resources keep depleting, potato production is able to meet the growing demand for food. Potato's popularity is due to its relatively short growing season: the very early varieties can be harvested in 50–60 days after planting. Potato is a flexible crop that is grown in various soil and climatic conditions: from northern Finland to the highlands of Ecuador and Kenya, in an area that spreads from 69°N to 50°S, starting from the sea level and rising up to 4,000 m above the sea level (m a.s.l.) (Hijmans and Spooner, 2001).

The potato is a light-demanding crop of the temperate climate, and the best conditions for its cultivation are in the areas with the average daily temperature above 5°C and below 21°C (Vos and Haverkort, 2007). The potato is grown in different agroecological zones of the globe: as a spring crop in the tropical highlands of Bolivia, Peru, and Mexico; as a summer crop in temperate zones of China, Europe, and America; as a winter crop at sea level of India, South China, and Bangladesh, and it is planted in spring and autumn in the Mediterranean region (Devaux et al., 2020;

DOI: 10.1201/9781003300564-15

Monneveux et al., 2014). The growing season of potatoes lasts for 70 days for the early varieties and up to 120–180 days for the late ripening ones. In many developed countries, potatoes are a cash crop that generates high profits. The economic effect depends on the region and cultivation technologies, the quality of seed material, and the productivity of the variety, optimal agricultural practices, and organization of production. For a potato plant, the most limiting environmental factors are the temperature and water-deficit stresses (Mihovilovich et al., 2014). In modern environment of unstable climate with frequent abnormally high or low temperatures and uneven rainfall, the tolerance of potatoes to abiotic stressors becomes an extremely urgent problem (Campbell et al., 2021; Handayani et al., 2019; Hill et al., 2021; Obidiegwu et al., 2015; Sharma et al., 2017).

15.2 EFFECT OF TEMPERATURE STRESS AND WATER DEFICIT ON THE GROWTH AND DEVELOPMENT OF POTATO

Cultivated potatoes originated in the Andean regions, where the climate is cool and the average daily temperature is 15°C–18°C (Mihovilovich et al., 2014), and on the coast of Chile and the islands in the Chilean archipelago with moderate summer temperatures, high air humidity, and abundant rainfall of 2,000–4,500 mm (Bukasov, 1971). One of the first descriptions of environmental factors necessary for the growth and development of potatoes was given in 1882 by I.A. Stebut in his fundamental work 'Basis of Field Crops and Their Improvement in Russia', in two volumes. According to the author, the cultivation of potatoes requires a sum of temperatures of 1,040°R–1,280°R (according to the Réaumur scale, i.e., 1,300°C–1,600°C); tubers begin to germinate at 3.2°R = 4°C; the first leaves appear at 9.2°R–10.4°R = 11.5°C–13°C; the flowering phase begins at an average air temperature of 11°R = 13.7°C, and 11.8°R = 14.7°C of soil (Stebut, 1882).

The sensitivity of potatoes to the temperature factor largely depends on the stage of development, that is, on physiological processes in the plant (Struik, 2007). Tubers start germinating at 3°C–5°C, but there is no root formation at such low temperature and, as a result, there is no normal shoot growth and emergence of sprouts. The minimum soil temperature required for sprouting is 7°C–8°C, and the optimal range is within the 18°C–25°C, at which sprouts appear in 12–13 days. At higher temperatures, the emergence of sprouts is delayed, for example, at 27°C–28°C they appear in 16–17 days (Potato production, 2011). The optimal air temperature for potato tubers' germination is 16°C–20°C, for sprouting and leaf area development it is 20°C–25°C, 32°C for shoot growth, 20°C–24°C for photosynthesis, 25°C for stolon initiation and growth, 15°C for tuber induction, 22°C for tuber initiation, and 14°C–22°C for tuber bulking (Struik, 2007). An increase of the nighttime temperature and high soil temperatures causes much more damage to crops than high daytime temperatures (Reynolds and Ewing, 1989b).

Potatoes cannot withstand negative temperatures (<0°C = freezing) and are sensitive to low temperatures (0°C–10°C = chilling). The Andean cultivated potato *Solanum tuberosum* subsp. *andigena* Hawkes (from here on, the classification of potato species corresponds to the system by Hawkes, 1990) is damaged at −2°C, while a more sensitive Chilean potato *S. tuberosum* L. is damaged at −1°C (Hijmans, 1998). Low temperatures slow down the aboveground plant parts and the process of tuberization. Varietal differences in response to low temperatures appear at 3°C–4°C (Li et al., 1979). At temperatures below 7°C, the shoot system growth stops almost completely (Alsmik et al., 1979). Changes in temperature regimes during the growing season are extremely unfavorable for potatoes (Mihovilovich et al., 2014).

The adverse effect of high temperatures on potatoes is manifested in the inhibition of photosynthesis and biomass accumulation, in suppression of tuberization and formation of abnormal tubers, in which the level of glycoalkaloids can increase and the dry matter content decrease (Levy and Veilleux, 2007). A response of potato to the temperature stress depends on the variety, plant development stage, and growing location (Chang et al., 2014; Condori et al., 2014; Levy, 1986; Rykaczewska, 2013; Struik, 2007; van Dam, et al., 1996). The optimal temperature for obtaining a high yield of potatoes is 20°C–24°C (Struik, 2007); it is 15°C–18°C for tuberization in temperate zone (Levy and Veilleux, 2007), 18°C–23°C for more southern regions, and 12°C–14°C for northern ones (Alsmik et al., 1979).

Varieties, landraces, and wild potato species show more variability in response to high temperatures (Campbell et al., 2021). Varietal differences, including earliness and unequal growth rate (different duration of development phases), determine sensitivity of potato plants to temperature. For instance, when plants are grown under conditions of contrasting photoperiods and different temperatures, potato variety Desire formed the largest number of tubers in a short photoperiod at 23°C, while an earlier maturing variety Spunta showed similar results in a long photoperiod at 15°C (van Dam et al., 1996). An increase in day/night temperature from 22°C/16°C to 30°C/20°C can reduce the tuber yield in Desire variety by 20%, as well as less starch and dry matter content (Hancock et al., 2014). At day/night temperatures of 30°C/21°C, tuberization of the early growing variety Superior slows down, the yield decreases by 30%, and with less number of large tubers (Kim and Lee, 2019). Among 117 clones of *S. tuberosum*, tolerance to high temperature was found in the clones with early tuberization and a long growing season, which ensured yields that were higher than those of clones with late tuberization or with early tuberization and a short growing season (Lyra et al., 2015).

The influence of temperature on tuberization depends on environmental factors determined by geographical parameters. For instance, in India, potato stems and leaves are growing quicker with an increase in day/night temperature from 20°C/10°C to 30°C/20°C, while the optimal night temperature for tuberization is 18°C–20°C (Singh et al., 2018).

Potatoes are water efficient compared to other crops (Vos and Haverkort, 2007), but have a weak root system and, therefore, are particularly demanding to soil conditions (Struik, 2007). The root mass of potatoes in relation to mass of the whole plant is less than that of many agricultural crops: the volume of soil encompassed by the root system of potatoes is almost 1.4 times less than that encompassed by the root system of barley, and 2.2 times less than that of sugar beet (Alsmik et al., 1979). The degree of potato plant sensitivity to soil moisture is tightly related to air and soil temperature, especially during the tuberization period (Alsmik et al., 1979).

The amount of precipitation required by potatoes during the growing season is 350–550 mm (Nasir and Toth, 2022) and up to 800 mm (Handayani et al., 2019), which is due to varietal differences, soil type, and climate. Drought adversely affects all physiological processes in plants, that is, the induction and development of the aboveground plant parts, photosynthesis, metabolites transportation, and yield formation. The degree of the negative impact of drought depends on its duration, the stage of plant development, and the ability of a variety to withstand unfavorable abiotic environmental factors (Obidiegwu et al., 2015; Soltys-Kalina et al., 2016; Vos and Haverkort, 2007). There is no consensus opinion about the relationship between the early maturity of a variety and drought tolerance. High temperatures in combination with a water deficit aggravated yield loss, especially in the late maturing varieties (Levy et al., 1990). On the other hand, late maturing varieties have been reported to produce root mass deeper and stem numbers greater than early maturing ones, thus early varieties could be more sensitive to drought at the end of the growing season (Nasir and Toth, 2022). The simultaneous impact of high temperatures and water deficiency has a more serious negative effect than each of the adverse factors has separately (Demirel et al., 2020; Levy, 1986).

Potatoes cultivated in South America (the center of the crop origin) and in the rest of the world represent a significant variety of forms that differ in the level of ploidy, adaptation to certain environmental conditions, and growing technologies. The response to stress factors (drought or frost) in potato varieties and native landraces depends to a large extent on the physiological conditions of plants, its development stage, growing location, soil type, and degree of stress (Ellis et al., 2020). The areas of distribution for potato wild relatives covered 16 countries in Northern and Southern Americas: Mexico, the USA, Costa Rica, Guatemala, Honduras, Panama, Argentina, Bolivia,

Brazil, Chile, Colombia, Ecuador, Paraguay, Peru, Uruguay, and Venezuela (Ellis et al., 2020; Hijmans and Spooner, 2001). Potato species are adapted to various natural climatic conditions, including those with a wide range related to m a.s.l., from at sea level to snow line, and from the territories of the Pacific coast to the highlands of the Andes. The wild species (particular, sections *Acaulia* Juz. and *Megistacroloba* Cárd. et Hawkes) exist and grow in very harsh climatic conditions of the highlands at the border of eternal snow. Species of *Cuneolata* Hawkes section can grow in dry cactus deserts and scrub across northern Peru to northwest Argentina and northern Chile (Hawkes, 1990). Such species as *S. medians* Bitt., *S. chancayense* Ochoa, and *S. mochiquense* Ochoa grow very well on the Pacific coast of Peru with poor rainfall, but abundant soil moisture during a short period of continuous fogs. Plants of *S. morelliforme* Bitt. et Muench are epiphytic, and they can grow in hollows and on tree trunks (Bukasov and Kameraz, 1972). The genetic diversity of potato species was determined by various climatic and ecological environments of their formation during evolution. However, potato genepool diversity was studied much better for resistance to pests and pathogens rather than tolerance to abiotic stresses (Handayani et al., 2019). In this regard, an analysis of experimental data on the reaction of wild and cultivated potato species on elevated or low temperatures and on water deficiency can be estimated as the first step toward the preparation and production of new modern potato varieties with improved tolerance to adverse climatic conditions for plant growth.

15.3 GENETIC RESOURCES OF FROST TOLERANCE IN POTATO

The first experiments to study frost tolerance in potato were performed by Klotzsch, who accidentally discovered frost-tolerant plants of wild species *S. utile* Klotzsch (= *S. demissum* Lindl.) at –3°C (–2°R = Réaumur scale). He produced interspecific hybrids, *S. utile* × *S. tuberosum*, and found that the progeny seedlings were very tolerant to repeated frosts up to –2°C (–1.5°R) without any visual damages (Bukasov, 1932). In another wild species, *S. commersonii*, the frost tolerance was reported in one of four studied accessions by Reddick in 1930 (Bukasov, 1932; Hawkes and Hjerting, 1969). Other studies of frost-tolerant tuber-forming species in the genus *Solanum* L. have been carried out, and the results published by many VIR scientists in 1930s (Bukasov and Kameraz, 1972). The highest level of frost tolerance was found in *S. acaule* plants, which withstood frosts down to –8°C and –10°C, while *S. commersonii* plants showed only higher temperature of frost tolerance at –3°C and –5°C (Table 15.1).

In 1968–1971, a VIR researcher M.A. Vavilova carried out very wide analysis and studied frost tolerance in 84 wild and cultivated species belonging to 20 sections. She found that plants of four wild species, *S. acaule, S. sanctae-rosae, S. chomatophilum*, and *S. commersonii*, can withstand temperature drop-down to –5.0°C for 6–8 hours with a total freezing time below 0°C up to 18 hours (Bukasov and Kameraz, 1972). The tolerance to frost of other six species, *S. boliviense, S. megistacrolobum, S. sogarandinum, S. toralapanum, S. multidissectum*, and *S. demissum*, was equally or slightly weaker, where some plants withstood a decrease in temperature only to –3.0°C to –4.0°C with the total duration of freezing up to 6–12 hours (Table 15.1).

Among other studied species, *S. juzepczukii*, representing the cultivated species, as well as three wild species *S. cañasense, S. leptophyes*, and *S. spegazzinii* were found to be more sensitive to frost. Finally, among the potato species studied by M.A. Vavilova, two cultivated species, *S. ajanhuiri* and *S. curtilobum*, together with one wild species, *S. sparcipilum*, were the most sensitive to frosts, and they withstood a temperature drop-down only to –2.0° to –3.0°C with duration of the freezing period no longer than 3–6 hours. In other study, the level of limiting freezing temperature below 0°C, at which plant death occurs, has been established for 60 various tuber-forming potato species (Li, 1977). In general, a higher level of tolerance to frost in species of sections *Acaulia* and *Megistacroloba*, as well as in plants of two potato species, *S. commersonii* and *S. vernei*, has been confirmed and reported (Table 15.1).

TABLE 15.1
Genetic Resources for Frost Tolerance in Wild Species of Potato, *Solanum*

Sections	Species	Temperature Tolerated by Plants	Temperature Killing Plants	Temperature Killing Plants after Acclimation[a]
Acaulia Juz.	*S. acaule* Bitt.	−5.0[b]; −8.0 to 10.0[c]	−4.5 to 6.5[d]	−9.0
	S. acaule subsp., *S aemulans* (Bitt. et Wittm.) Hawkes et Hjerting	−4.0[e]	No data	No data
Megistacroloba Cárd. et Hawkes	*S. sanctae-rosae* Hawkes	−4.0[e] to 5.0[b]	−5.5[d]	−5.5
	S. boliviense Dun.	−3.0 to 5.0[b]	−4.5[d]	−4.5
	S. megistacrolobum Bitt.	−3.0 to 5.0[b]	−3.5 to 5.0[d]	−5.5
	S. raphanifolium Cárd. et Hawkes	No data	−3.5 to 4.0[d]	No data
	S. sogarandinum Ochoa		−3.5[d]	No data
	S. toralapanum Cárd. et Hawkes		−4.0 to 5.0[d]	No data
Demissa Buk.	*S. demissum* Lindl.		−3.5 to 4.0[d]	No data
Commersoniana Buk.	*S. commersonii* Dun.	−2.8[e] to 5.0[b]	−4.0 to 5.4[e]	−11.5
Yungasensa Corr.	*S. chacoense* Bitt.	−2.8[e]	No data	No data
Conicibaccata Bitt.	*S. chomatophilum* Bitt.	−5.0[b]	−5.0[d]	−8.5
Tuberosa (Rydb.) Hawkes (wild species)	*S. cañasense* Hawkes	−3.0 to 4.0[b]	−3.5 to 4.5[d]	No data
	S. kurtzianum Bitt. et Wittm.	No data	−3.5[d]	−3.5
	S. leptophyes Bitt.	−2.0 to 4.0[b]	No data	No data
	S. microdontum Bitt.	−2.8[e]	−3.0[d]	−3.0
	S. multidissectum Hawkes	−3.0 to 5.0[b]	−3.5 to 4.0[d]	−8.5
	S. oplocense Hawkes	No data	−3.5[d]	−8.0
	S. sparcipilum (Bitt.) Juz. et Buk.	−2.0 to 3.0[b]	−3.5[d]	No data
	S. spegazzinii Bitt.	−2.0 to 4.0[b]	−3.0 to 3.5[d]	No data
	S. sucrense Hawkes	No data	−3.0[d]	−3.0
	S. ventirii Hawkes et Hjerting	No data	−3.5[d]	−3.5
	S. vernei Bitt. et Wittm.	−2.5 to 5.0[b]	−3.5 to 5.0[d]	−3.5
	S. verrucosum Schlechtd.	No data	−3.0[d]	−3.0
Tuberosa (cultivated species)	*S. ajanhuiri* Juz. et Buk.	−2.0 to 3.0[b]	−3.0 to 5.0[f]	No data
	S. curtilobum Juz. et Buk.	−2.0 to 3.0[b]; −5.0[c]	−3.0 to 5.0[f]	No data
	S. juzepczukii Juz. et Buk.	−3.0 to 4.0[b]; −5.0[c]	below −5.0[a]	No data
	S. stenotomum Juz. et Buk.	No data	−3.0 to 3.5[d]	−3.5
	S. tuberosum (Red Pontiac)	No data	−3.0[d]	−3.0
Pinnatisecta (Rydb.) Hawkes	*S. brachistotrichum* (Bitt.) Rydb.	No data	−3.5[d]	−3.5
	S. cardiophyllum Lindl.	No data	−2.5 to 3.0[d]	−3.0
	S. jamesii Torr.	No data	−3.0 to 3.5[d]	−3.0
	S. pinnatisectum Dun.	No data	−2.5 to 3.0[d]	−2. 5
	S. trifidum Corr.	No data	−3.0 to 3.5[d]	Death
Longipedicellata Buk.	*S. fendleri* Asa Gray	No data	−3.0[d]	−3.0
	S. hjertingii Hawkes	No data	−4.0[d]	No data
	S. polytrichon Rydb.	−2.8[e]	−3.0[d]	−6.0
	S. stoloniferum Schlechtd. et Bché.	No data	−3.0[d]	−3.0

Notes: [a]Chen and Li (1980a); [b]Bukasov and Kameraz (1972); [c]Bukasov (1932); [d]Li (1977); [e]Budykina et al. (1971); [f]Hijmans (1998).

According to Li et al. (1981), the species that can withstand −4.0°C should be estimated as frost tolerant, whereas those potato genotypes that can withstand only −2.5°C or above have to be designated as sensitive to frost. For most species (except five species, *S. sanctae-rosae*, *S. cañasense*, *S. sparcipilum*, *S. ajanhuiri*, and *S. juzepczukii*), the values of temperatures that plants can withstand and at which plant can die are quite close or identical (Table 15.1). Possible reasons may be related to genetic polymorphism of potato species, or might be simple differences in assessment methods for frost tolerance. In plants of species *S. acaule*, polymorphism was shown in differences between accessions for their ability to withstand negative temperatures in the range from −4.5°C to −10.0°C, while accessions of *S. vernei* can withstand temperatures in the range from −2.5°C to −5.0°C only (Table 15.1).

A polymorphism within five species, *S. alandiae* Cárd., *S. arnezii* Cárd., *S. berthaultii* Hawkes, *S. demissum*, and *S. sparcipilum*, has been established during experiments when plants were grown in subtropical areas of Northern India. Each species was found to contain plant material with clear damages ranging from 10% to 100%, in the period of January morning frosts (Luthra et al., 2007).

In a separate study, seedlings of 24 wild species were grown in containers, and the tolerance of plants to temperatures in the range from −2.1°C to −5.2°C was assessed in the phase of 5–6 leaves (Budykina et al., 1971). A clear differentiation of species was established and tolerance level (0%–10% damage from cooling at −2.1°C to −2.8°C) was shown in plants of seven species, *S. acaule*, *S. aemulans*, *S. sanctae-rosae*, *S. commersonii*, *S. megistracrolobum*, *S. chacoense*, and *S. polytrichon*. Plants highly tolerant to frost were typically identified in *S. acaule* (tolerated up to −5.2°), *S. aemulans,* and *S. sanctae-rosae* (up to −4.0°). The moderate level of frost tolerance was demonstrated in plants of five species, *S. commersonii*, *S. chacoense*, *S. megistracrolobum*, *S. vernei*, and *S. microdontum* (up to −2.8°C), which still was above average data. In general, Budykina et al. (1971) confirmed the results of other researchers indicating high tolerance to frost of species from sections *Acaulia* and *Megistacroloba*, and relatively moderate frost tolerance of the species in sections *Commersoniana* and *Demissa*.

Some accessions were found to be more sensitive to frost including species from sections *Yungasensa* (*S. chacoense*) and *Longipedicellata* (*S. polytrichon*), which demonstrated appropriate tolerance to temperatures below 0°C. However, authors suggested that it can be explained by genetic polymorphism within species and heterogeneous of seedling populations (Budykina et al., 1971).

The United States Potato Genebank collection was found to contain accessions of *S. albicans* Ochoa (section *Acaulia*) and *S. etuberosum* Lindl. (section *Etuberosa* Juz.) tolerant to high temperatures (Bamberg, 1994).

The ability to withstand negative temperatures increases in wild potato species after the acclimation period, or growing with a change in the temperature regime (de-acclimation), first at low temperatures (+2.0°C–+4.0°C), and then when the temperature rises up to 18°C–20°C (Chen and Li, 1980b; Vega and Palta, 2000). Plants of species *S. commersonii* were noted for its high tolerance to freezing and low temperatures, and the greatest acclimation ability. After growing within 15–20 days at the optimal acclimation temperature of +2.0°C, plants of *S. commersonii* withstood a temperature drop-down to −11.5°C (Table 15.1). Acclimation increases the frost tolerance in plants of other species, *S. acaule*, *S. megistacrolobum*, *S. chomatophilum*, *S. multidissectum*, *S. oplocense*, and *S. polytrichon* (Table 15.1). After an acclimation period, plants of *S. sanctae-rosae* and *S. toralapanum* demonstrated the best level of frost tolerance in the experiments of Vega and Palta (2000), when they withstood temperatures 1.6°C–2.0°C lower than those in the experiments by Chen and Li (1980a). Wild potato species also differ in their ability to withstand temperatures below 0°C after a short-term (within 12 hours) cultivation at +18°C (Vega and Palta, 2000).

15.4 GEOGRAPHY OF TOLERANT POTATO SPECIES AND GENETIC CONTROL OF FROST TOLERANCE

Frost tolerance was found in the species growing in the Andean highlands, in southern Peru, central and southern Bolivia, northern Chile, and northwestern Argentina (Palta and Li, 1979). Frost-tolerant *S. sanctae-rosae* and *S. boliviense* in *Megistacroloba* section grow in high mountain grassland or pastures, sandy and rocky places from 2,500 to 3,800 m a.s.l. (Hawkes, 1990). The areas of distribution of five frost-tolerant species, *S. acaule*, *S. megistacrolobum*, *S. toralapanum*, *S. ajanhuiri*, and *S. juzepczukii*, extend vertically to altitudes of 4,000–4,500 m a.s.l., and more, reaching the line of eternal snow. The habits of *S. acaule* plants and species in *Megistacroloba* sections are more typical with smaller rosettes and short stem (Figure 15.1).

A strong correlation was found to exist between the altitude of origin and tolerance to the temperature of −-5°C affecting the leaves of studied samples of eight species, *S. acaule*, *S. albicans*, *S. sanctae-rosae*, *S. multidissectum*, *S. alandiae*, *S. circaeifolium* Bitt., *S. bukasovii* Juz., and *S. marinasense* Vargas (Greaves and Wilson, 1987; Li et al., 1980; Smillie et al., 1983). A close relationship between the taxonomic species, geography of origin, altitude, and frost tolerance was shown during evaluation of 2,635 accessions of 101 potato species in a field trial experiment at Sturgeon Bay, Wisconsin, USA, in 1992 (Hijmans et al., 2003).

At the same time, notable is the frost tolerance in some samples of *S. vernei*, the upper limit of the distribution area of which is 3,500 m a.s.l., which was about 1,000 m below the growth limit of the most frost-tolerant species from sections *Acaulia* and *Megistacroloba* (Bukasov and Kameraz, 1972). In other study, plants of *S. commersonii* (CPC 3558), usually grown in lower woodlands level, up to an altitude of 400 m (Hawkes, 1990), had a high frost tolerance, at the level similar to those in high-altitude species (Greaves and Wilson, 1987). Plants of this species, *S. commersonii*, sometimes can produce erected habit of plants but generally they are decumbent or form lower rosette type of plants (Figure 15.2A).

In plants of hybrids and backcrosses involving two species contrasting in their reaction to cold exposure, that is, *S. commersonii* and *S. cardiophyllum* Lindl., frost tolerance and acclimation ability were identified as two independently inherited traits with polygenic, additive, and dominant genetic control (Stone et al., 1993, Valverde and Chen, 1999; Vega et al., 2003). In the BC_1 population originating from crosses between F_1 hybrids (*S. commersonii*×*S. cardiophyllum*) with the frost-tolerant parent *S. commersonii*, two QTLs were found associated with tolerance to temperature below 0°C and two other QTLs for the acclimation ability using RAPD and SSR markers. QTLs for each trait were detected in different genomic regions, and confirmed the independent genetic control of their phenotypic traits (Vega et al., 2003).

FIGURE 15.1 Wild species of potato, *Solanum*, in VIR germplasm collection. (A) *S. acaule* (k-11332); (B) *S. boliviense* (k-23188); and (C) *S. sanctae-rosae* (k-19341).

FIGURE 15.2 Wild species of potato, *Solanum*, in VIR germplasm collection. (A) *S. commersonii* (k-18975); and (B) *S. infundibuliforme* (k-12433).

15.5 GENETIC RESOURCES OF TOLERANCE TO HEAT AND DROUGHT IN POTATO

For screening potatoes for tolerance to high temperatures and drought, the developed techniques were based on changes in plant morphology and anatomy, on physiological and biochemical parameters (Handayani et al., 2019; Monneveux et al., 2014; Obidiegwu et al., 2015). The ability to grow and tuberize under unfavorable environmental conditions are two important traits for potato varieties that are tolerant to abiotic stresses. Screening and selection of breeding material tolerant to drought and high temperatures is carried out based on the phenotype tolerance index, which is based on the assessment of plants' ability to produce yield or other yield-related components under stress and in control non-stressed conditions (Cabello et al., 2013). The effect of drought on yield penalty can be different depending of germplasm accession, for example yield losses were noted lesser in landraces, while the cultivated varieties of potato showed significantly stronger reduction of tuber yield under drought (38% and 48%, respectively). However, landraces are characterized by a wider variability of the DSI (Drought Sensitivity Index), which reflects yield losses depending on drought intensity (Cabello et al., 2012).

In the field trials, potato plants were affected by not only the stress factor directly, but also plants under the influence of other additional stresses, such as salinity, micronutrient deficiency, soil compactness, and pathogens. All these unfavorable stressful factors disrupt plant growth, both above- and belowground, affect physiological processes, and thereby distort the accurate assessment of drought or heat effects on plants. Therefore, in order to identify the origin of tolerance to studying stress factors, it is necessary to assess accurately individual parameters that reflect changes in plants under the influence of a stress. At International Potato Center (CIP), in Lima (Peru), plant stress tolerance was assessed in both field trials and greenhouse experiments using more accurate and integrative measurements and image analysis software (Monneveux et al., 2014).

Experiments with heat and drought tolerance in potato were also performed in in vitro culture, sterile growth chambers, greenhouses, and fields, with whole intact plants and with individual organs, for example, roots, nodal cuttings, or detached leaves (Arvin and Donnelly, 2008; Handayani et al., 2019; Hetherington et al., 1983). Depending on the method and material, researchers can use various sets of parameters (Coleman, 2008; Hetherington et al., 1983; Midmore and Prange, 1991; Ibañez et al., 2021; Reynolds and Ewing, 1989a). Characteristics of stomata, photosynthesis, water use efficiency, relative water content, features of the root system, yield index, and biochemical parameters are of particular importance for drought, while heat tolerance is also associated with the stomata, photosynthesis, initiation of tuberization, and tuber bulking (Handayani et al., 2019). Genetic resources of heat tolerance were found among varieties, landraces, breeding lines, and cultivated and wild species of potatoes (Table 15.2).

One of the first studies of parameters for heat and drought tolerance in potato species was carried out by Krasnoshtein (1968) in the hot semi-desert zone of the Chelkar region in Kazakhstan. The field trial conditions were very extreme, the air temperature reached 40°C in daytime and 20°C at night, the amount of annual precipitation very low and varied in the range from 20.8 to 70.2 mm, and the relative humidity of the air was below 30%. Anatomical features and productivity of plants of wild and cultivated species and varieties of potato were studied only under irrigation. In such dry and hot conditions, high transpiration intensity (4.5–6 g) was recorded for plants of species *S. maglia*, *S. molinae*, *S. microdontum*, and *S. chacoense*. Differences in the size and density of stoma on the leaf underside in wild potato species were studied but no relationship was found between transpiration intensity and stoma size. High transpiration intensity was noted in both *S. molina* with large stomata (36 and 29 μm, length and width, respectively) and those of *S. chacoesne* with small stomata (16×18 μm). The largest number of small stomata (750 per 1 mm^2) was found in leaves of plants in *S. microdontum*, and few large stomata were discovered in *S. pinnatisectum* (271 stomata with size 35×20 μm). The stomata on leaves of plants in *S. demissum* were closed almost all day, while those of *S. molina* and *S. bulbocastanum* were widely open for a long time. The temperature threshold for the water-soluble protein coagulation in the leaves of wild potato species ranged from 48°C to 63°C. High tolerance to overheating (protein coagulation threshold above 58°C) was demonstrated in seven species, such as *S. maglia*, *S. molinae*, *S. kurtzianum*, *S. catarthrum*, *S. bulbocastanum*, *S. brachicarpum*, and *S. demissum*. The polymorphism in terms of protein coagulation temperature and stoma size was revealed within *S. chacoense* (Krasnoshtein, 1968).

According to Sergacheva (1954), some forms of *S. andigenum* were superior in heat tolerance to *S. tuberosum* in the conditions of the Aral Sea region under irrigation, which probably associated with a more powerful root system and a smaller leaf surface of strongly dissected leaves. The wild Chilean potato *S. molinae* is tolerant to high temperatures, but requires substantial moisture, while another wild potato from Chile *S. leptostigma* Juz. (syn. *S. tuberosum*) is more sensitive to heat.

The tolerance and sensitivity to heat and drought were assessed in wild and cultivated potato species in the conditions of southeastern Kazakhstan (Krasavin, 2009). In plants of six species, tolerant to heat and drought, *S. andigenum*, *S. berthaultii*, *S. demissum*, *S. vernei*, *S. chacoense*, and *S. tarijense*, the stomata on the leaves were open in the daytime, the leaves lost turgor only after 17–20 hours and restored within 3–4 hours, and protein coagulation occurred at 60°C–80°C (Krasavin, 2009).

The extensive analysis of sample collected from plants of 319 accessions covering 59 potato species was carried out for screening of their growth and tuberization ability at day/night temperature of 42°C/30°C (Reynolds and Ewing, 1989a). The intraspecific polymorphism of *S. chacoense* and other species for sensitivity to high temperatures was confirmed. It has been reported that the results of the nodal cutting tuberization assay do not always coincide with those of intact plants' tuberization, which confirms the importance of the adequate shoot growth of plants (Reynolds and Ewing, 1989a).

Assessments of growth, biomass accumulation, and photosynthetic parameters of the varieties, breeding lines, and wild and cultivated potato species have identified genotypes that were insensitive to elevated day/night temperatures of 33°C/25°C, respectively. These included plants from two species, *S. acaule* and *S. circaefolium*, varieties Norchip and Yungay, originated from the cross *S. tuberosum*×*S. andigena* (Midmore and Prange, 1991). Previously, Norchip variety was noted for the ability to form tubers at high temperatures of day and night (Reynolds and Ewing, 1989b). In experiments with varieties, landraces, and wild and cultivated potato species, the detached leaves from plants were subjected to a temperature of 37°C, or briefly to 41°C–42°C. Significant differences in the sensitivity of photosynthesis in the studied potato genotypes were established (Hetherington et al., 1983; Smillie et al., 1983). The most tolerant potato genotype with a lesser decrease in chlorophyll fluorescence was identified as the clone CH 140 of native potato

TABLE 15.2
Genetic Resources for Heat Tolerance in Wild Species of Potato, *Solanum*

Section	Species	Heat Tolerance Index	References
Acaulia	*S. acaule*	CO_2 assimilation and growth	Midmore and Prange (1991)
Olmosiana Ochoa	*S. olmosense* Ochoa	Chlorophyll fluorescence	Smillie et al. (1983)
Commersoniana	*S. commersonii*	Vigor of shoot growth. Electrolyte leakage assay	Reynolds and Ewing (1989a); Arivan and Donnely (2008)
Circaeifolia Hawkes	*S. circaeifolium*	CO_2 assimilation and growth	Midmore and Prange (1991)
Yungasensa	*S. chacoense*	Overheating tolerance Vigor of shoot growth Ability to tuberize Electrolyte leakage assay	Krasnoshtein (1968); Krasavin (2009); Reynolds and Ewing (1989a); Trapero-Mozos et al. (2018)
	S. tarijense	Chlorophyll fluorescence Overheating tolerance	Smillie et al. (1983); Krasavin (2009)
Maglia Bitt.	*S. maglia* Schlechtd.	Overheating tolerance	Krasnoshtein (1968)
Tuberosa (wild species)	*S. berthaultii*	Vigor of shoot growth. Ability to tuberize. Overheating tolerance	Reynolds and Ewing (1989a); Krasavin (2009)
	S. catarthrum	Overheating tolerance	Krasnoshtein (1968)
	S. gandarillasii Cárd	Overheating tolerance	Coleman (2008)
	S. kurtzianum	Overheating tolerance. Vigor of shoot growth. Ability to tuberize	Krasnoshtein (1968); Reynolds and Ewing (1989a)
	S. leptophyes	Electrolyte leakage assay	Trapero-Mozos et al. (2018)
	S. microdontum	High transpiration intensity Vigor of shoot growth	Krasnoshtein (1968); Reynolds and Ewing (1989a)
	S. spegazzinii	Ability to tuberize	Reynolds and Ewing (1989a)
	S. sucrense	Ability to tuberize	Reynolds and Ewing (1989a)
	S. vernei	Vigor of shoot growth. Overheating tolerance	Reynolds and Ewing (1989a); Krasavin (2009)
Tuberosa (cultivated species)	*S. molinae* Juz. (syn, *S. tuberosum)*	High transpiration intensity	Krasnoshtein (1968)
	S. phureja Juz. et Buk.	Chlorophyll fluorescence	Hetherington et al. (1983)
	Yungay *(tuberosum* × *andigena)*	CO_2 assimilation and growth	Midmore and Prange (1991)
	S. tuberosum Norchip	CO_2 assimilation and growth	Midmore and Prange (1991)
	S. tuberosum subsp. *tuberosum* (Chili) CH140	Chlorophyll fluorescence	Hetherington et al. (1983)
	LT1, LT5	CO_2 assimilation and growth	Midmore and Prange (1991)
Demissa	*S. brachycarpum* Corr.	Overheating tolerance. Vigor of shoot growth	Krasnoshtein (1968); Reynolds and Ewing (1989a)
	S. demissum	Overheating tolerance. Vigor of shoot growth. Ability to tuberize. Electrolyte leakage assay	Krasnoshtein (1968); Krasavin (2009); Reynolds and Ewing (1989a); Arivan and Donnely (2008)
Pinnatisecta	*S. jamesii*	Ability to tuberize. Electrolyte leakage assay	Reynolds and Ewing (1989a); Trapero-Mozos et al. (2018)
	S. stenophyllidium Bitt.	Electrolyte leakage assay	Trapero-Mozos et al. (2018)
	S. pinnatisectum	Electrolyte leakage assay	Trapero-Mozos et al. (2018)
Longipedicellata	*S. fendleri*	Vigor of shoot growth	Reynolds and Ewing (1989a)

(Continued)

TABLE 15.2 (*Continued*)
Genetic Resources for Heat Tolerance in Wild Species of Potato, *Solanum*

Section	Species	Heat Tolerance Index	References
	S. hertingii	Electrolyte leakage assay	Arivan and Donnely (2008)
	S. papita	Ability to tuberize	Reynolds and Ewing (1989a)
	S. stoloniferum	Vigor of shoot growth. Ability to tuberize	Reynolds and Ewing (1989a)
Polyadenia Buk. ex Correll	*S. polyadenium* Greenm.	High transpiration intensity	Krasnoshtein (1968)
Bulbocastana Hawkes (Rydb.)	*S. bulbocastanum* Dun.	Overheating tolerance. Vigor of shoot growth	Krasnoshtein (1968); Reynolds and Ewing (1989a)

from southern Chile (*S. tuberosum* subsp. *tuberosum*). Other following genotypes were also identified as highly tolerant to heat stress, including cultivated species *S. phureja*, six wild species, *S. chacoense*, *S. commersonii*, *S. demissum*, *S. hjertingii*, *S. tarijense*, and *S. olmosense*, as well as two cultivars, Norland and Bintje (Table 15.2).

The heat tolerance of five other species, *S. jamesii*, *S. stenophyllidium*, *S. pinnatisectum*, *S. leptophyes*, and *S. chacoense*, was shown in exposure of intact plants to heat stress temperatures of 40°C (Trapero-Mozos et al., 2018). Coleman (2008) reported that under high solar energy insolated together with drought, leaf temperatures of *S. gandarillasii* rose significantly above air temperatures that coincidence with results of leaf discs analyses for high tolerance of *S. gandarillasii* to heat stress. The collection at the United States Potato Genebank (NRSP-6) was also found to contain accessions of *S. guerreroense* Corr. (section *Demissa*), *S. mochicense* Ochoa (*Tuberosa* wild), and *S. morelliforme* (*Morelliformia*), which were more tolerant to high temperatures (Bamberg, 1994). In general, the results of repeated experiments confirm the heat tolerance results in wide range to *Solanum* species including *S. commersonii*, *S. chacoense*, *S. tarijense*, *S. berthaultii*, *S. kurtzianum*, *S. microdontum*, *S. vernei*, *S. demissum*, *S. brachycarpum*, *S. jamesii*, and *S. bulbocastanum* (Table 15.2).

In contrast, a relatively small number of drought-tolerant potato species were identified so far. According to Sergacheva (1954), the wild potato species *S. infundibuliforme* Phil. from section *Cuneolata* (Figure 15.2B) was found to show better tolerance to drought in the environment of semi-arid zone of the Aral Sea region. According to the report of Krasnoshtein (1968), indicators of drought tolerance of wild potato species were as follows: low transpiration intensity (2.4–3.6 g), small number of stomata, high concentration of cell sap, and small amplitude of water regime indicators (water content of leaf tissues) within a day. The plants of *S. bulbocastanum* and *S. pinnatisectum* were found to have the highest level of water regime, in which the water content was slightly decreased but water capacity was increased in leaf tissues in response to higher temperature (Krasnoshtein, 1968). A slight increase in water deficiency during a day was noted for plants of *S. bulbocastanum*, *S. demissum*, *S. chacoense*, *S. tarijense*, and *S. pinnatisectum* (Table 15.3).

Varieties, breeding lines, and accessions of potato cultivars were assessed for drought tolerance in a field test when grown on a desert coast of Peru (Cabello et al., 2012). *S. curtilobum* is the best genotype among cultivated species for tolerance to drought, but it slightly decreased the yield under a water deficiency. Another cultivated species, *S. juzepczukii*, produced a smaller yield under both water deficiency and irrigation, but the dry matter content in tubers was increased under stress (Table 15.3). Three accessions of *S. kurtzianum* tested for tolerance to moderate and severe water deficiency, demonstrated differences in morphological, anatomical, physiological, and biochemical parameters. One accession of *S. kurtzianum* was identified as drought-tolerant mainly explained through a rapid stomatal closure and lower stomatal density and size, and a better photoprotection (Ibañez et al., 2021). Field and laboratory experiments revealed differences between varieties,

TABLE 15.3
Genetic Resources for Drought Tolerance in Wild Species of Potato, *Solanum*

Sections	Species	Drought Tolerance Indicator	References
Acaulia	*S. acaule*	Electrolyte leakage assay	Arivan and Donnely (2008)
Yungasensa	*S. chacoense*	Tolerance to leaf wilting. Water use efficiency	Ekanayake and de Jong (1992); Krasnoshtein (1968)
	S. tarijense	Water use efficiency	Krasnoshtein (1968)
Tuberosa (wild)	*S. gandarillasii*	Low transpiration intensity	Coleman (2008)
	S. kurtzianum	Stomatal apparatus properties and photosynthesis indicators	Ibañez et al. (2021)
Tuberosa (cultivated)	*S. curtilobum*	Productivity and dry matter content	Cabello et al. (2012)
	S. juzepczukii	Dry matter content	Cabello et al. (2012)
	S. phureja P3-198	Growth and tuberization	Yang et al. (2019)
	S. stenotomum	Electrolyte leakage assay	Arivan and Donnely (2008)
Demissa	*S. demissum*	Electrolyte leakage assay. Water use efficiency	Arivan and Donnely (2008); Krasnoshtein (1968)
Pinnatisecta	*S. pinnatisectum*	Water use efficiency	Krasnoshtein (1968)
Bulbocastana	*S. bulbocastanum*	Water use efficiency	Krasnoshtein (1968)

landraces, and three wild species, *S. chacoense*, *S. raphanifolium* and *S. yungasense* Hawkes in sensitivity of leaves to a water deficit. This is because the stomata systems of *S. chacoense* was found to be the most tolerant to the stress (Ekanayake and de Jong, 1992). Tolerance and sensitivity to dehydration was studied in conditions of 35% polyethylene glycol (PEG) application for 24 hours. In such conditions, the isolated leaves from 10 varieties and germplasms, covering of 11 wild potato species, were tested. As a result, the detached leaves of varieties Alpha and Bintje showed better response and less water loss during the induced dehydration. Similar results were received in plants of three species, *S. acaule*, *S. demissum*, and *S. stenotonum*. Previously, similar dehydration test showed better adaptation of plants from two of these species, *S. acaule* and *S. demissum*, in both in vitro experiments and in greenhouse tests (Arvin and Donnelly, 2008).

The drought-tolerant genotype P3–198 from cultivated species *S. phureja* was selected in the progeny from crosses between contrasting diploid parental forms. The drought tolerance of *S. phureja* accession P3–198 was confirmed in experiments with the detached leaves for 0.5–2 hours on a medium with PEG as well as in field trial tests during 2 years in well-watered conditions (Control) and under drought (Yang et al., 2019).

15.6 GEOGRAPHIC DISTRIBUTION OF TOLERANT SPECIES AND GENETIC CONTROL OF HEAT AND DROUGHT TOLERANCE

The contradictory information has been found about the relationship between the geography of natural habitats, altitude, and tolerance to heat and drought in potato species. By analyzing the locations of natural growth of wild potato species, many authors note that heat- and drought-tolerant species originated from the areas with a hot and dry climate (Bukasov, 1971; Gorbatenko, 2006; Hawkes and Hjerting, 1969). According to Hawkes and Hjerting (1969), eight *Solanum* species, such as *S. infundibuliforme*, *S. chacoense*, *S. commersonii*, *S. tarijense*, *S. oplosense*, *S. spegazzinii*, *S. kurtzianum*, and *S. maglia*, growing in dry plateaus or high mountain slopes in southern Bolivia and northwestern Argentina, may serve as the original germplasms for breeding with improved tolerance to drought. Heat- and drought-tolerant genotypes of *S. chacoense* and *S. commersonii* grow in the arid regions of the northern Rio de la Plata basin. Wild potato species from sections

Conicibaccata and *Yungasensa*, as well as the Colombian cultivated species *S. estradae* and the Peruvian *S. hygrothermicum*, growing on the eastern slopes of the Andes in areas with annual total precipitation about 2,400 mm, were found to be more tolerant to high temperatures and humidity (Gorbatenko, 2006).

Ibañez et al. (2021) suggested that local adaptation of *S. kurtzianum* to natural habitat is taking place. Of the three *S. kurtzianum* accessions tested for tolerance to water deficiency, plants from environments that naturally have high evaporative demand (lower at an altitude of 1,100 m a.s.l.) was more tolerant than those from high altitudes (1,900 and 2,100 m a.s.l.).

However, a discussion of the results of testing more than 900 accessions from the CIP collection for tolerance and sensitivity to water deficit led Cabello et al. (2012) to a conclusion that differences in drought tolerance found in potato species and varieties were not directly related to their ecological and geographical origin. For instance, *S. curtilobum*, *S. ajanhuiri*, and *S. juzepczukii*, which are grown in the highlands of central Peru and Bolivia, were extremely different in drought tolerance. The authors believe that the relationship between drought tolerance and potato ploidy is more likely possible and, therefore, diploid species could be more sensitive to drought than polyploid potato genotypes.

Levy and Veilleux (2007) believe that tolerance to high temperature has polygenic control, particularly in multi-species hybrids with such genetic resources in their pedigree as in three species, *S. chacoense*, *S. berthaultii*, or *S. microdontum*. A search for QTLs associated with stress tolerance in potato was carried out in diploid and tetraploid populations (Campbell et al., 2021; Sharma et al., 2017; Schumacher et al., 2021). Based on the drought-tolerance index, eight QTLs were identified in the tetraploid population, of which three QTLs are autonomous and do not affect yield-relevant metabolism (i.e., do not affect the tuber yield and processes of starch synthesis and accumulation). QTL analyses confirmed the association of common abiotic stress responses with drought tolerance in potatoes due to the identification of candidate genes from signaling pathways for ethylene, abscisic acid, and jasmonic acid, that is, participants in the general stress response mechanism (Schumacher et al., 2021). It was found that chromosomes 1, 4, and 9 in the 06H1 diploid population contain QTLs associated with the potato yield at moderately high temperatures. The candidate gene encoding heat shock cognate 70 protein, *HSc70*, was reported with expression profile associated with yield stability at moderately high temperatures (Campbell et al., 2021). Different QTLs associated with changes in morphological and physiological parameters under drought conditions were identified in other diploid populations of hybrids obtained from three species, *S. phureja*, *S. vernei*, and *S. tuberosum*, or two ones, *S. tuberosum* and *S. goniocalyx* (Sharma et al., 2017). Most physiological parameters associated with drought tolerance in potato have low or moderate heritability (Monneveux et al., 2014; Sharma et al., 2017). Plant response to drought is more likely as a result of multiple processes governed by the coordinated action of many genes that control metabolic and signaling pathways (Schumacher et al., 2021).

15.7 PHYSIOLOGICAL MECHANISMS OF POTATO TOLERANCE TO EXTREME TEMPERATURES AND DROUGHT

For many years, the VIR germplasm collection of plant genetic resources was used in extensive studies of plant responses to drought, high and low temperatures, and salinity. The obtained results gave sufficient reason to make a conclusion about the non-specific (common) nature of plant adaptation to various types of stresses (Udovenko, 1995). The following uniform changes in physiological parameters under different stresses were established, including an increase of ion concentrations in tissues and the osmotic potential of cells, a decrease in the efficiency of cell transpiration and functional activity of DNA, a decrease in the pigment concentration, suppression of metabolite transport, etc. A fundamentally important conclusion was made about the general mechanisms of plant adaptation to various abiotic stresses, which represent a complex system of structures and functions acting at the cellular, organismal, and population levels (Udovenko, 1995). It has to be noted that most studies of mechanisms for tolerance to abiotic stresses was carried out on cereal, legume, vegetable, and fruit crops with reproductive organs for yield analysis. In contrast, in potatoes and related tuber-forming species, the major yield traits represented by vegetative parts such as tubes, where such species are obviously different. For the potato species in situ, seed propagation is a way of surviving under conditions that are unfavorable for vegetative propagation (Hawkes and Hjerting, 1969). In cultivated potatoes, high tuber yield can be combined with abundant berry formation, or with a complete lack of flowering. At the same time, numerous examples of the influence of environmental factors are known on the initiation and progression of generative and vegetative reproduction in cultivated and wild potatoes (Markarov et al., 2001).

The metabolic interaction between different shoots of the same plant and the interaction of vegetative organs with each other are poorly studied (Udovenko, 1995). The metabolic processes in the potato plant during the formation of generative and storage organs (tubers) under stress conditions have also been studied insufficiently. Among the diversity of potato species, various combinations of the photoperiodic reaction of flowering and tuberization are observed; the importance of the phytochrome system in the process of not only generative, but also vegetative reproduction in different wild potato species has been shown (Markarov et al., 2001). Obviously, the reaction of different potato species to unfavorable environmental conditions displays itself as a result of a complex interaction of exo- and endogenous factors in the process of realization of the developmental program embedded in the genotype.

At the cellular level, different types of potatoes were noted to have the ability to maintain the integrity of intracellular structures and stability in cytoplasmic osmoregulation under unfavorable temperatures and water deficiency. The significance of membrane stability for the tolerance of plants to elevated temperatures and drought has been noted in nine wild species as follows: *S. gandarillasii*, *S. jamesii*, *S. stenophyllidium*, *S. pinnatisectum*, *S. leptophyes*, *S. chacoense*, *S. commersonii*, *S. demissum*, and *S. hjertingii* (Arvin and Donnelly, 2008; Coleman, 2008; Trapero-Mozos et al., 2018). The property of cell membranes ensures tolerance of *S. demissum* to drought and temperatures −4°C and +37°C, *S. acaule* to drought and temperatures −4°C, and *S. commersonii* to temperatures −4°C and +37°C (Arvin and Donnelly, 2008). At the same time, these authors assessed the membrane stability in separated leaves of potato varieties and species and made a conclusion that there is a correlation with salt and drought tolerance and not with heat and drought, there being different mechanisms that provide tolerance to these stresses.

Frost-tolerant species *S. acaule* differs from more sensitive to frost *S. tuberosum* in its ability to maintain membrane integrity in cold conditions by changing the properties of membrane lipids (Li et al., 1981). Under cold stress, the membrane permeability of *S. tuberosum* decreases more significantly than that of *S. commersonii*, and the membrane permeability of *S. commersonii* increases upon acclimation (Fennell and Li, 1986). The structure of the RuBisCo protein (ribulose-1,5-bisphosphate carboxylase-oxygenase; catalyze CO_2 fixation during photosynthesis and of O_2 fixation during photorespiration) is more stable in *S. commersonii* than that in *S. tuberosum*, which plays an important role in the adaptation of wild species to temperatures below 0°C (Li et al., 1981). Negative temperatures have different effects on the cell structure and organelles integrity in *S. acaule* and in sensitive to frost *S. cardiophyllum* (Wu et al., 2019). After exposure to −5°C for 3 hours, the frost tolerance in plants of *S. acaule* was enhanced, while it was not changed in frost-sensitive plants of *S. cardiophyllum*. The subcellular characteristics related to the enhancement of freezing tolerance mainly include a decrease in the proportion of the vacuole to total cell volume, integrity of the biomembrane, and orderly arrangement of grana lamellae in photosystems due to the regulation of the lipid peroxidation system, permeability, and osmotic potential of cell membranes (Wu et al., 2019).

After acclimation, plants of *S. acaule* and *S. commersonii* were able to withstand temperatures down to −9.5°C and −12.0°C, while their content of soluble proteins, RNA, and phospholipids increases compared to control plants of *S. acaule* and variety Red Pontiac (Chen and Li, 1980a).

During cultivation at low temperatures of +4.0°C/+2.0°C day/night, potato plants of variety Zhengshu 6 were characterized by differential expression of many proteins involved in energy metabolism, photosynthesis, protein degradation, ribosome formation, signal transmission, cell movement, nitrogen metabolism, and other physiological processes. After exposure to low temperatures for 5–7 days, the content of soluble proteins, proline, and malondialdehyde (MDA), which is an indicator of membrane damage and a marker of oxidative stress, increased, and the content of chlorophyll decreased (Li et al., 2021). The acclimated plants of *S. acaule* had the MDA content 3–4 times lower, and the SOD (superoxide dismutase) content threefold higher than those in *S. cardiophyllum*; the proline content in *S. acaule* was five or more-fold higher than those in *S. cardiophyllum* (Wu et al., 2019).

According to Smillie et al. (1983), the tolerance to cold and high temperatures is represented by independent traits in wild potato species. However, the existence of a relationship between the ability to withstand temperatures of +41°C and 0°C was shown by assessing chlorophyll fluorescence parameter in potato varieties and landraces (Hetherington et al., 1983).

Palta and Li (1979) believe that the anatomical features of the leaves of wild potato species tolerant to temperatures below 0°C may be a sign of potential improvement in tolerance to water deficit. Plants of seven wild species, *S. acaule*, *S. commersonii*, *S. megistacrolobum*, *S. toralapanum*, *S. sanctae-rosae*, *S. multidissectum*, and *S. vernei*, were relatively sensitive to cold. They can die at −4.0°C or below, and the plants have two distinct layers of palisade parenchyma cells in leaves. However, other species, with death level reported at −3.5°C and above, were colder sensitive, and they had only one layer of palisade cells. This anatomical feature was inherited from the frost-tolerant parental form of *S. multidissectum* by hybrids [(*S. multidissectum*×*S. stenotomum*)×*S. phureja*], which could tolerate −4.0°C (Palta and Li, 1979). Frost-tolerant species had three times more stomata in the leaf epidermis than genotypes sensitive to frost, which probably promotes efficient gas exchange through more thicker palisade cell layer (Palta and Li, 1979).

Plants of frost-tolerant *S. toralapanum* which, like other species in section *Megistacroloba*, have a rosette form (Figure 15.1B and C) and were found in a landscape with globular cacti in Bolivia (Gorbatenko, 2006). Ross (Gorbatenko, 2006) divides drought-tolerant species into two groups: with normal and narrow developed leaflets and with narrow ones. The first group includes *S. tarijense*, *S. gandarillasii*, *S. pampasense*, *S. kurtzianum*, *S. sandemanii*, *S. acaule*, and *S. ochoae*, while the second group includes *S. spegazzinii*, *S. infundibuliforme*, *S. gourlayi*, and *S. xerophyllum* (Figures 15.1A and 15.2B). It is still to be found out whether morphological traits in wild potato species growing under conditions of heat and drought are combined with physiological features.

An extensive list of morphological and physiological traits in potatoes influenced by drought is offered in the review by Obidiegwu et al. (2015). The author notes that the information about the phenotypic manifestation of the effect of drought on cultivated forms of potatoes is contradictive. Such indicators as the yield index, root length, and stolon numbers increase or decrease under the influence of drought, depending on environmental conditions, soil type, stress duration and severity, and stage of plant development (Obidiegwu et al., 2015). In VIR germplasm collection, thousands of cereals, legume, vegetable, and fruit crop accessions were studied for tolerance to various extreme environmental factors, and promising genotypes were identified with combined tolerance to several abiotic stresses (Udovenko, 1995). Such potato species as *S. acaule*, *S. demissum*, *S. chacoense*, *S. commersonii*, *S. curtilobum*, and *S. juzepczukii* that show tolerance to several adverse factors are of particular interest for an 'in-depth' study of the mechanisms and genetic control of potato tolerance to adverse temperatures and moisture deficiency.

REFERENCES

Alsmik, P. I., Ambrosov, A. L., Vecher, A. S., Goncharik, M., and Mokronosov A.T. 1979. *Potato Physiology*. Moskow: Kolos. [In Russ.]

Arvin, M. J., and Donnelly, D. J. 2008. Screening potato cultivars and wild species to abiotic stresses using an electrolyte leakage bioassay. *Journal of Agricultural Science and Technology* 10: 33–42. Available at: http://jast.modares.ac.ir/article-23-7794-en.html.

Bamberg, J., 1994. *Elite Selections of Tuber-Bearing Solanum Species Germplasm: Based on Evaluations for Disease, Pest and Stress Tolerance*. Sturgeon Bay, WI: US Potato Genebank catalog.

Budykina, N., Drozdova, S., and Sinelnikova, V. 1971. Comparative tolerance of wild potato species to frost. *Bulletin of Applied Botany, Genetics and Plant Breeding* 46: 63–68. [In Russ.]

Bukasov, S. 1932. Frost tolerance in the potato. *Bulletin of Applied Botany of Genetics and Plant Breeding* 3: 287–297 [In Russ.]

Bukasov, S. M. 1971. Systematics of potato species of the *Tuberarium* (Dun.) Buk of the genus *Solanum* L. *Bulletin of Applied Botany, Genetics and Plant Breeding* 46: 3–45. [In Russ.]

Bukasov, S., and Kameraz, A. 1972. *Breeding and Production of Potato*. Kolos: Leningrad. [In Russ.]

Cabello, R., Mendiburu, F., Bonierbale, M., Monneveux, P., Roca, W., and Chujoy, E. 2012. Large-scale evaluation of potato improved varieties, genetic stocks and landraces for drought tolerance. *American Journal of Potato Research* 89: 400–410. doi: 10.1007/s12230-012-9260-5.

Cabello, R., Monneveux, P., Mendiburu, F., and Bonierbale, M. 2013. Comparison of yield based drought tolerance indices in improved varieties, genetic stocks and landraces of potato (*Solanum tuberosum* L.). *Euphytica* 193: 147–156 doi: 10.1007/s10681-013-0887-1.

Campbell, R., Ducreux, L., Mellado-Ortega, E., Hancock, R. D., and Taylor, M. A. 2021. Toward the design of potato tolerant to abiotic stress. In Dobnik, D., Gruden, K., Ramšak, Ž. and Coll, A. (eds), *Solanum Tuberosum. Methods in Molecular Biology*, vol. 2354, 387–399. New York, NY: Humana. doi: 10.1007/978-1-0716-1609-3_19.

Chang, D., Sohn, H., Cho, J., Im, J. S., Jin, Y. I., Do, G. R., et al. 2014. Freezing and frost damage of potato plants: A case study on growth recovery, yield response, and quality changes. *Potato Research* 57: 99–110 doi: 10.1007/s11540-014-9253-5.

Chen, H. H., and Li, P. 1980a. Biochemical changes in tuber-bearing *Solanum* species in relation to frost hardiness during cold acclimation. *Plant Physiology* 66: 414–421. doi: 10.1104/pp.66.3.414.

Chen, H. H., and Li, P. 1980b. Characteristics of cold acclimation and deacclimation in tuber bearing *Solanum* species. *Plant Physiology* 65: 1146–1148. doi: 10.1104/pp.65.6.1146.

Coleman, W. K. 2008. Evaluation of wild *Solanum* species for drought resistance: 1. *Solanum gandarillasii* Cardenas. *Environmental and Experimental Botany* 62: 221–230 doi: 10.1016/j.envexpbot.2007.08.007.

Condori, B., Hijmans, R. J., Ledent, J. F., and Quiroz, R. 2014. Managing potato biodiversity to cope with frost risk in the high Andes: a modeling perspective. *PloS One* 9: e81510. doi: 10.1371/journal.pone.0081510.

Demirel, U., Morris, W. L., Ducreux, L., Yavuz, C., Asim, A., Tindaset, I., et al. 2020. Physiological, biochemical, and transcriptional responses to single and combined abiotic stress in stress-tolerant and stress-sensitive potato genotypes. *Frontiers in Plant Science* 11: 169. doi: 10.3389/fpls.2020.00169.

Devaux, A., Goffart, J. P., Petsakos, A., Kromann, P., Gatto, M., Okello, J., et al. 2020. Global food security, contributions from sustainable potato agri-food systems. In Campos, H. and Ortiz, O. (eds), *The Potato Crop*, 3–35. Cham: Springer. doi: 10.1007/978-3-030-28683-5_1.

Ekanayake, I., and de Jong, J. 1992. Stomatal response of some cultivated and wild tuber-bearing potatoes in warm tropics as influenced by water deficits. *Annals of Botany* 70: 53–60. doi: 10.1093/oxfordjournals.aob.a088439.

Ellis, D., Salas, A., Chavez, O., Gomez, R., and Anglin, N. 2020. Ex situ conservation of potato [*Solanum* section *Petota* (Solanaceae)] genetic resources in genebanks. In Campos, H. and Ortiz, O. (eds), *The Potato Crop*, 109–138. Cham: Springer. doi: 10.1007/978-3-030-28683-5_4.

Fennell, A., and Li, P. H. 1986. Temperature response of plasma membranes in tuber-bearing *Solanum* species. *Plant Physiology* 80: 470–472. doi: 10.1104/pp.80.2.470.

Gorbatenko, L. 2006. *Potato Species Of South America (Ecology, Geography, Introduction, Systematics, Breeding Value)*. St. Petersburg: VIR, 456 pp. [In Russ.]

Greaves, J. A., and Wilson, J. M. 1987. Assessment of the frost sensitivity of wild and cultivated potato species by chlorophyll fluorescence analysis. *Potato Research* 30: 381–395. doi: 10.1007/BF02361917.

Hancock, R. D., Morris, W. L., Ducreux, L. J., Morris, J. A., Usman, M., Verrall, S. R., et al. 2014. Physiological, biochemical and molecular responses of the potato (*Solanum tuberosum* L.) plant to moderately elevated temperature. *Plant, Cell and Environment* 37: 439–450. doi: 10.1111/pce.12168.

Handayani, T., Gilani, S. A., and Watanabe, K. N. 2019. Climatic changes and potatoes: How can we cope with the abiotic stresses? *Breeding Science* 69: 545–563 doi: 10.1270/jsbbs.19070.

Hawkes J. G. 1990. *The Potato: Evolution, Biodiversity and Genetic Resources*. London: Belhaven Press, 259.

Hawkes, J. G., and Hjerting, J. P. 1969. *The Potatoes of Argentina, Brazil, Paraguay, and Uruguay*. Oxford: Clarendon Press, 525.

Hetherington, S. E., Smillie, R. M., Malagamba, P., and Huamán, Z. 1983. Heat tolerance and cold tolerance of cultivated potatoes measured by the chlorophyll-fluorescence method. *Planta* 159: 119–124. doi: 10.1007/BF00392981.

Hijmans, R. J. 1998. Estimating frost risk in potato production on the Altiplano. CIP program report 1997–98. *Natural Resource Management in the Andes*. Available at: https://cipotato.org/site/inrm/home/publicat/99cpa015.pdf (accessed April 12, 2022).

Hijmans, R. J., Jacobs, M., Bamberg, J. B., and Spooner, D. M. 2003. Frost tolerance in wild potato species: Assessing the predictivity of taxonomic, geographic, and ecological factors. *Euphytica* 130: 47–59 doi: 10.1023/A:1022344327669.

Hijmans, R. J., and Spooner, D. M. 2001. Geographic distribution of wild potato species. *American Journal of Botany* 88: 2101–2112. doi: 10.2307/3558435.

Hill, D., Nelson, D., Hammond, J., and Bell, L. 2021. Morphophysiology of potato (*Solanum tuberosum*) in response to drought stress: Paving the way forward. *Frontiers in Plant Science* 11: 597554. doi: 10.3389/fpls.2020.597554.

Ibañez, V. N., Kozub, P. C., González, C. V., Jerez, D. N., Masuelli, R. W., Berli, F. J., et al. 2021. Response to water deficit of semi-desert wild potato *Solanum kurtzianum* genotypes collected from different altitudes. *Plant Science* 308: 110911. doi: 10.1016/j.plantsci.2021.110911.

Kim, Y. U., and Lee, B. W. 2019. Differential mechanisms of potato yield loss induced by high day and night temperatures during tuber initiation and bulking: Photosynthesis and tuber growth. *Frontiers in Plant Science* 10: 300. doi: 10.3389/fpls.2019.00300.

Krasavin, V. F. 2009. *Potato Breeding in South-East of Kazakhstan*. Almaty: Oner. [In Russ.]

Krasnoshtein, R. G. 1968. *Potato Germplasm Collection Study in Semi-Arid Environment of Northern regions near Aral Sea*. PhD thesis. Leningrad: VIR. [In Russ.]

Levy, D. 1986. Genotypic variation in the response of potatoes (*Solanum tuberosum* L.) to high ambient temperatures and water deficit. *Field Crops Research* 15: 85–96 doi: 10.1016/0378-4290(86)90103-6.

Levy, D., Genizi, A., and Goldman, A. 1990. Compatibility of potatoes to contrasting seasonal conditions, to high temperatures and to water deficit: The association with time of maturation and yield potential. *Potato Research* 33: 325–334 doi: 10.1007/BF02359306.

Levy, D., and Veilleux, R.E. 2007. Adaptation of potato to high temperatures and salinity – a review. *American Journal of Potato Research* 84: 487–506. doi: 10.1007/BF02987885

Li, H., Luo, W., Ji, R., Xu, Y., Xu, G., Qiu, S., et al. 2021. A comparative proteomic study of cold responses in potato leaves. *Heliyon* 7: e06002. doi: 10.1016/j.heliyon.2021.e06002.

Li, P. H. 1977. Frost killing temperatures of 60 tuber-bearing *Solanum* species. *American Potato Journal* 54: 452–456. doi: 10.1007/BF02852059.

Li, P., Huner, N., Toivio-Kunnucan, M., Chen, H., and Palta, J. 1981. Potato freezing injury and survival, and their relationships to other stress. *American Potato Journal* 58: 15–29 doi: 10.1007/BF02855377.

Li, P. H., Palta, J. P., and Chen, H. H. 1979. Freezing stress in potato. In Lyons, J. M., et al. (eds), *Low Temperature Stress in Crop Plants: The Role of the Membrane*, 291–303. New York, NY: Academic Press.

Li, P. H., Palta, J. P., and Hawkes J. G. 1980. A Scientific note: Interrelationship between frost hardiness and elevation of genotype origin. *American Potato Journal* 57: 184–185. doi: 10.1007/BF02853873.

Luthra, S. K., Gopal, J., Manivel, P., Kumar, V., Singh, B.P., and Pandey, S. K. 2007. Screening of wild and cultivated species of potato for frost tolerance in north-central plains of India. *Potato J.* 34: 45–46.

Lyra, D., Ribeiro, G. H. M. R., de Figueiredo, I. C. R., Guedes, M. L., Carneiro, O. L. G., Pinto, C. A. B. P., et al. 2015. Tuber initiation, duration of the vegetative cycle and heat tolerance in potato genotypes. *Pesquisa Agropecuaria Brasileira* 50: 582–592. doi: 10.1590/S0100-204X2015000700008.

Markarov, A. M., Golovko, T. V., and Tabalenkova, G. N. 2001. *Tuber-Forming Plant Morphophysiology*. St Petersburg: Nauka. [In Russ.]

Midmore, D. J., and Prange, R. K. 1991. Sources of heat tolerance amongst potato cultivars, breeding lines, and *Solanum* species. *Euphytica* 55: 235–245 doi: 10.1007/BF00021244.

Mihovilovich, E., Carli, C., de Mendiburu, F., Hualla, V., and Bonierbale, M. 2014. *Tuber Bulking Maturity Assessment of Elite and Advanced Potato Clones Protocol*. Lima: International Potato Center (CIP), 43. doi: 10.4160/9789290604419.

Monneveux, P., Ramírez, D. A., Awais K. M., Raymundo, R. M., Loayza, H., and Quiroz, R. 2014. Drought and heat tolerance evaluation in potato (*Solanum tuberosum* L.). *Potato Research* 57: 225–247 doi: 10.1007/s11540-014-9263-3.

Nasir, M. W., and Toth, Z. 2022. Effect of drought stress on potato production: A review. *Agronomy* 12: 635. doi: 10.3390/agronomy12030635.

Obidiegwu, J. E., Bryan, G. J., Jones, H. G., and Prashar, A. 2015. Coping with drought: Stress and adaptive responses in potato and perspectives for improvement. *Frontiers in Plant Science* 6: 542. doi: 10.3389/fpls.2015.00542.

Palta, J. P. and Li, P. H. 1979. Frost hardiness in relation to leaf anatomy and natural distribution of *Solanum* species. *Crop Science* 19: 665–671. doi: 10.2135/cropsci1979.0011183X001900050031x.

Potato Production: Planting Through Harvest. 2011. USAID-Inma. Available at: https://pdf.usaid.gov/pdf_docs/pnadz110.pdf (accessed April 12, 2022).

Reynolds, M. P., and Ewing, E. E. 1989a. Heat tolerance in tuber bearing *Solanum* species: A protocol for screening. *American Potato Journal* 66: 63–74. doi: 10.1007/BF02854425.

Reynolds, M. P., and Ewing, E. E. 1989b. Effects of high air and soil temperature stress on growth and tuberization in *Solanum tuberosum. Annals of Botany* 64: 241–247. doi: 10.1093/oxfordjournals.aob.a087837.

Rykaczewska, K. 2013. The impact of high temperature during growing season on potato cultivars with different response to environmental stresses. *American Journal of Plant Sciences* 4: 2386–2393. doi: 10.4236/ajps.2013.412295.

Schumacher, C., Thümecke, S., Schilling, F., Köhl, K., Kopka, J., Sprenger, H., et al. 2021. Genome-wide approach to identify quantitative trait loci for drought tolerance in tetraploid potato (*Solanum tuberosum* L.). *International Journal of Molecular Sciences* 22: 6123. doi: 10.3390/ijms22116123.

Sergacheva, A. P. 1954. *Biological Characterization of Potato Cultivars and Species in Irrigation Conditions of Northern Regions near Aral Sea*. PhD thesis. Leningrad: VIR. [In Russ.]

Sharma, S, Pandey, S., Muthamilarasan, M., Chaudhry, V., Dulani, P., Prasad, M. 2017. Genomics resources for abiotic stress tolerance in Solanaceae crops. In Chakrabarti, S. K., et al. (eds), *The Potato Genome*, 195–216. doi: 10.1007/978-3-319-66135-3_12.

Singh, B., Tomar, M., Dutt, S., Raigond, P., and Singh, B. P. 2018. Potato physiology in relation to crop yield. In Abbas, Z., et al. (eds), *Emerging Trends of Plant Physiology for Sustainable Crop Production*, 1–24. Waretown, NJ: Apple Academic Press. doi: 10.1201/b22338.

Smillie, R. M., Hetherington, S. E., Ochoa, C., and Malagamba, P. 1983. Tolerances of wild potato species from different altitudes to cold and heat. *Planta* 159: 112–118. doi: 10.1007/BF00392980.

Soltys-Kalina, D., Plich, J., Strzelczyk-Żyta, D., Śliwka, J., and Marczewski, W. 2016. The effect of drought stress on the leaf relative water content and tuber yield of a half-sib family of 'Katahdin'-derived potato cultivars. *Breeding Science* 66: 328–331 doi: 10.1270/jsbbs.66.328.

Stebut, I. A. 1882. *Basis of Field Crops and their Improvement in Russia*. Vol. 2. Moscow. Available at: http://elib.cnshb.ru/books/free/0385/385131 [In Russ.]

Stone, J., Palta, J., Bamberg, J., Weiss, L, Harbage, J. 1993. Inheritance of freezing resistance in tuber-bearing Solanum species: Evidence for independent genetic control of nonacclimated freezing tolerance and cold acclimation capacity. *Proceedings of the National Academy of Sciences of the USA* 90: 7869–7873 doi: 10.1073/pnas.90.16.7869.

Struik, P. C. 2007. Responses of the potato plant to temperature. In Vreugdenhill, D. (ed), *Potato Biology and Biotechnology. Advances and Perspectives*, 367–393. Oxford: Elsevier.

Trapero-Mozos, A., Ducreux, L., Bita, C. E., Morris, W., Wiese, C., Morris, J. A., et al. 2018. A reversible light- and genotype-dependent acquired thermotolerance response protects the potato plant from damage due to excessive temperature. *Planta* 247: 1377–1392. doi: 10.1007/s00425-018-2874-1.

Udovenko, G. V. 1995. Plant tolerance to abiotic stresses. In Udovenko, G. V. and Shevelukha, V. S. (eds), *Physiological Principles of Plant Breeding. Theoretical Basis of Plant Breeding*, vol. 2, 293–352. St. Petersburg: VIR. [In Russ.]

Valverde, R., and Chen, H. 1999. Genetic analysis of frost hardiness traits in tuber-bearing *Solanum* species. *Journal of Plant Biology* 42: 174–180.

Van Dam, J., Kooman, P. L. and Struik, P. C. 1996. Effects of temperature and photoperiod on early growth and final number of tubers in potato (*Solanum tuberosum* L.). *Potato Research* 39: 51–62. doi: 10.1007/BF02358206.

Vega, S., del Rio, A., Jung, G., Bamberg, J. and Palta J. P. 2003. Marker-assisted genetic analysis of non-acclimated freezing tolerance and cold acclimation capacity in a backcross *Solanum* population. *American Journal of Potato Research* 80: 359–369. doi: 10.1007/BF02854247.

Vega, S. and Palta, J. 2000. Variability in the rate of cold acclimation and deacclimation among tuber-bearing *Solanum* (potato) species. *Journal of the American Society for Horticultural Science* 125: 205–211. doi: 10.21273/JASHS.125.2.205.

Vos, J. and Haverkort, A. J. 2007. Water availability and potato crop performance. In Vreugdenhill, D. (ed), *Potato Biology and Biotechnology. Advances and Perspectives*, 333–352. Oxford: Elsevier.

Wu, Q., He, T., Liu, H., Luo, X., Yin, W., Chenet, E., et al. 2019. Cell ultrastructure and physiological changes of potato during cold acclimation. *Canadian Journal of Plant Science* 99: 873–884. doi: 10.1139/cjps-2019-0092.

Yang, X., Liu, J., Xu, J., Duan, S., Wang, Q., Li, G., et al. 2019. Transcriptome profiling reveals effects of drought stress on gene expression in diploid potato genotype P3–198. *International Journal of Molecular Sciences* 20: 852. doi: 10.3390/ijms20040852.

16 Metabolomics for Plant Stress Amelioration

The Status of Cereal Crops

Tanzeel Bashir, Syed Anam Ul Haq, and Amjad M. Husaini
Sher-e-Kashmir University of Agricultural Sciences & Technology of Kashmir (SKUAST-K)

CONTENTS

16.1 INTRODUCTION

Metabolomics is a relatively recent "omics" tool for comprehending complicated biological systems and analyzing plant metabolomes. Using metabolomics, molecules smaller than 1,500 Da can be analyzed and compared (Wishart, 2008). Metabolomics studies the compositions and interactions of metabolites in living systems by identifying and quantifying them (Khakimov et al., 2014). Depending on the goal of a research study metabolomics can be classified either as targeted or untargeted. The absolute measurement of one or a few metabolites from a list of predetermined known compounds is the subject of targeted metabolomics. As a result, the focused approach is extremely sensitive and quantitative, making it ideal for tracking metabolites connected to certain stressors. The relative abundances of hundreds to thousands of all known metabolites are evaluated using the discovery-based method known as targeted metabolomics. On the other hand, the untargeted method can measure unknown metabolites using mass spectrometric characteristics.

Metabolites that code for specific loci are used as biomarkers in association studies. Metabolite concentrations in plant tissues can reveal the genetic basis of these concentrations and to analyze metabolome quantitative trait loci (mQTL). mQTL can also reveal new associations between metabolomic pathways, structural genes, and agronomically important traits. Including as many metabolites as possible when carrying out extensive analyses is crucial. It is vital to examine small components such as sterols, vitamins, and phenolics, as well as major ones such as carbohydrates or proteins, when examining how gene alterations affect the composition of a cereal crop.

Plants' adaptive responses to stress have been studied extensively using metabolomics. Studying the synthesis of certain metabolites under various stresses is crucial for understanding how plants adapt to unfavorable environments. Numerous metabolome studies have been carried out over the past 10 years to look into changes in metabolite concentrations in response to various biotic and abiotic stress conditions. New molecules and metabolic pathways that accumulate in response to

DOI: 10.1201/9781003300564-16

various stressors are discovered through metabolomic research (Wen et al., 2014). *Oryza sativa*, *Triticum aestivum*, *Zea mays*, and *Sorghum bicolor* have all been researched using metabolomic techniques to investigate their stress responses (Ashraf et al., 2017; Husaini et al., 2010; Wani et al., 2021). Metabolomics research also contributes to better our understanding of different metabolic pathways (Wani and Husaini, 2021a).

16.2 CEREAL METABOLOMICS

Cereal metabolomics involves different qualitative and quantitative assessments of the metabolites from cereal plants. In the 1950s, ^{14}C labelling was used on wheat plants to understand the regulation of proteins, carbohydrates, and energy expenditure systems (McConnell, 1959). The role of metabolomics in cereal science expanded significantly in the 20th century due to the development of better analytical techniques, improvements in analyzing metabolites, and the subsequent interpretation of highly complex metabolomic datasets, especially in relation to biotic and abiotic stresses (Table 16.1).

TABLE 16.1
Application of Metabolomics Techniques for Examining Crop Responsiveness to Biotic and Abiotic Stresses

Crop	Treatment/ Stress Factor	Technique	Significant Metabolic Changes Response	References
Oryza sativa	SUB1A gene presence or absence/ flooding	1H NMR	• In submergence-tolerant SUB1A-containing rice, carbohydrate consumption is restricted and amino acids synthesized during submergence. • In rice without SUB1A gene there are elevated levels of glutamine, glutamate, leucine, isoleucine, threonine, and valine.	Barding et al. (2012), Yang et al. (2019), Casartelli et al. (2018), Jones et al. (2011), Norvienyeku et al. (2021), Suharti et al. (2016)
	Low temperature	LC–MS/ MS	• In response to low-temperature stress there is significant change in metabolites involved in glutathione and inositol phosphate metabolism, and biosynthesis of phenylpropanoids.	
	Water deficiency	GC–MS	• Enhanced accumulation of N-rich metabolites and reduction in malate, glyceric acid, and glyceric acid-3-phosphate in resistant cultivars. • Greater fold increase in the oligosaccharide raffinose in sensitive genotype.	
	Magnaporthe grisea	NMR, GC/ LC–MS/ MS	• There is strong association of alanine, malate, glutamine, proline, and cinnamate with fungal penetration.	
		UPLC, QTOF, HPLC	• There is strong positive correlation of saponin bayogenin 3-*O*-cellobioside with blast resistance.	
	Rhizoctonia solani	GC–MS	• There is a significant increase in glyceric, mucic and jasmonic acid in the resistant genotype.	
Triticum aestivum	Low nitrogen	UPLC–QTOF	• Flavonoids and their derivatives such as iso-vitexin, iso-orientin, methylisoorientin-2″-*O*-rhamnoside accumulate during mild nitrogen stress.	Zhang et al. (2017), Vincent et al. (2012), Guo et al. (2020)

(Continued)

TABLE 16.1 (*Continued*)
Application of Metabolomics Techniques for Examining Crop Responsiveness to Biotic and Abiotic Stresses

Crop	Treatment/ Stress Factor	Technique	Significant Metabolic Changes Response	References
	Stagonospora nodorum	GC–MS	• There is a significant increase in TCA cycle intermediates as an attempt to generate ATP in response to the collapse of photosynthesis caused by effector protein SnToxA.	
	Drought	UPLC–QTOF–MS	• Resistant variety (HX10) exhibited higher accumulation of amino acids, alkaloids, organic acids, and flavonoids under drought.	
Zea mays	*Fusarium verticillioides*	LC–HRM	• Infection causes changes in level of *N*-hydroxynamic acid amides, flavonoids, and amino acids l-tryptophan and tyrosine.	Ciasca et al. (2020), Marti et al. (2013), Erb, Flors et al. (2009), Erb, Gordon-Weeks et al. (2009), Christensen et al. (2021)
	Herbivory	UHPLC–TOF–MS	• Upon infestation the levels of 1,3-benzoxazin-4-ones, phospholipids, *N*-hydroxycinnamoyl tyramines, azelaic acid, and tryptophan are elevated in maize leaves.	
	Diabrotica virgifera *Spodoptera littoralis*	HPLC–DAD	• Enhanced accumulation of DIMBOA (2,4-dihydroxy-7-methoxy-1,4-benzoxazin-3-one) in the plant following the infestation.	
	Heat stress/*C. heterostrophus*	UHPLC–HRMS	• *p*-Coumaric acid deficiency is associated with heat-induced vulnerability of maize to *C. heterostrophus*.	
Hordeum vulgare	Salt stress	GC–MS	• Accumulation of more proline and putrescine in salt-sensitive line. • Higher concentration of hexose phosphate and tricarboxylic acid cycle intermediate in the salt-tolerant line.	Patterson et al. (2009), Chen et al. (2022), Han et al. (2020), Bollina et al. (2011), Kumaraswamy et al. (2011), Karre et al. (2017), Cao et al. (2017), Swarcewicz et al. (2017)
	Drought stress	LC/MS	• Significant changes are observed in relation to ABC transporters, amino acid metabolism, carbohydrate metabolism, and lipid metabolism.	
	Drought stress	UHPLC–QTOF–MS	• In response to drought stress there are significant changes in phenolic metabolites.	
	F. graminearum	GC–MS, LC–MS	• Significant changes occur in metabolite levels such as flavonoids, fatty acids, phenylpropanoids, linolenic acid, *p*-coumaric, and sinapic acids in response to infection.	
	Gibberella zeae	LC–MS	• Significant changes in metabolites belonging to phenolic acids, fatty acids, and flavonoids occur on infection.	
	Fusarium graminearum	HPLC, LC–HRMS	• Accumulation of secondary metabolites from the hydroxycinnamic acid (HCAA), jasmonic acid, and phenylpropanoid pathways takes place.	

(*Continued*)

TABLE 16.1 (*Continued*)
Application of Metabolomics Techniques for Examining Crop Responsiveness to Biotic and Abiotic Stresses

Crop	Treatment/ Stress Factor	Technique	Significant Metabolic Changes Response	References
	Salinity	GC–MS, LC–MS	• Elevated levels of abscisic acid in roots. • Elevated salicylic acid levels were associated with an increase in leaf chlorophyll content.	
	Drought	GC–MS	• Identification of compounds as potential drought tolerance biomarkers.	
Sorghum bicolor	*Burkholderia andropogonis*	LC–MS	• Dynamic and complex network of sorghum defense resources focused at *B. andropogonis*, enabling the plant to deploy an enhanced defensive capability for disease suppression.	Mareya et al. (2019), Pavli et al. (2013), Sheflin et al. (2019)
	Water deficiency	GC–MS	• Differences in metabolite content was observed in leaves and roots. In leaves there were significant changes in myo-inositol and l-asparagine content, while in roots d-(+) trehalose and d-mannitol predominated.	
	Low nitrogen	GC–MS, LC–MS	• Under reduced N conditions there is a strong correlation with salicylic acid (SA) and phenylalanine, a precursor to SA, demonstrating the limited metabolic capability for defensive response.	

The primary components of cereal grains are starch, carbohydrates, lipids, proteins, dietary fiber, and minerals. However, minor metabolites such as phenolic acids, sterols, and flavonoids significantly impact the grain's characteristics. These low-molecular-weight phytochemicals have positive health impacts due to their antioxidant, antiproliferative, and radical scavenging activities (Madhujith and Shahidi, 2007). Because of their potential role in resistance to biotic and abiotic stressors, polyphenols have recently attracted a lot of attention among these (Amarowicz et al., 2007; Manach et al., 2004). Cereal phenolics can be found in free or conjugated forms. Several projects have examined the phytochemical makeup of cereals within the HEALTHGRAIN program (Andersson et al., 2008; Li et al., 2008; Nystrom et al., 2008; Shewry et al., 2008; Ward et al., 2008). These investigations have shown that the phytochemical content might significantly vary depending on the growth environment and geographic location even within the same grain variety. Hereunder we discuss the crop-wise metabolomic studies carried out in some important cereal crops cultivated worldwide.

16.2.1 *Oryza sativa* (Rice)

Natural catastrophes can significantly impact agricultural output and quality, including drought, extremely high temperatures, and flooding. Understanding how these stresses affect the metabolism of crops gives us insights into the basic systems that can affect survival. An untargeted metabolomics investigation on two aus-type and drought-intolerant irrigated varieties generated useful information about drought-responsive molecules linked to tolerance (Casartelli et al., 2018). Metabolomic profiling of root and shoot samples at tillering stage demonstrated a generally larger concentration of nitrogen-rich metabolites in the shoots of the resistant cultivars. There was a greater reduction of malate, glyceric acid, and glyceric acid-3-phosphate in the roots of aus-type cultivars. However, both the shoots and roots of the sensitive genotypes displayed a greater fold increase in the oligosaccharide raffinose. In a study by Yang et al. (2019), two different rice varieties belonging to japonica and indica group were exposed to low temperature (LT), normal temperature (NT), and a transition from LT to NT stress and tissues were examined at various times during germination. The study led to the identification of 730 metabolites, among seven significantly changed in response to low-temperature stress, making them important metabolites. These seven essential metabolites play a role in the metabolism of glutathione and inositol phosphate, as well as the biosynthesis of amino acids, and phenylpropanoids.

In all rice-growing regions of the world the fungus *Magnaporthe grisea* causes significant production losses (Shahriar et al., 2020; Wani and Husaini, 2021b). The reaction of primary metabolites to the infection process of *Oryza sativa* with the compatible and incompatible strains of *M. grisea* has been studied using metabolomics (Jones et al., 2011). The metabolomics approach using NMR and GC/LC-MS/MS showed that alanine along with malate, glutamine, proline, and cinnamate demonstrated a strong association between the time of fungal penetration into the leaf and the divergence of metabolite profiles in each interaction. Norvienyeku et al. (2021) investigated how several rice cultivars respond to *M. grisea* challenges. The level of saponin bayogenin 3-*O*-cellobioside in infected rice was strongly associated with its resistance characteristics. Engineering saponin expression in cereal crops presents an alluring and sustainable disease management strategy, due to the link between bayogenin 3-*O*-cellobioside levels and blast resistance.

Rhizoctonia solani is another wide host range pathogen causing necrotic lesions on rice plants that ultimately affects its growth (Anderson et al., 2017). In a study by Suharti et al. (2016), infection with *R. solani* significantly increased levels of adenosine diphosphate (ADP) and glyceric, mucic, and jasmonic acids in the 32R genotype. In 29S genotype, the plant response to *R. solani* infection involved inosine monophosphate. *R. solani* infection in 32R are linked to the activation of metabolic plant activities such as respiration, photorespiration, pectin synthesis, and lignin buildup, while in 29S the infection is thought to be related to nitrogen metabolism.

16.2.2 *Hordeum vulgare* (Barley)

Barley is the fourth-most significant cereal crop in terms of global production. Due to high fiber content and antioxidants [particularly (1,3;1,4)-glucan, phenolic acids, flavonoids, and tocols] present in the barley grain and their health-promoting benefits, two-thirds of the world's hordeum production is used as animal feed, one-third is used to manufacture malt for beer and distilled beverages, and 2%–3% is used for human consumption (Baik and Ullrich, 2008; Idehen et al., 2017). A thorough phytochemical investigation of ten distinct barley lines cultivated in one area as part of the HEALTHGRAIN program was carried out by Andersson et al. (2008). The study examined dietary fiber, phytosterols, phenolic acids, tocols, folates, and alkylresorcinols using different metabolomics techniques. Another study used GC–MS-based metabolomics approach to demonstrate how *H. vulgare* cultivars differing in salinity tolerance respond to salt stress (Patterson et al., 2009). The research demonstrated a propensity for the more salt-sensitive *H. vulgare* line to accumulate more proline and putrescine. Hexose phosphate and tricarboxylic acid cycle intermediate concentrations were higher in the salt-tolerant line.

Zhao et al. (2021) compared the metabolite profiles of *H. vulgare* seedlings exposed to different abiotic stresses viz. osmotic, metal, and nutrient deficiency stresses to identify the common and specific metabolites, which may be used as the markers for diagnosing certain abiotic stress that plants are suffering. Han et al. (2020) identified 118 phenolic metabolites affected by drought stress using a metabolomics approach in barley. Chen et al. (2022) identified a total of 187 metabolites in barley during seed germination in response to salt stress using LC/MS approach. These metabolites were mainly related to ABC transporters, amino acid metabolism, carbohydrate metabolism, and lipid metabolism.

Numerous researchers have used metabolomics approach to identify the metabolites that relate to the plant's defense mechanism against *F. graminearum* infection in barley plants (Bollina et al., 2011). Resistant and susceptible genotypes were examined for their metabolomic profiles, and 115 metabolites were putatively identified as being connected to resistance. These include flavonoids, fatty acids, phenylpropanoids, linolenic acid, *p*-coumaric, and sinapic acids. Another study used GC–MS to analyze the metabolite fingerprints of *H. vulgare* whole seeds, endosperms, and embryos during industrial malting (Gorzolka et al., 2012). This study identified prospective metabolic markers for a particular developmental stage that may be applied in industrial process control.

Untargeted LC–MS analysis was employed in another barley metabolomic study to explain the impact of *Gibberella zeae* fungus (Kumaraswamy et al., 2011). Both susceptible and resistant barley genotypes were used and 161 metabolites were identified as resistance-related (RR) and/or pathogenicity-related (PR) metabolites. Fifty-three out of 161 metabolites were tentatively identified, with the majority coming from phenolic acids, fatty acids, and flavonoids. In another similar study following infection with *Fusarium graminearum*, Karre et al. (2017) found the accumulation of a specific group of induced secondary metabolites from the hydroxycinnamic acid (HCAA), jasmonic acid, and phenylpropanoid pathways.

16.2.3 *Zea mays* (Maize)

The evaluation of transgenic lines, comparison of transgenic lines to non-transgenic lines, and the impact of abiotic stressors are the three main areas of attention in maize. Utilizing integrated transcriptomic, proteomic, and metabolomic studies of two GMO (Bt maize and herbicide-tolerant Roundup Ready maize) and corresponding non-GMO parental maize lines, phenotypic alterations brought on by genetic modifications, and environmental factors were examined (Barros et al., 2010). Another maize study examined the roots and leaves of resistant and susceptible maize for herbivore-induced metabolites (Marti et al., 2013). This study demonstrated that the concentrations of 1,3-benzoxazin-4-ones, phospholipids, *N*-hydroxycinnamoyltyramines, azelaic acid, and tryptophan were significantly increased on the infestation of maize leaves, while only minor changes occurred in the metabolome of the roots. Erb Flors et al. (2009) and Erb Gordon-Weeks et al. (2009) examined how the aboveground organs of maize responded to the belowground attack by *Diabrotica virgifera virgifera* in terms of their ability to defend themselves against *Spodoptera littoralis*. They also tracked the buildup of defensive metabolites when *Spodoptera* challenged the leaves. Using HPLC–DAD, the metabolites discovered in these damaged leaves were measured. The amount of DIMBOA (2,4-dihydroxy-7-methoxy-1,4-benzoxazin-3-one) in the plant increased following the infestation. After *Fusarium verticillioides* infection of seedlings, metabolomics alterations of two maize recombinant inbred lines with different phenotypes were examined with regard to disease severity. Numerous changes were found in metabolites from various metabolic pathways, including three *N*-hydroxynamic acid amides, five flavonoids, and two amino acids (l-tryptophan and tyrosine) (Ciasca et al., 2020). Using targeted and untargeted metabolomics, Christensen et al. (2021) looked at the molecular reactions of maize to abiotic, biotic, and combinatorial stress. The study concluded that *p*-coumaric acid deficiency might be a factor in the heat-induced vulnerability of maize to *Cochliobolus heterostrophus*.

According to the study using NMR-based metabolomic profiling of maize under salt stress amino acids, trace organic acids, and sugar levels in the shoots and roots of control and treated maize lines were clearly different (Gavaghan et al., 2011). Another study discovered 37 metabolite indicators of salt-induced osmotic stress tolerance after profiling the metabolomes of 266 maize inbred lines using an LC–MS-based untargeted metabolomics technique (Liang et al., 2021). Similarly, Witt et al. (2012) identified numerous compounds that showed conserved responses to drought using a metabolomics approach.

In order to understand the molecular mechanisms underlying the response to gradually dropping temperatures at the vegetative stage and to identify metabolic responses that could be extrapolated from growth chamber conditions to early sowing in the field, researchers (Urrutia et al., 2021) used untargeted metabolomic and proteomic approaches on a collection of genetically diverse maize hybrids grown in a growth chamber and in the field. The investigation identified a number of potential metabolites and proteins that may be tested as biomarkers to predict the response to early sowing in the field from a growth chamber experiment. In order to investigate the impact of high temperatures on photosynthetic physiological traits and metabolome, two transgenic wheat lines, PC27 and PC51, were used. In comparison to wild type, about 13 different metabolites, including 1 organic acid, 6 amino acids, 4 sugars, and 2 polyols, were identified under non-stress treatment; under high-temperature stress, 25 different metabolites, including 6 organic acids, 12 amino acids, 4 sugars, and 3 polyols, were identified (Qi et al., 2017).

16.2.4 *Triticum aestivum* (Wheat)

One of the most extensively cultivated major grain crops worldwide is wheat (*Triticum aestivum* L.). Water deficiency has a major impact on the expression and accumulation of the grain proteome, the content of storage proteins, and the gluten quality. Using an untargeted approach with ultra-performance liquid chromatography–mass spectrometry (UPLC–MS), Guo et al. (2020) investigated the profiles of metabolites produced by wheat seedlings in response to drought stress from the cross between drought-tolerant genotype (HX10) and drought-sensitive genotype (YN211). HX10 displayed better growth indices than YN211 under drought stress. Following treatment for drought stress, HX10 accumulated more phenolics than YN211, whereas YN211 had roughly 13 times as much thymine.

Zhang et al. (2017) performed a large-scale untargeted metabolomic analysis of two wheat cultivars under different induced nitrogen levels using ultra-performance liquid chromatography–quadrupole time-of-flight mass spectrometry (UPLC–QTOF) in order to investigate the impact of low nitrogen stress on various metabolites of the flag leaf of wheat. Eleven secondary metabolites that could serve as indicators of mild nitrogen stress were found. Chemical tests revealed that flavonoids and their associated derivatives, such as iso-vitexin, iso-orientin, methylisoorientin-2″-*O*-rhamnoside, etc., made up the majority of the detected secondary metabolites.

The proteomics and metabolomics approach was utilized by Vincent et al. (2012) to examine the reaction of wheat to exposure to the SnToxA effector protein released by the fungal pathogen *Stagonospora nodorum* during infection. During a 72-hour time frame, wheat leaves treated with SnToxA showed differential abundances of 91 distinct acidic and basic proteins and 101 metabolites. Significant increases in tricarboxylic acid (TCA) cycle intermediates indicate an effort by the plant to produce ATP and reducing equivalents in response to the collapse of photosynthesis caused by SnToxA, which indicate significant metabolic perturbations in central carbon metabolism. This was confirmed by the finding that SnToxA increased the activity of the TCA cycle enzyme malate dehydrogenase. This approach shows the complementary nature of proteomics and metabolomics in examining effector–host interactions. It adds credence to the idea that necrotrophic pathogens, such as *S. nodorum*, may leverage host cell death processes already in place to facilitate pathogen development and induce disease.

16.2.5 *Sorghum bicolor* (Sorghum)

Sorghum is the fifth most productive cereal crop in the world, and is used in the production of sugar, animal feed, and lignocellulosic biomass for bioenergy (Bakari et al., 2022). It is a viable option for development of an affordable biofuel due to its high yield of biomass and drought tolerance. Crop development initiatives to boost the production of biofuel feedstocks benefit from a deeper understanding of the metabolic responses to low nitrogen (N) levels. In a study by Sheflin et al. (2019), roots were collected at two time points for metabolomics from 10 distinct sorghum cultivars field-grown under low and full N environments. Both sampling dates revealed changed metabolic profiles in the roots of plants cultivated in low N environments, including metabolites crucial for N storage and aromatic amino acid production. When *Pseudomonas* abundance and root metabolites were correlated, it was found that there was a strong negative link with salicylic acid (SA) under full N conditions but not under reduced N circumstances. Reduced phenylalanine, a precursor to SA, was found in the roots of plants under N stress, further demonstrating the limited metabolic capability for defensive response in low N environments.

In a study conducted by Pavli et al. (2013), young sorghum seedlings cultivated under controlled conditions were exposed to drought stress for 7 days using two concentrations of polyethylene glycol. Gas chromatography and mass spectrometry (GC–MS) were used to profile metabolites in the leaves and roots of stressed and control plants. The sugars d-mannose, d-glucose, isomaltose, fructose, and sucrose as well as myo-inositol and l-asparagine showed the most significant changes in leaves. Similarly sugars including d-glucose, fructose, sucrose, and d-(+)trehalose, as well as d-mannitol were the most significant molecules in roots.

Bacterial leaf stripe disease caused by *Burkholderia andropogonis* is one of the main diseases affecting *Sorghum bicolor* economically. Aspects of the molecular pathways underlying the pathogenic interaction between *S. bicolor* and *B. andropogonis* were recently identified using metabolomics (Mareya et al., 2020). The findings showed a dynamic and complex network of sorghum defense resources focused at *B. andropogonis*, enabling the plant to deploy an enhanced defensive capability for disease suppression (Mareya et al., 2019).

16.3 FUTURE PERSPECTIVE

There is a need to integrating metabolomics with genetic approaches as it can help researchers understand how plants regulate their genetics in relation to metabolomics. Reverse genetics coupled with metabolomics techniques in metabolomics-assisted breeding can be very useful. This strategy can help crop improvement projects in generating high-yielding, stress-tolerant, and climate-adapted crop varieties more efficiently. Cutting-edge technologies like single-cell metabolomics can aid in pathway elucidation and quantification at the single-cell level. Single-probe MS technology is a recent technological advancement, which can revolutionize in situ-focused metabolomic investigations on real-time in single cells. The complementary use of proteomics and metabolomics in examining complex molecular interactions can further help in understanding the regulatory role of these metabolic products. Activity-based protein profiling (ABPP) is an upcoming technology in chemical proteomics that uses mass spectrometry analysis to identify 'unknown' labels and validate the targets (Husaini et al., 2018). It can be highly effective in better understanding biological processes and metabolomes in underutilized crops, even if there is limited information about their genomes and transcriptomes (Husaini et al., 2022).

REFERENCES

Amarowicz, R., Zegarska, Z., Karamac, M., and Kosinska, A. 2007. Antioxidant and radical scavenging activities of a barley crude extract and its fraction. *Czech Journal of Food Sciences* 25(2): 73.

Anderson, J. P., Sperschneider, J., Win, J., Kidd, B., Yoshida, K., Hane, J., Saunders, D. G., and Singh, K. B. 2017. Comparative secretome analysis of Rhizoctonia solani isolates with different host ranges reveals unique secretomes and cell death inducing effectors. *Scientific Reports* 7(1): 1–13.

Andersson, A. A., Lampi, A.-M., Nystrom, L., Piironen, V., Li, L., Ward, J. L., Gebruers, K., Courtin, C. M., Delcour, J. A., and Boros, D. 2008. Phytochemical and dietary fiber components in barley varieties in the HEALTHGRAIN diversity screen. *Journal of Agricultural and Food Chemistry* 56(21): 9767–9776.

Ashraf, H., Murtaza, I., Nazir, N., Wani, A. B., Naqash, S., and Husaini, A. M. 2017. Nutritional profiling of pigmented and scented rice genotypes of Kashmir Himalayas. *Journal of Pharmacognosy and Phytochemistry* 6(6): 910–916.

Baik, B.-K., and Ullrich, S. E. 2008. Barley for food: Characteristics, improvement, and renewed interest. *Journal of Cereal Science* 48(2): 233–242.

Bakari, H., Ruben, Z. F., Roger, D. D., Cedric, D., Guillaume, P., Pascal, D., Philippe, M., and Gwendoline, C. 2022. Sorghum (*Sorghum bicolor* L. Moench) and Its Main Parts (By-Products) as Promising Sustainable Sources of Value-Added Ingredients. *Waste and Biomass Valorization* 14(4): 1023–1044.

Barding Jr, G. A., Fukao, T., Béni, S., Bailey-Serres, J., and Larive, C. K. 2012. Differential metabolic regulation governed by the rice SUB1A gene during submergence stress and identification of alanylglycine by 1H NMR spectroscopy. *Journal of Proteome Research* 11(1): 320–330.

Barros, E., Lezar, S., Anttonen, M. J., Van Dijk, J. P., Röhlig, R. M., Kok, E. J., and Engel, K. H. 2010. Comparison of two GM maize varieties with a near-isogenic non-GM variety using transcriptomics, proteomics and metabolomics. *Plant Biotechnology Journal* 8(4): 436–451.

Bollina, V., Kushalappa, A. C., Choo, T. M., Dion, Y., and Rioux, S. 2011. Identification of metabolites related to mechanisms of resistance in barley against Fusarium graminearum, based on mass spectrometry. *Plant Molecular Biology* 77(4): 355–370.

Cao, D., Lutz, A., Hill, C. B., Callahan, D. L., and Roessner, U. 2017. A quantitative profiling method of phytohormones and other metabolites applied to barley roots subjected to salinity stress. *Frontiers in Plant Science* 7: 2070.

Casartelli, A., Riewe, D., Hubberten, H. M., Altmann, T., Hoefgen, R., and Heuer, S. 2018. Exploring traditional aus-type rice for metabolites conferring drought tolerance. *Rice* 11(1): 1–16.

Chen, Y., Wang, J., Yao, L., Li, B., Ma, X., Si, E., Yang, K., Li, C., Shang, X., and Meng, Y. 2022. Combined proteomic and metabolomic analysis of the molecular mechanism underlying the response to salt stress during seed germination in barley. *International Journal of Molecular Sciences* 23(18): 10515.

Christensen, S. A., Alborn, H. T., Block, A. K., and Chamberlain, C. A. 2021. Metabolomics by UHPLC-HRMS reveals the impact of heat stress on pathogen-elicited immunity in maize. *Metabolomics* 17(1): 1–11.

Ciasca, B., Lanubile, A., Marocco, A., Pascale, M., Logrieco, A. F., and Lattanzio, V. M. 2020. Application of an integrated and open source workflow for LC-HRMS plant metabolomics studies. Case-control study: Metabolic changes of maize in response to Fusarium verticillioides infection. *Frontiers in Plant Science* 11: 664.

Erb, M., Flors, V., Karlen, D., De Lange, E., Planchamp, C., D'Alessandro, M., Turlings, T. C., and Ton, J. 2009. Signal signature of aboveground-induced resistance upon belowground herbivory in maize. *The Plant Journal* 59(2): 292–302.

Erb, M., Gordon-Weeks, R., Flors, V., Camañes, G., Turlings, T. C., and Ton, J. 2009. Belowground ABA boosts aboveground production of DIMBOA and primes induction of chlorogenic acid in maize. *Plant Signaling & Behavior* 4(7): 639–641.

Gavaghan, C. L., Li, J. V., Hadfield, S. T., Hole, S., Nicholson, J. K., Wilson, I. D., Howe, P. W., Stanley, P. D., and Holmes, E. 2011. Application of NMR-based metabolomics to the investigation of salt stress in maize (*Zea mays*). *Phytochemical Analysis* 22(3): 214–224.

Gorzolka, K., Lissel, M., Kessler, N., Loch-Ahring, S., and Niehaus, K. 2012. Metabolite fingerprinting of barley whole seeds, endosperms, and embryos during industrial malting. *Journal of Biotechnology* 159(3): 177–187.

Guo, X., Xin, Z., Yang, T., Ma, X., Zhang, Y., Wang, Z., Ren, Y., and Lin, T. 2020. Metabolomics response for drought stress tolerance in chinese wheat genotypes (*Triticum aestivum*). *Plants* 9(4): 520.

Han, Z., Ahsan, M., Adil, M. F., Chen, X., Nazir, M. M., Shamsi, I. H., Zeng, F., and Zhang, G. 2020. Identification of the gene network modules highly associated with the synthesis of phenolics compounds in barley by transcriptome and metabolome analysis. *Food Chemistry* 323: 126862.

Husaini, A. M., Haq, S. A. U., and Jiménez, A. J. L. 2022. Understanding saffron biology using omics- and bioinformatics tools: Stepping towards a better Crocus phenome. *Molecular Biology Report* 49(6): 5325–5340. doi: 10.1007/s11033-021-07053-x.

Husaini, A. M., Morimoto, K., Chandrasekar, B., Kelly, S., Kaschani, F., Palmero, D., Jiang, J., Kaiser, M., Ahrazem, O., and Overkleeft, H. S. 2018. Multiplex fluorescent, activity-based protein profiling identifies active α-glycosidases and other hydrolases in plants. *Plant Physiology* 177(1): 24–37. doi: 10.1104/pp.18.00250.

Husaini, A. M., Srivastava, D. K., Zainul Abdin, M., and Kumar, A. T. 2010. Transgenic technology for engineering ROS and abiotic stress-tolerant crops. In Ahmad, P. and Umar, S. (eds), *Oxidative Stress: Role of Antioxidants in Plants*, 91–128. New Delhi: Studium Press India Pvt Ltd.

Idehen, E., Tang, Y., and Sang, S. 2017. Bioactive phytochemicals in barley. *Journal of Food and Drug Analysis* 25(1): 148–161.

Jones, O. A., Maguire, M. L., Griffin, J. L., Jung, Y.-H., Shibato, J., Rakwal, R., Agrawal, G. K., and Jwa, N.-S. 2011. Using metabolic profiling to assess plant-pathogen interactions: An example using rice (*Oryza sativa*) and the blast pathogen Magnaporthe grisea. *European Journal of Plant Pathology* 129(4): 539–554.

Karre, S., Kumar, A., Dhokane, D., and Kushalappa, A. C. 2017. Metabolo-transcriptome profiling of barley reveals induction of chitin elicitor receptor kinase gene (HvCERK1) conferring resistance against Fusarium graminearum. *Plant Molecular Biology* 93(3): 247–267.

Khakimov, B., Bak, S., and Engelsen, S. B. 2014. High-throughput cereal metabolomics: Current analytical technologies, challenges and perspectives. *Journal of Cereal Science* 59(3): 393–418.

Kumaraswamy, G. K., Bollina, V., Kushalappa, A. C., Choo, T. M., Dion, Y., Rioux, S., Mamer, O., and Faubert, D. 2011. Metabolomics technology to phenotype resistance in barley against Gibberella zeae. *European Journal of Plant Pathology* 130(1): 29–43.

Li, L., Shewry, P. R., and Ward, J. L. 2008. Phenolic acids in wheat varieties in the HEALTHGRAIN diversity screen. *Journal of Agricultural and Food Chemistry* 56(21): 9732–9739.

Liang, X., Liu, S., Wang, T., Li, F., Cheng, J., Lai, J., Qin, F., Li, Z., Wang, X., and Jiang, C. 2021. Metabolomics-driven gene mining and genetic improvement of tolerance to salt-induced osmotic stress in maize. *New Phytologist* 230(6): 2355–2370.

Madhujith, T., and Shahidi, F. 2007. Antioxidative and antiproliferative properties of selected barley (Hordeum vulgarae L.) cultivars and their potential for inhibition of low-density lipoprotein (LDL) cholesterol oxidation. *Journal of Agricultural and Food Chemistry* 55(13): 5018–5024.

Manach, C., Scalbert, A., Morand, C., Rémésy, C., and Jiménez, L. 2004. Polyphenols: Food sources and bioavailability. *The American Journal of Clinical Nutrition* 79(5): 727–747.

Mareya, C. R., Tugizimana, F., Di Lorenzo, F., Silipo, A., Piater, L. A., Molinaro, A., and Dubery, I. A. 2020. Adaptive defence-related changes in the metabolome of Sorghum bicolor cells in response to lipopolysaccharides of the pathogen Burkholderia andropogonis. *Scientific Reports* 10(1): 1–17.

Mareya, C. R., Tugizimana, F., Piater, L. A., Madala, N. E., Steenkamp, P. A., and Dubery, I. A. 2019. Untargeted metabolomics reveal defensome-related metabolic reprogramming in Sorghum bicolor against infection by Burkholderia andropogonis. *Metabolites* 9(1): 8.

Marti, G., Erb, M., Boccard, J., Glauser, G., Doyen, G. R., Villard, N., Robert, C. A. M., Turlings, T. C., Rudaz, S., and Wolfender, J. L. 2013. Metabolomics reveals herbivore-induced metabolites of resistance and susceptibility in maize leaves and roots. *Plant, Cell & Environment* 36(3): 621–639.

McConnell, W. 1959. Studies on wheat plants using carbon-14 labelled compounds: X. the incorporation of glutamic acid-l-c14. *Canadian Journal of Biochemistry and Physiology* 37(8): 933–936.

Norvienyeku, J., Lin, L., Waheed, A., Chen, X., Bao, J., Aliyu, S. R., Lin, L., Shabbir, A., Batool, W., and Zhong, Z. 2021. Bayogenin 3-O-cellobioside confers non-cultivar-specific defence against the rice blast fungus *Pyricularia oryzae*. *Plant Biotechnology Journal* 19(3): 589–601.

Nystrom, L., Lampi, A.-M., Andersson, A. A., Kamal-Eldin, A., Gebruers, K., Courtin, C. M., Delcour, J. A., Li, L., Ward, J. L., and Fras, A. 2008. Phytochemicals and dietary fiber components in rye varieties in the HEALTHGRAIN diversity screen. *Journal of Agricultural and Food Chemistry* 56(21): 9758–9766.

Patterson, J. H., Newbigin, E., Tester, M., Bacic, A., and Roessner, U. 2009. Metabolic responses to salt stress of barley cultivars, Sahara and Clipper, which differ in salinity tolerance. *Journal of Experimental Botany* 60(14): 4089–4103.

Pavli, O. I., Vlachos, C. E., Kalloniati, C., Flemetakis, E., and Skaracis, G. N. 2013. Metabolite profiling reveals the effect of drought on sorghum ('*Sorghum bicolor*' L. Moench) metabolism. *Plant Omics* 6(6): 371–376.

Qi, X., Xu, W., Zhang, J., Guo, R., Zhao, M., Hu, L., Wang, H., Dong, H., and Li, Y. 2017. Physiological characteristics and metabolomics of transgenic wheat containing the maize C4 phosphoenolpyruvate carboxylase (PEPC) gene under high temperature stress. *Protoplasma* 254(2): 1017–1030.

Shahriar, S. A., Imtiaz, A. A., Hossain, M. B., Husna, A., and Eaty, M. N. K. 2020. Rice blast disease. *Annual Research & Review in Biology* 35(1): 50–64.

Sheflin, A. M., Chiniquy, D., Yuan, C., Goren, E., Kumar, I., Braud, M., Brutnell, T., Eveland, A. L., Tringe, S., and Liu, P. 2019. Metabolomics of sorghum roots during nitrogen stress reveals compromised metabolic capacity for salicylic acid biosynthesis. *Plant Direct* 3(3): e00122.

Shewry, P. R., Piironen, V., Lampi, A.-M., Nystrom, L., Li, L., Rakszegi, M., Fras, A., Boros, D., Gebruers, K., and Courtin, C. M. 2008. Phytochemical and fiber components in oat varieties in the HEALTHGRAIN diversity screen. *Journal of Agricultural and Food Chemistry* 56(21): 9777–9784.

Suharti, W. S., Nose, A., and Zheng, S.-H. 2016. Metabolomic study of two rice lines infected by Rhizoctonia solani in negative ion mode by CE/TOF-MS. *Journal of Plant Physiology* 206: 13–24.

Swarcewicz, B., Sawikowska, A., Marczak, Ł., Łuczak, M., Ciesiołka, D., Krystkowiak, K., Kuczyńska, A., Piślewska-Bednarek, M., Krajewski, P., and Stobiecki, M. 2017. Effect of drought stress on metabolite contents in barley recombinant inbred line population revealed by untargeted GC–MS profiling. *Acta Physiologiae Plantarum* 39(8): 1–16.

Urrutia, M., Blein-Nicolas, M., Prigent, S., Bernillon, S., Deborde, C., Balliau, T., Maucourt, M., Jacob, D., Ballias, P., and Bénard, C. 2021. Maize metabolome and proteome responses to controlled cold stress partly mimic early-sowing effects in the field and differ from those of Arabidopsis. *Plant, Cell & Environment* 44(5): 1504–1521.

Vincent, D., Du Fall, L. A., Livk, A., Mathesius, U., Lipscombe, R. J., Oliver, R. P., Friesen, T. L., and Solomon, P. S. 2012. A functional genomics approach to dissect the mode of action of the Stagonospora nodorum effector protein SnToxA in wheat. *Molecular Plant Pathology* 13(5): 467–482.

Wani, A. B., and Husaini, A. M. 2021a. Biotechnology and molecular breeding in rice improvement. In Kearns, H. N. (ed), *Oryza Sativa: n and Uses Production, Cultivatio*, 1–38. Nova Science Publishers, New York.

Wani, A. B., and Husaini, A. M. 2021b. Conventional approaches to rice improvement. In Kearns, H. N. (ed), *Oryza sativa: Production, Cultivation and Uses*, 31–70. Nova Science Publishers, New York.

Wani, A. B., Husaini, A. M., Wani, S. M., Mir, S. A., Shafi, F., and Jabeen, N. 2021. Cold stress and the role of signalling hormones: A preliminary study on cold-tolerant high-altitude Himalayan rice genotypes. *Pharma Innovation* 10(3): 692–699. doi: 10.22271/tpi.2021.v10.i3j.5859.

Ward, J. L., Poutanen, K., Gebruers, K., Piironen, V., Lampi, A.-M., Nyström, L., Andersson, A. A., Boros, D., Rakszegi, M., and Bedő, Z. 2008. The HEALTHGRAIN cereal diversity screen: Concept, results, and prospects. *Journal of Agricultural and Food Chemistry* 56(21): 9699–9709.

Wen, W., Li, D., Li, X., Gao, Y., Li, W., Li, H., Liu, J., Liu, H., Chen, W., and Luo, J. 2014. Metabolome-based genome-wide association study of maize kernel leads to novel biochemical insights. *Nature Communications* 5(1): 1–10.

Wishart, D. S. 2008. Applications of metabolomics in drug discovery and development. *Drugs in R & D* 9(5): 307–322.

Witt, S., Galicia, L., Lisec, J., Cairns, J., Tiessen, A., Araus, J. L., Palacios-Rojas, N., and Fernie, A. R. 2012. Metabolic and phenotypic responses of greenhouse-grown maize hybrids to experimentally controlled drought stress. *Molecular Plant* 5(2): 401–417.

Yang, M., Yang, J., Su, L., Sun, K., Li, D., Liu, Y., Wang, H., Chen, Z., and Guo, T. 2019. Metabolic profile analysis and identification of key metabolites during rice seed germination under low-temperature stress. *Plant Science* 289: 110282.

Zhang, Y., Ma, X.-M., Wang, X.-C., Liu, J.-H., Huang, B.-Y., Guo, X.-Y., Xiong, S.-P., and La, G.-X. 2017. UPLC-QTOF analysis reveals metabolomic changes in the flag leaf of wheat (*Triticum aestivum L.*) under low-nitrogen stress. *Plant Physiology and Biochemistry* 111: 30–38.

Zhao, H., Ni, S., Cai, S., and Zhang, G. 2021. Comprehensive dissection of primary metabolites in response to diverse abiotic stress in barley at seedling stage. *Plant Physiology and Biochemistry* 161: 54–64.

17 Applications of "Omics" Technologies in Plant Responses to Combined Drought and Heat Stress

Trends and Future Perspectives

Rudo Ngara
University of the Free State

Stephen Chivasa
Durham University

CONTENTS

17.1 INTRODUCTION

Plants encounter a wide range of abiotic stresses, which limit growth and development (Levitt, 1980a, b; Hale et al., 1987). In agricultural ecosystems, crop yields may be reduced (Boyer, 1982), and food security threatened (IPCC 2007; Nelson et al., 2009). As a direct response to the negative impact of climate change on crop production, systems biology research (Yuan et al., 2008) on plant responses to abiotic stresses as well as the development of climate-resilient crops is increasing. Consequently, plant responses to individually occurring abiotic stresses such as drought, salinity, extreme temperatures, nutrient deficiency, and heavy metal toxicity have been extensively described and reviewed (Shao et al., 2007; Cramer et al., 2011, Ngara and Ndimba, 2014b; Ngara et al., 2021; Yang et al., 2021). Other studies have mined "omics" data to identify target genes, proteins, or metabolites for use in developing transgenic plants with improved resilience to individual stresses (Park et al., 2021).

Under natural field conditions, however, plants often encounter multiple stresses concurrently, leading to positive and/or negative impacts on plant growth than each of the individual stresses (Mittler, 2006; Mittler and Blumwald, 2010; Suzuki et al., 2014; Zandalinas et al., 2018). Plant responses to stress combinations are presumably complex, and thought to involve selective, additive, antagonistic, or new sets of regulatory mechanisms (Rivero et al., 2022). Therefore, transgenic approaches for developing crop varieties with improved resilience against single stresses could be

DOI: 10.1201/9781003300564-17

less effective in protecting plants against stresses occurring simultaneously (Mittler and Blumwald, 2010). As such, interest in studying plant responses to a combination of abiotic stresses as a new form of stress is growing (Suzuki et al., 2014; Zandalinas et al., 2018).

Among these combined stresses, drought and heat are most likely to co-occur in the field, causing negative interactions (Mittler, 2006) and increased plant vulnerability to their harsh combined effect (Suzuki et al., 2014). Climate models are also predicting increased occurrences of drought and heat waves (Gan et al., 2016), making their combined effect an imminent threat to crop productivity. Although the negative impact of climate change and global warming may vary with geographical regions (Kerr, 2007; Nelson et al., 2009), the drought-prone sub-Saharan African countries are envisaged to be worst hit by these dry and hot spells (IPCC, 2007; Gan et al., 2016). In many African countries, agriculture is the cornerstone of economies, and recurring drought and heat episodes may lead to famines and increased malnutrition for the already marginalized population groups. Therefore, there is a need to produce crops that can withstand combined drought and heat stress, which is prevalent in arid and semi-arid tropics. In this chapter, we review some of the "omics" studies on plant responses to combined drought and heat stress in *Arabidopsis*, tomato, and cereals. Potential targets for use in transgenic or gene-editing plant transformation platforms for improving combined stress tolerance are discussed.

17.2 COMPLEXITY OF PLANT RESPONSES TO COMBINED DROUGHT AND HEAT STRESSES

Adequate water supply and optimum physiological temperatures are essential requirements for normal plant growth and development (Taiz and Zeiger, 2010). For example, water is a universal solvent for biochemical processes, while optimum temperatures are required for maintaining the structure and function of lipid membranes, proteins, and enzymes, which are essential in preserving cell integrity and metabolism (Taiz and Zeiger, 2010). Under conditions of water deficits and/or high temperature stress, cellular homeostasis is disrupted, and plants alter their morphology, physiology, and metabolism to counteract the negative effects of stress and promote survival. General plant responses to individually applied drought (Levitt 1980a; Bray, 1993, 1997; Xiong and Zhu, 2002; Farooq et al., 2009) and heat (Levitt, 1980b; Kotak et al., 2007; Wahid et al., 2007; Nievola et al., 2017) stresses, including crosstalk between stresses (Knight and Knight, 2001; Wang et al., 2003; Chinnusamy et al., 2004) have been extensively described elsewhere, and will not be considered here. Although some plant responses are common between different types of stress, others are not. For example, the increased accumulation of transcripts and proteins of heat shock proteins (HSPs) and a range of antioxidant enzymes may be common in plants exposed to drought and heat (Wang et al., 2003).

However, plants may also recruit distinct and at times antagonistic physiological (Mittler, 2006) and molecular (Sewelam et al., 2020) responses towards the individually applied stresses. For instance, stomatal conductance generally decreases during periods of water deficits, preventing excessive transpiration water loss (Levitt, 1980a). In contrast, plants under heat stress maintain low leaf surface temperatures through transpiration cooling that is facilitated by high stomatal conductance (Levitt, 1980b). Such contrasting stomatal responses have been reported in tobacco (*Nicotiana tabacum*) (Rizhsky et al., 2002), *Arabidopsis* (*Arabidopsis thaliana*) (Rizhsky et al., 2004), and tomato (*Solanum lycopersicum*) (Nankishore and Farrell, 2016) in response to individually applied drought and heat stresses. Consequently, when subjected to a combination of drought and heat stress, these plants may require unique sets of defense mechanisms to survive the new physiological and metabolic strains exerted by the stress combination (Rizhsky et al., 2002, 2004; Mittler and Blumwald, 2010).

Alternatively, depending on the timing of onset and intensity of the individual stresses within the combination, and/or the ultimate effect of the stress combination, plants may utilize one of the responses to an individual stress, or even a blend of the two (Rivero et al., 2022). For this reason, Mittler (2006) suggests that stress combinations be regarded as new forms of stress in future studies. Zandalinas et al. (2018) have already given an excellent review on physiological, molecular, and metabolic responses of *Arabidopsis*, tobacco, and other plants to stress combinations, including drought and heat. Below is a discussion of some of the "omics" studies of *Arabidopsis*, tomato, and cereals to a combination of drought and heat stress.

17.3 "OMICS" STUDIES UNDER COMBINED DROUGHT AND HEAT STRESS

As with many other abiotic factors, drought and heat stresses modulate plant physiology and metabolism to various extents depending on the magnitude and duration of the stress, timing of onset within the plant's life cycle, and the plasticity of plant genomes in response to the changing environment (Rizhsky et al., 2004). Some of these changes have been captured in numerous "omics" studies, where transcripts, proteins, and/or metabolites are extracted from different treatment groups, identified, quantified, and analyzed for stress-induced changes. Examples of such studies are listed in Table 17.1. As expected with global "omics" workflows, only snapshots of cellular responses are captured at any given time point, making multifactorial time-course "omics" experiments relevant in understanding the complex changes in metabolism under stress. Other reviews have also highlighted potential applications of multi-omics (genomics, transcriptomics, proteomics, metabolomics, ionomics, and phenomics) technologies in crop breeding strategies (Yang et al., 2021; Zenda et al., 2021; Bhardwaj et al., 2022) and can be accessed for further reading.

TABLE 17.1
Examples of "Omics" Studies of Plants under Combined Drought and Heat Stress

"Omics" Technology	Plant Species	Plant Tissues	Techniques Used	References
Transcriptomics				
	Tobacco (*Nicotiana tabacum*)	-	cDNA arrays	Rizhsky et al. (2002)
	Arabidopsis (*Arabidopsis thaliana*)	Leaves	DNA chip analysis	Rizhsky et al. (2004)
	Durum wheat (*Triticum turgidum* subsp. *durum*)	Leaves	cDNA-AFLP, semi quantitative RT-PCR	Rampino et al. (2012)
	Sorghum (*Sorghum bicolor*)	Leaves	Microarrays, qRT PCR	Johnson et al. (2014)
	Soybean (*Glycine max*)	Leaves	RNA sequencing, qRT PCR	Wang et al. (2018)
	Lentil (*Lens culinaris*)	Leaves	RNA sequencing	Hosseini et al. (2021)
	Tomato (*Solanum lycopersicum*)	Leaves	Small RNA deep sequencing, degradome sequencing, qRT PCR	Zhou et al. (2020)
	Arabidopsis (*Arabidopsis thaliana*)	Rosettes	RNA sequencing	Sewelam et al. (2020)
Proteomics				
	Soybean (*Glycine max*)	Leaves	2D-DIGE, MALDI–TOF MS	Das et al. (2016)
	Wild barley (*Hordeum spontaneum*)	Leaves	2D-DIGE, MALDI–TOF MS	Ashoub et al. (2015)

(Continued)

TABLE 17.1 (*Continued*)
Examples of "Omics" Studies of Plants under Combined Drought and Heat Stress

"Omics" Technology	Plant Species	Plant Tissues	Techniques Used	References
Metabolomics				
	Arabidopsis (*Arabidopsis thaliana*)	Leaves	GC–MS	Rizhsky et al. (2004); Sewelam et al. (2020)
	Tobacco (*Nicotiana tabacum*)	Leaves, Roots	HPLC–MS	Dobra et al. (2010)
Phenomics				
	Wheat (*Triticum aestivum*)	Whole plants	Structural "red–green–blue" chlorophyll fluorescence and thermal infrared imaging systems	Abdelhakim et al. (2021)

Note: cDNA, complementary DNA; AFLP, amplified fragment length polymorphism; 2D-DIGE, two-dimensional difference in gel electrophoresis; GC–MS, gas chromatography–mass spectrometry; HPLC–MS, high-performance liquid chromatography–mass spectrometry; MALDI–TOF MS, matrix-assisted laser desorption/ionization–time of flight mass spectrometry; qRT PCR, quantitative real-time polymerase chain reaction.

In one of the first transcriptomics studies of *Arabidopsis* (*Arabidopsis thaliana*) plants subjected to drought, heat, and their combination, leaf transcripts were modulated in response to the three stress treatments (Rizhsky et al., 2004). For example, 454 and 318 transcripts were specifically up- and downregulated only in response to the stress combination, highlighting differences in plant response to individual stress factors versus their combination. The increased expression of HSPs is well-documented in plants under a range of stress treatments including drought and heat (Vierling, 1991; Wang et al., 2004; Al-Whaibi, 2011). In the study, *Arabidopsis* leaves exhibited some unique expression patterns of these molecular chaperones and their regulators of heat shock transcription factors (HSFs) in response to the three stresses. Other transcripts specifically elevated in response to the combined heat and drought stress included those of proteases, lipid biosynthetic enzymes, starch degrading enzymes, signal transduction proteins, membrane channel proteins, and antioxidant enzymes.

Likewise, Sewelam et al. (2020) conducted a comprehensive transcriptome analysis of *Arabidopsis* plants under single, double, and triple combinations of NaCl-induced salinity, mannitol-induced osmotic, and heat stresses. Of the 27,655 genes identified in the study, 3,529 were modulated (≥ twofold change) by at least one of the seven treatments. Further analysis of the transcriptome changes induced by individual heat and osmotic stresses highlighted 281 genes whose expression was not modified by salt stress, but largely upregulated by the osmotic stress and downregulated by heat. While it is difficult to identify the actual causes of such conflicting effects of the two stresses on gene expression, the authors highlighted the probable effects of antagonistic physiological strains exerted by the individual stresses on plants. Among the top 50 genes with opposite expression profiles in response to the osmotic and heat stresses were transcripts of LEAs, receptor-like kinases, glutathione *S*-transferases, and WRKY transcription factors (Sewelam et al., 2020). Further functional studies are, however, required to validate the contributions of such differentially expressed genes in single and multifactorial abiotic stress combinations.

Among the cereals, sorghum (*Sorghum bicolor*) is better adapted to hot and dry conditions when compared to maize (*Zea mays*), wheat (*Triticum aestivum*), and rice (*Oryza sativa*) (House, 1985; Doggett, 1988). Despite its prime position as a model cereal in abiotic stress biology (Ngara and Ndimba, 2014a), sorghum transcriptomics studies on combined drought and heat stress are still limited (Johnson et al., 2014), while considerable research has been carried out in response to individually applied drought stress (Ngara et al., 2021). As previously reported for *Arabidopsis* (Rizhsky et al., 2004), leaf transcripts of sorghum exposed to combined drought and heat stress exhibited unique expression changes when compared to the individual stresses (Johnson et al., 2014). For example, transcripts of some HSPs, signaling proteins, ion transporters, transcription factors, LEAs, and polyamine biosynthetic enzymes were only modulated in response to the combined stress treatment. Similarly, a cDNA-AFLP (amplified fragment length polymorphism)-based study of durum wheat (*Triticum turgidum* subsp. *durum*) identified genes that were specifically upregulated in response to the combined drought and heat stress (Rampino et al., 2012). These included genes with putative functions in response to stress, transcriptional regulation, receptor-related roles, and DNA integration.

Although microRNAs (miRNA) are key regulatory elements during plant stress adaptation (Szymanski et al., 2003; Guleria et al., 2011), studies on their expression profiles in response to combined drought and heat stress are still limited (Zhou et al., 2020). Zhou et al. (2020) conducted miRNA sequencing and degradome sequencing of tomato (*Solanum lycopersicum*) leaf samples in response to drought, heat, and their combination. The study identified a total of 765 miRNAs, of which 335 were conserved and 430 novel miRNAs. Of these, 257 conserved and 254 novel miRNAs were shared in the four libraries of the control, drought, heat, and combined stress treatment groups, highlighting common plant responses across different abiotic stresses. However, the differential expression of miRNAs varied with treatments, with some being unique to the combined drought and heat stress treatment group. Overall, the results indicated complex gene regulatory systems in tomato, where a single gene could be regulated by multiple miRNAs, several genes being targets of a single miRNA, and in some cases, a single gene being targeted by the same miRNA across treatment groups. When the expression levels of four miRNAs and their four target genes were validated using quantitative real-time-polymerase chain reaction (qRT PCR), it was observed that increased accumulation of a miRNA sequence in a treatment group relative to the control correlated with a reciprocal decline of its target gene and vice versa. For example, under drought, heat, and combined stress, sly-miR398b-3p-stu and PC-56-5p levels decreased, while those of their respective target genes *SODCC.5* and *22.7 kDa class IV heat shock protein–like* increased (Zhou et al., 2020). Generally, the identified differentially expressed target genes had putative functions in transcription, transcription factor activity, sequence-specific DNA binding, regulation of transcription, nucleus, and DNA binding, highlighting some of the important gene modulations during stress response.

Some stress-induced transcriptome changes contribute towards alterations in protein abundances for further gene regulatory and stress signaling functions as well as for the direct protection of cell components against stress damage (Shinozaki and Yamaguchi-Shinozaki, 1996; Seki et al., 2003). Such changes in protein dynamics can be quantified using a range of gel-based and non-gel-based proteomics methods (Monteoliva and Albar, 2004; Barkla et al, 2013). Global proteomics studies under individual drought and heat stresses and their combination have also demonstrated that plants respond to these three stress treatments using common and unique mechanisms (Ashoub et al., 2015; Das et al., 2016). For example, two-dimensional difference in gel electrophoresis (2D DIGE) coupled with matrix-assisted laser desorption/ionization–time-of-flight–time-of-flight mass spectrometry (MALDI-TOF-TOF MS) revealed that the leaf proteome of wild barley (*Hordeum spontaneum*) exposed to water limitation, heat, and their combination exhibited some common and stress-specific differentially expressed proteins (Ashoub et al., 2015). The study identified a total of 20, 8, and 5 proteins whose abundances were specifically modulated in response to the drought, heat, and combined stress, respectively. However, the five uniquely differentially expressed proteins under combined drought and heat stress were uncharacterized and with unknown functions (Ashoub et al., 2015). These results further highlight the possible existence of many untapped plant genes with important functions in response to stress combinations.

Changes in the plant transcriptome and proteome under stressful conditions ultimately result in alterations in metabolic processes, and the differential accumulation of a myriad of metabolites including soluble sugars, amino acids, proline, and glycine betaine (Ashraf and Foolad, 2007;

Taiz and Zeiger, 2010; Hayat et al., 2012). Although the accumulation levels may differ with plant species, genotype, and/or tissue type (McNeil et al., 1999; Chen and Murata, 2002; Goche et al., 2020), as well as the nature and severity of the stress (Sewelam et al., 2020), some stress-induced metabolites serve in osmoregulatory, signaling, and protective functions (Taiz and Zeiger, 2010). For example, proline is widely reported to accumulate in drought-stressed plants including *Arabidopsis* (Rizhsky et al., 2004; Sewelam et al., 2020), sorghum (Fadoul et al., 2018; Goche et al., 2020), wheat (Qaseem et al., 2019; Sattar et al., 2020), and tobacco (Dobra et al., 2010), and its multifunctional roles under stress have been extensively described (Ashraf and Foolad, 2007; Taiz and Zeiger, 2010; Hayat et al., 2012).

However, there is no consensus on proline metabolism and accumulation patterns under heat stress alone or when heat is combined with drought across different plant species. For instance, in bread wheat (Qaseem et al., 2019; Sattar et al., 2020) and tomato (Raja et al., 2020), endogenous proline contents were elevated under conditions of drought, heat, and combined stress when compared to the control, with the combined stress treatment inducing the highest amount of the amino acid. In contrast, high levels of proline were only detected in drought-stressed *Arabidopsis* plants, with negligible amounts recorded under heat and the combined stress (Rizhsky et al., 2004). The absence of proline in heat-stressed *Arabidopsis* plants either individually or in combination with drought could be attributed to its toxic nature and/or that of its metabolic intermediate pyrroline-5-carboxylate (Rizhsky et al., 2004; Suzuki et al., 2014). Therefore, the role played by proline under individual stresses of drought and heat, together with their combination may differ between plant species and also nature of the stress. For example, in another study, 200 mM mannitol-induced osmotic stress activated proline biosynthesis and accumulation in *Arabidopsis* leaves, but at a much lower magnitude than when the osmotic stress was combined with heat. In some cases, proline and other osmoprotectants may be replaced by sugars such as sucrose, which accumulate under severe stresses or in stress combinations (Rizhsky et al., 2004; Suzuki et al., 2014). Therefore, the range of osmoprotectants accumulating under combined stresses needs further investigation in specific plant families or species prior to their incorporation in plant transformation programs for improved stress resilience (Sewelam et al., 2020).

17.4 EXPLOITING "OMICS" DATA FOR DEVELOPING RESILIENT CROPS

While large-scale datasets obtained from proteomics and transcriptomics analyses provide key insights into stress-adaptive responses of plants, the ultimate goal is to use this information to develop stress-resilient crops. The overriding challenge is selection of targets from hundreds of differentially expressed genes and/or proteins. Clearly the proteome and/or metabolome associated with optimal growth under normal conditions has to be reconfigured to enable stress survival. Resource allocation across cellular metabolic networks is changed to prioritize cellular processes promoting stress survival. Global changes in gene and/or protein expression underpin the adjustments to resource allocation. Thus, the bulk of differentially expressed genes and/or proteins are not central to surviving stress, but rather a reflection of the reset in metabolic homeostasis. Therefore, identifying gene targets with functional roles in stress survival is a monumental challenge.

In multiple stresses, there are differentially expressed genes and/or proteins unique to each specific stress and others that overlap between the stresses. It stands to reason that gene and/or proteins common between the stresses applied individually could hold in part the key targets important in the adaptive process. Johnson et al. (2014) applied drought stress and heat stress separately on sorghum in addition to a combination of the stresses. They identified a total of 438 differentially expressed genes common to all three treatments. There were 2,043 genes that were unique to the drought and heat stress combination. Which of these two groups could hold the key genes for multiple stress resilience? Despite the increasing research on combined stresses in the last 5 years (Table 17.1), there is currently no sufficient experimental data to provide a simple answer, but the logic for the former seems more convincing.

Nevertheless, priority should be given to transcription factors and receptor-like proteins, which could be critical regulatory hubs (Bechtold et al., 2016) controlling multiple stress-survival processes. In an extensive study examining *Arabidopsis* response to single, double, and triple combinations of drought, heat, and viral disease stresses (Prasch and Sonnewald, 2013), 11 differentially expressed genes were found at the intersection of all the multifactorial treatments. Among the 11 candidates, two were transcription factors, Rap2.9 and G-BOX BINDING FACTOR 3 (GBF3). A later study by Ramegowda et al. (2017) provided genetic evidence supporting a central role for GBF3 in drought stress tolerance. *Arabidopsis* loss-of-function *gbf3* mutants have increased sensitivity to salinity stress and osmotic stress, while transgenic *Arabidopsis* overexpressing *Arabidopsis thaliana* or *Eleusine coracana* (finger millet) *GBF3* have increased tolerance to both stresses (Ramegowda et al., 2017). Furthermore, virus-induced silencing of *Zea mays GBF3* reduced photosynthesis and increased cell damage under drought stress (Ramegowda et al., 2017). Although currently there are no reports on the role of GBF3 in combined drought and heat stresses, it is clear that selection of transcription factors as targets for crop improvement is likely to result in success.

17.5 CONCLUSION

Indeed, current projections in global warming and climate change scenarios point to the increased likelihood of combined drought and heat stress occurring in the tropics and a projected decline in crop yield. While "omics" research on plant responses to stress combinations is still in its infancy, some large datasets of stress-responsive genes/proteins/metabolites have been generated. However, other data mining projects of various published "omics" studies are required to extract and present common and unique trends across experiments. In that way, potential universal target elements could be identified for use in generating crops with improved resilience to stress combinations. Nevertheless, HSPs, LEAs, antioxidants, compatible solutes, transcription factors, and miRNAs could be worthwhile targets for further functional validation studies prior to their incorporation in breeding programs.

REFERENCES

Abdelhakim, L. O. A., Rosenqvist, E., Wollenweber, B., Spyroglou, I., Ottosen, C. O., and Panzarova, K. 2021. Investigating combined drought- and heat stress effects in wheat under controlled conditions by dynamic image-based phenotyping. *Agronomy* 11: 364. doi: 10.3390/agronomy11020364.

Al-Whaibi, M. M. 2011. Plant heat-shock proteins: A mini review. *Journal of King Saud University - Science* 23: 139–150. doi: 10.1016/j.jksus.2010.06.022.

Ashoub, A., Baeumlisberger, M., Neupaertl, M., Karas, M., and Bruggemann. W. 2015. Characterization of common and distinctive adjustments of wild barley leaf proteome under drought acclimation, heat stress and their combination. *Plant Molecular Biology* 87: 459–471 doi: 10.1007/s11103-015-0291-4.

Ashraf, M., and Foolad, M. R. 2007. Roles of glycine betaine and proline in improving plant abiotic stress resistance. *Environmental and Experimental Botany* 59: 206–216. doi: 10.1016/j.envexpbot.2005.12.006.

Barkla, B. J., Vera-Estrella, R., and Pantoja, O. 2013. Progress and challenges for abiotic stress proteomics of crop plants. *Proteomics* 13: 1801–1815. doi: 10.1002/pmic.201200401.

Bechtold, U., Penfold, C. A., Jenkins, D. J., Legaie, R., Moore, J. D., Lawson, T., et al. 2016. Time-series transcriptomics reveals that AGAMOUS-LIKE22 affects primary metabolism and developmental processes in drought-stressed arabidopsis. *Plant Cell* 28: 345–366. doi: 10.1105/tpc.15.00910.

Bhardwaj, A., Devi, P., Chaudhary, S., Rani, A., Jha, U. C., Kumar, S., et al. 2022. 'Omics' approaches in developing combined drought and heat tolerance in food crops. *Plant Cell Reports* 41: 699–739 doi: 10.1007/s00299-021-02742-0.

Boyer, J. S. 1982. Plant productivity and environment. *Science* 218(4571): 443–448. doi: 10.1126/science.218.4571.443.

Bray, E. A. 1993. Molecular responses to water deficit. *Plant Physiology* 103: 1035–1040. doi: 10.1104/pp.103.4.1035.

Bray, E. A. 1997. Plant responses to water deficit. *Trends in Plant Science* 2: 48–54. doi: 10.1016/S1360-1385(97)82562-9.

Chen, T. H., and Murata, N. 2002. Enhancement of tolerance of abiotic stress by metabolic engineering of betaines and other compatible solutes. *Current Opinion in Plant Biology* 5: 250–257. doi: 10.1016/S1369-5266(02)00255-8.

Chinnusamy, V., Schumaker, K., and Zhu, J. K. 2004. Molecular genetic perspectives on cross-talk and specificity in abiotic stress signalling in plants. *Journal of Experimental Botany* 55: 225–236. doi: 10.1093/jxb/erh005.

Cramer, G. R., Urano, K., Delrot, S., Pezzotti, M., and Shinozaki, K. 2011. Effects of abiotic stress on plants: A systems biology perspective. *BMC Plant Biology* 11: 163. doi: 10.1186/1471-2229-11-163.

Das, A., Eldakak, M., Paudel, B., Kim, D. W., Hemmati, H., Basu, C., et al. 2016. Leaf proteome analysis reveals prospective drought and heat stress response mechanisms in soybean. *BioMed Research International* 2016: 6021047. doi: 10.1155/2016/6021047.

Dobra, J., Motyka, V., Dobrev, P., Malbeck, J., Prasil, I. T., Haisel, D., et al. 2010. Comparison of hormonal responses to heat, drought and combined stress in tobacco plants with elevated proline content. *Journal of Plant Physiology* 167: 1360–1370 doi: 10.1016/j.jplph.2010.05.013.

Doggett, H. 1988. *Sorghum*, 2nd. ed. New York, NY: Longman Scientific & Technical.

Fadoul, H. E., El Siddig, M. A., Abdalla, A. W. H., and El Hussein, A. A. 2018. Physiological and proteomic analysis of two contrasting *Sorghum bicolor* genotypes in response to drought stress. *Australian Journal of Crop Science* 12: 1543–1551. doi: 10.21475/ajcs.18.12.09.PNE134.

Farooq, M., Wahid, A., Kobayashi, N., Fujita, D., and Basra, S. M. A. 2009. Plant drought stress: effects, mechanisms and management. *Agronomy for Sustainable Development* 29: 185–212 doi: 10.1051/agro:2008021.

Gan, T. Y., Ito, M., Hulsmann, S., Qin, X., Lu, X. X. Liong, S. Y., et al. 2016. Possible climate change/variability and human impacts, vulnerability of drought-prone regions, water resources and capacity building for Africa. *Hydrological Sciences Journal* 61: 1209–1226 doi: 10.1080/02626667.2015.1057143.

Goche, T., Shargie, N. G., Cummins, I., Brown, A. P., Chivasa, S. and Ngara, R. 2020. Comparative physiological and root proteome analyses of two sorghum varieties responding to water limitation. *Scientific Reports* 10: 11835. doi: 10.1038 s41598-020-68735-3.

Guleria, P., Mahajan, M., Bhardwaj, J., and Yadav, S. K. 2011. Plant small RNAs: Biogenesis, mode of action and their roles in abiotic stresses. *Genomics, Proteomics and Bioinformatics* 9: 183–199 doi: 10.1016/S1672-0229(11)60022-3.

Hale, M. G., Orcutt, D. M., and Thompson, L. K. 1987. *The Physiology of Plants under Stress*. New York, NY: Wiley.

Hayat, S., Hayat, Q., Alyemeni, M. N., Wani, A. S., Pichtel, J., and Ahmad, A. 2012. Role of proline under changing environments: A review. *Plant Signalling and Behaviour* 7: 1456–1466. doi: 10.4161/psb.21949.

Hosseini, S. Z., Ismaili, A., Nazarian-Firouzabadi, F., Fallahi, H., Nejad, A. R., and Sohrabi, S. S. 2021. Dissecting the molecular responses of lentil to individual and combined drought and heat stresses by comparative transcriptomic analysis. *Genomics* 113: 693–705 doi: 10.1016/j.ygeno.2020.12.038.

House, L. R. 1985. *A Guide to Sorghum Breeding*, 2nd ed. Patancheru: International Crops Research Institute for the Semi-Arid Tropics.

IPCC. 2007. Summary for policymakers. In Parry, M. L., et al. (eds), *Climate Change: Impacts, Adaptation and Vulnerability, Contribution of Working Group II to the Fourth Assessment Report of the Intergovernmental Panel on Climate Change*, 7–22. Cambridge: Cambridge University Press.

Johnson, S. M., Lim, F. L., Finkler, A., Fromm, H., Slabas, A. R., and Knight, M. R. 2014. Transcriptomic analysis of *Sorghum bicolor* responding to combined heat and drought stress. *BMC Genomics* 15: 456. doi: 10.1186/1471-2164-15-456.

Kerr, R. A. 2007. Global warming is changing the world. *Science* 316(5822): 188–190. doi: 10.1126/science.316.5822.188.

Knight, H., and Knight, M. R. 2001. Abiotic stress signalling pathways: Specificity and cross-talk. *Trends in Plant Science* 6: 262–267. doi: 10.1016/S1360-1385(01).01946-X.

Kotak, S., Larkindale, J., Lee, U., von Koskull-Doring, P., Vierling, E. and Scharf, K. D. 2007. Complexity of the heat stress response in plants. *Current Opinion in Plant Biology* 10: 310–316. doi: 10.1016/j.pbi.2007.04.011.

Levitt, J. 1980a. Responses of plants to environmental stresses: Water, radiation, salt, and other stresses. In T. T. Kozlowski (ed.) *Physiological Ecology* (607 pages), 2nd ed., vol. II. New York, NY: Academic Press.

Levitt, J. 1980b. Responses of plants to environmental stresses: Chilling, freezing, and high temperature stress. In T. T. Kozlowski (ed.) *Physiological Ecology* (497 pages), 2nd ed., vol. 1. New York, NY: Academic Press.

McNeil, S. D., Nuccio, M. L., and Hanson, A. D. 1999. Betaines and related osmoprotectants. Targets for metabolic engineering of stress resistance. *Plant Physiology* 120: 945–949 doi: 10.1104/pp.120.4.945.

Mittler, R. 2006. Abiotic stress, the field environment and stress combination. *Trends in Plant Science* 11: 15–19 doi: 10.1016/j.tplants.2005.11.002.

Mittler, R., and E. Blumwald. 2010. Genetic engineering for modern agriculture: Challenges and perspectives. *Annual Review of Plant Biology* 61: 443–462. doi: 101146/annurev-arplant-042809-112116.

Monteoliva, L., and Albar, J. P. 2004. Differential proteomics: An overview of gel and non-gel based approaches. *Briefings in Functional Genomics and Proteomics* 3: 220–239. doi: 10.1093/bfgp/3.3.220.

Nankishore, A., and Farrell, A. D. 2016. The response of contrasting tomato genotypes to combined heat and drought stress. *Journal of Plant Physiology* 202: 75–82. doi: 10.1016/j.jplph.2016.07.006.

Nelson, G. C., Rosegrant, M., Koo, J., Robertson, R., Sulser, T., Zhu, T., et al. 2009. *Climate Change. Impact on Agriculture and Costs of Adapation.* Washington, DC: International Food Policy Research Institute.

Ngara, R., Goche, T., Swanevelder, D. Z. H., and Chivasa, S. 2021. Sorghum's whole-plant transcriptome and proteome responses to drought stress: A review. *Life* 11: 704. doi: 10.3390/life11070704.

Ngara, R., and Ndimba, B. K. 2014a. Model plant systems in salinity and drought stress proteomics studies: A perspective on arabidopsis and sorghum. *Plant Biology* 16: 1029–1032. doi: 10.1111/plb.12247.

Ngara, R., and Ndimba, B. K. 2014b. Understanding the complex nature of salinity and drought-stress response in cereals using proteomics technologies. *Proteomics* 14: 611–621. doi: 10.1002/pmic.201300351.

Nievola, C. C., Carvalho, C. P., Carvalho, V., and Rodrigues. E. 2017. Rapid responses of plants to temperature changes. *Temperature* 4: 371–405. doi: 10.1080/23328940.2017.1377812.

Park, S. I., Kwon, H. J., Cho, M. H., Song, J. S., Kim, B. G., Baek, J., et al. 2021. The *OsERF115/AP2EREBP110* transcription factor is involved in the multiple stress tolerance to heat and drought in rice plants. *International Journal of Molecular Sciences* 22: 7181. doi: 10.3390/ijms22137181.

Prasch, C. M., and Sonnewald, U. 2013. Simultaneous application of heat, drought, and virus to arabidopsis plants reveals significant shifts in signaling networks. *Plant Physiology* 162: 1849–1866. doi: 10.1104/pp.113.221044.

Qaseem, M. F., Qureshi, R., and Shaheen, H. 2019. Effects of pre-anthesis drought, heat and their combination on the growth, yield and physiology of diverse wheat (*Triticum aestivum* L.) genotypes varying in sensitivity to heat and drought stress. *Scientific Reports* 9: 6955. doi: 10.1038.s41598-019-43477-z.

Raja, V., Qadir, S. U., Alyemeni, M. N., and Ahmad, P. 2020. Impact of drought and heat stress individually and in combination on physio-biochemical parameters, antioxidant responses, and gene expression in *Solanum lycopersicum 3. Biotech* 10: 208. doi: 10.1007/s13205-020-02206-4.

Ramegowda, V., Gill, U. S., Sivalingam, P. N., Gupta, A., Gupta, C., Govind, G., et al. 2017. GBF3 transcription factor imparts drought tolerance in *Arabidopsis thaliana. Scientific Reports* 7: 9148. doi: 10.1038/s41598-017-09542-1.

Rampino, P., Mita, G., Fasano, P., Borrelli, G. M., Aprile, A., Dalessandro, G., et al. 2012. Novel durum wheat genes up-regulated in response to a combination of heat and drought stress. *Plant Physiology and Biochemistry* 56: 72–78. doi: 10.1016/j.plaphy.2012.04.006.

Rivero, R. N., Mittler, R., Blumwald, E., and Zandalinas, S. I. 2022. Developing climate-resilient crops: Improving plant tolerance to stress combination. *Plant Journal* 109: 373–389 doi: 10.1111/tpj.15483.

Rizhsky, L., Liang, H. J., and Mittler, R. 2002. The combined effect of drought stress and heat shock on gene expression in tobacco. *Plant Physiology* 130: 1143–1151. doi: 10.1104/pp.006858.

Rizhsky, L., Liang, H. J., Shuman, J., Shulaev, V., Davletova, S., and Mittler, R. 2004. When defense pathways collide. The response of arabidopsis to a combination of drought and heat stress. *Plant Physiology* 134: 1683–1696 doi: 10.1104/pp.103.033431.

Sattar, A., Sher, A., Ijaz, M., Ul-Allah, S., Rizwan, M. S., Hussain, M., et al. 2020. Terminal drought and heat stress alter physiological and biochemical attributes in flag leaf of bread wheat. *PLoS One* 15: e0232974. doi: 10.1371/journal.pone.0232974.

Seki, M., Kamei, A., Yamaguchi-Shinozaki, K., and Shinozaki, K. 2003. Molecular responses to drought, salinity and frost: Common and different paths for plant protection. *Current Opinion in Biotechnology* 14: 194–199 doi: 10.1016/S0958-1669(03)00030-2.

Sewelam, N., Brilhaus, D., Brautigam, A., Alseekh, S., Fernie, A. R., and Maurino, V. G. 2020. Molecular plant responses to combined abiotic stresses put a spotlight on unknown and abundant genes. *Journal of Experimental Botany* 71: 5098–5112 doi: 10.1093/jxb/erra250.

Shao, H. B., Jiang, S. Y., Li, F. M., Chu, L. Y., Zhao, C. X., Shao, M. A., et al. 2007. Some advances in plant stress physiology and their implications in the systems biology era. *Colloids and Surfaces B-Biointerfaces* 54: 33–36. doi: 10.1016/J.colsurfb.2006.05.011.

Shinozaki, K., and Yamaguchi-Shinozaki, K. 1996. Molecular responses to drought and cold stress. *Current Opinion in Biotechnology* 7: 161–167. doi: 10.1016/s0958-1669(96)80007-3.

Suzuki, N., Rivero, R. M., Shulaev, V., Blumwald, E., and Mittler, R. 2014. Abiotic and biotic stress combinations. *New Phytologist* 203: 32–43. doi: 10.1111/nph.12797.

Szymanski, M., Barciszewska, M. Z., Zywicki, M., and Barciszewski, J. 2003. Noncoding RNA transcripts. *Journal of Applied Genetics* 44: 1–19.

Taiz, L., and Zeiger, E. 2010. *Plant Physiology*, 4th ed. Sunderland, MA: Sinauer Associates Inc. Publishers.

Vierling, E. 1991. The roles of heat shock proteins in plants. *Annual Review of Plant Physiology and Plant Molecular Biology* 42: 579–620. doi: 10.1146/annurev.pp.42.060191.0030551.

Wahid, A., Gelani, S., Ashraf, M., and Foolad, M. R. 2007. Heat tolerance in plants: An overview. *Environmental and Experimental Botany* 61: 199–223. doi: 10.1016/j.envexpbot.2007.05.011.

Wang, L. B., Liu, L. J., Ma, Y. L., Li, S., Dong, S. K., and Zu, W. 2018. Transcriptome profilling analysis characterized the gene expression patterns responded to combined drought and heat stresses in soybean. *Computational Biology and Chemistry* 77: 413–429. doi: 10.1016/j.combiolchem.2018.09.012.

Wang, W., Vinocur, B. and Altman, A. 2003. Plant responses to drought, salinity and extreme temperatures: Towards genetic engineering for stress tolerance. *Planta* 218: 1–14 doi: 10.1007/s00425-003-1105-5.

Wang, W. X., Vinocur, B., Shoseyov, O., and Altman, A. 2004. Role of plant heat-shock proteins and molecular chaperones in the abiotic stress response. *Trends in Plant Science* 9: 244–252. doi: 10.1016/j.tplants.2004.03.006.

Xiong, L., and Zhu, J. K. 2002. Molecular and genetic aspects of plant responses to osmotic stress. *Plant Cell and Environment* 25: 131–139. doi: 10.1046/j.1365-3040.2002.00782.x.

Yuan, J. S., Galbraith, D. W., Dai, S. Y., Griffin, P., and Stewart, C. N. 2008. Plant systems biology comes of age. *Trends in Plant Science* 13: 165–171. doi: 10.1016/j/tplants.2008.02.003.

Yang, Y. D., Saand, M. A., Huang, L. Y., Abdelaal, W. B., Zhang, J., Wu, Y., et al. 2021. Applications of multi-omics technologies for crop improvement. *Frontiers in Plant Science* 12: 563953. doi: 10.3389/fpls.2021.563953.

Zandalinas, S. I., Mittler, R., Balfagon, D., Arbona, V., and Gomez-Cadenas, A. 2018. Plant adaptations to the combination of drought and high temperatures. *Physiologia Plantarum* 162: 2–12. doi: 10.1111/ppl.12540.

Zenda, T., Liu, S. T., Dong, A. Y., Li, J., Wang, Y. F., Liu, X. Y., et al. 2021. Omics-facilitated crop improvement for climate resilience and superior nutritive value. *Frontiers in Plant Science* 12: 774994. doi: 10.3389/fpls.2021.774994.

Zhou, R., Yu, X. Q., Ottosen, C. O., Zhang, T. L., Wu, Z., and Zhao, T. M. 2020. Unique miRNAs and their targets in tomato leaf responding to combined drought and heat stress. *BMC Plant Biology* 20: 107. doi: 10.1186/s12870-020-2313-x.

18 Genome Editing for Multiple Stress Tolerance

Laxmipreeya Behera, K.A. Pawan, and Shraddha Bhaskar Sawant
Odisha University of Agriculture and Technology

S. Sanghamitra and C. Parmeswaran
ICAR-National Rice Research Institute

C.S. Kailash
Odisha University of Agriculture and Technology

Bijayalaxmi Sahoo and Iteeshree Nayak
ICAR-National Rice Research Institute

CONTENTS

18.1 INTRODUCTION

Genome editing is rapidly advancing and has the potential to become the dominant technology for managing abiotic stress such as drought, salinity, extreme heat and cold, and heavy metal contamination in soils, as well as biotic stress such as insects, bacteria, virus, fungus, and nematodes, all of which have a significant impact on crop production. The clustered regularly interspaced short palindromic repeat (CRISPR)–Cas (CRISPR-associated) system is one of the most efficient, convenient, and time-saving approaches to effective gene manipulation. The Cas9 protein, an RNA-guided DNA endonuclease, causes a site-specific double-stranded break, which triggers a DNA-repair mechanism that results in genome modification. Drought, salt, temperature, and heavy metals are all abiotic stressors that represent a severe threat to crop productivity and result in significant yield reductions around the world. The most sustainable and environmental-friendly strategy to dealing with this dilemma is to breed resistant cultivars against these pressures. Agriculture and food security are threatened by climate change (Hasegawa et al., 2018). Excessive greenhouse gas emissions are to blame for the frequent occurrences of high temperatures and drought stress in crops. Wheat,

DOI: 10.1201/9781003300564-18

rice, and maize yields are expected to decrease by 6%, 10%–20%, and 21%–31%, respectively, with a 1°C increase in ambient temperature. Precision genome engineering for particular crop attributes has become possible thanks to advancements in genome editing technology. The choice of candidate genes, on the other hand, is crucial to the achievement of the desired features. These genes can be divided into two groups: structural genes and regulatory genes. Structural genes code for proteins that give direct stress tolerance, whereas regulatory genes govern the expression of other genes involved in various cellular processes. In order to achieve stress tolerance, *cis*-regulatory sequences are also required. We acclaim utilizing clustered regularly interspaced short palindromic repeat (CRISPR)/CRISPR-associated protein 9 (Cas9) system to target these regulatory and/or structural genes, as well as the *cis*-regulatory sequences, as a reliable, effective, and realistic way of developing climate-resilient crop varieties. We also highlight how CRISPR/Cas-mediated promoter targeting could be used to create new quantitative trait loci for abiotic stress tolerance. These genome editing technologies are expected to contribute to not only the development of novel plant varieties that are resistant to diverse abiotic challenges, but also public acceptance of these goods in the future.

18.2 UNDERSTANDING AND FUNCTIONAL GENE ANALYSIS THROUGH CRISPR/CAS

Understanding the function of abiotic trait associated gene is a difficult task. Adaptation to particular abiotic stress is influenced by interactions between mechanisms of numerous signaling, regulatory and metabolic networks (Mickelbart et al., 2015) governed by wide range of genes (Table 18.1). Abiotic stress-resistant T genes have been found and put into crop plants by molecular breeders. The study of drought response is well-understood by mitogen-activated protein kinases3 (slmapk3) mutant gene developed by CRISPR/Cas in tomato (Wang et al., 2017) and stress-ABA-activated protein kinase2 mutant in case of rice (Lou et al., 2017). CRISPR-Cas9 was utilized to generate mutants (cbfs double and triple mutants) in *Arabidopsis* under cold stress to examine the involvement of C-repeat binding factors. To enhance drought tolerance in maize, the CRISPR-Cas9 technique was utilized to boost the expression level of the *ARGOS8* gene (negatively control ethylene response); the promoter of *ARGOS8* was altered to GOS2. Under drought circumstances in the field, these mutants increased grain yields. Furthermore, overexpressing TaCP and SPCP2 in *Arabidopsis* improved drought tolerance. Abiotic stress tolerance has been observed in plants that overexpress the melatonin production genes. Targeted editing of the *TMS5* gene in hybrid rice has resulted in the rapid production of temperature-sensitive breeding lines. After selecting yield-related genes during domestication operations, plant breeding activities may have reduced T gene alleles. Furthermore, whole genome duplication events have occurred in plants, and a high percentage of genes are represented by multigene families with functional redundancy. Often, knocking out a single gene does not result in any/desired phenotype, making it difficult to determine its function. More number of genes can be targeted simultaneously utilizing the CRISPR-Cas9 system because to the ease of creation and high efficiency of sgRNAs, overcoming the problem of functional redundancy of genes. In model plants and agricultural crop, multiplex genome editing has been effectively implemented (Mao et al., 2013). Such techniques can help researchers figure out what role many functionally redundant genes play in a biological process such as abiotic stress response. Another strategy could be to use HDR-mediated gene targeting to pyramid/stack numerous genes involved in a stress response pathway or regulatory network. CRISPR-Cas9 technologies could be used to target genes implicated in the stress-related gene regulatory network, signal transduction, and metabolite production in order to create stress-tolerant crop plants. Crop development programs rely on the availability of wild germplasm and genetic variability in crop plants. However, both basic and practical research, particularly in crop plants, is hampered by a lack of natural germplasm, genetic diversity, and mutant collections. Genome editing techniques allow for the production of such variants in the genomes of crop plants, allowing for the circumvention of these constraints. SgRNA libraries targeting practically all genes can be constructed for any plant species due to their small size and ease of design, and they can be used to generate genome-scale point mutations and gene knockouts. Large-scale genetic screening in agricultural and non-model plants can be aided by the availability of such collections. SgRNA libraries can also be used to modify non-coding genomic elements, making it easier to find gene regulatory sequences. Recent research has demonstrated that the CRISPR-Cas9 system can perform potent negative and positive shortlisting screens in humans (Shalem et al., 2014; Wang et al., 2014). Even though these resources have not yet been created in plants, they are expected to do so in the not-too-distant future. Screening mutants with altered abiotic stress responses in plants could improve gene function analysis and the production of stress-tolerant agricultural kinds. The CRISPR-Cas9 system's capabilities can be expanded by using

TABLE 18.1
List of Abiotic Stress-Related Genes Edited through Genome Editing

Sr. No.	Gene Modified	Trait	Target Plant	Trait Improvement	Reference
1	*Bentazon-sensitive lethal (BEL)*	Herbicide tolerance	*Oryza sativa* Japonica cv. Nipponbare	Bantazone and sulfonylurea tolerant	Xu et al. (2014)
2	*Acetolactate synthase (ALS)*	Herbicide tolerance	*Oryza sativa* Japonica cv. Nipponbare	Bispyribac sodium tolerant	Sun et al. (2016)
3	*OsRAV2*	Salt tolerance		Salt stress responsive	Duan et al. (2017)
4	*OsNramp5*	Cadmium uptake	Indica cv. Huazhan, Longke 638s	Low cadmium accumulation	Tang et al. (2017)
5	*OsAnn3*	Cold tolerance		Loss of function: reduced cold tolerance	Shen et al. (2017)
6	*TIFY1a, TIFY1b*	Cold tolerance	*Oryza sativa* Japonica cv. Nipponbare	Enhance cold tolerance	Huang et al. (2017)
7	*SAPK2*	Drought stress	Japonica	ABA-mediated seed dormancy, drought tolerance	Lou et al. (2017)
8	*OsARM1*	Arsenic stress	Japonica	Tolerant/sensitive to arsenic	Wang et al. (2017)
9	*CCd7*	Strigolactone biosynthesis	*Oryza sativa* Japonica cv. Nipponbare	Increased tiller no. and reduced plant height	Butt et al. (2018)
10	*CCx2*	Cadmium stress	Japonica cv. Nipponbare	Grain cadmium accumulation	Hao et al. (2018)
11	*RR22*	Salinity	Japonica cv. WPB106	Enhanced tolerance	Zhang et al. (2019a)
12	*Semi-rolled leaf 1,2*	Drought tolerance	Restorer rice lines (GXU16, GXU20, GXU28)	Leaf rolling, increased panicle number, grain number per panicle, and yield	Liao et al. (2019)
13	*VPI*	Seed dormancy	Japonica cv. Dongjin	Fast seed germination in mutant lines	Jung et al. (2019)
14	*ACS*	Phosphate deficiency	Japonica cv. Nipponbare	Low phosphate tolerance and lateral root elongation	Lee et al. (2019)
15	*PRP1*	Cold response	Indica cv. GXU2-44	Enhanced cold tolerance	Nawaz et al. (2019)
16	*NACO41*	Salt response	Japonica cv. Nipponbare	Salt sensitivity plant	Angon et al. (2023)

(Continued)

TABLE 18.1 (*Continued*)
List of Abiotic Stress-Related Genes Edited through Genome Editing

Sr. No.	Gene Modified	Trait	Target Plant	Trait Improvement	Reference
17	*PPa6*	Salinity response	Japonica cv. Kitaake	Lowered alkaline tolerance and dwarf plant height	Wang et al. (2019)
18	*PIN5b* *GS3* *MYB 30*	Yield and cold tolerance	Japonica cv. Nipponbare	Increased in panicle length, and enlarged grain size	Zhu et al. (2020)
19	*DST1*	Drought and salinity tolerance	Indica cv. MTU1010	Improved drought and salt tolerance and grain yield	Santosh Kumar et al. (2020)
20	*ERA1*	Drought tolerance	Japonica cv. Nipponbare	Increased primary root growth	Ogata et al. (2020)
21	*PYL9*	Drought tolerance and grain yield		Enhanced drought tolerance, 1,000 grain weight, grain length, grain width, and yield per panicle. Decreased flag leaf length and width	Usman et al. (2020)

various Cas9 protein variants. For example, employing dCas9 (catalytically inactive Cas9), CRISPR interference (CRISPRi) can be utilized to disrupt gene function (Gilbert et al., 2013). To improve transcriptional suppression, several effector domains, such as KRAB/SID, have been coupled with dCas9. Through use of paired Cas9 nickases improves the specificity of mutagenesis significantly. The addition of transcriptional activation domains to dCas9, such as VP16/VP64, can boost target gene expression and allow for abiotic stress tolerance phenotyping. Using a large number of sgRNAs for a single gene promoter has been proven in a few experiments to have a high activation impact (Mali et al., 2013). A synthetic transcriptional activator or repressor based on CRISPR has been proven to influence the transcription of target endogenous genes in plants (Piatek et al., 2015). The role of methylation or distinct chromatin states in abiotic stress responses/adaptation can be better understood by tethering dCas9 to epigenetic modifiers. A similar approach might be used to regulate or fine-tune the expression of stress-responsive genes.Cas9 protein reassembly can be used to create inducible CRISPR-Cas9 systems that allow for spatially precise changes. The incorporation of light-inducible domains, CIB1 and CRY2, to create TALENS has been effectively shown. To avoid unwanted pleiotropic effects, conditional (stress-inducible and/or tissue-specific) promoters can be used to induce Cas9 and sgRNA expression. Mutations in an AP2/ERF family transcription factor gene, ABA-INSENSITIVE 4, important in abiotic stress responses, were induced using a custom-designed zinc finger nuclease and a heat-shock promoter (Osakabe et al., 2010). There was a high frequency (up to 3%) of gene mutations that resulted in the expected phenotypes.

18.3 EXPLOITATION OF GENOME EDITING TOOLS FOR SUSTAINABLE MANAGEMENT OF BIOTIC STRESS

Biological stress is a significant limitation to agricultural progress and global food security. Diseases limit crop yields by impairing plant growth and development, as well as lowering the quality of agricultural goods in the field and during storage. Diseases reduce crop output by 20%–40% globally, according to insufficient statistical data. Plants are sensitive to a wide range of diseases, including fungus, bacteria, and viruses, which cause significant economic losses; improving plant resistance is critical for altering crop production to meet world population growth (http://www.fao.org/news/story/en/item/280489/icode/). When used correctly, disease management methods that rely on resistant cultivars and agrochemicals are usually very efficient. However, because many plant diseases have a high evolutionary potential, emerging genotypes resistant to the resistance gene or the phytosanitary agent might emerge quickly through mutation or recombination. When this happens, disease control strategies can soon become ineffective as novel genotypes become more common through natural selection and spread to other locations, resulting in control failure across wide geographic areas. For many years, scientists have been trying to figure out how plants interact with populations of bacteria, fungus, and other microbes. Biotic stressors, such as viral, fungal, and bacterial infections, account for 20%–40% of global agricultural output losses (Savary et al., 2012).

Advances in genome editing tools have opened up new avenues for improving crop resistance. The CRISPR/Cas system has been used to address a variety of agricultural difficulties in recent years, including improving biotic stress resistance. CRISPR/Cas technologies have primarily been used to combat virus infection, with efforts to improve fungal and bacterial disease resistance following Arora and Narula (2017). Destructive fungal and oomycete diseases have reduced crop yield significantly, and these pathogens continue to pose a danger to global food security. Although chemical and cultural controls have been used for crop protection, these involve continuous costs and time and fungicide resistance among plant pathogens has been increasingly reported. Production of disease-resistant cultivars is the most effective technique to protect crops against plant diseases (Sánchez-Martín and Keller, 2019). Traditional breeding methods, on the other hand, are time consuming and labor expensive. The CRISPR/Cas9 system has recently been used to improve disease resistance in a variety of crops, including rice, cacao, wheat, tomato, and grape. Plant disease treatment can be made more ecologically friendly and frequently complementary by developing plant kinds with in-built disease resistance through breeding (Table 18.2). Plant breeding for disease resistance often involves identifying plants with significant disease resistance features, growing and developing breeding candidates in a disease-friendly environment, and then selecting disease resistance individuals that retain yield. According to Schenke and Cai (2020), there are numerous techniques for using the CRISPR/Cas system to study plant disease resistance: (1) knocking out genes that code for susceptibility factors (e.g., MLO; a mildew resistance locus O) (Nekrasov et al., 2017); (2) *cis*-element deletion, alteration, or introduction in promoters (Oliva et al., 2019); (3) using HDR to introduce particular mutations in coding areas; (4) modifying amino acids in plant surface receptor proteins to avoid pathogen effectors produced from the pathogen (e.g., AtBAK1) (Li et al., 2016). Negative regulators of plant defense responses (e.g., TcNPR3) (Fister et al., 2018) can be knocked out, and central regulators of defense responses can be modified (e.g., BnWRKY70) (Sun et al., 2018). Three ways are used to implement these strategies: first one is delivering a ribonucleoprotein (RNP) complex consisting of the Cas protein and sgRNA into plant cells for DNA-free plant genome editing via polyethylene glycol (PEG)-mediated protoplast transformation or second, ballistic delivery of RNP complex into plant embryo cells or through *Agrobacterium*-mediated transformation (AMT) (Fister et al., 2018). PTI is activated quickly in general when PAMPs are recognized by pattern recognition receptors (PRRs). The CRISPR/Cas9 system was first discovered in bacteria and archaea as an RNA-mediated adaptive defense response against viral invasion. CRISPR technology has opened up new possibilities for identifying essential defense-related genes in plants and improving disease resistance in crops in order to assure food safety and long-term agricultural development.

18.3.1 Fungal Resistance

Mildew, smut, rust, rot, and a variety of other syndromes are caused by fungal infections. These diseases not only result in significant yield losses each year around the world, but they also affect the quality of the harvested crops. Furthermore, mycotoxigenic fungi are a big concern because they produce secondary metabolites known as mycotoxins, which cause serious health problems in humans and animals when they consume contaminated food or feed. RNA-guided Cas9 endonuclease has been used to target MLO loci in three different plant species: bread wheat, tomato, and grapevine. MLO codes for a plasma membrane-localized protein with seven transmembrane

TABLE 18.2
List of Biotic Stress-Related Genes Edited through Genome Editing

Sr. No.	Gene Modified	Promoter Used	Target Plant	Effector Protein	Strategy of Transformation	Trait Improvement	Reference
				BACTERIA			
1.	OsSWEET14	AtU6	*O. sativa* Zhonghua	SpCas9	*Agrobacterium* EHA105	Disruption of *OsSWEET14* in the zhobghua 11 background confer strong resistance to African Xoo strain AXO1947 and Asian Xoo strain PXO86	Zeng et al. (2020)
2.	DMR6	2XCaMV35S OsU6 CaMV35S	Banana	SpCas9	*Agrobacterium* EHA105	Musadmr6 mutants of banana showed enhanced	Tripathi et al. (2020)
3.	EBEs of OsSWEET14 gene	OsU3, U6	Super Basmati (Pakistan's indigenous rice variety *O. sativa)*	SpCas9	Biolistic	Resistance in super basmati rice lines against Xoo strain carrying AvrXa7 TALE	Zafar et al. (2020)
4.	EBE of three promoters, i.e., SWEET11, SWEET13, and SWEET14	U6	O. sativa	SpCas9	*Agrobacterium*	Enhanced broad-spectrum disease resistance to *Xanthomonas oryzae* pv. *oryzae*	Oliva et al. (2019)
5.	EBE of OsSWEET11 and OsSWEET14	SWEET11, SWEET14	O. sativa	SpCas9	*Agrobacterium*	Enhanced broad-spectrum disease resistance to *Xanthomonas oryzae* pv. *oryzae*	Xu et al. (2019)
6.	S1JAZ2	Ubiquitin	*Solanum lycopersicum*	SpCas9	*Agrobacterium*	Resistance against bacterial speak caused by *Pseudomonas syringae*, DC3000	Talakayala et al. (2022)
7.	CaLOB1	CaMV35S	*Citrus paradisi*	SpCas9	*Agrobacterium* infiltration	Resistance *Xanthomonas citri* subsp. to citrus canker	Jia et al. (2017)
				FUNGUS			
1.	StDMR6-1 and StDMR6-2	-	*Solanum tuberosum* L	SpCas9	*Agrobacterium* C58	Editing confer increased late blight resistance	Kieu et al. (2021)
2.	VvPR4b	AtU6	*Vitis vinifera*	SpCas9	*Agrobacterium* GV3101	Editing decreases downy mildew (*Plasmopara viticola*) resistance in grapevine	Li et al. (2020)
3.	PMR4	AtU6	*Solanum lycopersicum* L	SpCas9	*Agrobacterium* AGL1	Knockdown of SLPMR4 enhances resistance against	Santillan-Martinez et al. (2020)
4.	VvWRKY52	AtU3d, AtU3b, AtU6-1, AtU6-29	*Vitis vinifera*	SpCas9	*Agrobacterium* EHA105	Resistance to gray mold (*Botrytis cinerea*)	Talakayala et al. (2022)

(*Continued*)

TABLE 18.2 (*Continued*)
List of Biotic Stress-Related Genes Edited through Genome Editing

Sr. No.	Gene Modified	Promoter Used	Target Plant	Effector Protein	Strategy of Transformation	Trait Improvement	Reference
5.	Clpsk1	AtU6	Watermelon (*Citrullus lanatus*)	SpCas9	*Agrobacterium* EHA105	Loss-of-function rendered watermelon seedlings more resistant to infection by *Fusarium oxysporum* f. sp. *niveum*	Zhang et al. (2020)
6.	MLO6, DMR		*Vitis vinifera*	SpCas9	*Agrobacterium*	Resistance against grape wine powdery mildew (*Erysiphe necator*) and downy mildew	Talakayala et al. (2022)
7.	Gh14-3-3d	AtU3	Cotton	SpCas9	*Agrobacterium* VIGS-mediated GV3101	Resistance to *Verticillium dahliae* infestation	Zhang et al. (2018)
8.	TcNPR3, a suppressor of the defense response	AtU6-26	*Theobroma cacao*	SpCas9	*Agrobacterium* infiltration	Increased resistance to infection with the cacao pathogen *Phytophthora tropicalis*	Fister et al. (2018)
9.	Solyc08g075770/ genomic region	CaMV35S	Tomato	SpCas9	Agrobacterium GV3101	Enhanced susceptibility to disease *Fusarium oxysporum* f.sp. *lycopersici*	Prihatna et al. (2018)
10.	alEPIC8/cds region	Hsp70, HAM34 RPL41	Papaya	SpCas9	*Agrobacterium* EHA105	Increased resistance against *P. palmivora*	Ahmad et al. (2020)
11.	StMlo1	pAGM4723	*Solanum lycopersicm*	SpCas9	*Agrobacterium*	Resistance against *Oidium neolycopersici* (tomato powdery mildew)	Nekrasov et al. (2017)
12.	TaEDR1	Pjit163-Ubi-Cas9, TaU6	*Triticum aestivum*	SpCas9	Biolistic bombardment	Resistance against *Blumeriagraminis* f.sp. *Triticii* (powdery mildew)	Zhang et al. (2017)
				VIRUS			
1.	gRNAs – pCVA and pCVB of Cabbage leaf curl virus (CaLCuV)	CaMV35S, AtU6	*Arabidopsis*	SpCas9	*Agrobacterium* GV3101	Demonstrated gemini viral-mediated VIGE for genome editing	Yin and Qiu (2019)
	CLCuMuV C1 (Rep) and IR	35S, U6	*Nicotiana benthamiana*	SpCas9	*Agrobacterium* GV2260	Resistance to cotton leaf curl multan virus	
2.	TMV (Tobacco mosaic virus) SRBDSV (Southern rice black-streaked dwarf virus) RSMV (rice stripe mosaic virus)	CaMV35S, OsU6, AtU6	Tobacco and *Oryza sativa*	LshCas13a	*Agrobacterium* GV3101	Resistance to viruses, i.e., TMV, SRBDSV, RSMV	Zhang et al. (2019b)

(*Continued*)

TABLE 18.2 (*Continued*)
List of Biotic Stress-Related Genes Edited through Genome Editing

Sr. No.	Gene Modified	Promoter Used	Target Plant	Effector Protein	Strategy of Transformation	Trait Improvement	Reference
3.	ChiLCV	U3, U6 and CaMV35S	*Nicotiana benthamiana*	SpCas9	*Agrobacterium*-transient assay GV3103	Resistant to ChiLCV reduced viral accumulation	Roy et al. (2019a)
4.	ORF1, 2, 3 and IR of BSV	OsU6	*Musa* spp. (banana)	SpCas9	*Agrobacterium* EHA105	Generated resistant banana plants against endogenous banana streak virus	Tripathi et al. (2021)
5.	Four WDV-specific sgRNAs – MP, CP, Rep/Rep, IR/WDV	ZmUbi1, CaMV35S	Barley	SpCas9	*Agrobacterium* C58C1	Resistant to wheat dwarf virus	Kis et al.
6.	AC2 and AC3 of African cassava mosaic virus	CaMV35S, T7, U6	Cassava	SpCas9	*Agrobacterium* LBA4404	Failed to show clear disease resistance against ACMV in cassava	Mehta et al. (2019)
7.	P3-membrane protein, CI-cytoplasmic inclusion body, Nib RNA-dependent RNA polymerase	U10, AtU6	*Solanum tuberosum* (Potato)	LshCas13a	*Agrobacterium* GV3101	Resistance to potato virus Y disease	Zhan et al. (2019)
8.	TuMV/CP	CaMV35S	*Nicotiana benthamiana*	LshCas13a	*Agrobacterium* GV3101	Resistance to leaf curl diseases	Aman et al. (2018)

domains that are found in monocots and dicots alike. It had previously been revealed that MLO genes were susceptibility (S) genes, and that homozygous loss-of-function mutants in barley, *Arabidopsis*, and tomato had considerably improved resistance to powdery mildew. Bread wheat plants with CRISPR/Cas9 mutations in one of the three MLO homoalleles (TaMLO-A1) showed increased resistance to *Blumeria graminis* f. sp. *tritici* infection, demonstrating the importance of TaMLO genes in powdery mildew disease once again (Wang et al., 2014). SlMlo1, previously recognized as the most significant of 16 SlMlo genes, was targeted at two places in tomato, yielding a 48-bp deletion. To generate CRISPR/Cas cassette-free individuals, the modified plants were self-pollinated. This new non-transgenic variety, "Tomelo," was fully resistant to *Oidium neolycopersici*. Off-target research also revealed no influence on genomic areas other than the SlMlo1 locus (Nekrasov et al., 2017). The molecular feasibility of VvMLO7 deletion in grapevine protoplasts has been proven using CRISPR/Cas9 RNP, although no plants have been regenerated (Malnoy et al., 2016). CRISPR/Cas9 components were introduced into cocoa leaves to target the non-expressor of pathogenesis-related 3 (NPR3) gene, an immune system suppressor, and resulted in leaves that were more resistant to *Phytophthora tropicalis*. CRISPR/Cas9 was used to create rice plants resistant to rice blast disease by interrupting the *OsERF922* and *OsSEC3A* genes (Ma et al., 2018). Ossec3a mutant plants with a mutation in a putative subunit of a complex involved in exocytosis showed a pleiotropic phenotype that included enhanced *Magnaporthe oryzae* resistance, greater salicylic acid (SA) content, and upregulation of pathogenesis- and SA-related genes, but also shorter stature (Ma et al., 2018). Homologous recombination of donor DNA or split marker (for fungi) was traditionally thought to be one of the classic strategies for generating mutants in fungi and oomycetes, but this method has the drawback of producing a small number of mutants (Mcleod et al., 2008). As a result, combining the donor DNA or split marker with the CRISPR system was attempted and proved to be highly effective in efficiency. Fungal or human codon optimized Cas9 and Cpf1 sequences have been utilized to genetically edit fungus and oomycetes utilizing the CRISPR system. A constitutive fungal/oomycete promoter (e.g., ptef1, pgpd, pENO1, ptrpC, and pHam34) and terminator control expression of the CRISPR (Cas9 or Cpf1) gene, which is coupled with an SV40 or synthetic nuclear localization signal. The promoters of heterologous U6 Pol III, SNR52 promoter, 5s rRNA and tRNA for fungi, and RPL41 promoter for oomycetes, especially *Phytophthora* species, control the SgRNA or crRNA guide sequence with a CRISPR target (Schuster and Kahmann, 2019). There are four common approaches for filamentous fungi: (1) exogenous DNA introduction into protoplasts formed by cell wall lysing enzymes using PEG solution (protoplast-mediated transformation, PMT) (Nødvig et al., 2015); (2) co-cultivation of *Agrobacterium tumefaciens* for vector DNA transfer (AMT) (Zou et al., 2020); (3) reversible membrane permeabilization caused by localized electric pulses (electroporation, EP); and (4) tungsten particles coated with DNA, accelerated to a high velocity, and injected into cells (biolistic transformation, BT) (Wang, 2018). Drug resistance markers such as hygromycin-B-phosphotransferase (hph, hygromycin resistance), phleomycin resistance protein (Ble, phleomycin resistance) (Zhang et al., 2016), or nourseothricin *N*-acetyl transferase (NAT, nourseothricin resistance) and auxotrophic markers such as uridine and arginine, pyrG (van Leeuwe et al., 2019) and argB, have been widely used in fungi. *P. capsici*'s oxathiapiprolin resistance gene (PcMuORP1) was lately investigated as a new selective marker and shown to be highly effective in *Phytophthora* species transformation. Disease control is difficult due to fungi's diverse lives and great genetic flexibility, which allows them to adapt to new hosts, break resistance (R) gene-mediated resistance, and evolve resistance to fungicides. The CRISPR/Cas9 system has allowed for efficient and precise targeted mutagenesis in plants to improve resistance to fungal infections. Susceptibility genes and negative regulators involved in the defense mechanism for plant disease resistance have received a lot of attention (Yin and Qiu, 2019). Rice resistance against *Magnaporthe oryzae*-caused rice blast disease has been improved by employing the CRISPR/Cas9 system to target negative regulators associated with the defensive mechanism. The ethylene-responsive factor in rice, known as APETELA2/ethylene response factor (AP2/ERF)-type transcription factor (OsERF922), was knocked down by RNAi, resulting in lower abscisic acid buildup and hence enhanced disease resistance against *M. oryzae* (Liu et al., 2012). Using CRISPR/Cas9, researchers boosted resistance to rice blast and bacterial blight by inserting mutations into the coding areas of the thermosensitive male sterile gene *TMS5*, proline-rich protein *Pi21* gene, and bacterial blight resistance a recessive gene *Xa13*. This is due to the fact that knocking out recessive genes *Pi21* and *Xa13* resulted in rice blast and bacterial blight resistance, respectively (Li et al., 2019). CRISPR/Cas9 was used to create a novel chromosomal rearrangement strategy in soybean (Nagy et al., 2021). This study artificially created chimeric paralogs of two NLR gene clusters called Rpp1-like and Rps1, which were linked to resistance in *Phakopsora pachyrhizi* and *Phakopsora sojae*, respectively, and demonstrated a possible advanced approach for managing plant disease through artificially creating chimeric paralogs. Oomycete diseases have been successfully reduced using CRISPR/Cas technologies. The PAMP-triggered immunity repressor AtERF019 (ethylene-responsive factor 19 gene) was knocked out in *A. thaliana* using CRISPR/Cas9, which boosted resistance to *Phytophthora parasitica*, a model organism for *Phytophthora* species (Lu et al., 2020). Genome editing was carried out in cacao leaves and cotyledon cells using *Agrobacterium*-mediated temporary transformation to introduce CRISPR/Cas9 components that targeted non-expressor pathogenesis-related 3 (TcNPR3), a suppressor of the defensive response. The TcNPR3-deficient leaves had higher levels of defense genes, including as pathogenesis-related (PR) genes, and were more resistant to *Phytophthora tropicalis*' black pod disease (Fister et al., 2018). EDR1 (enhanced disease resistance 1) inhibits resistance to the powdery mildew fungus *Erysiphe cichoracearum* in *Arabidopsis*, but has only a minor impact on plant growth.

18.3.2 Bacterial Resistance

Only a few hundred of the world's bacterial species are involved in crop damage, which often manifests as various signs of disease (Schloss and Handelsman, 2004). Controlling phytopathogenic bacteria is difficult, owing to undiscovered asymptomatic infections and a lack of appropriate agrochemicals. In general, bacteriological plant management relies on genetic resistance, agronomic methods, and biocontrol agents to prevent and exclude the pathogen (Kerr, 2016). Rice has been mutated using CRISPR/Cas9 to gain resistance to the bacterial blight disease caused by the proteobacterium *Xanthomonas oryzae* pv. *Oryzae* (Zhou et al., 2015). OsSWEET13 is a susceptibility (S) gene that codes for a sucrose transporter that plays a role in the plant–pathogen interaction. PthXo2, an effector protein produced by *X. oryzae*, causes OsSWEET13 expression in the host and, as a result, susceptibility. In prior work using the TALEN method to mutate the OsSWEET14 promoter, the loss of this gene caused the *X. oryzae* effector unable to bind OsSWEET14, resulting in disease resistance (Li et al., 2012). CRISPR/Cas9 has been used in two recent studies to create citrus plants that are resistant to citrus bacterial canker (CBC). The most common disease among commercial cultivars is CBC, which is caused by *Xanthomonas citri* subsp. *citri* (Xcc). In the first study by Jia et al. (2016), canker-resistant mutants were created in Duncan grapefruit by altering the PthA4 effector-binding sites in the promoter of the *lateral organ boundaries 1* (*CsLOB1*) gene. Four days after inoculation with Xcc, mutant lines demonstrated a reduction in typical canker symptoms, but no other phenotypic changes were observed. Complete deletion of the EBEPthA4 sequence from both CsLOB1 alleles resulted in an increase in CBC resistance. Furthermore, no changes in plant development were found after the CsLOB1 promoter was modified. To facilitate their agronomic application in CBC prevention, additional efforts will be necessary to create non-transgenic canker-resistant citrus types. Type III effectors are secreted into the plant cell during the bacterial infection process. For disease development, these effectors typically disrupt the host's defensive pathways and/or activate the *S* genes. As a result, CRISPR/Cas9-mediated gene editing of *S* genes and negative regulators of plant innate immune response is a good target site for improving plant resistance. In rice, knocking down the endogenous *OsSWEET11* gene, which is a TALE target, led to dramatically increased resistance to *X. oryzae* pv. *oryzae* (Xoo) without affecting pollen formation. TALE-binding elements (EBEs) in the promoter regions of OsSWEET11 and OsSWEET14 were also targeted, and rice lines with indels gave substantial resistance to most Xoo strains. Citrus canker disease resistance was conferred by genome editing of the promoter region of the *S* gene *CsLOB1*, which is the target of TALE from *X. citri* pv. *citri* (Xcc) (Peng et al., 2017). Although DMR6 is required for downy mildew resistance in *A. thaliana*, CRISPR/SpCas9 deletion of its homolog SIDMR6–1 in tomato made the plant resistant to a variety of bacterial diseases, including *P. syringae*, *P. capsici*, and *Xanthomonas* spp. Furthermore, the CRISPR/SpCas9-generated SlJAZ2jas tomato germplasm provided resistance to *P. syringae* pv. *tomato* DC3000 without affecting the plant's defense response to the necrotrophic fungal disease *Botrytis cinerea*.

18.3.3 Viral Resistance

Plant viral infections are challenging to control due to the fast evolution of viruses and the presence of insect vectors. Transgenic production of virus proteins or RNAs has been frequently employed to improve plant virus resistance over the last three decades, and the resulting resistance is known as pathogen-derived resistance. RNAi generated by double-stranded RNA has recently been regarded an effective approach for conferring virus resistance to plants. Genome editing technology gives plant virus researchers a new weapon in their armory. Plant viruses in the Geminiviridae family cause worldwide crop losses in Cucurbitaceae, Euphorbiaceae, Solanaceae, Malvaceae, and Fabaceae families, among others (Zaidi et al., 2016). The first two investigations examined 43 putative sgRNA/Cas9 target sites in coding and non-coding sections of the BSCTV genome, focusing on resistance to geminiviruses, beet severe curly top virus (BSCTV), and bean yellow dwarf virus (BeYDV) in model plants such as *N. benthamiana* and *Arabidopsis* (Baltes et al., 2015). All of the sgRNA/Cas9 constructions reduced virus accumulation in inoculated leaves to varied degrees, although *Nicotiana* and *Arabidopsis* plants with the greatest levels of Cas9 and sgRNA expression showed better resistance to virus infection. sgRNA molecules put into *N. benthamiana* plants through a tobacco rattle virus (TRV) vector persistently overexpress the Cas9 protein (Ali et al., 2015). SgRNAs targeted the viral capsid protein, the replication protein's RCRII motif, and the intergenic region of the tomato yellow leaf curl virus (TYLCV) (IR). Turnip mosaic virus (TuMV) RNA genome manipulation using an RNA-guided ribonuclease (Aman et al., 2018). Two targets in the green fluorescent protein (GFP) area, one in the helper component proteinase silencing suppressor (HC-Pro) and another in the coat protein region were selected. The most effective viral interference was reported with CRISPR RNA editing of the HC-Pro and GFP2 genes, which resulted in TuMV replication and dissemination in tobacco leaves being inhibited. Moreover, because Cas13 has the innate ability to convert pre-CRISPR RNA into functional CRISPR RNA, this alternative method could significantly increase the multiplex targeting of numerous viral mRNA (Aman et al., 2018). eIF4E and its isoforms are translation initiation factors in plants that are involved in the multicomponent translation complex that recruits ribosomes to the 5′ untranslated regions (UTRs) of mRNAs and are also essential for viral protein translation. A loss-of-function mutation in the *eIF(iso)4E* gene leads to TuMV resistance, according to genetic screening for *Arabidopsis* mutants with increased resistance to Turnip mosaic virus (TuMV), an RNA virus. Furthermore, CRISPR/Cas9 deletion of *Arabidopsis eIF(iso)4E* resulted in TuMV resistance but had no effect on plant vigor. As a result, plant *eIF4E* genes are likely to be suitable targets for genome editing to generate broad-spectrum viral resistance. CVYV (cucumber vein yellowing virus), ZYMV (zucchini yellow mosaic virus), and PRSV-W resistance was found in cucumber *eif4e* mutants with mutations at two locations in the *eIF4E* gene produced by CRISPR/Cas9 (papaya ring spot mosaic virus-W) (Chandrasekaran et al., 2016). Only nCBP-1 (new cap-binding protein-1) and nCBP-2 (novel cap-binding protein-2) are associated with VPgsin cassava (viral genome-linked proteins). Only the ncbp-1/ncbp-2 double mutants produced using CRISPR/Cas9 consistently demonstrated delayed and attenuated symptoms following CBSV infection (cassava brown streak virus). CRISPR/Cas13a (C2c2) looks to be an even more potent editing technology that allows single-stranded RNA to be modified directly. In tobacco, CRISPR/Cas13a (C2c2) targeting the viral RNA genome resulted in TuMV resistance. In plant-–virus interactions, the eukaryotic translation initiation factor 4E (eIF4E), also known as a cap-binding protein, has been shown to be a susceptibility factor. Innate immunity against several potyviruses (viruses belonging to the Potyviridae family) is induced when the *eIF4E* gene is disrupted in various plant species. Cucumber plants with eIF4E-edited CRISPR/SpCas9-mediated resistance to ZYMV, PRSMV-W, and cucumber vein yellowing virus, for example, were resistant to ZYMV, PRSMV-W, and cucumber vein yellowing virus (CVYV). Furthermore, using the CRISPR/SpCas9 system, mutation of eIF4E alleles in cassava and *A. thaliana* reduced cassava brown streak disease symptoms and turnip mosaic virus (TuMV) infection, respectively. CRISPR/SpCas9 was also utilized to modify *eIF4G* in rice, resulting in resistance to the rice tungro spherical virus (RTSV) (Macovei et al., 2018). The cytidine base editor recently inserted a C>G conversion (N176K) into the wild-type eIF4E1 in *A. thaliana*, affording resistance to clover yellow vein virus (ClYVV). SpCas9 and sgRNA constructs targeting the Rep (replication-associated protein) gene and the intergenic region (IR) in beet severe curly top virus (BSCTV) and bean yellow dwarf virus (BeYDV) viral genomes were transformed into *N. benthamiana* and *A. thaliana*, respectively. The plants that resulted were highly resistant to the virus in question. Similarly, *N. benthamiana* and tomato plants expressing SpCas9 and sgRNA specific for tomato yellow leaf curl virus (TYLCV) coat protein or Rep sequences showed considerable virus resistance. More research found that the CRISPR/SpCas9 system developed wheat dwarf virus (WDV) resistance in barley and banana streak virus (BSV) resistance in bananas by targeting MP, coat protein, and any other conserved areas in viral genomes. Cas proteins (such as Cas13a and FnCas9) that target RNA molecules rather than DNA can also directly target RNA viruses. In *N. benthamiana*, for example, Aman and colleagues used CRISPR/Cas13a to target and destroy

TuMV viral RNA (Aman et al., 2018). CRISPR/FnCas9 was also employed by Zhang and colleagues to degrade the viral genomes of cucumber mosaic virus (CMV) and tobacco mosaic virus (TMV) in transgenic *N. benthamiana* and *A. thaliana* plants, respectively. Resistance to potato virus Y (PVY) has been successfully developed in tobacco, as well as resistance to rice stripe mosaic virus (RSMV) and southern rice black-streaked draught virus (SRBSDV) in rice.

18.3.4 Insect/Pest Management

Insect pests, like other biotic stress factors, have raised severe worries about yield losses, putting agricultural output at risk. Plant trait modification for crop enhancement began with breeding techniques and progressed to genetic engineering. Stringent regulatory standards for risk assessment, as well as a lack of social acceptance for genetically modified crops around the world, have prompted researchers to consider other options. Genome engineering, also known as genome editing, is a revolutionary breeding technology that can alter the genomes of plants, animals, microorganisms, and humans. Several gene editing strategies are being used, with variants appearing on a regular basis. The efficacy of numerous editing techniques, ranging from endonucleases to CRISPR/Cas, in several areas of plant growth, development, and stress reduction has been discovered by scientists. The creation of tools and procedures, such as the construction of binary vectors, to make genome engineering easier is being pursued categorically. In this chapter, we will look at a variety of genome editing technologies that can be used to develop insect pest resistance in plants and are currently being employed. This chapter focuses on major options available for precise editing for pest management in plants, given the success of genome editing in plants and its ability to counteract biotic stress factors. This chapter is unique in that it presents groundbreaking research on the application of genome editing for insect control, not just in plants to discourage insects but also in target pests to diminish their ability to attack plants (Tyagi et al., 2020). CRISPR/Cas9, the most recent revolutionary technology for genome editing, is based on RNA-guided engineered nucleases that hold great promise because of their specificity, simplicity, efficiency, and versatility in addressing key challenges posed by other genome editing tools. CRISPR systems evolved as an adaptive immune system for archaea and bacteria, and are divided into two classes based on the effector molecules used. According to the architecture of CRISPR/Cas (CRISPR-associated) loci and the kind of Cas proteins with crRNA for CRISPR interference, they are further divided into six distinct types (IVI). A strategy to target detoxification genes in polyphagous pests can be a promising option. Knockdown of insecticidal detoxification genes like gossypol-inducing cytochrome P450 resulted in insect susceptibility. CRISPR/Cas9 knockdown of CYP6AE gene cluster in the polyphagous pest *H. armigera* proved the role of these enzymes in detoxification of various toxic phytochemicals. The capacity to target genes that could impair chemical communication and mating partner recognition is another degree of insect management that genome editing can provide. These are just two of the many components that go into creating successful plant–insect relationships. In insects, olfactory receptors (ORs) are vital for recognizing the odorant of the host plant as well as the odorant of the mating partner. A mutation in the *Or83b* gene in *Drosophila* altered the selection of the egg-laying site (host) and reduced olfactory detecting capabilities. Similarly, CRISPR/Cas9 knockout of the *Orco* (olfactory receptor coreceptor) gene in *Spodopthera litura* resulted in distraction in mating partner selection as well as loss of identity toward host plants, resulting in anosmia. Adoption of such technologies will be a potential option to keep insect pests away from the crops and prevent insect damage.

18.4 MULTIPLEX GENOME EDITING TO DEVELOP MULTIPLE STRESS TOLERANCE

Several research have described several gene knockouts in various plant species for distinct reasons in this area. The use of CRISPR/Cas9 in multiplexing two sgRNAs to delete ssDNA viruses broadened the scope of virus biological control in plants (Ali et al., 2015). Using a multiplex CRISPR/Cas9 system to alter the effector-binding elements (EBEs) of the bacterial blight TAL effector for two susceptibility genes, *OsSWEET11* and *OsSWEET14*, in rice cv. Kitaake (harboring the defeated OsSWEET13) generated a multiplex knockout to develop broad-spectrum disease resistance in rice (Xu et al., 2017).

Additional sequences must be inserted for further processing following transcription using single polycistronic multiple gRNA techniques. Cas12a, formerly known as Cpf1, a newer member of the Cas protein family, may be able to overcome these constraints (Zetsche et al., 2015). Cas12a is modest in size, requires a short crRNA (42 nt) for synthesis and multiplexing, and contains a necessary PAM sequence that allows it to target T-rich areas. Furthermore, the Cas12a multiplexing system outperforms conventional DNA editing technologies. Point mutations at the target site are caused by fusing dCas with cytosine deaminase (CBE) or adenosine deaminase (ABE) and sgRNAs. A single nucleotide change may be enough to change gene structure and gain a desired function (Komor et al., 2016; Eid et al., 2018).

Multiple herbicide-resistant point mutations in rice were induced. Xu et al. have developed a toolkit that combines high-fidelity dCas9 variants coupled to a tandemly arrayed tRNA-modified sgRNA architecture in rice to boost the effectiveness of this technology by two- to fivefold (Xu et al., 2019). Furthermore, the fusing of dCas enzymes to effector domains, such as CRISPR-mediated inhibitor (CRISPRi) and activator (CRISPRa), allowed for directed transcriptional regulation. CRISPR/Cas9 transcriptional activation and repression of *Arabidopsis* genes toolbox for cloning multiplex (Lowder et al., 2017). The researchers were able to edit five sites successfully; however, there were no editing events at the third site, which was assumed to be due to a lack of CG content. The gamma-aminobutyric acid (GABA) shunt metabolic pathway in tomatoes has been engineered and manipulated. They targeted GABA-TP1, GABA-TP2, GABA-TP3, CAT9, and SSADH, five critical genes in the system.

18.5 FUTURE ASPECTS

Genome editing is steadily and regularly becoming a method of choice for elucidating gene function as well as its translational application in laboratories all around the world. They're being used in a variety of crop development efforts to combat both biotic and abiotic stressors. Because of their great specificity, efficiency, cost, and time effectiveness, editing strategies have evolved as useful tools. Despite their success in combating diseases in economically important crops, their utility in insect management has not been fully exploited. Despite the intellectual effort put toward developing insect pest resistance methods in both insects and plants, the results have been disappointing. Though insect editing is a fascinating prospect, it necessitates caution in the selection of features that must be environmentally benign in order to avoid disrupting the food chain. In contrast to other pressures, the major hurdle has been the lack of target gene availability. As a result, experts around the world must focus on identifying resistance sources that might serve as a foundation for insect management. On the basis of their appropriateness for reducing off-target impacts on target insects, a number of novel targets are being considered.

An essential hint that needs to be addressed in a path-breaking manner is extensive phenotyping and enrichment of the resistance gene pool. This necessitates a review of available germplasm, including wild relatives of certain crops for pest response, as well as the use of multi-omics techniques to identify stress-responsive genes. The nitty-gritty of pest management utilizing genome editing involves multiplex editing of such discovered resistance genes using high-throughput transformation techniques. With such research already popular, targeted mutations leading to the transformation of sensitive plants into ones that can handle their pests aren't far away in space or time. These factors, together with regulatory considerations for gene-edited crops, could lead to the technology's success, not just in terms of scientific advancement but also in terms of societal acceptance. Crop protection through genetic modification offers a promising option with no clear impact on human health or the environment in an era characterized by political and societal pressure to limit the use of pesticides. Disease resistance is widely recognized as the most potential application of CRISPR/Cas9 technology in agriculture, and genome editing is one of the genetic levers that can be used. This is due to three key factors: To begin with, scientific understanding of the molecular mechanisms underpinning a variety of pathosystems has progressed far enough to allow for the suggestion of genes that could be altered to attain resistance. Second, disease resistance may often be established by modifying a single gene, which is less technically difficult. This is related to metabolic pathway modification, where a single gene can have an all-or-nothing effect, but different from abiotic stress tolerance, where multiple genes must be tweaked in a coordinated manner to make incremental gains. Third, because the inactivation of susceptibility genes leads to disease resistance, targeted mutagenesis, the sole use of CRISPR/Cas9 technology yet mastered with respect to crops, is easily applicable to disease resistance. The first problem is to show that the promises made in proofs of concept in constrained environments can be kept in real-world situations. Field testing are also required for an accurate assessment of the modified crops' agronomic fitness. The second issue is the disease resistance's long-term viability and agronomic management. This difficulty must be taken seriously in order to persuade a population that is typically skeptical of technology. The answers are the same for introgressed resistance genes discovered in the genetic variability of the species. Durability is not a specific aspect of resistance genes obtained by genome editing, and the answers are the same for introgressed resistance genes discovered in the genetic variability of the species. The use of CRISPR/Cas9 technology in the stacking process can also be beneficial. In contrast to the introduction of conventional resistance genes, the technology not only avoids genetic drag on neighboring regions, which could have negative agronomic consequences, but also allows multiplexing, or the simultaneous use of several sgRNAs targeting different genes, to create multiple resistances in a single generation. Multiplexing gets increasingly difficult as ploidy levels rise, as seen in the above example of hexaploid wheat. The third task is to overcome the current technical limitations of targeted mutagenesis in crop plants and to execute genuine genome editing. True genome editing adds specified base alterations at one or more precise sites in a gene, whereas targeted mutagenesis introduces random mutations (usually short insertions or deletions) at a predetermined site of a gene, generally resulting to loss-of-function. In comparison to other GE technologies, the CRISPR/Cas system is now the most widely used technology for targeted genomic editing, having been developed and applied in a large number of host plants and plant pathogens for dissecting the molecular mechanisms underlying plant–pathogen interactions and improving host resistance to bacteria, fungi, oomycetes, DNA viruses, and RNA viruses. The CRISPR/Cas system is useful for making gene loss-of-function and gain-of-function mutants, as well as researching plant–pathogen interactions and decreasing the harm caused by damaging infections in agricultural applications. Finally, the CRISPR/Cas system and its variants offer a fresh way to investigate the complicated field of plant–pathogen interactions. We expect CRISPR technologies to play a significant role in decoding the interaction between plant and pathogen and building durable and broad-spectrum disease-resistant plants in the future, given the constant changes in agricultural production activities and plant disease systems.

18.6 LIMITATIONS

There are, however, a few constraints that must be addressed. Because crop plants have a longer life cycle than model plants, precision and frequency of mutant production are important considerations. As a result, more study is needed into quick and easy approaches to confirm the resistance phenotype by transient production of CRISPR/Cas9 components. More basic challenges, such as a lack of selectable markers, limit genome editing in fungi/oomycetes. Before using CRISPR/Cas9 genome editing tools, consider the potential adverse effects of editing or altering plant susceptibility genes. The susceptibility gene(s) are known to be associated to plant growth and fitness, as well as having pleiotropic effects. This is also a very common occurrence after artificially increasing plant immunity by knocking off a negative regulator. Although multiple studies have shown that

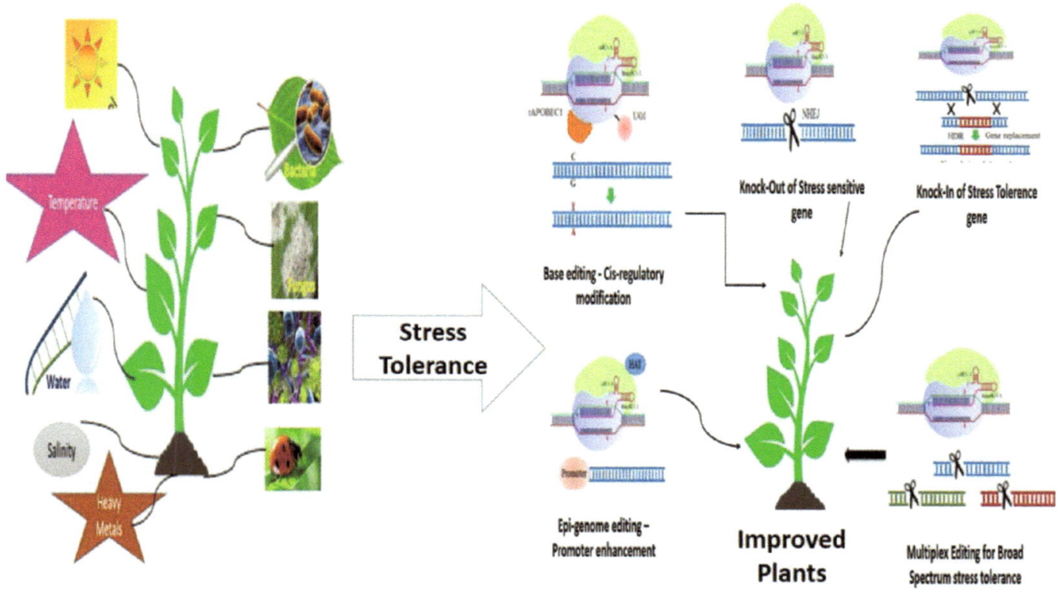

FIGURE 18.1 Genome editing and stress tolerance in plant.

deletion or mutation of these genes has no harmful influence on plant growth or health, more research and testing is required. Furthermore, the CRISPR/Cas9 system's ability to introduce off-target mutations is a crucial factor to address. An off-target mutation is an unfavorable outcome that results in non-specific changes in the genomic sequences, affecting protein structure and/or function (Figure 18.1).

REFERENCES

Ahmad, S., Wei, X., Sheng, Z., Hu, P., and Tang, S. 2020. CRISPR/Cas9 for development of disease resistance in plants: recent progress, limitations and future prospects. *Briefings in Functional Genomics* 19(1): 26–39.

Ali, Z., Abulfaraj, A., Idris, A., Ali, S., Tashkandi, M., and Mahfouz, M. M. 2015. CRISPR/Cas9-mediated viral interference in plants. *Genome Biology* 16(1): 1–11. doi: 10.1186/s13059-015-0799-6.

Aman, R., Ali, Z., Butt, H., Mahas, A., Aljedaani, F., et al. 2018. RNA virus interference via CRISPR/Cas13a system in plants. *Genome Biology* 19(1): 1–9. doi: 10.1186/s13059-017-1381-1.

Angon, P. B., Mondal, S., Akter, S., Sakil, M. A., and Jalil, M.A. 2023. Roles of CRISPR to mitigate drought and salinity stresses on plants. *Plant Stress* 8, 100169. https://doi.org/10.1016/j.stress.2023.100169.

Arora, L., and Narula, A. 2017. Gene editing and crop improvement using CRISPR-Cas9 system. *Frontiers in Plant Science* 8: 1932. doi: 10.3389/fpls.2017.01932.

Baltes, N. J., Hummel, A. W., Konecna, E., Cegan, R., Bruns, A. N., Bisaro, D. M., and Voytas, D. F. 2015. Conferring resistance to geminiviruses with the CRISPR–Cas prokaryotic immune system. *Nature Plants* 1: 15145. doi: 10.1038/nplants.2015.145.

Butt, H., Jamil, M., Wang, J. Y., Al-Babili, S., and Mahfouz, M. 2018. Engineering plant architecture via CRISPR/Cas9-mediated alteration of strigolactone biosynthesis. *BMC Plant Biology*, 18(1): 1–9.

Chandrasekaran, J., Brumin, M., Wolf, D., Leibman, D., Klap, C., Pearlsman, M., et al. 2016. Development of broad virus resistance in non-transgenic cucumber using CRISPR/Cas9 technology. *Molecular Plant Pathology* 17(7): 1140–1153. doi: 10.1111/mpp.12375.

Duan, Y. B., Li, J., Qin, R. Y., Xu, R. F., Li, H., Yang, Y. C., Ma, H., Li, L., Wei, P. C., and Yang, J. B. 2016. Identification of a regulatory element responsible for salt induction of rice OsRAV2 through ex situ and in situ promoter analysis. *Plant Molecular Biology* 90: 49–62.

Eid, A., Alshareef, S., and Mahfouz, M. M. 2018. CRISPR base editors: genome editing without double-stranded breaks. *Biochemical Journal* 475(11): 1955–1964. doi: 10.1042/BCJ20170793.

Fister, A. S., Landherr, L., Maximova, S. N., and Guiltinan, M. J. 2018. Transient expression of CRISPR/Cas9 machinery targeting TcNPR3 enhances defense response in *Theobroma cacao. Frontiers in Plant Science* 9: 268. doi: 10.3389/fpls.2018.00268.

Gilbert, L. A., Larson, M. H., Morsut, L., Liu, Z., Brar, G. A., Torres, S. E., et al. 2013. CRISPR-mediated modular RNA-guided regulation of transcription in eukaryotes. *Cell* 154(2): 442–451. doi: 10.1016/j.cell.2013.06.044.

Hao, X., Zeng, M., Wang, J., Zeng, Z., Dai, J., Xie, Z., Yang, Y., Tian, L., Chen, L., and Li, D. 2018. A node-expressed transporter OsCCX2 is involved in grain cadmium accumulation of rice. *Frontiers in Plant Science*, 9: 476.

Hasegawa, T., Fujimori, S., Havlík, P., Valin, H., Bodirsky, B. L., Doelman, J. C., Fellmann, T., Kyle, P., Koopman, J. F., Lotze-Campen, H., and Mason-D'Croz, D. 2018. Risk of increased food insecurity under stringent global climate change mitigation policy. *Nature Climate Change*, 8(8): 699–703.

Huang, X., Zeng, X., Li, J., and Zhao, D. 2017. Construction and analysis of tify1a and tify1b mutants in rice (Oryza sativa) based on CRISPR/Cas9 technology. *Journal of Agricultural Biotechnology*, 25(6): 1003–1012.

Jia, H., Orbovic, V., Jones, J. B., and Wang, N. 2016. Modification of the PthA4 effector binding elements in Type I Cs LOB 1 promoter using Cas9/sg RNA to produce transgenic Duncan grapefruit alleviating XccΔpthA4: dCs LOB 1.3 infection. *Plant Biotechnology Journal* 14(5): 1291–1301. doi: 10.1111/pbi.12495.

Jia, H., Zhang, Y., Orbović, V., Xu, J., White, F. F., Jones, J. B., and Wang, N. 2017. Genome editing of the disease susceptibility gene Cs LOB 1 in citrus confers resistance to citrus canker. *Plant Biotechnology Journal* 15(7): 817–823.

Jung, Y. J., Lee, H. J., Bae, S., Kim, J. H., Kim, D. H., Kim, H. K., Nam, K. H., Nogoy, F. M., Cho, Y. G., and Kang, K. K. 2019. Acquisition of seed dormancy breaking in rice (Oryza sativa L.) via CRISPR/Cas9-targeted mutagenesis of OsVP1 gene. *Plant Biotechnology Reports* 13: 511–520.

Kerr, A. 2016. Biological control of crown gall. *Australasian Plant Pathology* 45(1): 15–18. doi: 10.1007/s13313-015-0389-9.

Kieu, N. P., Lenman, M., Wang, E. S., Petersen, B. L., and Andreasson, E. 2021. Mutations introduced in susceptibility genes through CRISPR/Cas9 genome editing confer increased late blight resistance in potatoes. *Scientific Reports* 11(1): 1–12.

Komor, A. C., Kim, Y. B., Packer, M. S., Zuris, J. A., and Liu, D. R. 2016. Programmable editing of a target base in genomic DNA without double-stranded DNA cleavage. *Nature* 533(7603): 420–424. doi: 10.1038/nature17946.

Lee, H. Y., Chen, Z., Zhang, C., and Yoon, G. M. 2019. Editing of the OsACS locus alters phosphate deficiency-induced adaptive responses in rice seedlings. *Journal of Experimental Botany* 70(6): 1927–1940.

Li, L., Kim, P., Yu, L., Cai, G., Chen, S., Alfano, J. R., and Zhou, J. M. 2016. Activation-dependent destruction of a co-receptor by a *Pseudomonas syringae* effector dampens plant immunity. *Cell Host & Microbe* 20(4): 504–514.doi: 10.1016/j.chom.2016.09.007.

Li, M. Y., Jiao, Y. T., Wang, Y. T., Zhang, N., Wang, B. B., Liu, R. Q., Yin, X., Xu, Y., and Liu, G.T. 2020. CRISPR/Cas9- mediated VvPR4b editing decreases downy mildew resistance in grapevine (Vitis vinifera L.). *Horticulture Research* 7, 1–11. https://doi.org/10.1038/s41438-020-00371-4

Li, S., Shen, L., Hu, P., Liu, Q., Zhu, X., Qian, Q., et al. 2019. Developing disease-resistant thermosensitive male sterile rice by multiplex gene editing. *Journal of Integrative Plant Biology* 61(12): 1201–1205.doi: 10.1111/jipb.12774.

Li, T., Liu, B., Spalding, M. H., Weeks, D. P., and Yang, B. 2012. High-efficiency TALEN-based gene editing produces disease-resistant rice. *Nature Biotechnology* 30(5): 390–392. doi: 10.1038/nbt.2199.

Liao, S., Qin, X., Luo, L., Han, Y., Wang, X., Usman, B., Nawaz, G., Zhao, N., Liu, Y., and Li, R. 2019. CRISPR/Cas9-induced mutagenesis of semi-rolled leaf1, 2 confers curled leaf phenotype and drought tolerance by influencing protein expression patterns and ROS scavenging in rice (*Oryza sativa* L.). *Agronomy* 9(11): 728.

Liu, D., Chen, X., Liu, J., Ye, J., and Guo, Z. 2012. The rice ERF transcription factor OsERF922 negatively regulates resistance to Magnaporthe oryzae and salt tolerance. *Journal of Experimental Botany* 63(10): 3899–3911.

Lou, D., Wang, H., Liang, G., and Yu, D. 2017. OsSAPK2 confers abscisic acid sensitivity and tolerance to drought stress in rice. *Frontiers in Plant Science* 8: 993.

Lowder, L. G., Paul, J. W., and Qi, Y. 2017. Multiplexed transcriptional activation or repression in plants using CRISPR-dCas9-based systems. In Kaufmann, K. and Mueller-Roeber, B. (eds), *Plant Gene Regulatory Networks*, 167–184. New York, NY: Humana Press.

Lu, W., Deng, F., Jia, J., Chen, X., Li, J., Wen, Q., et al. 2020. The Arabidopsis thaliana gene AtERF019 negatively regulates plant resistance to *Phytophthora parasitica* by suppressing PAMP-triggered immunity. *Molecular Plant Pathology* 21(9): 1179–1193.doi: 10.1111/mpp.12971.

Ma, J., Chen, J., Wang, M., Ren, Y., Wang, S., Lei, C., and Cheng, Z. 2018. Disruption of OsSEC3A increases the content of salicylic acid and induces plant defense responses in rice. *Journal of Experimental Botany* 69(5): 1051–1064. doi: 10.1093/jxb/erx458.

Mali, P., Esvelt, K. M., and Church, G. M. 2013. Cas9 as a versatile tool for engineering biology. *Nature Methods* 10(10): 957–963.

Macovei, A., Sevilla, N. R., Cantos, C., Jonson, G. B., Slamet-Loedin, I., Čermák, T., et al. 2018. Novel alleles of rice eIF4G generated by CRISPR/Cas9-targeted mutagenesis confer resistance to Rice tungro spherical virus. *Plant Biotechnology Journal* 16(11): 1918–1927. doi: 10.1111/pbi.12927.

Malnoy, M., Viola, R., Jung, M. H., Koo, O. J., Kim, S., Kim, J. S., et al. 2016. DNA-free genetically edited grapevine and apple protoplast using CRISPR/Cas9 ribonucleoproteins. *Frontiers in Plant Science* 7. doi: 10.3389/fpls.2016.01904.

Mao, Y., Zhang, H., Xu, N., Zhang, B., Gou, F., and Zhu, J. K. 2013. Application of the CRISPR–Cas system for efficient genome engineering in plants. *Molecular Plant* 6(6): 2008–2011. doi: 10.1093/mp/sst121.

Mcleod, A., Fry, B. A., Zuluaga, A. P., Myers, K. L., and Fry, W. E. 2008. Toward improvements of oomycete transformation protocols. *Journal of Eukaryotic Microbiology* 55(2): 103–109. doi: 10.1111/j.1550-7408.2008.00304.x.

Mehta, D., Stürchler, A., Anjanappa, R. B., Zaidi, S. S. E. A., Hirsch-Hoffmann, M., Gruissem, W., and Vanderschuren, H. 2019. Linking CRISPR-Cas9 interference in cassava to the evolution of editing-resistant geminiviruses. *Genome Biology* 20(1): 1–10.

Mickelbart, M. V., Hasegawa, P. M., and Bailey-Serres, J. 2015. Genetic mechanisms of abiotic stress tolerance that translate to crop yield stability. *Nature Reviews Genetics* 16(4): 237–251. doi: 10.1038/nrg3901.

Nagy, E. D., Stevens, J. L., Yu, N., Hubmeier, C. S., LaFaver, N., Gillespie, M., et al. 2021. Novel disease resistance gene paralogs created by CRISPR/Cas9 in soy. *Plant Cell Reports* 40(6): 1047–1058.doi: 10.1007/s00299-021-02678-5.

Nawaz, G., Han, Y., Usman, B., Liu, F., Qin, B., and Li, R. 2019. Knockout of OsPRP1, a gene encoding proline-rich protein, confers enhanced cold sensitivity in rice (Oryza sativa L.) at the seedling stage. *3 Biotech* 9: 1–18.

Nekrasov, V., Wang, C., Win, J., Lanz, C., Weigel, D., and Kamoun, S. 2017. Rapid generation of a transgene-free powdery mildew resistant tomato by genome deletion. *Scientific Reports* 7(1): 1–6. doi: 10.1038/s41598-017-00578-x.

Nødvig, C. S., Nielsen, J. B., Kogle, M. E., and Mortensen, U. H. 2015. A CRISPR-Cas9 system for genetic engineering of filamentous fungi. *PloS One* 10(7): e0133085.

Ogata, T., Ishizaki, T., Fujita, M., and Fujita, Y. 2020. CRISPR/Cas9-targeted mutagenesis of OsERA1 confers enhanced responses to abscisic acid and drought stress and increased primary root growth under non-stressed conditions in rice. *PloS One* 15(12): e0243376.

Oliva, R., Ji, C., Atienza-Grande, G., Huguet-Tapia, J. C., Perez-Quintero, A., Li, T., et al. 2019. Broad-spectrum resistance to bacterial blight in rice using genome editing. *Nature Biotechnology* 37(11): 1344–1350. doi: 10.1038/s41587-019-0267-z.

Osakabe, K., Osakabe, Y., and Toki, S. 2010. Site-directed mutagenesis in Arabidopsis using custom-designed zinc finger nucleases. *Proceedings of the National Academy of Sciences* 107(26): 12034–12039. doi: 10.1073/pnas.1000234107.

Peng, A., Chen, S., Lei, T., Xu, L., He, Y., Wu, L., et al. 2017. Engineering canker-resistant plants through CRISPR/Cas9-targeted editing of the susceptibility gene Cs LOB 1 promoter in citrus. *Plant Biotechnology Journal* 15(12): 1509–1519. doi: 10.1111/pbi.12733.

Piatek, A., Ali, Z., Baazim, H., Li, L., Abulfaraj, A., Al-Shareef, S., et al. 2015. RNA-guided transcriptional regulation in planta via synthetic dC as9-based transcription factors. *Plant Biotechnology Journal* 13(4): 578–589. doi: 10.1111/pbi.12284.

Prihatna, C., Barbetti, M. J., and Barker, S. J. 2018. A novel tomato fusarium wilt tolerance gene. *Frontiers in Microbiology* 9: 1226.

Roy, A., Zhai, Y., Ortiz, J., Neff, M., Mandal, B., Mukherjee, S. K., and Pappu, H.R. 2019. Multiplexed editing of a begomovirus genome restricts escape mutant formation and disease development. *PLoS One* 14(10): e0223765.

Sánchez-Martín, J., and Keller, B. 2019. Contribution of recent technological advances to future resistance breeding. *Theoretical and Applied Genetics* 132(3): 713–732. doi: 10.1007/s00122-019-03297-1.

Santillán Martínez, M. I., Bracuto, V., Koseoglou, E., Appiano, M., Jacobsen, E., Visser, R. G., Wolters, A. M. A., and Bai, Y. 2020. CRISPR/Cas9-targeted mutagenesis of the tomato susceptibility gene PMR4 for resistance against powdery mildew. *BMC Plant Biology* 20(1): 1–13.

Santosh Kumar, V. V., Verma, R. K., Yadav, S. K., Yadav, P., Watts, A., Rao, M. V., and Chinnusamy, V. 2020. CRISPR-Cas9 mediated genome editing of drought and salt tolerance (OsDST) gene in indica mega rice cultivar MTU1010. *Physiology and Molecular Biology of Plants* 26: 1099–1110.

Savary, S., Ficke, A., Aubertot, J. N., and Hollier, C. 2012. Crop losses due to diseases and their implications for global food production losses and food security. *Food Security* 4(4): 519–537. doi: 10.1007/s00203-017-1426-6.

Schenke, D., and Cai, D. 2020. Applications of CRISPR/Cas to improve crop disease resistance: beyond inactivation of susceptibility factors. *Iscience* 23(9): 101478.

Schloss, P. D., and Handelsman, J. 2004. Status of the microbial census. *Microbiology and Molecular Biology Reviews* 68(4): 686–691. doi: 10.1128/MMBR.68.4.686-691.2004.

Schuster, M., and Kahmann, R. 2019. CRISPR-Cas9 genome editing approaches in filamentous fungi and oomycetes. *Fungal Genetics and Biology* 130: 43–53.

Shalem, O., Sanjana, N. E., Hartenian, E., Shi, X., Scott, D. A., Mikkelsen, T. S., et al. 2014. Genome-scale CRISPR-Cas9 knockout screening in human cells. *Science* 343(6166): 84–87. doi: 10.1126/science.1247005.

Shen, C., Que, Z., Xia, Y., Tang, N., Li, D., He, R., and Cao, M. 2017. Knock out of the annexin gene OsAnn3 via CRISPR/Cas9-mediated genome editing decreased cold tolerance in rice. *Journal of Plant Biology* 60: 539–547.

Sun, Q., Lin, L., Liu, D., Wu, D., Fang, Y., Wu, J., and Wang, Y. 2018. CRISPR/Cas9-mediated multiplex genome editing of the BnWRKY11 and BnWRKY70 genes in *Brassica napus* L. *International Journal of Molecular Sciences* 19(9): 2716.doi: 10.3390/ijms19092716.

Sun, Y., Zhang, X., Wu, C., He, Y., Ma, Y., Hou, H., Guo, X., Du, W., Zhao, Y., and Xia, L. 2016. Engineering herbicide-resistant rice plants through CRISPR/Cas9-mediated homologous recombination of acetolactate synthase. *Molecular Plant* 9(4): 628–631.

Talakayala, A., Ankanagari, S., and Garladinne, M. 2022. CRISPR-Cas genome editing system: A versatile tool for developing disease resistant crops. *Plant Stress* 3: 100056. https://doi.org/10.1016/j.stress.2022.100056.

Tang, L., Mao, B., Li, Y., Lv, Q., Zhang, L., Chen, C., He, H., Wang, W., Zeng, X., Shao, Y., and Pan, Y. 2017. Knockout of OsNramp5 using the CRISPR/Cas9 system produces low Cd-accumulating indica rice without compromising yield. *Scientific Reports* 7(1): 14438.

Tripathi, L., Ntui, V. O., and Tripathi, J. N. 2020. CRISPR/Cas9-based genome editing of banana for disease resistance. *Current Opinion in Plant Biology* 56: 118–126.

Tripathi, L., Ntui, V. O., Tripathi, J. N., and Kumar, P. L. 2021. Application of CRISPR/Cas for diagnosis and management of viral diseases of banana. *Frontiers in Microbiology* 11: 609784.

Tyagi, S., Kesiraju, K., Saakre, M., Rathinam, M., Raman, V., Pattanayak, D., and Sreevathsa, R. 2020. Genome editing for resistance to insect pests: an emerging tool for crop improvement. *ACS Omega* 5(33): 20674–20683. doi: 10.1021/acsomega.0c01435.

Usman, B., Nawaz, G., Zhao, N., Liao, S., Liu, Y., and Li, R. 2020. Precise editing of the OsPYL9 gene by RNA-guided Cas9 nuclease confers enhanced drought tolerance and grain yield in rice (*Oryza sativa* L.) by regulating circadian rhythm and abiotic stress responsive proteins. *International Journal of Molecular Sciences* 21(21): 7854.

van Leeuwe, T. M., Arentshorst, M., Ernst, T., Alazi, E., Punt, P. J., and Ram, A. F. 2019. Efficient marker free CRISPR/Cas9 genome editing for functional analysis of gene families in filamentous fungi. *Fungal Biology and Biotechnology* 6(1): 1–13. doi: 10.1186/s40694-019-0076-7.

Wang, P. 2018. Two distinct approaches for CRISPR-Cas9-mediated gene editing in Cryptococcus neoformans and related species. *Msphere* 3(3): e00208–18. doi: 10.1128/mSphereDirect.00208-18.

Wang, L., Chen, L., Li, R., Zhao, R., Yang, M., Sheng, J., and Shen, L. 2017. Reduced drought tolerance by CRISPR/Cas9-mediated SlMAPK3 mutagenesis in tomato plants. *Journal of Agricultural and Food Chemistry* 65(39): 8674–8682.

Wang, W., Xue, Z., Miao, J., Cai, M., Zhang, C., Li, T., ... and Liu, X. 2019. PcMuORP1, an oxathiapiprolin-resistance gene, functions as a novel selection marker for Phytophthora transformation and CRISPR/Cas9 mediated genome editing. *Frontiers in Microbiology* 10: 2402.

Wang, Y., Cheng, X., Shan, Q., Zhang, Y., Liu, J., Gao, C., and Qiu, J. L. 2014. Simultaneous editing of three homoeoalleles in hexaploid bread wheat confers heritable resistance to powdery mildew. *Nature Biotechnology* 32(9): 947–951. doi:10.1038/nbt.2969.

Xu, L., Zhao, L., Gao, Y., Xu, J., and Han, R. 2017. Empower multiplex cell and tissue-specific CRISPR-mediated gene manipulation with self-cleaving ribozymes and tRNA. *Nucleic Acids Research* 45(5): e28–e28.

Xu, R., Li, H., Qin, R., Wang, L., Li, L., Wei, P., and Yang, J. 2014. Gene targeting using the Agrobacterium tumefaciens-mediated CRISPR-Cas system in rice. *Rice* 7: 1–4.

Xu, W., Song, W., Yang, Y., Wu, Y., Lv, X., Yuan, S., et al. 2019. Multiplex nucleotide editing by high-fidelity Cas9 variants with improved efficiency in rice. *BMC Plant Biology* 19(1): 1–10. doi: 10.1186/s12870-019-2131-1.

Xu, Z., Xu, X., Gong, Q., Li, Z., Li, Y., Wang, S., Yang, Y., Ma, W., Liu, L., Zhu, B., and Zou, L. 2019. Engineering broad-spectrum bacterial blight resistance by simultaneously disrupting variable TALE-binding elements of multiple susceptibility genes in rice. *Molecular Plant* 12(11): 1434–1446.

Yin, K., and Qiu, J. L. 2019. Genome editing for plant disease resistance: applications and perspectives. *Philosophical Transactions of the Royal Society B* 374(1767): 20180322. doi: 10.1098/rstb.2018.0322.

Zafar, K., Khan, M. Z., Amin, I., Mukhtar, Z., Yasmin, S., Arif, M., Ejaz, K., and Mansoor, S. 2020. Precise CRISPR-Cas9 mediated genome editing in super basmati rice for resistance against bacterial blight by targeting the major susceptibility gene. *Frontiers in Plant Science* 11: 575.

Zaidi, S. S. E. A., Tashkandi, M., Mansoor, S., and Mahfouz, M. M. 2016. Engineering plant immunity: using CRISPR/Cas9 to generate virus resistance. *Frontiers in Plant Science* 7: 1673. doi: 10.3389/fpls.2016.01673.

Zetsche, B., Gootenberg, J. S., Abudayyeh, O. O., Slaymaker, I. M., Makarova, K. S., Essletzbichler, P., et al. 2015. Cpf1 is a single RNA-guided endonuclease of a class 2 CRISPR-Cas system. *Cell* 163(3): 759–771. doi: 10.1016/j.cell.2015.09.038.

Zhan, X., Zhang, F., Zhong, Z., Chen, R., Wang, Y., Chang, L., Bock, R., Nie, B., and Zhang, J. 2019. Generation of virus-resistant potato plants by RNA genome targeting. *Plant Biotechnology Journal* 17(9): 1814–1822.

Zhang, A., Liu, Y., Wang, F., Li, T., Chen, Z., Kong, D., Bi, J., Zhang, F., Luo, X., Wang, J., and Tang, J., 2019a. Enhanced rice salinity tolerance via CRISPR/Cas9-targeted mutagenesis of the OsRR22 gene. *Molecular Breeding* 39: 1–10.

Zhang, C., Meng, X., Wei, X., and Lu, L. 2016. Highly efficient CRISPR mutagenesis by microhomology-mediated end joining in *Aspergillus fumigatus*. *Fungal Genetics and Biology* 86: 47–57.doi: 10.1016/j.fgb.2015.12.007.

Zhang, M., Liu, Q., Yang, X., Xu, J., Liu, G., Yao, X., Ren, R., Xu, J., and Lou, L. 2020. CRISPR/Cas9-mediated mutagenesis of Clpsk1 in watermelon to confer resistance to *Fusarium oxysporum* f. sp. niveum. *Plant Cell Reports* 39: 589–595.

Zhang, T., Zhao, Y., Ye, J., Cao, X., Xu, C., Chen, B., An, H., Jiao, Y., Zhang, F., Yang, X., and Zhou, G., 2019b. Establishing CRISPR/Cas13a immune system conferring RNA virus resistance in both dicot and monocot plants. *Plant Biotechnology Journal* 17(7): 1185.

Zhang, Y., Bai, Y., Wu, G., Zou, S., Chen, Y., Gao, C., and Tang, D. 2017. Simultaneous modification of three homoeologs of Ta EDR 1 by genome editing enhances powdery mildew resistance in wheat. *The Plant Journal* 91(4): 714–724.doi: 10.1111/tpj.13599.

Zhang, Z., Ge, X., Luo, X., Wang, P., Fan, Q., Hu, G., Xiao, J., Li, F., and Wu, J. 2018. Simultaneous editing of two copies of Gh14-3-3d confers enhanced transgene-clean plant defense against Verticillium dahliae in allotetraploid upland cotton. *Frontiers in Plant Science* 9: 842.

Zhou, J., Peng, Z., Long, J., Sosso, D., Liu, B. O., Eom, J. S., et al. 2015. Gene targeting by the TAL effector PthXo2 reveals cryptic resistance gene for bacterial blight of rice. *The Plant Journal* 82(4): 632–643. doi: 10.1111/tpj.12838.

Zhu, H., Li, C., and Gao, C. 2020. Applications of CRISPR–Cas in agriculture and plant biotechnology. *Nature Reviews Molecular Cell Biology* 21(11): 661–677.

Zou, Z., Liu, F., Selin, C., and Fernando, W. D. 2020. Generation and characterization of a virulent *Leptosphaeria maculans* isolate carrying a mutated AvrLm7 gene using the CRISPR/Cas9 system. *Frontiers in Microbiology* 11: 1969. doi: 10.3389/fmicb.2020.01969.

Index